STRONG COUPLING GAUGE THEORIES IN THE LHC PERSPECTIVE

Proceedings of KMI-GCOE Workshop

STRONG COUPLING GAUGE THEORIES IN THE LHC PERSPECTIVE

Proceedings of KMI-GCOE Workshop

Nagoya University, Nagoya, Japan 4 – 7 December 2012

Editors

Yasumichi Aoki
Toshihide Maskawa
Koichi Yamawaki

Nagoya University, Japan

World Scientific

NEW JERSEY • LONDON • SINGAPORE • BEIJING • SHANGHAI • HONG KONG • TAIPEI • CHENNAI

Published by

World Scientific Publishing Co. Pte. Ltd.
5 Toh Tuck Link, Singapore 596224
USA office: 27 Warren Street, Suite 401-402, Hackensack, NJ 07601
UK office: 57 Shelton Street, Covent Garden, London WC2H 9HE

British Library Cataloguing-in-Publication Data
A catalogue record for this book is available from the British Library.

STRONG COUPLING GAUGE THEORIES IN THE LHC PERSPECTIVE (SCGT12)
Proceedings of KMI-GCOE Workshop

ISBN 978-981-4566-24-7

Printed in Singapore by Fuisland Offset Printing (S) Pte Ltd

PREFACE

A series of the Nagoya workshops on SCGT (Strong Coupling Gauge Theories) started in 1988, with its title "New Trends in Strong Coupling Gauge Theories (SCGT 88)", which was motivated by a couple of our new proposals in mid 80's: The "Hidden Local Symmetries" formulating massive composite vector bosons as gauge bosons, and the "Walking Technicolor" which, having a large anomalous dimension near unity, predicted a Technidilaton as a light composite Higgs. The latter model was based on the scale-invariant strong gauge dynamics, with the coupling being almost constant near criticality on the fixed point. After yet another new proposal of the "Top Quark Condensate" having even larger anomalous dimension, the second meeting SCGT 90 followed, hoping that the origin of the electroweak symmetry breaking would be soon revealed by the SSC (Superconducting Super Collider), which was unfortunately cancelled in 1993.

After the SSC collapse, the Nagoya SCGT series still continued in 1996, 2002, 2006, 2009 and 2012 (SCGT12Mini), through the present one, SCGT 12. In December 4-7, 2012, we organized "Strong Coupling Gauge Theories in the LHC Perspective (SCGT 12)" at "Sakata-Hirata Hall" of Nagoya University, a new building in commemoration of the mentors of four Nobel Laureates from Nagoya University, M. Kobayashi and T. Maskawa for physics and R. Noyori and O. Shimomura for chemistry.

The SCGT 12 was held in an exciting time after the Higgs discovery at LHC on July 4, 2012, which we have been awaiting for about a quarter century since the first meeting in 1988: We are now at the stage where the strong coupling gauge theories for the electroweak symmetry breaking having large anomalous dimension, as identified as "New Trends" in the SCGT 88, can really be tested by the experiments. Also the QCD at extreme conditions will have new features, having some resemblance to the above "new trends" and is facing a new stage in the era of LHC.

From the very beginning of the series, the tradition of the SCGT workshops has been a unique combination of the phenomenological and the model-building studies with the theoretical ones including lattice and stringy approaches. In SCGT 12 we had experimentalists, phenomenologists, model builders, lattice experts, string theorists, and so on, strongly correlated in a packed space and time box. During 4 days sessions we had 92 participants (24 from abroad), with 42 talks in the main program and 17 presentations at poster sessions. Topics were broad, not just conformal gauge dynamics but various aspects of strong coupling gauge theories including QCD in extreme conditions, in a wide variety of approaches covering lattice studies, holography, effective theories, and so on. Also several experimental talks were presented. We do believe that these presentations included in the Proceedings are very useful for those who are interested in the strong coupling gauge theories in the revolutionary era of LHC.

The workshop was sponsored by the Kobayashi-Maskawa Institute for the Origin of Particles and the Universe (KMI), Nagoya University and the Nagoya University Global COE Program "Quest for Fundamental Principles in the Universe: from Particles to the Solar System and the Cosmos". It was financially supported by the JSPS Grant-in-Aid for Scientific Research (S): "Symmetry Breakings and the Gauge Dynamics", Daiko Foundation and Nagoya University Foundation. We would like to express our sincere thanks to these organizations for generous supports.

We would also like to thank all the members of the International Advisory Committee, the Organizing Committee, the Local Organizing Committee, and the SCGT 12 Secretariat and the GCOE support stuff office, and the graduate students at Nagoya University for their devoted contributions to make the workshop successful.

September, 2013

Editors
Yasumichi Aoki
Toshihide Maskawa
Koichi Yamawaki

WORKSHOP ORGANIZATION

International Advisory Committee

T. W. Appelquist (Yale)
W. A. Bardeen (FNAL)
R. S. Chivukula (Michigan State)
J. Ellis (Kings College/CERN)
H. Georgi (Harvard)
B. Grinstein (UC San Diego)
G. 't Hooft (Utrecht)
Y. Iwasaki (KEK/Tsukuba)
M. Kobayashi (KEK/JSPS/Nagoya)
J. Kogut (DOE/Maryland)
T. Kugo (YITP Kyoto)
J. Kuti (UC San Diego)
V. A. Miransky (Western)
M. Rho (Saclay)
E. H. Simmons (Michigan State)
D. T. Son (Washington)

Organizing Committee

President: T. Maskawa (KMI)

Chairperson: K. Yamawaki (KMI)
Co-chairperson: M. Harada (Nagoya)
M. Tanabashi (KMI/Nagoya)

K-I. Aoki (Kanazawa)
Y. Aoki (KMI)
Y. Kikukawa (Tokyo)
R. Kitano (Tohoku)
K. I. Kondo (Chiba)
T. Onogi (Osaka)
T. Sakai (Nagoya/KMI)
A. Shibata (KEK)
I. Sogami (Kyoto Sangyo)

Local Organizing Committee

Y. Akamatsu (KMI)
Y. Aoki (KMI)
T. Aoyama (KMI)
M. Harada (Nagoya)
M. Hayakawa (Nagoya)
J. Jia (KMI)
M. Kurachi (KMI)
N. Maekawa (KMI/Nagoya)
S. Matsuzaki (Kyoto Sangyo)
K. Nagai (KMI)
S. Nojiri (Nagoya/KMI)
C. Nonaka (KMI/Nagoya)
H. Ohki (KMI)
E. Rinaldi (KMI)
T. Sakai (Nagoya/KMI)
M. Shigemori (KMI)
M. Tanabashi (KMI/Nagoya)
K. Tobe (Nagoya/KMI)
K. Yamawaki (KMI)
T.Yamazaki (KMI)

Sponsorship

Kobayashi-Maskawa Institute for the Origin of Particles and the Universe (KMI), Nagoya University.
Nagoya University Global COE Program "Quest for Fundamental Principles in the Universe from Particles to the Solar System and the Cosmos".
Daiko Foundation.
JSPS Grant-in-Aid for Scientific Research (S) No. 22224003.
Nagoya University Foundation.

SCGT 12
KMI-GCOE Workshop on
Strong Coupling Gauge Theories in the LHC Perspective
December 4-7, 2012
Nagoya

CONTENTS

Conformal Window and Correlation Functions in Lattice Conformal QCD*

Y. Iwasaki

University of Tsukuba,
Tsukuba, 305-8577, Japan
**E-mail: iwasaki@ccs.tsukuba.ac.jp*
www.ccs.tsukuba.ac.jp/people/iwasaki

We discuss various aspects of Conformal Field Theories on the Lattice. We mainly investigate the $SU(3)$ gauge theory with N_f degenerate fermions in the fundamental representation, employing the one-plaquette gauge action and the Wilson fermion action.

First we make a brief review of our previous works on the phase structure of lattice gauge theories in terms of the gauge coupling constant and the quark mass. We thereby clarify the reason why we conjecture that the conformal window is $7 \leq N_f \leq 16$.

Secondly, we introduce a new concept, "conformal theories with IR cutof" and point out that any numerical simulation on a lattice is bounded by an IR cutoff Λ_{IR}. Then we make predictions that when N_f is within the conformal window, the propagator of a meson $G(t)$ behaves at large t, as $G(t) = c \exp(-m_H t)/t^\alpha$, that is, a modified Yukawa-type decay form, instead of the usual exponential decay form $\exp(-m_H t)$, in the small quark mass region. This holds on an any lattice for any coupling constant g, as far as g is between 0 and g^*, where g^* is the IR fixed point. We verify that numerical results really satisfy the predictions for the $N_f = 7$ case and the $N_f = 16$ case.

Thirdly, we discuss small number of flavors ($N_f = 2 \sim 6$) QCD at finite temperatures. We point out theoretically and verify numerically that the correlation functions at $T/T_c > 1$ exhibit the characteristics of the conformal function with IR cutoff, an exponential decay with power correction.

Investigating our numerical data by a new method which we call the "local-analysis" of propagators, we observe that the $N_f = 7$ case and the $N_f = 2$ at $T \sim 2\,T_c$ case are similar to each other, while the $N_f = 16$ case and the $N_f = 2$ at $T = 10^2 \sim 10^5 Tc$ cases are similar to each other.

Further, we observe our data are consistent with the picture that the $N_f = 7$ case and the $N_f = 2$ at $T \sim 2\,T_c$ case are close to the meson unparticle model. On the other hand, the $N_f = 16$ case and the $N_f = 2$ at $T = 10^2 \sim 10^5\,Tc$ cases are close to a free state in the $Z(3)$ twisted vacuum. All results are consistent with naive physical intuition and give clues for long standing issues at high temperatures such as why the free energy at high temperatures does not reach the Stefan-Boltzmann ideal gas limit even at $T = 100\,T_c$.

Keywords: Conformal Theories, Conformal Window, IR cutoff, Unparticle

---

*Based on the work in Collaboration with K.-I. Ishikawa, Yu Nakayama and T. Yoshie

1. Introduction and Summary

Recently there has been a lot of interest in finding a conformal or near-conformal theory, which could play a key role in constructing a theory Beyond the Standard Model (BSM) of particle physics.[1] Non-perturrbative investigation of non-supersymmetric conformal theories in four dimensional space-time is also ardently desired.

We already described in the abstract partly the objectives, the target, the method and the results of this work. In order to avoid to repeat the same things, we will skip what was described in the abstract and will describe remaining important matters.

The strategy we take is as follows: We define continuum conformal theories as the continuum limit of lattice theories. This is a constructive field theory approach. First, we propose the properties of continuum conformal theories from knowledge of the continuum super–symmetric conformal theories and RG argument.[2] Then, we conjecture the properties of lattice theories.

Lattice theories are defined on a Euclidian four-dimensional lattice with lattice size N_x, N_y, N_z, N_t and lattice spacing a. For simplicity, we assume $N_x = N_y = N_z = N$ and $N_t = rN$, with r a constant. The continuum limit of the theory is defined by $a \to 0$ with $N \to \infty$ keeping $L = N\,a$ constant. When $L =$ finite, the continuum limit defines a theory with an IR cutoff.

We propose that the propagator of a meson $G(t)$ behaves at large t, as $G(t) = c\, \exp{(-m_H t)}/t^{\alpha}$ for a small enough quark mass. We note that this form of the propagator in the coordinate representation corresponds to a cut instead of a pole in the momentum representation.

After we have verified that numerical results satisfy the predictions for the $N_f = 7$ case and the $N_f = 16$ case, we propose a new method of analysis of propagators which we call the "local-analysis of propagator": Writing the propagator of a meson as $G(t) = c\, \exp{(-m(t)t)}/t^{\alpha(t)}$, investigate the t dependence of $m(t)$ and $\alpha(t)$. The $m(t)$ and $\alpha(t)$ are reflected by the evolution of RG transformation flow and therefore contain useful information of the dynamics.

Carefully investigating the $m(t)$ and $\alpha(t)$, we observe that our data are consistent with the picture that the $N_f = 7$ theory is close to the meson unparticle model, while the $N_f = 16$ theory is close to free fermion state. We also observe the similarity between the large N_f case and the small N_f at $T \geq T_c$ case.

2. Stage and Tools

We employ the Wilson quark action and the standard one-plaquette gauge action. For fermion field an anti-periodic boundary condition is imposed in the time direction, otherwise periodic boundary conditions. The theory is defined by two parameters; the bare coupling constant g_0 and the bare quark mass m_0.

We usually use, instead of the bare gauge coupling constant g_0 and the bare quark mass m_0, β defined by $\beta = 6/g_0^2$ and $K = 1/2(m_0 a + 4)$, which is called the hopping parameter.

In order to investigate properties of the theory in the continuum limit, one has to first clarify the phase structure of lattice QCD at zero temperature to identify a UV fixed point and/or an IR fixed point. When $N_F \leq 16$, the point $g_0 = 0$ and $m_0 = 0$ is a UV fixed point. Therefore a theory governed by this fixed point is an asymptotically free theory. We restrict ourselves to the theory defined by this UV fixed point in this articles.

When there is an IR fixed point[3] at finite coupling constant g on the massless line which starts from the UV fixed point, the long distance behavior is determined by the IR fixed point. This defines a conformal theory. The minimum number of flavors N_f for the existence of such an IR fixed point is denoted by N_f^c. Usually $N_f^c \leq N_f \leq 16$ is called the conformal window.

The local meson operator is denoted by $G_H(t)$. When the theory is in the confining phase, it decays exponentially at large t

$$G_H(t) = c\, \exp(-m_H t) \tag{1}$$

for a free boundary, where m_H is the mass of the ground state hadron contained in the channel H.

The boundary conditions we take imply

$$G(t) = c\,(\exp(-m_H t) + \exp(-m_H(L_t - t))) \tag{2}$$

for $t \to L_t/2$. From this behavior we extract the energy of the ground state m_H in the channel H.

The quark mass is defined through Ward-Takahashi identities

$$m_q = \frac{\langle 0|\nabla_4 A_4|\mathrm{PS}\rangle}{2\langle 0|P|\mathrm{PS}\rangle}, \tag{3}$$

where P is the pseudo-scalar density and A_4 the fourth component of the local axial vector current. Renormalization constants are suppressed, since this expression is enough for our purpose.

3. Phase Diagram: A Brief Survey of Our Previous Works

In our earlier article,[4] we investigated the phase structure of the SU(3) and SU(2) gauge theory with various numbers of flavors N_f of Wilson fermions in the fundamental representation, and conjectured that the conformal window is $7 \leq N_F \leq 16$ for the SU(3) and $3 \leq N_F \leq 10$ for the SU(2), based on numerical results for various numbers of flavors.

Although we would like to extract the phase diagram at infinite volume and at zero temperature, we have to make simulations at finite volume N^3 and at finite $N_t = rN$. Therefore the phase diagram is a three dimensional space parameterized by, g_0, m_0 and N. Finally we have to take the limit N infinity.

3.1. $N_f \leq 6$

When N is finite, the phase diagram is given as in Fig. 1(a). In 1996, we studied[5] the finite temperature transition of QCD for N_F =2, 3, 6. Our results implies that the transition in the chiral limit is of first order for $N_f = 3$ and 6. The order of the chiral transition in the $N_f = 2$ case is more subtle. We will discuss this issue later.

At $K = 0$ which corresponds to the pure gauge theory, the phase transition is of first order. Thereby the transition line connects between the critical point at $K = 0$ and the chiral transition point at $K = K_c$.

As lattice size N increases, the phase transition points both at quenched QCD and at massless QCD move towards larger β as shown in Fig. 1(b). Eventually in the limit $N \to \infty$, all the region is covered with the confinement phase as shown in Fig. 3(a).

3.2. $7 \leq N_f \leq 16$

When N is finite, we found that the phase diagram is given as in Fig. 2(a). Due to the lack of chiral symmetry the structure of phase diagram is complicate.

The salient facts are the following:

First, the massless line which starts the UV fixed point moves towards smaller β and hits the bulk transition. The massless line belongs to a deconfining phase all through the region.

Secondly, in the confinement phase which exists at small β region there is no massless line. When we decrease the quark mass from the quenched QCD, we en-

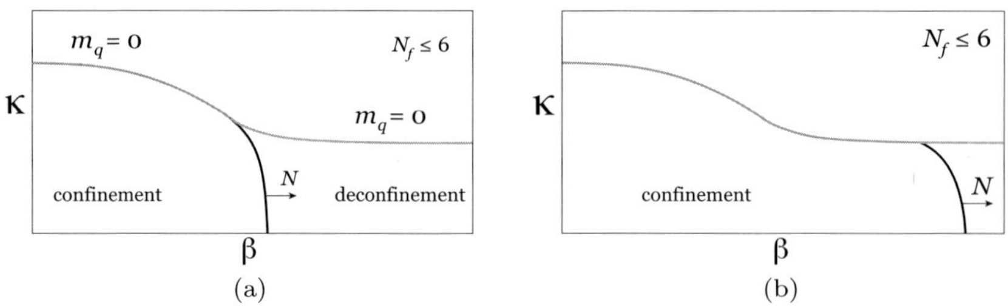

Fig. 1. Phase diagram for $N_f \leq 6$. (a) For not so large N. (b) For larger N.

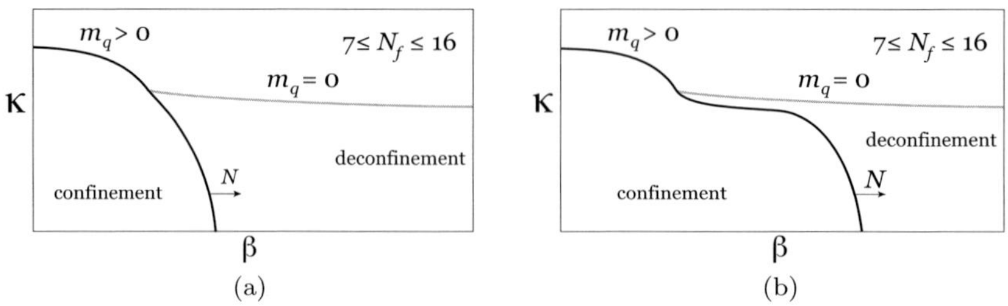

Fig. 2. Phase diagram for $7 \leq N_f \leq 16$. (a) For not so large N. (b) For larger N.

counter a bulk transition at finite quark mass. Thus this confinement phase is irrelevant to the continuum theory.

As lattice size N increases the phase diagram changes as shown in Fig. 2(b) and Fig. 3(b). It should be noted that there is still no massless line which belongs to a confinement phase. This implies that even in the continuum limit, the massless line does not belong to the confinement phase. On the other hand, if there would be no IR fixed point, the massless line should belong to a confining phase in the continuum limit.

More generally, consider a set of actions of the Wilson quark action and improved gauge actions. There is no chiral non-symmetric phase on the massless line in this set of actions. Assume that the quark confinement is realized for a some large lattice with some action which belongs to the set of actions. Then make a renormalization transformation towards the IR region. The IR behavior does not change following the idea of RG transformation. For the change of factor s, the linear size of the lattice becomes $1/s$ and the gauge coupling constant g becomes larger. Repeat this procedure. Then finally the lattice size becomes similar to that we investigated the phase structure. Since the IR behavior should not change, there must exist the confinement phase on a small lattice also. However, there is no confinement region on a small lattice within the set of actions. Thus we reach contradictory. Therefore we may conclude the quark is not confined in this set of actions.

From this reasoning we conjecture that the conformal window is $7 \leq N_f \leq 16$.

4. Strategy for Identification of the Conformal Window

The logics in the previous work is an indirect way to identify the conformal window. Certainly more direct ways are desirable. The most direct and solid way is to identify the IR fixed point. However, for small N_f the IR fixed point exists deep in strong coupling region, and it becomes difficult to identify the IR fixed point.

An alternative way is to find directly the characteristics of conformal theories. This is the way we take in this work.

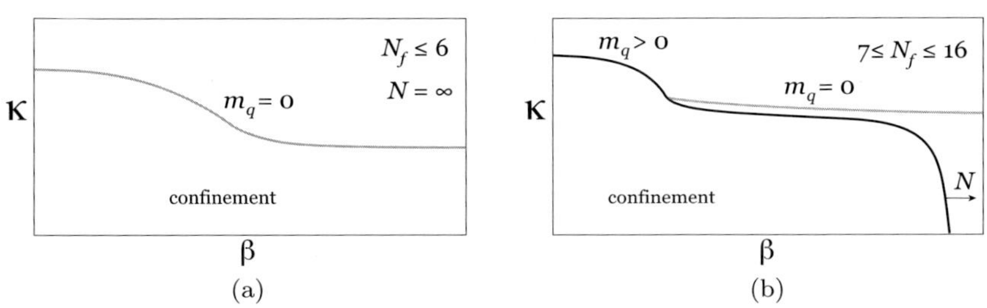

Fig. 3. Phase diagra for large N. (a) For $N_f \leq 6$. (b) For $7 \leq N_f \leq 16$.

5. Conformal Field Theories

In this section we address ourselves to a basic and fundamental issue of what happens when the number of flavor N_f is within the conformal window.

The strategy we take is as follows: We define continuum conformal theories as the continuum limit of lattice theories. First, we propose the structure of continuum conformal theories from knowledge of the continuum super–symmetric conformal theories and RG argument. Then, we conjecture the structure of conformal theories on the lattice.

5.1. *Continuum Conformal Field Theories*

We discuss separately the $L = \infty$ case and the $L =$ finite case, since there exists a clear difference between the two cases.

5.1.1. *The case* $L = \infty$

This defines a continuum theory on R^4 Euclidean four-dimensional space, which corresponds to the flat infinite space at zero temperature. In this case the IR cutoff Λ_{IR} is zero.

Therefore the relevant quantity is only the running coupling constant $g(\mu)$ and there are no physical quantities which have physical dimensions.

The "conformal region" exists only on the massless line. All other region is the "confinement region". The phase structure is shown in Fig. 4(a). In super-symmetric theories this kind phase structure is known. The shaded region in the strong coupling region does not exist in the $\beta - m_q$ plane. When the phase structure is described in terms of $\beta - K$, the corresponding phase belongs to a region for Wilson doubles (Aoki phase). Therefore when it is mapped to the $\beta - m_q$ planes, the region corresponding to the shaded one does not exist.

The propagator on the massless quark line takes the form

$$G(t) = c\,\frac{1}{t^{\alpha}}, \tag{4}$$

since, as mentioned, there is no physical quantities with physical dimensions.

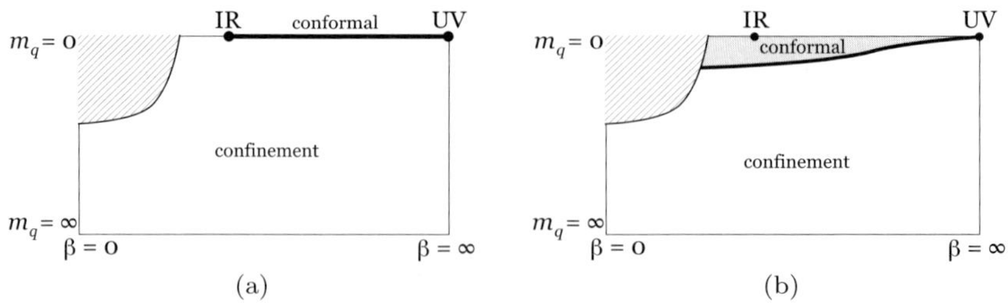

Fig. 4. Phase diagra for large $N = \infty$. (a) $\Lambda_{IR} = 0$. (b) $\Lambda_{IR} =$ finite

If we take the coupling constant $g = g^*$ at UV scale μ,

$$\alpha(t) = 3 - 2\gamma^*, \tag{5}$$

with γ^* being the anomalous mass dimension γ at $g = g^*$. The theory is scale invariant (and shown to be conformal invariant within perturbation theory.[6] See also e.g. Nakayama[7] and references therein from AdS/CFT approach). In general, the power α varies with t:

$$\alpha = \alpha(t). \tag{6}$$

When t is small, the correlation function is governed by the UV fixed point. There the "meson state" is a free quark and an anti-quark state. Thus the power is given by

$$\alpha(t) = 3. \tag{7}$$

When t is large, it is governed by the IR fixed point. The exponent is given by

$$\alpha(t) = 3 - 2\gamma^*. \tag{8}$$

The exponent of the power α in IR is given by the renormalization equation or derived just by dimension counting.

5.1.2. *The case $L =$ finite*

This defines a continuum theory on T^4 Euclidean four-dimensional space, which corresponds to a compact T^3 space at finite temperature. In this case the IR cutoff $\Lambda_{IR} = 1/L$ is finite.

The renormalization transformation to IR does stop to evolve at the scale of IR cutoff Λ_{IR}. Thus the relevant physical quantities we may choose are

$$\Lambda_{CFT}, \Lambda_{IR}, m_H \tag{9}$$

Here Λ_{CFT} is a scale parameter for the transition region from UV to IR, and m_H is a generic mass of a meson such as the pion and the ρ meson.

Now we make a proposal. The propagator of a meson $G(t)$ behaves at large t, as

$$G(t) = c\frac{\exp(-mt)}{t^\alpha}, \tag{10}$$

that is, a modified Yukawa-type decay form in the region

$$m_H \leq c\Lambda_{IR}, \tag{11}$$

instead of the usual behavior of the propagator in the "confinement region",

$$G(t) = c\exp(-mt). \tag{12}$$

Here Λ_{IR} is an IR cutoff, m_H is a generic mass of a hadron such as the pion mass., and c is a constant of order 1. The conformal region is depicted in Fig. 4(b). The relation Eq. (11) reduces to

$$m_q \leq c\,(1/L)^{(1+\gamma^*)}, \tag{13}$$

using the relation[8]

$$m_H = c\, m_q^{1/(1+\gamma^*)}. \tag{14}$$

The phase structure with $L =$ finite smoothly approaches the case $L = \infty$, as L varies to ∞. The form of the propagator also smoothly becomes

$$G(t) = c\frac{1}{t^\alpha}, \tag{15}$$

in the massless limit.

As is the case in the $\Lambda_{IR} = 0$ case, the power α varies with t:

$$\alpha = \alpha(t). \tag{16}$$

When t is small,

$$\alpha(t) = 3. \tag{17}$$

On the other hand, when t is large ($t \gg 1/\Lambda_{CFT}$), the behavior of $\alpha(t)$ is complicate: When m is small enough and if $m\,t \ll 1$, the power is given by

$$\alpha(t) = 3 - 2\gamma^*. \tag{18}$$

If we take larger t such that $m\,t \gg 1$, the $\alpha(t)$ takes a value which depends on the dynamics of the theory. We give explicit examples when we discuss some models.

The concept of "conformal theory with IR cutoff" and "conformal region" determined by equations (10) and (11) are our key proposals.

5.2. *Conformal Field Theories on the Lattice*

We are able to perform numerical simulations only on a finite size lattice. Thus the system is automatically bounded by the IR cutoff $\Lambda_{IR} \sim 1/(Na)$. In this case, the relevant physical parameters we may use are Λ_{CFT}, Λ_{IR} and m_H.

Our main target is first to verify the transition of a meson propagator from an exponential decay form to a modified Yukawa-type, that is, an exponential form with power correction given by Eq. (10), at the critical hadron mass given by Eq. (11).

When the lattice size N_1 increased to $N_2 = s\,N_1$ with β being fixed, then

$$m_H^{critical}(\beta, N_2) = 1/s\, m_H^{critical}(\beta, N_1). \tag{19}$$

The critical hadron mass decreases. Thus we have to carefully choose the simulation parameters in order to find the conformal region and the Yukawa-type propagators.

6. Numerical Simulations

6.1. *Algorithm and Parameters*

We have made simulations mainly for the cases $N_f = 7$ and $N_f = 16$, which are, we think, the boundaries of the conformal window. The algorithm we employ is the blocked HMC algorithm[9] for $N_f = 2\,N$ and the RHMC algorithm[10] for $N_f = 1$, in

Table 1. Parameters of Simulations

Nf	β	K	mq	Nf	β	K	mq	Nf	β	K	mq
7	6.0			16	11.5			2	6.5		
		0.1400	0.22			0.126	0.22			0.146	0.035
		0.1446	0.084			0.127	0.19		10.0		
		0.1452	0.062			0.130	0.10			0.135	0.028
		0.1459	0.045			0.1315	0.055		15.0		
		0.1472	0.006			0.13322	0.003			0.130	0.046

the case $N_f = 2N + 1$. The parameters we have simulated are listed in Table 1. (The parameters for $N_f = 2$ at high temperatures are also listed.) We have chosen the hopping parameter K in such a way that the quark mass takes the value from 0.25 to 0.0. It might be worthwhile to mention that we are able to take K which corresponds to a negative quark mass crossing the zero quark mass, since the phase is a deconfining phase and therefore the pion mass is not zero there. The coupling constant of the IR fixed point for $N_f = 16$ is $\beta = 11.48$ in one-loop approximation. This is the reason why we choose $\beta = 11.5$ for $N_f = 16$.

The lattice sizes we choose are $8^3 \times 32, 16^3 \times 64$ and 24×96. The run-parameters are chosen such that the acceptance of the global metropolis test is about 70%. The statistics is in general 1,000 MD trajectories for thermalization and 1,000 MD trajectories for measurement.

7. Numerical Results

Let us first discuss the results for $N_f = 7$ case on a $16^3 \times 64$ lattice. Fig. 5 shows the effective mass plots for pions with the local-sink local-source (black squares), local-sink doubly-smeared-source (red circles) and local-sink doubly-wall-source (green triangles) cases. The effective masses in the relatively large quark mass case shown on the left panel, $m_q = 0.25 (K = 0.1400)$, clearly show the plateau at $t = 22 \sim 31$. On the other hand, in the $m_q = 0.045$ $(K = 0.1459)$ case shown on the right panel, there is no plateau up to $t = 31$. The effective mass is slowly decreasing.

The effective masses for all cases with $m_q \leq 0.084$ $(K \geq 0.1446)$ exhibit similar behavior to $K = 0.1459$. We have conjectured that propagators decay exponentially with power correction. Thus, to verify the conjecture, we fit the local-local data with the Yukawa-type for the fitting range $t = [15 : 31]$. The fits for $K = 0.1459$ and $K = 0.1452$ are depicted in Fig. 6. The both fits well reproduce the data with range $t = [15 : 31]$ with $\alpha = 0.8(1)$.

In the case of $N_f = 16$, also, the effective masses on a $16^3 \times 64$ lattice exhibit the modified Yukawa-type for all cases when $K \geq 0.130$. The results of the fit to the local-local data for $K = 0.130$ and $K = 0.1315$ with the Yukawa-type for the fitting range $t = [15 : 31]$ are plotted in Fig. 7. The fits well reproduce the data. However, if we carefully compare the fits of $N_f = 7$ and $N_f = 16$ cases, the fit of $N_f = 16$ case at large t is not so good as $N_f = 7$ case. We will discuss this point later.

The data on a $24^3 \times 96$ lattice for both cases of $N_f = 7$ with $K \geq 0.1459$ and $N_f = 16$ with $K \geq 0.130$ exhibit the modified Yukawa type decay.

Thus we have verified that the propagators decay with the Yukawa-type decay form for the $N_f = 7$ and $N_f = 16$ cases, for the small quark mass. In term of the pion mass the critical mass can be expressed as $m_H \leq c\,\Lambda_{IR}$. Here m_H is the pion mass and we take $\Lambda_{IR} = 2\pi\,(N^3 \times N_t)^{1/4}$. There is no *a priori* definition of Λ_{IR} in the case of asymmetric lattices. This is one of choices. With this definition, we estimate the constant c is $c \sim 2.16$ for $N_f = 7$, and $c \sim 1.94$ for $N_f = 16$.

8. Local-Analysis of Propagator

As we have verified our conjecture that the propagators decay with the Yukawa-type decay form for the small quark mass, the next issue we would like to solve is what kind of theory is defined. To do so, we propose a new method of analysis of propagators which we call the "local-analysis" of propagators: Writing the propagator of

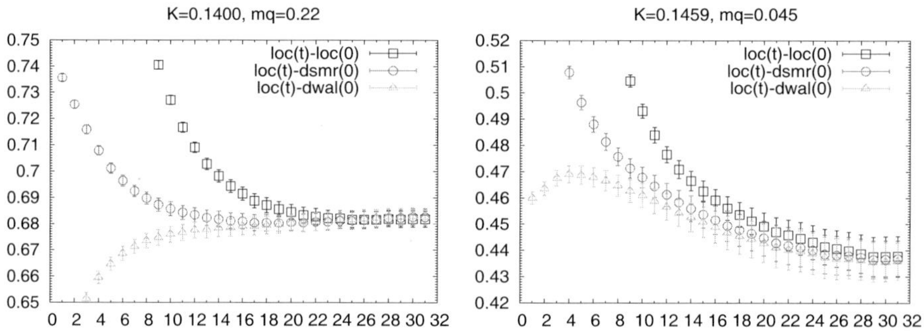

(a) (b)

Fig. 5. The asymptotic behavior of propagators in the $N_f = 7$ case on a $16^3 \times 64$ lattice at $\beta = 6.0$. (a) Exponential decay form.(b) Modified Yukawa-tye decay form.

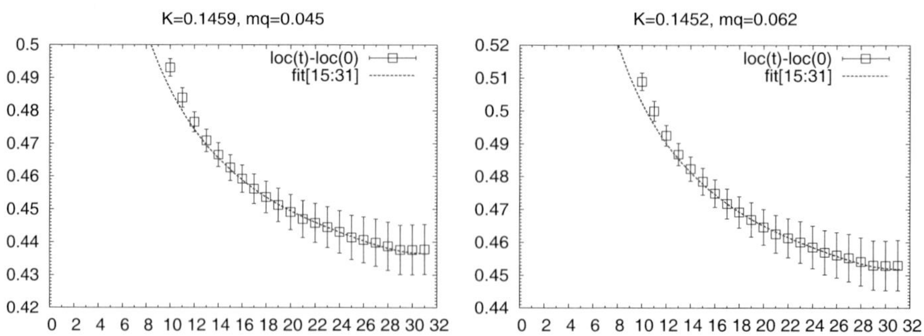

(a) (b)

Fig. 6. Two examples of modified Yukawa-type decay form in the $N_f = 7$ case on a $16^3 \times 64$ lattice at $\beta = 6.0$ (a) K=0.1459. (b) K=0.1452.

a meson as

$$G(t) = c \exp(-m(t)t)/t^{\alpha(t)} \tag{20}$$

investigate the t dependence of $m(t)$ and $\alpha(t)$. The $m(t)$ and $\alpha(t)$ are reflected by the evolution of RG transformation flow and therefore contain useful information of the dynamics.

We depict $m(t)$ and $\alpha(t)$ in Figs. 9$\sim$ 19 for various cases.

First, at small t (we disregard the $\alpha(t)$ and $m(t)$ at $t = 1$ and 2, since they are affected by the boundary effect),

$$\alpha(t) \sim 3.0 \tag{21}$$

and

$$m(t) \sim 2\,m_q \tag{22}$$

are well satisfied, when $m_q \geq 1/N$.

As t increases, $m(t)$ increases from $2\,m_q$ to m_H. The behavior of $\alpha(t)$ is more complex and it reflects the dynamics.

In the $N_f = 7$ case on the $16^3 \times 64$ lattice, the $\alpha(t)$ at large t is relatively stable and it seems that $\alpha(t)$ shows a plateau at $t = 15 \sim 31$, although it is noisy. The statics of the data on the 24×96 lattice is not enough to see the $\alpha(t)$, as it requires more statistics than the Yukawa-type fit with a wider range of t. We are storing configurations.

The t dependence of $\alpha(t)$ in the $N_f = 16$ case is quite different from that of the $N_f = 7$ case. In the $N_f = 16$ case, although there is no plateau structure in $\alpha(t)$, there is a shoulder with a value around 1.5 at $t = 12 \sim 24$. on the $16^3 \times 64$ lattice. On the 24×96 lattice (the statistics is enough, since the $N_f = 16$ case is much faster than the $N_f = 7$ case), the t dependence of $\alpha(t)$ is different each other. In the case $K = 0.130$, the shoulder becomes very wide down to $t \sim 32$. In the case $K = 0.1315$ the shoulder is clearly seen for $t = 12 \sim 25$. on the other hand,

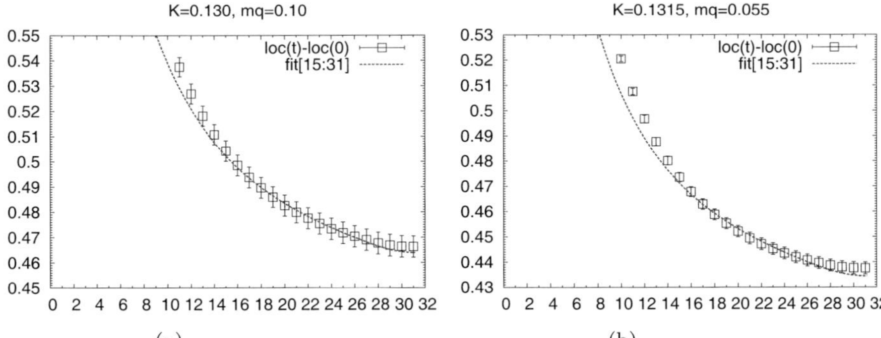

Fig. 7. Two examples of modified Yukawa-type decay form in the $N_f = 16$ case on a $16^3 \times 64$ lattice at $\beta = 11.5$. (a) K=0.130. (b) K=0.1315.

the shoulder disappears and the plateau at $\alpha \sim 0.5$ is seen for $t \geq 24$ in the case $K = 0.13322$.

In order to find what kind of theory is defined, we will investigate some models which we consider are good candidates to compare with the theory. To do so, we find it is better to discuss conformal theories at high temperatures before that. Therefore we examine conformal theories at high temperatures in the next section.

9. Conformal Field Theories at Finite Temperature

Now let us consider lattice QCD at high temperatures for small N_f $(2 \leq N_f \leq 6)$ where the chiral phase transition occurs at T_c. Note that the system is bounded by the infrared cutoff $\Lambda_{IR} = 1/T$. The continuum limit of the theory at finite temperature is defined by $a \to 0$ with $N \to \infty$ and $N_t \to \infty$ keeping $Na = L$ and $N_t a = 1/T$ constant. Taking the limit $L \to \infty$, the gauge theory at finite temperature T is defined.

We first consider the case that the renormalized quark mass is zero. Then the renormalized coupling constant is the only relevant variable in the theory. A running coupling constant $g(\mu;T)$ at temperature T can be defined as in the case of $T = 0$. The following discussion can be applied to any method of the definition of the running coupling constant $g(\mu;T)$. One way is the Wilson RG method.

Although in several literatures (see e.g. Ref. 11) the running coupling constant $g(\mu;T)$ was discussed, it was mainly the UV behavior. Up to now it seems that it has not been discussed systematically the IR behavior.

In the UV region the running coupling constant at finite T is essentially as the same as that of $T = 0$. Since the theory is asymptotically free, at small g region the following relation

$$g^2(\mu;T) = g^2(\mu;T=0) + cg^4(\mu;T=0), \tag{23}$$

with c constant, is satisfied. The constant c depends on both of the schemes of $g(\mu;T=0)$ and $g(\mu;T)$ in addition to N_f. It might be worthwhile to stress that the leading term is universal in the limit $g \to 0$.

However, in the IR region $g(\mu;T)$ may be quite different from $g(\mu;T=0)$, since the IR cut-off Λ_{IR} in the time direction is $1/T$. This is quite different from the $T = 0$ case.

Furthermore, the fact that when $T/T_c > 1$, the quark is not confined, implies that the running coupling constant $g(\mu;T)$ cannot be arbitrarily large.This means that there is a fixed point in the β function when $T/T_c > 1$.

Let us increase the temperature gradually from $T = 0$.

As far as $T < T_c$, the β function is negative all through g. As the temperature is increased further, the form of β function will changes as in Fig. 8: (a) When $T > T_c$ but $T \sim T_c$, the β function changes the sign from negative to positive at large g. (b) When $T \gg T_c$ it changes the sign at small g.

The T/T_c dependence of the β functions resembles the N_f dependence[3] for large N_f: $N_f = 16 \Leftrightarrow T/T_c \gg 1$ and $N_f = 7 \Leftrightarrow T/T_c \geq 1$.

9.1. *Numerical Results at High Temperatures*

We have made simulation in the $N_f = 2$ case on a $16^3 \times 64$ lattice for several sets of parameters. Let us discuss the three cases listed in Table 1: $\beta = 6.5$ corresponds to $T \sim 2\,T_c$, $\beta = 10.0$ to $T \sim 100\,T_c$, and $\beta = 15.0$ to $T \sim 10^5\,T_c$.

We clearly see that the propagators in the three cases decay with the modified Yukawa type decay form. Therefore we make the local-analysis of the propagators. The $m(t)$ and $\alpha(t)$ are depicted in Figs. 14, 15 and 16.

10. Correspondence between N_f and T

We observe the t dependence of $m(t)$ and $\alpha(t)$ in the case $T \sim 2T_c$ are very similar to those in the $N_f = 7$ case. On the other hand, the t dependence in the $T \sim 100T_c$ and $T \sim 10^5 T_c$ cases are very similar to those in the $N_f = 16$ case. We plot in Figs. 17, 18 and 19, the large N_f data on the right panel and the $N_f = 2$ at high temperature data on the left panel in order to compare them directly.

The similarities between them are excellent. We could anticipate this from the similarities of the β functions: $N_f = 16 \Leftrightarrow T/T_c \gg 1$ and $N_f = 7 \Leftrightarrow T/T_c \geq 1$.

We point out a possibility that there is a correspondence between the large N_f with Λ_{IR} and small N_f at high temperature T. Both of the large N_f with Λ_{IR} and the small N_f at high temperature T are described by two parameters in addition to N and N_t; β and m_q. In order to have a correspondence we have to map from one set of the parameters to the other set. Although it is too premature to discuss the details, the correspondence between N_f and T would be very intriguing and very useful. Although the flavor number N_f is a discrete integer, the temperature T is a continuous number. Thus the properties of large N_f could be deduced from those of small N_f at high temperatures with this correspondence.

11. Models

We investigate a few models to compare the results with the models.

(1) Meson unparticle model

(2) Fermion unparticle model

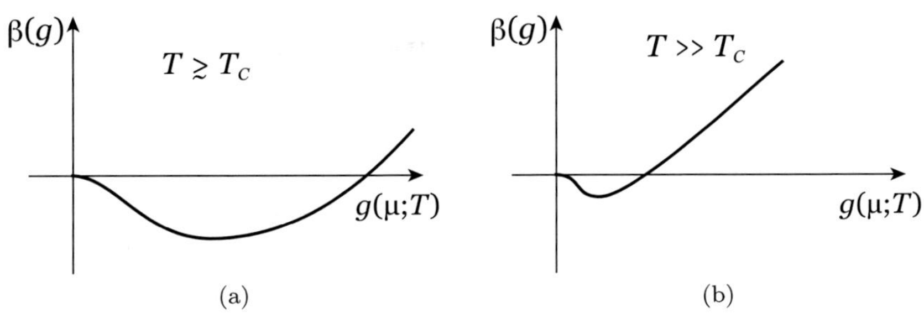

Fig. 8. β function with $T =$ constant. (a) $T \geq T_c$. (b) $T \gg T_c$.

The meson unparticle model and fermion unparticle model are motivated by the soft-wall model in AdS/CFT correspondence.[12]

(1) The soft-wall model predicts the form of the propagator with scale dimension Δ in the momentum space as

$$\langle O(p)O(-p)\rangle = \frac{1}{(p^2+m^2)^{2-\Delta}} \ . \tag{24}$$

In this case, when $m\,t \ll 1$, $\alpha(t) = 3 - 2\,\gamma^*$. This is model independent and universal. On the other hand, when $m\,t \gg 1$, $\alpha(t) = 2 - \gamma^*$ for $t \gg \Lambda_{CFT}^{-1}$. This is model dependent.

(2) Another possible scenario is to treat $\bar{\psi}\gamma_5\psi(x)$ as the non-bound state of un-fermions. The soft-wall model predicts the form of the propagator with scale dimension Δ_f in the momentum space as

$$\langle \Psi(p)\bar{\Psi}(-p)\rangle = (p^\mu\gamma_\mu + m)\frac{1}{(p^2+m^2)^{\frac{5}{2}-\Delta_f}} \ . \tag{25}$$

In this case, when $m\,t \ll 1$, $\alpha(t) = 3 - 2\,\gamma^*$. On the other hand, when $m\,t \gg 1$, $\alpha(t) = 1.5 - \gamma^*$ for $t \gg \Lambda_{CFT}^{-1}$. This is model dependent.

12. What Kind of Theories are Defined?

First let us consider the $N_f = 16$ case. We expect that the $N_f = 16$ case is close to a free fermion state. After the conference we further investigate this issue, and find that it is indeed close to a free fermion state, but the vacuum is the $Z(3)$ twisted vacuum. We also find that $N_f = 2$ case at $T \sim 100\,T_c$ and $T \sim 10^5\,T_c$ is close to a free fermion state in the $Z(3)$ twisted vacuum. The approach to a free state is very slow. Investigation in detail of propagators reveals that the difference cannot be attributed to the fermion unparticle model. This is different what we thought at the conference.

Next let us consider the $N_f = 7$ case. The t dependence of $\alpha(t)$ is very similar to the $N_f = 2$ case at $T \sim 2\,T_c$. The $\alpha(t)$ shows a plateau at large t ($t = 15 \sim 31$). We conclude from this fact that the $N_f = 7$ theory is close to the meson unparticle model.

Now we estimate γ^* for $N_f = 7$, using the formula we have derived. The value of $\alpha(t)$ at plateau($t = 15 \sim 31$) is 0.8(1) for $K = 0.1452$ and $K = 0.1459$. Applying the formula $\alpha(t) = 2 - \gamma^*$, we have $2 - \gamma^* = 0.8(1)$. Thus $\gamma^* = 1.2(1)$. Although the estimate should be much more refined, taking the continuum limit, we may expect that γ^* is order of unity.

13. Solutions for Long Standing Issues at High Temperatures

There are long standing issues at high temperatures such as:

The free energy of quark-gluon plasma state does not reach that of the Stefan-Boltzmann ideal gas limit even at $T/T_c = 100$.[13]

Wave functions of "meson" just above T_c can be obtained, although quarks are not confined.

What is the order of the chiral phase transition in the $N_f = 2$ case ? This is theoretically and phenomenologically important issue.

We concentrate the issue of the order of the chiral phase transition in the $N_f = 2$ case in this report.

Pisarski and Wilczek[14] mapped $N_f = 2$ QCD at high temperature to the three dimensional sigma model and pointed out that if $U_A(1)$ symmetry is not recovered at the chiral transition temperature, the chiral phase transition of QCD in the $N_f = 2$ case is 2nd order with exponents of the three dimensional $O(4)$ sigma model.

The $O(4)$ scaling relation was first tested for staggered quarks.[15] For the Wilson quarks it was shown that the chiral condensate satisfies remarkably the $O(4)$ scaling relation, with the RG improved gauge action and the Wilson quark action[16] and with the same gauge action and the clover-improved Wilson quark action[17] (See, for example, Fig. 6 in Ref. 16). It was also shown for staggered quarks the scaling relation is satisfied in the $N_f = 2 + 1$ case,[18] extending the region from $T/T_c > 1$ adopted in[16] and[17] to the region including $T/T_c < 1$. These results imply the transition is second order.

However, recently, it was shown that the expectation value of the chiral susceptibility $\chi_\pi - \chi_\sigma$ is zero[19] in thermodynamic limit when the $SU(2)$ chiral symmetry is recovered under the assumptions we will discuss below. This is consistent with that the $U_A(1)$ symmetry is recovered, which implies the transition is 1st order according to.[14] Apparently the two conclusions are in contradiction.

Here we revisit this issue with the new insight of conformal field theories with an IR cutoff. It is assumed in Ref. 19 that the vacuum expectation value of mass-independent observable is an analytic function of m_q^2, if the chiral symmetry is restored. However, in the conformal region the propagator of a meson behaves as Eq. (3) and the relation between the m_H and the m_q is given by the hyper-scaling relation[8]

$$m_H = c\, m_q^{1/(1+\gamma*)},$$

with γ^* the anomalous mass dimension. This anomalous scaling implies m_H is not analytic in terms of m_q^2 and the analyticity assumption does not hold. It should be noted that the Ward-Takahashi Identities in[19] are proved in the thermodynamic limit and therefore the numerical verification of the hyperscaling in the limit will be decisive. We stress however that the hyper-scaling is theoretically derived with the condition of the existence of the IR fixed point and multiplicative renormalization of m_q. We believe that this violation of the analyticity assumption resolves the apparent discrepancy as also mentioned in[19] as a viable possibility.

For the other issues given in the beginning of this section, we only mention that the slow approach to the free fermion state in the $Z(3)$ twisted vacuum is related to the slow approach of the free energy of quark-gluon plasma state to the Stefan-Boltzmann ideal gas limit, and when $T \sim T_c$, "mesons" are meson unparticles, similar to meson particles in some aspects. We will investigate these points further.

14. Discussion and Conclusions

First of all, our conjectures for the "conformal theories with an IR cutoff" have been verified in the cases of $N_f = 7$ and $N_f = 16$. This fact is consistent with the assertion that the conformal window is $7 \leq N_f \leq 16$. We have further verified the "conformal theories with IR cutoff" in the case $T/T_c > 1$ for $N_f = 2$. We have stressed that IR cutoff is inherent with simulations on a lattice and QCD at high temperatures.

The micro-analysis of propagators is proposed and is shown to be very useful. Even in the confining phase the t dependence of $\alpha(t)$ contains information of RG flow from UV to IR. If it is affected by an IR fixed point, it will appear as remnant in the $\alpha(t)$. If it is not affected, it is difficult to estimate γ^* correctly from the scaling rule, Eq. (5.1.2).

By the "micro-analysis" of the propagator, we observe "$N_f = 7$" and "$T \sim T_c$" are similar to each other, and "$N_f = 16$" and "$T \gg T_c$" are similar to each other. Using the correspondence between "N_f" and "T" that our data suggest, we conclude our data are consistent with the picture that "$N_f = 7$" and "$T \sim T_c$" are close to the meson unparticle model, and estimate $\gamma^* = 1.2(1)$.

Surely a lot of things theoretically and numerically should be done. In particular, the continuum limit of propagators in the large N_f case should be estimated in order to obtain with more confident what kind of theories are defined. To do so, we have to investigate finite size effects and the β dependence of propagators. In parallel with these efforts, physics implications of unparticles such as the effects in the early universe should be much more deepened.

Acknowledgments

I would express gratitude to K.-I. Ishikawa, Yu Nakayama and T. Yoshie for a stimulating and fruitful collaboration. I also would like to thank T. Yanagida for making a chance to start this collaboration. I thank K. Kanaya for his help in preparing the manuscript. I would like to thank the organizer of SCGT 2013 for giving the chance to talk on this subject. The calculations were performed using HA-PACS computer at CCS, University of Tsukuba and SR16000 at KEK. I would like to thank members of CCS and KEK for their strong support for this work.

References

1. J. Giedt, PoS(Lattice 2012); E. Neil, PoS(Lattice 2011)009; L. Del Debbio, PoS(Lattice 2010)004; and references quoted.
2. K.G. Wilson and J. Kogut, Phys. Reports 12C (1974) 75.
3. T. Banks and A. Zaks, Nucl. Phys. **B196** (1982) 189.
4. Y. Iwasaki, K. Kanaya, S. Kaya, S. Sakai, and T. Yoshié, Phys. Rev. **D69** (2004) 014507.
5. Y. Iwasaki, K. Kanaya, S. Kaya, S. Sakai, and T. Yoshié, Phys. Rev. **D54** (1996) 7010.
6. J. Polchinski, Nucl. Phys. B **303**, 226 (1988).
7. Y. Nakayama, Int. J. Mod. Phys. A **25** (2010) 4849.

8. V.A. Miransky, Phys. Rev. **D59** (1999) 105003; L. Del Debbio and R. Zwicky, Phys. Rev. **D82** 014502.
9. M. Hayakawa, et al., PoS LATTICE **2010** (2010) 325.
10. M. A. Clark and A. D. Kennedy, Phys. Rev. Lett. **98** (2007) 051601.
11. O. Kaczmarek, F. Karsch, F. Zantow and P. Petreczky, Phys. Rev. **D70**(2004) 074505.
12. G. Cacciapaglia, G. Marandella and J. Terning, JHEP 0902, 049 (2009).
13. Sz. Borsanyi, et al., arXiv:1204.6184
14. R, Pisarski and F. Wilczek, Phys. Rev. **29** (1984) 338.
15. F. Karsch, Phys. Rev. D49 (1994) 3791; F. Karsch and E. Laermann, ibid. D50 (1994) 6954.
16. Y. Iwasaki, K. Kanaya, S. Kaya, and T. Yoshié, Phys. Rev. Lett. **78** (1997) 179.
17. A. Ali Khan et al. (CP-PACS Collaboration), Phys. Rev. **D63** (2000) 034502.
18. S. Ejiri et al., Phys. Rev. **D80**(2009) 094505; A. Bazavov et al., Phys. Rev. **D85**(2012) 054503.
19. S. Aoki, H. Fukaya, Y. Taniguchi, arXiv:1209.2061.

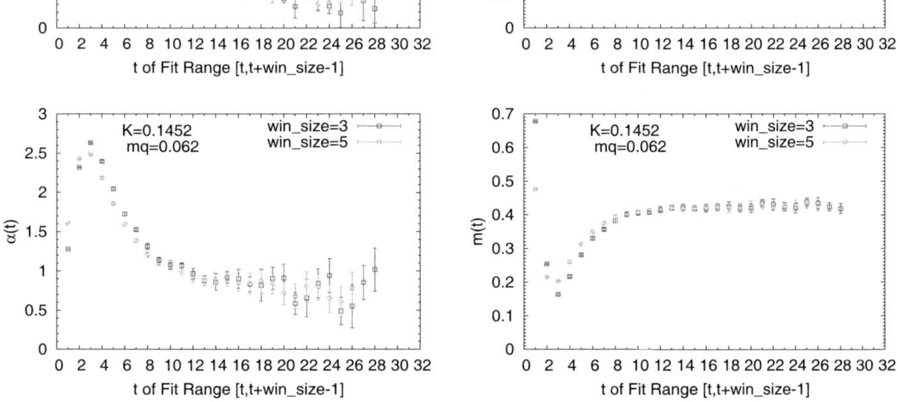

Fig. 9. $\alpha(t)$ and $m(t)$; $N_f = 7$ on a lattice $16^3 \times 64$ at $K = 0.1446$ and $K = 0.1452$.

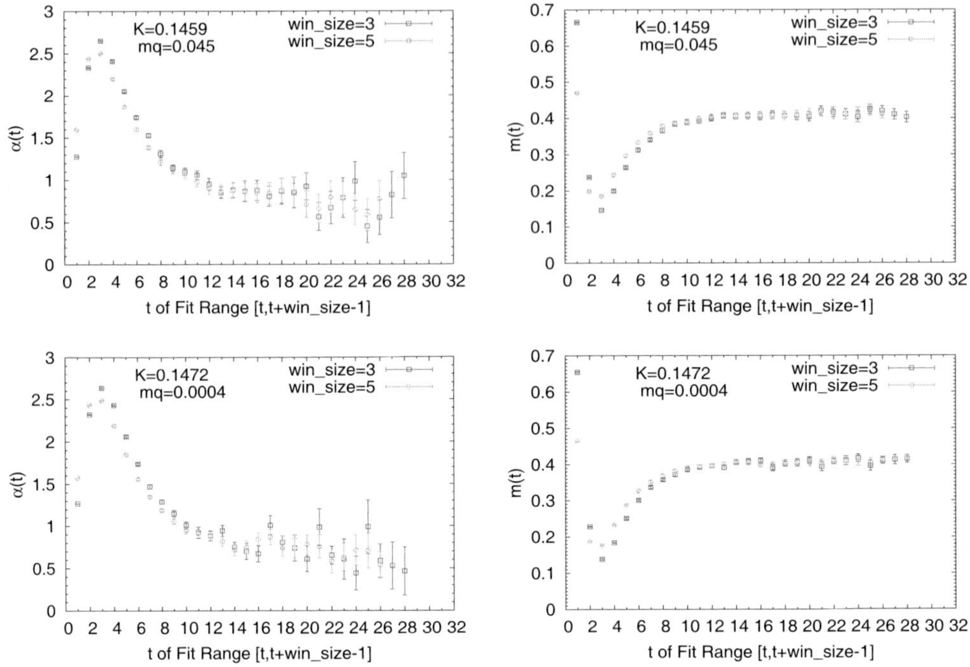

Fig. 10. $\alpha(t)$ and $m(t)$; $N_f = 7$ on a lattice $16^3 \times 64$ at $K = 0.1459$ and $K = 0.1472$.

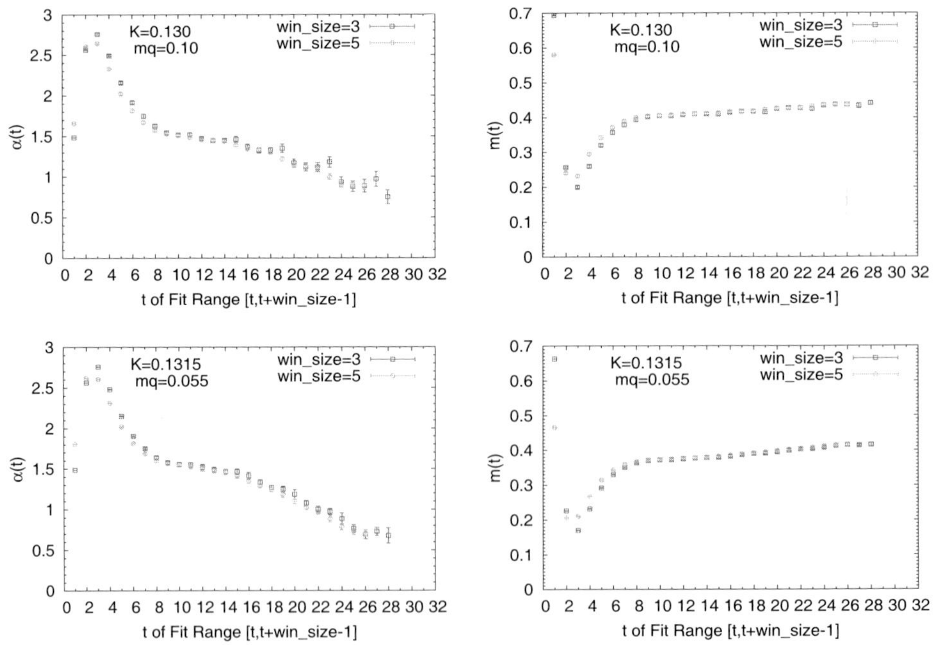

Fig. 11. $\alpha(t)$ and $m(t)$; $N_f = 16$ on a lattice $16^3 \times 64$ at $K = 0.130$ and $K = 0.1315$.

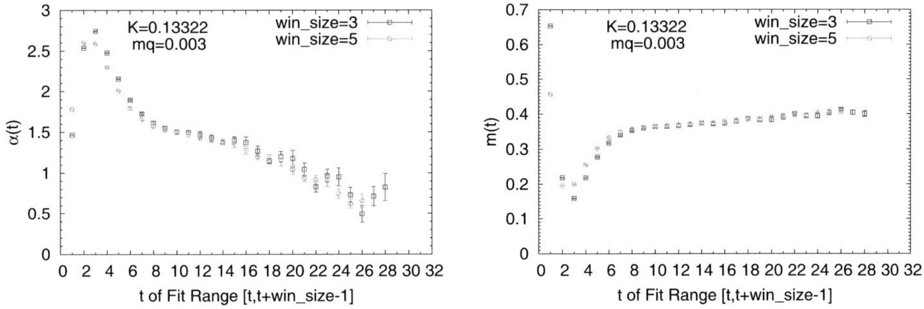

Fig. 12. $\alpha(t)$ and $m(t)$; $N_f = 16$ on a lattice $16^3 \times 64$ at $K = 0.13322$.

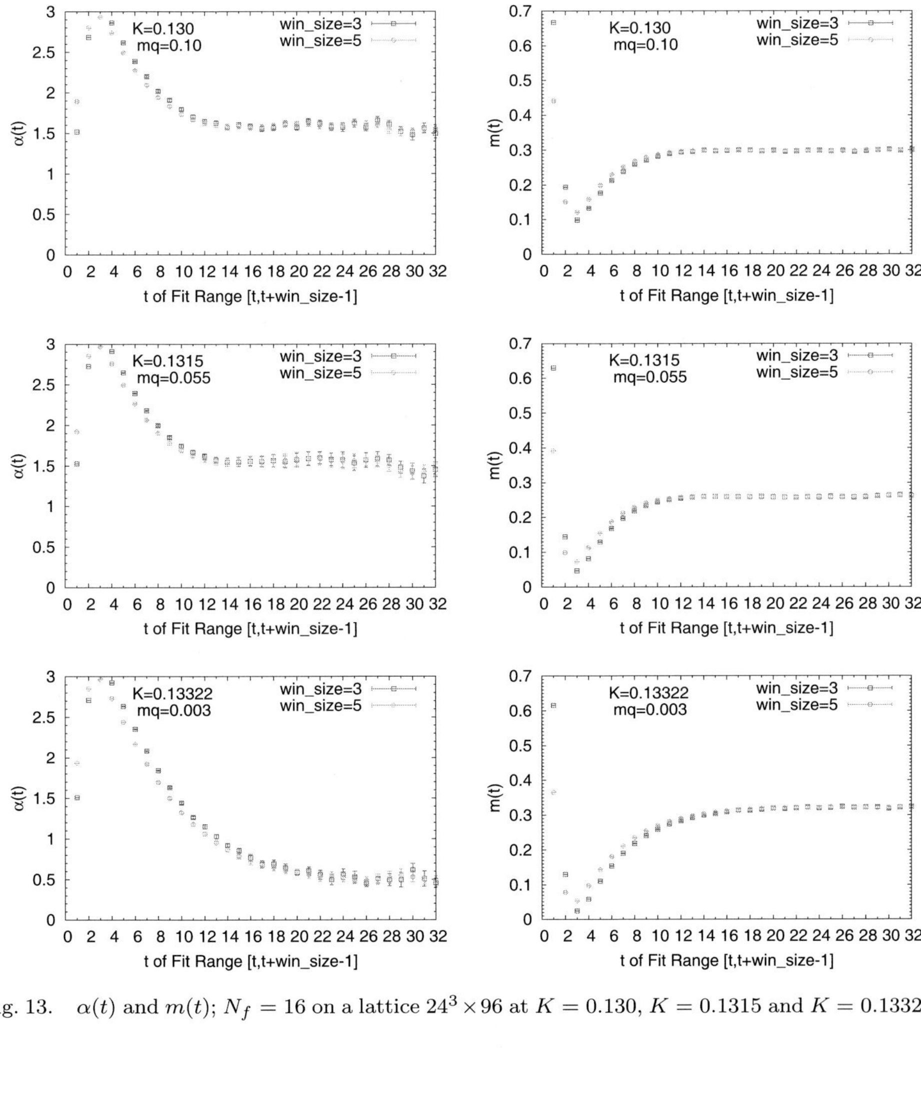

Fig. 13. $\alpha(t)$ and $m(t)$; $N_f = 16$ on a lattice $24^3 \times 96$ at $K = 0.130$, $K = 0.1315$ and $K = 0.13322$.

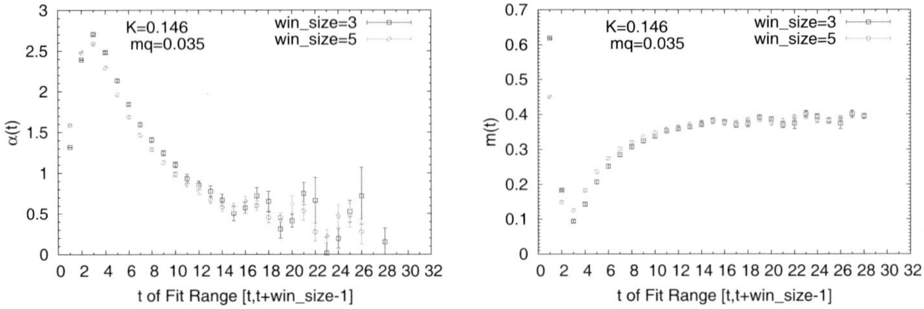

Fig. 14. $\alpha(t)$ and $m(t)$; $N_f = 2$ on a lattice $16^3 \times 64$ at $\beta = 6.5$ and $K = 0.146$.

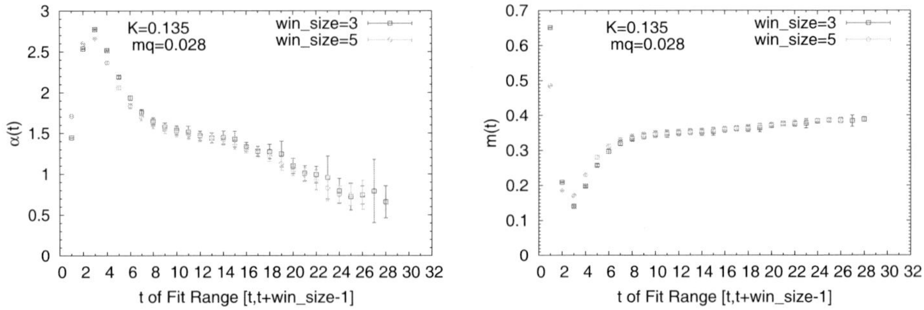

Fig. 15. $\alpha(t)$ and $m(t)$; $N_f = 2$ on a lattice $16^3 \times 64$ at $\beta = 10.0$ and $K = 0.135$.

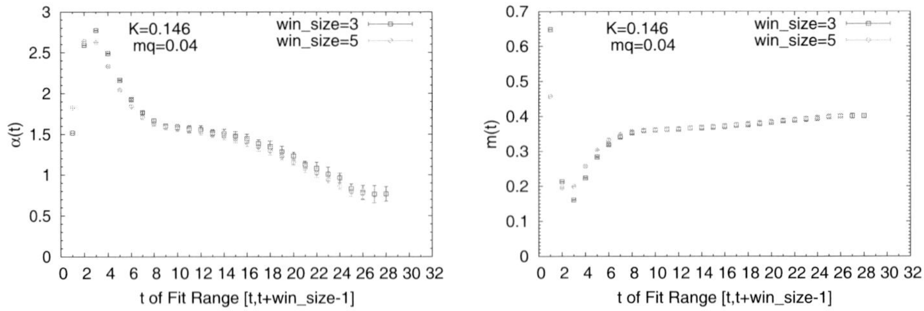

Fig. 16. $\alpha(t)$ and $m(t)$; $N_f = 2$ on a lattice $16^3 \times 64$ at $\beta = 15.0$ and $K = 0.130$.

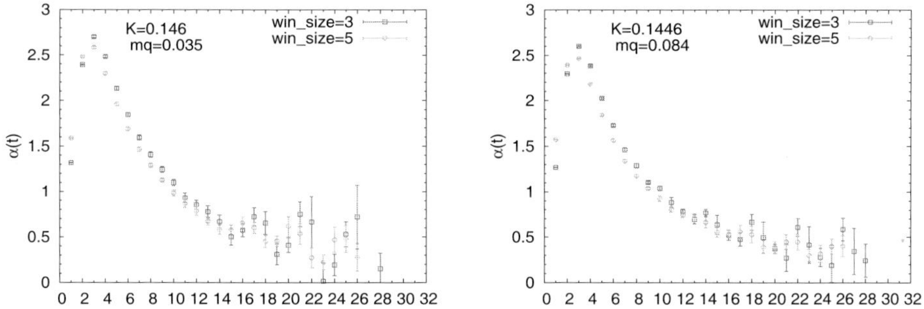

(a) (b)

Fig. 17. The exponent $\alpha(t)$ on a lattice $16^3 \times 64$.(a) $N_f = 2T \sim 2\,T_c$.(b) $N_f = 7; K = 0.1446$.

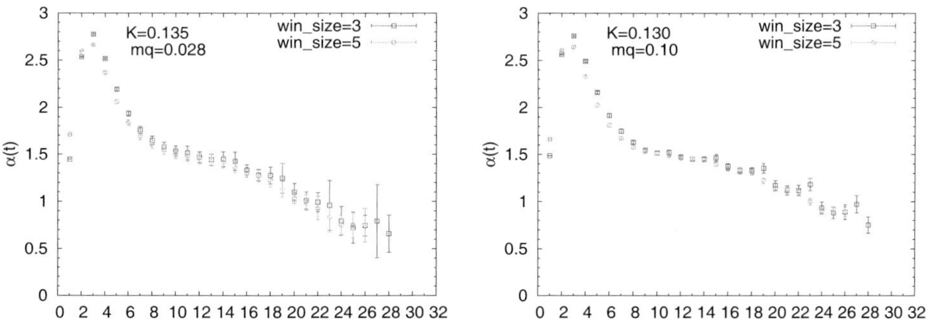

(a) (b)

Fig. 18. The exponent $\alpha(t)$ on a lattice $16^3 \times 64$. (a) $N_f = 2; T \sim 100\,T_c$. (b) $N_f = 16; K = 0.130$.

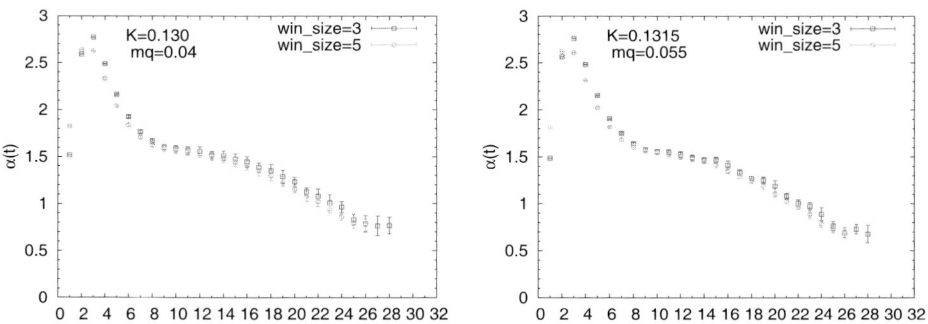

(a) (b)

Fig. 19. The exponent $\alpha(t)$ on a lattice $16^3 \times 64$. (a) $N_f = 2; T \sim 10^5\,T_c$. (b) $N_f = 16; K = 0.1315$.

Can the Higgs Impostor Hide Near the Conformal Window?

Zoltán Fodor

Department of Physics, University of Wuppertal
Gaußstrasse 20, D-42119, Germany
E-mail: fodor@bodri.elte.hu

Kieran Holland

University of the Pacific, Stockton, and Albert Einstein Center, ITP, University of Bern
3601 Pacific Ave, Stockton CA 95211, USA, and Sidlerstrasse 5, Bern CH-3012, Switzerland
E-mail: kholland@pacific.edu

Julius Kuti*

Department of Physics 0319, University of California, San Diego
9500 Gilman Drive, La Jolla, CA 92093, USA
E-mail:jkuti@ucsd.edu

Dániel Nógrádi

Institute for Theoretical Physics, Eötvös University
H-1117 Budapest, Hungary
E-mail:nogradi@bodri.elte.hu

Christopher Schroeder

Physical Sciences Directorate, Lawrence Livermore National Laboratory
Livermore, California 94550, USA
E-mail:chris.schroeder@gmail.com

Chik Him Wong

Department of Physics 0319, University of California, San Diego
9500 Gilman Drive, La Jolla, CA 92093, USA
E-mail:rickywong@physics.ucsd.edu

We present a minimal candidate model for the composite Higgs mechanism with a fermion doublet in the two-index symmetric (sextet) representation of the SU(3) color gauge group.[3] The model is known to be close to the conformal window with a small β-function.[4] Chiral symmetry breaking with vacuum condensate becomes the origin of electroweak symmetry breaking from compositeness. Close to the conformal window, a light scalar state with Higgs quantum numbers is emerging. Associations with the recently observed 126 GeV Higgs boson, or with a dilaton state from broken scale invariance, are subjects of ongoing investigations.

*Speaker at the conference

Keywords: Lattice simulations; Electroweak sector; Composite Higgs; Conformal.

1. Introduction

The discovery of the Higgs-like resonance at 126 GeV at the Large Hadron Collider[1,2] provides the first insight into the origin of electroweak symmetry breaking (EWSB) in the standard model. The minimal realization of EWSB is implemented by introducing an SU(2) doublet scalar Higgs field whose vacuum expectation value, v = 246 GeV, sets the electroweak scale. This simple solution should be regarded as a parametrization rather than a dynamical explanation of EWSB. In particular, the mass-squared parameter of the light Higgs has to be finely tuned, leading to the well-known hierarchy problem. Searching for a deeper dynamical explanation, and resolving the shortcomings of the minimal standard model with its elementary Higgs doublet new physics beyond the Standard Model (BSM) in the framework of some new strongly-interacting gauge theory with a composite Higgs mechanism is an idea which was outside experimental reach when it was first introduced as an attractive BSM scenario.[5–12] The original framework has been considerably extended by new explorations of the multi-dimensional theory space in fermion flavor number, the choice of color gauge group, and fermion representation.[13–22] Systematic and non-perturbative lattice studies play an important role in studies of this extended theory space.[23–64]

One example is the light dilaton as a pseudo-Goldstone particle of spontaneous breaking of scale invariance that has been featured in recent phenomenological discussions as one possible interpretation of the discovery.[65–68] Nearly conformal gauge theories serve as theoretical laboratories for realistic implementations of this scenario.[11,19,65,69–76,99,100] Unfortunately, a credible realization of the idea as a strongly interacting BSM gauge theory is still lacking. We investigate here a candidate theory with a fermion flavor doublet in the two-index symmetric (sextet) representation of the SU(3) color gauge group close to the conformal window, if it is hiding a light Higgs-like scalar state with or without dilaton-like interpretation.

The sextet force and a new fermion doublet driving electroweak symmetry breaking was introduced in QCD a long time ago by Marciano.[15] Early pioneering lattice work, limited to the quenched approximation at that time, investigated the sextet fermion representation.[16] The main difference in the model we investigate here is the introduction of a new SU(3) gauge force not associated with QCD gluons and motivated by ideas of compositeness in BSM scenarios. After chiral symmetry breaking we find three massless Goldstone pions in the spectrum providing the minimal realization of the Higgs mechanism, just like in the original technicolor idea.[5,6] The important new ingredient is the sextet representation of the fermion doublet which brings the model very close to the conformal window as indicated in a recent paper.[4] The accuracy of the very small nearly vanishing β-function in difficult simulations could not resolve the existence of a conformal fixed point gauge coupling from the alternative slowly walking scenario. When combined with our observation of chiral symmetry breaking (χSB) reported here for small fermion mass deformations, the overall consistency of all simulations is resolved if the sextet model is close to the conformal window with a very small non-vanishing β-function (see, also[51,52]). In this case the model exhibits the simplest composite Higgs mechanism and leaves open the possibility of a light scalar state with quantum numbers of the Higgs impostor emerging as the pseudo-Goldstone dilaton state from spontaneous symmetry breaking of scale invariance. Even if scale symmetry breaking is entangled with χSB without dilaton interpretation, a light Higgs-like scalar state can emerge from the new force close to the conformal window. [a]

In section 2 we will outline the computational strategy and the simulation set-up including the important treatment of finite size effects. In section 3 results on the chiral condensate are presented with extrapolation to the massless fermion limit. The spectrum is presented in Section 4 and compared with the χSB hypothesis. Section 5 will describe our ongoing Higgs project to determine the scalar 0^{++} mass spectrum when disconnected diagrams are included in the calculations. Closely related to the dilaton interpretation, we also outline in Section 5 the important role of the non-perturbative gluon condensate and our strategy for investigating it using the gradient flow on gauge field configurations.[77]

2. Computational strategy and finite volume analysis

Probing χSB, we extrapolate the spectrum to infinite volume at fixed fermion mass m. In large volumes the leading finite size corrections are exponentially small and dominated by the lowest state of the spectrum which has pion quantum numbers. From the mass spectrum, extrapolated to infinite volume, we can probe the pattern of χSB when small fermion mass deformations are simulated close to the massless limit. Our results are consistent with the χSB hypothesis.

2.1. *The algorithm and simulation results*

We have used the tree-level Symanzik-improved gauge action for all simulations reported in this paper. The conventional $\beta = 6/g^2$ lattice gauge coupling is defined as the overall factor in front of the well-known terms of the Symanzik lattice action. Its values are $\beta = 3.20$ and $\beta = 3.25$ for our simulations. The link variables in the staggered fermion matrix were exponentially smeared with two stout steps;[78] the precise definition of the staggered stout action was given in.[79] The RHMC algorithm was deployed in all runs. The fermion flavor doublet requires rooting in the algorithm. For molecular dynamics time evolution we applied multiple time scales[80] and the Omelyan integrator.[81] Our error analysis of hadron masses used correlated fitting with double jackknife procedure on the covariance matrices.[82] The

---

[a] Our very recent follow-up work on the model, not reported here, strongly indicates the emergence of a light scalar with 0^{++} quantum numbers not far from the EWSB scale.[83]

time histories of the fermion condensate, the plaquette, and correlators were used to monitor autocorrelation times in the simulations.

We have simulation results at $\beta = 3.2$ in the fermion mass range m $= 0.003 - 0.010$ on $24^3 \times 48$, $28^3 \times 56$, and $32^3 \times 64$ lattices. Five fermion masses at m $= 0.003, 0.004, 0.005, 0.006, 0.008$ are used in most fits. A very large and expensive $48^3 \times 96$ run was added recently at m $= 0.003$ to control finite size effects. We also have new simulation results at $\beta = 3.25$ in the mass range m $= 0.004 - 0.008$ on $24^3 \times 48$, $28^3 \times 56$, and $32^3 \times 64$ lattices. In our report, only some limited results are shown at this gauge coupling.

2.2. *Finite size effects*

Infinite-volume extrapolations of the lowest state in the spectrum with pion quan-

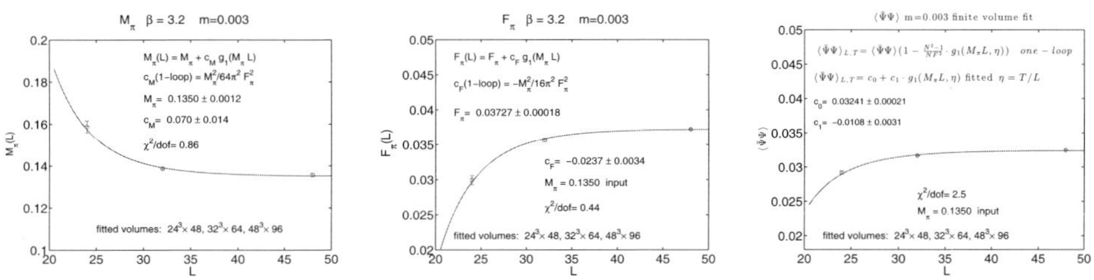

Fig. 1. (Color online). Finite volume dependence at the lowest fermion mass for $\beta = 3.2$. The form of $\widetilde{g}_1(\lambda, \eta)$ is a complicated infinite sum which contains Bessel functions and requires numerical evaluation.[84] Since we are not in the chiral log regime, the prefactor of the $\widetilde{g}_1(\lambda, \eta)$ function was replaced by a fitted coefficient. The leading term of the function $\widetilde{g}_1(\lambda, \eta)$ is a special exponential Bessel function $K_1(\lambda)$ which dominates in the simulation range.

tum numbers, the related F_π, and the condensate $\langle\overline{\psi}\psi\rangle$ are shown in Figure 1 where $\widetilde{g}_1(\lambda, \eta)$ describes finite volume corrections from the exchange of the lightest pion state with $\lambda = M_\pi L$ and lattice aspect ratio $\eta = T/L$, similarly to what was introduced in.[85] The fitting procedure approximates the leading treatment of the pion which wraps around the finite volume, whether in chiral perturbation theory (χpt), or in Lüscher's non-perturbative finite size analysis.[86] This equivalence relaxes the requirement on the fitted parameters c_M,c_F,c_1 to agree with 1-loop χPT as long as the pion is the lightest state dominating the finite volume corrections.

The infinite-volume limits of M_π, F_π, and $\langle\overline{\psi}\psi\rangle$ for $m = 0.003$ at $\beta = 3.2$ were determined self-consistently from the fitting procedure. Similar fits were applied to other composite states. The value of M_π in the fit on the left side plot of Figure 1 was determined from the highly non-linear fitting function and used as input in the other two fits. Based on the fits at $m = 0.003$, the results are within one percent of the infinite-volume limit at $M_\pi L = 5$. In the fermion mass range $m \geq 0.004$ the condition $M_\pi L > 5$ is reached at $L = 32$. Although it will require high precision runs to test, we expect less than one percent residual finite size effects in the $32^3 \times 64$ runs for $m \geq 0.004$. Based on these observations, we will interpret the results from the $32^3 \times 64$ runs for $m \geq 0.004$ as infinite-volume behavior in mass deformed chiral perturbation theory.

3. The chiral condensate

Our simulations show that the chiral condensate $\langle\overline{\psi}\psi\rangle$ is consistent with χSB and remains non-vanishing in the massless fermion limit. It has the infinite-volume spectral representation,

$$\langle\overline{\psi}\psi\rangle = -2m \cdot \int_0^\Lambda d\lambda \frac{\rho(\lambda)}{m^2 + \lambda^2}\,, \tag{1}$$

which is UV-divergent when the cutoff Λ is taken to infinity. The divergences are isolated by writing the integral of the spectral representation in twice subtracted form,[87]

$$\langle\overline{\psi}\psi\rangle = -2m \cdot \int_0^\mu d\lambda \frac{\rho(\lambda)}{m^2 + \lambda^2} - 2m^5 \cdot \int_\mu^\Lambda \frac{d\lambda}{\lambda^4} \frac{\rho(\lambda)}{m^2 + \lambda^2} + c_1(a) \cdot m + c_3(a) \cdot m^3\,. \tag{2}$$

The first integral in Eq. (2) isolates the infrared part and recovers the well-known relation $\langle\overline{\psi}\psi\rangle = -\pi\rho(0)$ in the $m \to 0$ limit.[88] The linear fermion mass term $c_1(a) \cdot m$ is a quadratically divergent UV contribution $\approx a^{-2} \cdot m$ with lattice cutoff a. There is also a very small third-order UV term $c_3(a) \cdot m^3$ with logarithmic divergence which is hard to detect for small m.

IR finite contributions to the condensate from the chiral Lagrangian are connected at the low energy scale μ with the first integral in Eq. (2). In the chiral expansion of the condensate there is an m-independent constant term which is proportional to BF^2, a linear term proportional to $B^2 \cdot m$, a quadratic term $\sim B^3F^{-2} \cdot m^2$, and higher order terms, in addition to logarithmic corrections generated from chiral loops. The expansion in the fermion mass is expressed in terms of low energy constants of chiral perturbation theory, like B and F.[89] We used two independent methods for the determination of the chiral condensate in the massless fermion limit. In the first method fits were made directly to $\langle\overline{\psi}\psi\rangle$ with constant and linear terms in the fitted function. The third order term is hard to detect within the accuracy of the data and its inclusion somewhat increases the error on the intercept. The result is shown on the left side plot of Figure 2.

We also studied the subtracted chiral condensate operator defined with the help of the connected part χ_{conn} of the chiral susceptibility χ,

$$\left[1 - m_{\rm v} \frac{d}{dm_{\rm v}}\right] \langle\overline{\psi}\psi\rangle\Big|_{m_{\rm v}=m} = \langle\overline{\psi}\psi\rangle - m \cdot \chi_{con}\,,$$

$$\chi = \frac{d}{dm}\langle\overline{\psi}\psi\rangle = \chi_{con} + \chi_{disc}\,, \quad \chi_{con} = \frac{d}{dm_{\rm v}}\langle\overline{\psi}\psi\rangle_{pq}\Big|_{m_{\rm v}=m}. \tag{3}$$

The derivatives d/dm and $d/dm_{\rm v}$ are taken at fixed gauge coupling β. The derivative $d/dm_{\rm v}$ is defined in the partially quenched functional integral of $\langle\overline{\psi}\psi\rangle_{pq}$ with respect to the valence mass $m_{\rm v}$ and the limit $m_{\rm v} = m$ is taken after differentiation. The

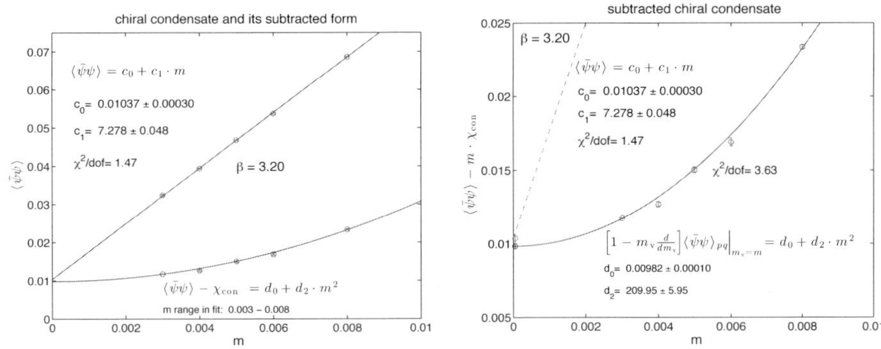

Fig. 2. (Color online). The chiral condensate and its reduced form with subtracted derivative (both have to converge to the same chiral limit) are shown on the left side plot with linear fit to the condensate. The data without derivative subtraction cannot detect higher order fermion mass terms with significant accuracy. The fit to the reduced form with subtracted derivative is defined in the text and shown on the right side plot. The value of d_0 at $m = 0$ is consistent with the direct determination of c_0 from the chiral limit of $\langle\bar\psi\psi\rangle$. The consistency is reassuring since the two results are derived from independent determinations. For $m = 0.003$ data from infinite-volume extrapolation were used in the fit. As we explained earlier, at higher m values the largest volume $32^3 \times 64$ runs were used for the condensate and its derivative subtraction.

subtraction of the derivative term significantly reduces the dominant UV-divergent linear part of the $\langle\bar\psi\psi\rangle$ condensate without changing the intercept in the $m = 0$ limit. Once the derivative term is subtracted, the non-perturbative IR contribution is better exposed. The two independent determinations give consistent non-vanishing fit results in the massless chiral limit as shown on the right side plot of Figure 2.

The two independent determinations of the non-vanishing condensate are consistent [b] but differ from the GMOR[90] relation $\langle\bar\psi\psi\rangle = 2BF^2$, if B and F are taken from chiral fits of Section 3. The value of $2B$ is determined in lattice units from the pion spectrum using the leading $M_\pi^2 = 2B \cdot m$ relation. We find the numerical value $2B = 6.35(21)$ as shown on the left side plot of Figure 3. F is determined from the pseudoscalar correlator which satisfies the PCAC relation. We find in lattice units the numerical value $F = 0.0279(4)$ from the the right side plot of Figure 3 with $2BF^2 = 0.0049(2)$. Both sides of the GMOR relation are sensitive to cutoff effects in B and F at bare lattice coupling $\beta = 3.2$. Our preliminary fits based on staggered chiral perturbation theory indicate that cutoff and two-loop effects modifying the continuum values of B and F are likely sources of the discrepancy.[83] Some increase in the cutoff and two-loop sensitive values of B and F, which is the observed trend, would bring the two sides of the GMOR relation in agreement.

[b] Our new determination of the condensate from the mode number distribution of the fermion spectrum[83] is consistent with the analyses reported here.

4. Goldstone spectrum and resonance predictions

Spectrum calculations in a gauge theory with massless fermions in the chiral limit require important and difficult lattice extrapolations which include extrapolation from finite lattice size to infinite volume, extrapolation to the massless fermion limit, and extrapolation in lattice spacing to the continuum. All three issues are addressed. The strategy of finite size corrections was explained in Section 2 and it will be applied here. Extrapolation from finite fermion masses will be used consistent with χSB. As a first step to address the removal of finite lattice spacing, we will compare the Goldstone and non-Goldstone pion spectra at two different lattice spacings to probe the restoration of taste symmetry for staggered fermions as the lattice spacing is decreased.

4.1. *The Goldstone pion and F_π*

The chiral Lagrangian describes the low energy theory of Goldstone pions and non-Goldstone pions in the staggered lattice fermion formulation. It will be used as an effective tool probing the χSB hypothesis at finite fermion masses including extrapolation to the massless chiral limit. Close to the chiral limit, the pion spectrum

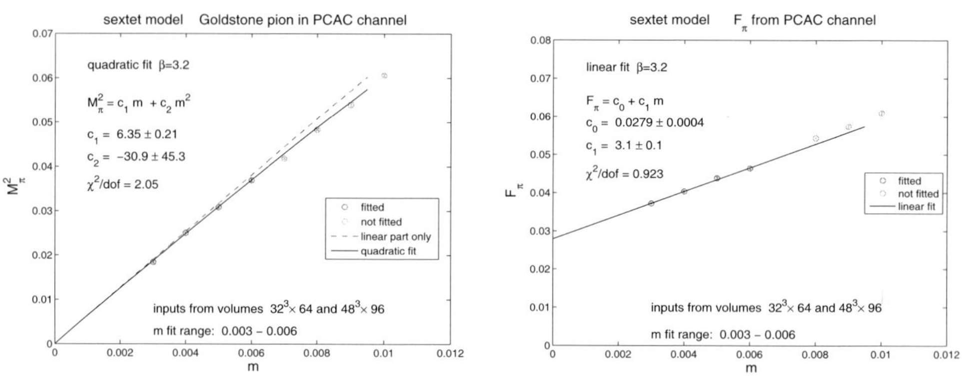

Fig. 3. (Color online). Polynomial fits from the analytic mass dependence of the chiral Lagrangian without logarithmic loop corrections are shown for the Goldstone pion and F_π. The dashed line on the left side plot for the Goldstone pion shows the leading linear contribution. The data point at $m = 0.003$ is determined from the $48^3 \times 96$ lattice and the other fitted points are based on the $32^3 \times 64$ runs.

and the pion decay constant F_π are organized in powers of the fermion mass m which is an input parameter in the simulations. Chiral log corrections to the polynomial terms are generated from pion loops.[91] Their analysis will require an extended dateset with high statistics.

In Section 2 we presented results of infinite-volume extrapolations. The effects are largest at $m = 0.003$ in our dataset and the infinite-volume limits of M_π and F_π were shown for $m = 0.003$ for fixed lattice cutoff and bare coupling $\beta = 3.2$. Similar fits were applied to the chiral condensate and composite states in the spectrum at

$m = 0.003$. Based on the analysis at $m = 0.003$, we determined that the infinite-volume limit is reached at $M_\pi L = 5$ within one percent accuracy. It is expected that similar or better accuracy is reached for $M_\pi L \geq 5$ at higher m values in all states of the spectrum. In the fermion mass range $m \geq 0.004$ the condition $M_\pi L > 5$ is reached at $L = 32$. Based on these observations, in fits to the observed pion spectrum and F_π we will use infinite-volume extrapolation at $m = 0.003$ and treat the $32^3 \times 64$ runs for $m \geq 0.004$ as if the volume were infinite.

In Figure 3 we used the local pion correlator with noisy sources to extract M_π and F_π. The correlator is tagged as the PCAC channel since the PCAC relation, based on axial Ward identities, holds for this correlator and F_π can be directly determined from the residue of the pion pole. The other staggered meson states and correlators we use are defined elsewhere.[92] For example, what we call the non-Goldstone scPion and the f_0 meson are identified in correlator I of Table 1 in an earlier publication.[92] Similarly, the non-Goldstone i5Pion is from correlator VII, the non-Goldstone ijPion is from correlator VIII, and the rho and A1 mesons are from correlator III of Table 1.[92] We measure the Goldstone pion in two different ways, with one of them defined above and the other is correlator II of Table 1 in the already cited publication.[92] For baryon states in the sextet fermion representation, not presented here, we use our own construction of correlators which are different from the baryon correlators of the cited paper.[92]

Based on the analytic fermion mass dependence of the chiral Lagrangian, and using the lowest four fermion masses, good polynomial fits were obtained for M_π and F_π as shown in Figure 3 with fitting functions

$$M_\pi^2 = 2B \cdot m + p_M \cdot m^2 \ , \qquad F_\pi = F + p_F \cdot m \ . \tag{4}$$

The parameters B and F are defined in the two leading terms of the chiral Lagrangian.[91] In this simple fitting procedure B is set from the fit to M_π^2 and F is set independently from the fit to F_π. Their role in the GMOR relation was discussed earlier in Section 3. The fit parameters p_M and p_F describe the respective leading power corrections in m to the pion mass and decay constant. Limited to one lattice spacing only, the accuracy of our dataset is not sufficient for the robust determination of chiral log corrections in Eq. (4). It will provide an important consistency check for B and F in our future analysis of an extended dataset. This analysis requires rooted and partially quenched staggered perturbation theory at finite lattice spacing for simultaneous fits of M_π and F_π with a consistent pair of cutoff-dependent F and B values.[93,94]

Our preliminary analysis of the dataset at $\beta = 3.2$ is consistent with staggered perturbation theory including chiral logs, but the cutoff dependent corrections to the F and B parameters require further testing at weaker gauge couplings and a set of valence fermion masses. We made the first step in this direction by adding a new run set to our database at $\beta = 3.25$. In Figure 4 we show taste-breaking effects in two pion spectra for comparison. Reduction in taste breaking is significant at $\beta = 3.25$ with smaller lattice spacing. Our staggered perturbation theory analysis will be presented in a longer follow-up report.[83]

4.2. *Taste breaking in the non-Goldstone pion spectrum*

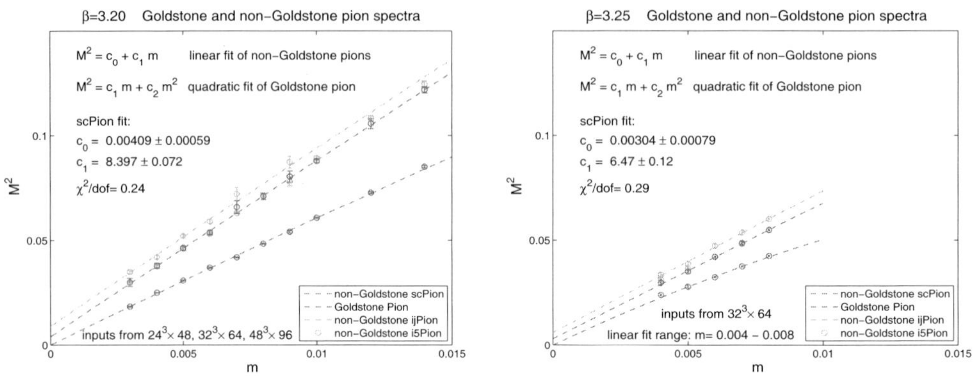

Fig. 4. (Color online). The left side plot in the figure is the spectrum at $\beta = 3.2$. It shows the polynomial fit of the Goldstone pion (magenta points). The red points are the non-Goldstone scPion data covering the green i5Pion data with complete degeneracy. The slightly split ijPion is shown with cyan color. The right side plot in the figure is the spectrum at $\beta = 3.25$. In identical notation it displays the improvement in taste splitting with a considerably less taste-broken spectrum when plotted on the same scale.

The non-Goldstone pion spectra, quite different from the one found in QCD, are shown at $\beta = 3.2$ in the left side plot of Figure 4 using standard notation, introduced earlier. The non-Goldstone i5Pion is split from the Goldstone pion and remains exactly degenerate with the non-Goldstone scPion, a similar feature in QCD. The new feature is the mass dependence of the split between the Goldstone pion and the non-Goldstone i5Pion with non-parallel slopes of the fitting functions. The non-Goldstone ijPion is further split from the i5Pion with a small mass-independent offset. Although taste breaking effects appear substantial on the scale of the plot, they are comparable with those from the HISQ action when the lattice spacings are matched.[95] The trends of the splits, particularly the fan-out structure and the lack of parallel equi-spaced splits with a constant slope determined by B is characteristic of gauge models as they get close to the conformal window. A very small residual mass at $m = 0$ is consistent with fits for the non-Goldstone pion states and decreases as we lower the lattice spacing with the weaker coupling at $\beta = 3.25$. This is shown on the right side plot of Figure 4 which exhibits a similar structure for the same pion states as the left side plot in the figure but on a significantly more collapsed scale. Taste breaking is reduced considerably. It will be interesting to conduct a full analysis of all data on the finer lattice scale, closer to the continuum limit, and compare with the results presented here on the coarser lattice scale.[83]

4.3. *The ρ and A_1 parity partner states*

It is useful and important to investigate the chiral limit of composite hadron states separated by a gap from the Goldstone and non-Goldstone pion spectra. The baryon mass gap in the chiral limit can provide further evidence for χSB but our preliminary results are not shown here. Hadron masses of parity partners also provide important information with split parity masses in the chiral limit. This is particularly helpful not only to confirm χSB but to obtain a first estimate on the S parameter for probing the model against electroweak precision tests.[96] As an example, we will briefly review our results for the ρ meson state and its parity partner, the A_1 meson. Particularly interesting is the $\rho - \mathrm{A}_1$ mass splitting with parity violation. Figure 5 shows fits to the ρ meson and its A_1 parity partner. The left side plot is a

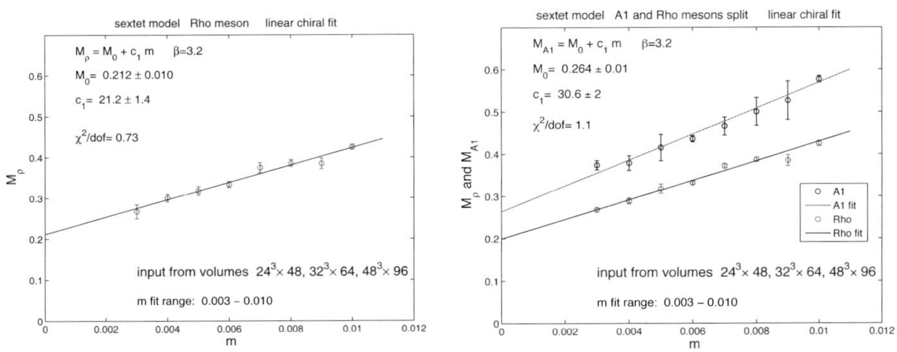

Fig. 5. (Color online). Linear fit to the ρ meson mass is shown on the left side plot of the figure. The right side plot shows the linear fit to the A_1 meson superimposed on the ρ meson plot. The parity split is quite visible with varying size errors in the fitted m range.

linear fit to the ρ meson with a non-vanishing mass at $m = 0$, consistent with χSB. The right side plot shows the linear fit to the A_1 meson. Both states extrapolate to non-vanishing masses in the chiral limit.

5. The scalar spectrum of the ongoing sextet Higgs project

With χSB in the massless fermion limit of the sextet model, its relevance for the realization of the composite Higgs mechanism is self-evident. The three Goldstone pions of the model have the perfect match for providing the longitudinal components of the $W^{\pm}$ and Z bosons. The remaining most important issues are: (1) to calculate the mass of the 0^{++} state when the disconnected part of correlator I in Table 1 of Ref.[92] is included; (2) associations with the recently observed 126 GeV Higgs boson, or with a dilaton state from broken scale invariance; (3) a more precise determination of the running coupling for which we will deploy our new method based on the gradient flow of the gauge field in finite volume.[97] We will briefly summarize the first and second issues.

5.1. *The f_0 state in the 0^{++} channel*

Figure 6 shows the fermion mass dependence of the f_0 meson without including the disconnected part of correlator I in Table 1 of the cited Ref.[92] The non-Goldstone scPion and f_0 are parity partner states in this correlator. Strictly speaking this is the isovector a_0 meson which is not effected by the disconnected annihilation diagram. The linear fit from the connected diagram is shown in Figure 6. It has a non-zero intercept in the chiral limit with a mass more than five times F so it corresponds to the heavy a_0 state and not a Higgs candidate.

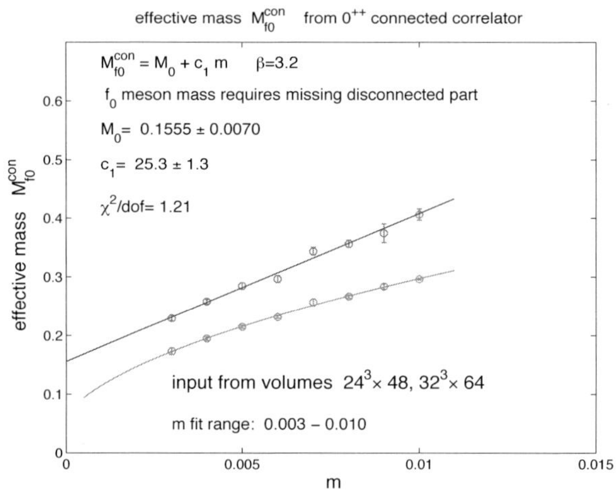

Fig. 6. (Color online). The linear fit is shown to the mass of the 0^{++} f_0 meson from the connected part of correlator I in Table 1 of the paper cited earlier.[92] For comparison, the scPion which is the parity partner of the f_0 meson in the correlator is replotted with its fit from Figure 4 (magenta color). In the continuum limit, the mass of the non-Goldstone scPion will vanish and the f_0 state could become light close to the conformal window. The disconnected part of the correlator is required to resolve this issue.

The quantum numbers of the f_0 meson should match that of the flavor singlet 0^{++} state in the full staggered correlator including the disconnected part (often called hairpin diagrams). Flavor-singlet correlators have fermion-line connected and fermion-line disconnected contributions from hairpin diagrams. To evaluate disconnected quark loops with zero momentum, we need to sum over propagators from sources at each spatial location for a given time slice. To avoid the very costly $\mathcal{O}(V)$ inversions to compute all-to-all propagators in lattice terminology, random sources have to be used with noise reduction. Close to the conformal window it is reasonable to expect that the disconnected diagram will dramatically reduce the f_0 mass when the chiral limit is taken, different from the σ particle of QCD.

Using the technology outlined above, our recent and ongoing work strongly indicates the emergence of a light f_0 scalar state with 0^{++} quantum numbers not far from the EWSB scale.[83] The full f_0 state including the disconnected diagrams could perhaps act as the Higgs impostor after additional mass shift from electroweak corrections.[101]

A very interesting further challenge is the existence of two types of distinct 0^{++} scalar mesons. One of them is the composite fermion state and the other is the scalar glueball with the same quantum number. If the glueball has a low mass these two types of states can mix with a spectrum of light scalar mesons. To resolve this issue requires a well-chosen variational operator set to disentangle the scalar states. This is part of our ongoing investigations.

5.2. *The Higgs particle and the dilaton*

Although the sextet gauge theory with massless fermions is scale invariant on the classical level, from the scale anomaly the dilatation current has a non-vanishing divergence which leads to a well-known partially conserved dilatation current (PCDC) relation among three non-perturbative quantities,

$$m_\sigma^2 \simeq -\frac{4}{f_\sigma^2}\langle 0|\left[\Theta_\mu^\mu(0)\right]_{NP}|0\rangle\,, \tag{5}$$

where m_σ is the dilaton mass, f_σ is its decay amplitude, and $[\Theta_\mu^\mu]_{NP} = \frac{\beta(\alpha)}{4\alpha}[G^a_{\mu\nu}G^{a\mu\nu}]_{NP}$ designates the non-perturbative trace of the energy-momentum tensor, built from gauge fields in the massless fermion limit.[65,73,76] Although $\alpha(\mu)$ and $G^a_{\mu\nu}G^{a\mu\nu}$ depend on the renormalization scale μ, the trace of the energy-momentum tensor is scheme independent after renormalization.

Predictions for m_σ close to the conformal window depend on the behavior of f_σ and the gluon condensate $\left[G^a_{\mu\nu}G^{a\mu\nu}\right]_{NP}$. There are two different expectations about the limit of the gluon condensate to f_σ ratio when the conformal window is approached. In one interpretation, the right-hand side of Eq. (5) is predicted to approach zero in the limit, so that the dilaton mass $m_\sigma^2 \simeq (N_f^c - N_f)\cdot\Lambda^2$ would parametrically vanish when the conformal limit is reached. The Λ scale is defined where the running coupling becomes strong to trigger χSB. The formal parameter $N_f^c - N_f$ with the non-physical (fractional) critical number of fermions vanishes when the conformal phase is reached.[73] In an alternate interpretation the right-hand side ratio of Eq. (5) remains finite in the limit and a residual dilaton mass is expected when scaled with $f_\sigma \simeq \Lambda$.[65,76]

It is important to note that there is no guarantee, even with a very small β-function near the conformal window, for the realization of a light dilaton in association with the light scalar particle. Realistic BSM models have not been built with parametric tuning close to the conformal window. Only non-perturbative lattice calculations can explore the dynamical properties and characteristics of the scalar particle. To shed some more light on near-conformal behavior and the related dilaton, we are currently investigating the lattice determination of the nonperturbative energy-momentum tensor from the gradient flow on gauge field configurations.[77]

Acknowledgments

Julius Kuti would like to thank Koichi Yamawaki and the organizers for kind hospitality at the most enjoyable SCGT2012 workshop of the Kobayashi-Maskawa Institute. This work was supported by the DOE under grant DE-SC0009919, by the NSF under grants 0704171 and 0970137, by the EU Framework Programme 7 grant (FP7/2007-2013)/ERC No 208740, by OTKA under grant OTKA-NF-104034, and by the Deutsche Forschungsgemeinschaft grant SFB-TR 55. The simulations were performed using USQCD computational resources at Fermilab and JLab. Further support was provided by the UCSD GPU cluster funded by DOE ARRA Award ER40546. Some of the simulations used allocations from the Extreme Science and Engineering Discovery Environment (XSEDE), which is supported by National Science Foundation grant number OCI-1053575. In addition, some computational resources were used at the University of Wuppertal, Germany. We are grateful to Kalman Szabo and Sandor Katz for their code development building on Wuppertal gpu technology.[98] KH wishes to thank the Institute for Theoretical Physics and the Albert Einstein Center for Fundamental Physics at Bern University for their support.

References

1. G. Aad *et al.* [ATLAS Collaboration], [arXiv:1207.7214 [hep-ex]].
2. S. Chatrchyan *et al.* [CMS Collaboration], [arXiv:1207.7235 [hep-ex]].
3. Z. Fodor, K. Holland, J. Kuti, D. Nogradi, C. Schroeder and C. H. Wong, Phys. Lett. B **718**, 657 (2012) [arXiv:1209.0391 [hep-lat]].
4. T. DeGrand, Y. Shamir and B. Svetitsky, arXiv:1201.0935 [hep-lat].
5. S. Weinberg, Phys. Rev. D **19**, 1277 (1979).
6. L. Susskind, Phys. Rev. D **20**, 2619 (1979).
7. S. Dimopoulos, L. Susskind, Nucl. Phys. **B155**, 237-252 (1979).
8. E. Eichten, K. D. Lane, Phys. Lett. **B90**, 125-130 (1980).
9. E. Farhi and L. Susskind, Phys. Rept. **74**, 277 (1981).
10. B. Holdom, Phys. Lett. **B150**, 301 (1985).
11. K. Yamawaki, M. Bando, K. -i. Matumoto, Phys. Rev. Lett. **56**, 1335 (1986); Phys. Lett. B **178**, 308 (1986).
12. T. Appelquist, L. C. R. Wijewardhana, Phys. Rev. **D36**, 568 (1987).
13. W. E. Caswell, Phys. Rev. Lett. **33**, 244 (1974).
14. T. Banks and A. Zaks, Nucl. Phys. B **196**, 189 (1982).
15. W. J. Marciano, Phys. Rev. D **21**, 2425 (1980).
16. J. B. Kogut, J. Shigemitsu and D. K. Sinclair, Phys. Lett. B **145**, 239 (1984).
17. T. Appelquist, M. Piai, R. Shrock, Phys. Rev. **D69**, 015002 (2004).
18. F. Sannino, K. Tuominen, Phys. Rev. **D71**, 051901 (2005).
19. D. D. Dietrich, F. Sannino, K. Tuominen, Phys. Rev. **D72**, 055001 (2005).
20. M. A. Luty, T. Okui, JHEP **0609**, 070 (2006).
21. D. D. Dietrich, F. Sannino, Phys. Rev. **D75**, 085018 (2007).
22. M. Kurachi, R. Shrock, JHEP **0612**, 034 (2006).
23. Z. Fodor, K. Holland, J. Kuti, D. Nogradi and C. Schroeder, Phys. Lett. B **681**, 353 (2009).
24. Z. Fodor, K. Holland, J. Kuti, D. Nogradi and C. Schroeder, PoS **LAT2010**, 060 (2010) [arXiv:1103.5998 [hep-lat].

25. Z. Fodor, K. Holland, J. Kuti, D. Nogradi, C. Schroeder, Phys. Lett. B **703**, 348 (2011).
26. Z. Fodor, K. Holland, J. Kuti, D. Nogradi, C. Schroeder and C. H. Wong, PoS **LAT2011**, 073 (2011) arXiv:1205.1878 [hep-lat].
27. T. Appelquist, G. T. Fleming and E. T. Neil, Phys. Rev. Lett. **100**, 171607 (2008).
28. T. Appelquist, G. T. Fleming and E. T. Neil, Phys. Rev. D **79**, 076010 (2009),
29. T. Appelquist, G. T. Fleming, M. F. Lin, E. T. Neil and D. A. Schaich, Phys. Rev. D **84**, 054501 (2011) [arXiv:1106.2148 [hep-lat]].
30. T. Appelquist *et al.*, Phys. Rev. Lett. **104**, 071601 (2010).
31. A. Deuzeman, M. P. Lombardo, E. Pallante, Phys. Lett. **B670**, 41-48 (2008).
32. A. Deuzeman, M. P. Lombardo, E. Pallante, Phys. Rev. **D82**, 074503 (2010).
33. A. Deuzeman, M. P. Lombardo, T. N. da Silva and E. Pallante, arXiv:1111.2590 [hep-lat].
34. A. Hasenfratz, Phys. Rev. **D80**, 034505 (2009).
35. A. Hasenfratz, Phys. Rev. **D82**, 014506 (2010).
36. A. Cheng, A. Hasenfratz and D. Schaich, arXiv:1111.2317 [hep-lat].
37. X. -Y. Jin, R. D. Mawhinney, PoS **LAT2009**, 049 (2009).
38. X. -Y. Jin, R. D. Mawhinney, PoS **LATTICE2010**, 055 (2010).
39. Y. Aoki, T. Aoyama, M. Kurachi, T. Maskawa, K. -i. Nagai, H. Ohki, A. Shibata, K. Yamawaki, and T. Yamazaki, arXiv:1202.4916 [hep-lat].
40. S. Catterall, F. Sannino, Phys. Rev. **D76**, 034504 (2007).
41. S. Catterall, J. Giedt, F. Sannino, J. Schneible, JHEP **0811**, 009 (2008).
42. A. J. Hietanen, J. Rantaharju, K. Rummukainen, K. Tuominen, JHEP **0905**, 025 (2009).
43. A. J. Hietanen, K. Rummukainen, K. Tuominen, Phys. Rev. **D80**, 094504 (2009).
44. L. Del Debbio, B. Lucini, A. Patella, C. Pica, A. Rago, Phys. Rev. **D82**, 014510 (2010).
45. F. Bursa, L. Del Debbio, L. Keegan, C. Pica, T. Pickup, Phys. Lett. **B696**, 374-379 (2011).
46. L. Del Debbio, R. Zwicky, Phys. Rev. **D82**, 014502 (2010).
47. L. Del Debbio, B. Lucini, A. Patella, C. Pica and A. Rago, arXiv:1111.4672 [hep-lat].
48. Y. Shamir, B. Svetitsky, T. DeGrand, Phys. Rev. **D78**, 031502 (2008).
49. T. DeGrand, Y. Shamir, B. Svetitsky, Phys. Rev. **D82**, 054503 (2010).
50. T. DeGrand, Phys. Rev. D **84**, 116901 (2011) [arXiv:1109.1237 [hep-lat]].
51. J. B. Kogut, D. K. Sinclair, Phys. Rev. **D81**, 114507 (2010).
52. J. B. Kogut and D. K. Sinclair, Phys. Rev. D **84**, 074504 (2011) [arXiv:1105.3749 [hep-lat]].
53. E. Bilgici, A. Flachi, E. Itou, M. Kurachi, C. -J D. Lin, H. Matsufuru, H. Ohki, T. Onogi, Phys. Rev. **D80**, 034507 (2009).
54. E. Itou, T. Aoyama, M. Kurachi, C. -J. D. Lin, H. Matsufuru, H. Ohki, T. Onogi, E. Shintani, PoS **LATTICE2010**, 054 (2010).
55. N. Yamada, M. Hayakawa, K. -I. Ishikawa, Y. Osaki, S. Takeda, S. Uno, PoS **LAT2009**, 066 (2009).
56. M. Hayakawa, K. -I. Ishikawa, Y. Osaki, S. Takeda, S. Uno, N. Yamada, [arXiv:1011.2577 [hep-lat]].
57. R. V. Gavai, Nucl. Phys. **B269**, 530 (1986).
58. N. Attig, B. Petersson, M. Wolff, Phys. Lett. **B190**, 143 (1987).
59. S. Meyer, B. Pendleton, Phys. Lett. **B241**, 397-402 (1990).
60. P. H. Damgaard, U. M. Heller, A. Krasnitz, P. Olesen, Phys. Lett. **B400**, 169-175 (1997).

61. S. -y. Kim, S. Ohta, Phys. Rev. **D46**, 3607-3617 (1992).
62. F. R. Brown, H. Chen, N. H. Christ, Z. Dong, R. D. Mawhinney, W. Schaffer, A. Vaccarino, Phys. Rev. **D46**, 5655-5670 (1992).
63. Y. Iwasaki, K. Kanaya, S. Kaya, S. Sakai, T. Yoshie, Phys. Rev. **D69**, 014507 (2004).
64. Y. Iwasaki, arXiv:1212.4343 [hep-lat].
65. S. Matsuzaki and K. Yamawaki, Phys. Rev. D **86**, 035025 (2012) [arXiv:1206.6703 [hep-ph]].
66. J. Ellis and T. You, arXiv:1207.1693 [hep-ph].
67. I. Low, J. Lykken and G. Shaughnessy, arXiv:1207.1093 [hep-ph].
68. D. Elander and M. Piai, arXiv:1208.0546 [hep-ph].
69. W. A. Bardeen, C. N. Leung and S. T. Love, Phys. Rev. Lett. **56**, 1230 (1986).
70. B. Holdom and J. Terning, Phys. Lett. B **187**, 357 (1987); Phys. Lett. B **200**, 338 (1988).
71. V. A. Miransky and K. Yamawaki, Phys. Rev. D **55**, 5051 (1997).
72. W. D. Goldberger, B. Grinstein and W. Skiba, Phys. Rev. Lett. **100**, 111802 (2008).
73. T. Appelquist and Y. Bai, Phys. Rev. D **82**, 071701 (2010).
74. B. Grinstein and P. Uttayarat, JHEP **1107**, 038 (2011).
75. O. Antipin, M. Mojaza and F. Sannino, Phys. Lett. B **712**, 119 (2012).
76. M. Hashimoto and K. Yamawaki, Phys. Rev. D **83**, 015008 (2011).
77. M. Luscher, JHEP **1008**, 071 (2010) [arXiv:1006.4518 [hep-lat]].
78. C. Morningstar and M. J. Peardon, Phys. Rev. D **69**, 054501 (2004).
79. Y. Aoki, Z. Fodor, S. D. Katz, K. K. Szabo, JHEP **0601**, 089 (2006).
80. C. Urbach, K. Jansen, A. Shindler, U. Wenger, Comput. Phys. Commun. **174**, 87-98 (2006).
81. T. Takaishi, P. de Forcrand, Phys. Rev. **E73**, 036706 (2006).
82. L. Del Debbio, L. Giusti, M. Luscher, R. Petronzio and N. Tantalo, JHEP **0702**, 082 (2007) [hep-lat/0701009].
83. Z. Fodor, K. Holland, J. Kuti, D. Nogradi, and C. H. Wong, unpublished (2013).
84. J. Gasser, H. Leutwyler, Phys. Lett. **B184**, 83 (1987).
85. H. Leutwyler, Phys. Lett. B **189**, 197 (1987).
86. M. Luscher, Commun. Math. Phys. **104**, 177 (1986).
87. H. Leutwyler, A. V. Smilga, Phys. Rev. **D46**, 5607-5632 (1992).
88. T. Banks and A. Casher, Nucl. Phys. B **169**, 103 (1980).
89. J. Bijnens, J. Lu, JHEP **0911**, 116 (2009). [arXiv:0910.5424 [hep-ph]].
90. M. Gell-Mann, R. J. Oakes and B. Renner, Phys. Rev. **175**, 2195 (1968).
91. J. Gasser and H. Leutwyler, Annals Phys. **158**, 142 (1984).
92. N. Ishizuka, M. Fukugita, H. Mino, M. Okawa, A. Ukawa, Nucl. Phys. **B411**, 875-902 (1994).
93. C. Aubin and C. Bernard, Phys. Rev. D **68**, 034014 (2003).
94. C. Aubin and C. Bernard, Phys. Rev. D **68**, 074011 (2003).
95. A. Bazavov *et al.* [HotQCD Collaboration], PoS LATTICE **2010**, 169 (2010).
96. M. E. Peskin and T. Takeuchi, Phys. Rev. D **46**, 381 (1992).
97. Z. Fodor, K. Holland, J. Kuti, D. Nogradi and C. H. Wong, JHEP **1211**, 007 (2012).
98. G. I. Egri, Z. Fodor, C. Hoelbling, S. D. Katz, D. Nogradi and K. K. Szabo, Comput. Phys. Commun. **177**, 631 (2007).
99. S. Matsuzaki and K. Yamawaki, arXiv:1209.2017 [hep-ph].
100. O. Antipin, S. Di Chiara, M. Mojaza, E. Molgaard and F. Sannino, arXiv:1205.6157 [hep-ph].
101. R. Foadi, M. T. Frandsen and F. Sannino, arXiv:1211.1083 [hep-ph].

Understanding Finite-Volume Effects in Deformed CFTs on the Lattice

L. Del Debbio*, B. Lucini⋆, A. Patella◇,♣, C. Pica♡, A. Rago◇

* *School of Physics & Astronomy, The University of Edinburgh, Edinburgh EH9 3JZ, UK.*
⋆ *College of Science, Swansea University, Singleton Park, Swansea SA2 8PP, UK.*
◇ *School of Computing and Mathematics, University of Plymouth, Plymout PL4 8AA, UK.*
♣ *CERN, Physics Department, CH-1211 Geneva 23, Switzerland.*
♡ *CP*[3]*-Origins & DIAS, University of Southern Denmark, Campusvej 55, DK-5230 Odense M.*

Four-dimensional gauge theories with sufficiently many fermion flavors can provide explicit examples of theories with an infra-red fixed point (IRFP). In recent years the existence of IRFPs has been the subject of intense scrutiny using numerical simulations of gauge theories defined on a lattice. Being able to disentangle the signatures of the fixed point from lattice artefacts is the main challenge in order to reach robust conclusions. In this work, we discuss preliminary results on the finite volume effects for the SU(2) gauge theories with two fermions in the adjoint representation.

Keywords: Infrared fixed points, lattice simulations

1. Introduction

The existence of a conformal window[1,2] for nonsupersymmetric gauge theories is an interesting theoretical problem, which has potentially important consequences for model-building beyond the Standard Model.[3–5] This issue has been actively studied in recent years by numerical simulations, see e.g. Ref.[6] and references therein for a recent summary of methodologies and results. The existence of an infrared fixed point (IRFP) can be tested by studying the scaling of the spectrum with the mass of the fermionic matter fields.[7–9] Unfortunately the scaling is distorted in actual simulations by finite volume effects. The latter need to be understood in detail, and brought under control, in order to get a quantitative understanding of the long distance dynamics of these theories. Early studies suggest that the SU(2) gauge theory with $n_f = 2$ Dirac fermions in the adjoint representation has an IRFP. This theory was introduced as a candidate for dynamical electroweak symmetry breaking under the name of Minimal Walking Technicolor (MWT).[10] Extensive investigations have confirmed the presence of a bias in the spectrum computed at finite volume, see e.g. Ref.[11] and references therein. In this work we give an overview of the most recent simulations that we performed in order to tame the finite volume effects in the spectrum of MWT.

2. Lattice simulations

Simulations are performed using the RHMC algorithm described in Ref.[12] Early results from our investigations, and their interpretation, have been presented in a series of papers.[13–15] Our latest results have been presented in Ref.,[16] and a first account of the size of finite-volume effects was discussed in Ref.[11] In the latter paper we discussed in detail the possible sources of finite-volume effects, and presented preliminary estimates of the lattice sizes needed to reduce them at the percent level. The finite extent of the lattice in the temporal direction is responsible for poor plateaux in the effective masses, and hence a poor signal when trying to fit for the masses in the spectrum. One way to address this problem was explored in Ref.,[16] where smeared sources were used to minimize the contamination from excited states in the lattice correlators. In this work large temporal extensions are used in order to always have robust fitting ranges for the determination of the masses in the spectrum, and the focus will be on the effects due to the finite spatial volume of our lattices.

Simulations are performed at a single value of the bare gauge coupling, $\beta = 2.25$, which is the value used in previous studies. We choose two values of the fermion bare mass $am_0 = -1.05, -1.15$, corresponding to pseudoscalar masses $aM_{\mathrm{PS}} \approx 1.19, 0.64$ respectively; finite volume effects are investigated by comparing results obtained for different volumes. At the lightest quark mass, we also performed simulations with twisted boundary conditions.[17] Results obtained from lattices with different boundary conditions should agree when the volume of the system is large compared to the typical correlation length of the system, which is determined by the inverse of the lightest mass in the spectrum. The new lattices generated for this work are summarized in Table 1.

Table 1. List of lattices used in this study. The bare mass, number of trajectories, and lengths of the trajectories are reported. Values for the average plaquette, and the eigenvalue of the Dirac operator are shown, together with their respective integrated autocorrelation time.

lattice	V	$-am_0$	N_{traj}	t_{traj}	$\langle P \rangle$	τ	λ	τ_λ
E0	64×8^3	1.15	810	3	0.66536(22)	3.6(1.2)	0.2005(58)	1.42(31)
F0	64×12^3	1.15	530	1.5	0.66601(15)	1.93(59)	0.2054(40)	1.54(43)
G0	64×16^3	1.15	1500	1.5	0.665992(61)	2.32(46)	0.2116(16)	2.38(48)
D3	64×24^3	1.15	2387	1.5	0.665927(26)	3.92(79)	0.21478(70)	1.70(24)
H0	64×32^3	1.15	2541	1.5	0.665946(30)	3.37(62)	0.2115(12)	1.06(12)
J0	80×48^3	1.15	2200	1.5	0.665943(17)	5.1(1.2)	0.2237(17)	0.637(58)
B2	24×12^3	1.05	7819	1	0.647633(70)	6.79(99)	1.4936(51)	5.80(78)
G1	64×16^3	1.05	2648	1.5	0.647645(48)	4.63(96)	1.4389(36)	1.26(14)
D4	64×24^3	1.05	4000	1.5	0.647695(37)	3.56(53)	1.3906(45)	0.722(54)
K0	48×32^3	1.05	3590	1.5	0.647680(30)	4.28(74)	1.3708(49)	0.632(45)
TW0	64×8^3	1.15	565	1.5	0.66665(22)	2.8(1.0)	0.5557(96)	0.85(18)
TW1	64×12^3	1.15	741	1.5	0.66590(11)	2.96(96)	0.2709(48)	1.93(50)
TW2	64×16^3	1.15	1162	1.5	0.665990(61)	2.91(73)	0.2484(17)	6.2(2.2)
TW3	64×24^3	1.15	2701	1.5	0.665912(35)	4.63(95)	0.21840(88)	2.43(37)

Interestingly the twist in the boundary conditions induces a change of sign in the finite-volume corrections to the pseudoscalar meson. The large volume limit is reached from different directions, and the comparison of the twisted and untwisted simulations yields a visually clear indication of the approach to a regime where finite volume effects are negligible.

The observables that have been considered in this study are the masses of the mesonic flavor non-singlet states, their decay constants, and the masses of the flavor-singlet mesons and glueballs. The masses and decays constant are extracted as usual from the large-time behaviour of correlators of fields with the desired quantum numbers. The fermion bilinear operators used here have been defined in Ref.[15] For the glueball spectrum we use the variational method discussed in Ref.[18]

3. Results

Let us first address the issue of thermal effects. The effective mass in the pseudoscalar channel is shown in Fig. 1. It is clear for the plot that a time extent $L/a \geq 64$ is sufficient in order to identify a fitting range that is free of contaminations from higher states at the current level of statistical precision for all spatial volumes. The dependence of the mass on the spatial volume is also visible in this figure: the value of the plateaux changes significantly as the thermodynamic limit is approached. In the rest of this work we will assume that we have sufficiently large time extensions, and will not discuss thermal effects.

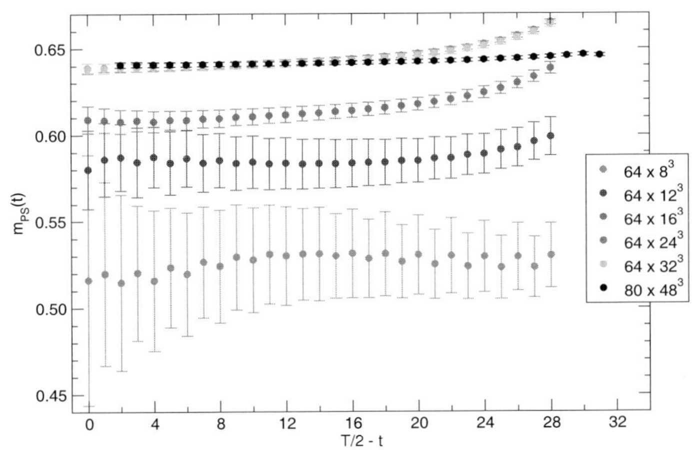

Fig. 1. Plateaux for the effective mass in the pseudoscalar channel for the lighter fermion mass. Different curves correspond to different spatial volumes.

Results for the mesonic spectrum at the higher value of the fermion mass are summarized in Fig. 2(a). Note that the effective masses extracted from different correlators with the same quantum numbers agree nicely, and that the finite volume effects are small for $L/a \geq 16$. In all channels we can perform a robust extrapolation to the thermodynamic limit; the results of this extrapolation are shown in the same figure. The mesonic spectrum for the lighter mass is reported in Fig. 2(b). The

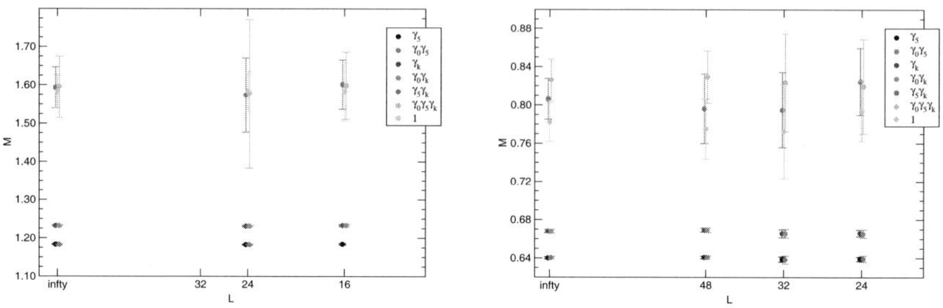

(a) (b)

Fig. 2. Summary of the mesonic spectrum. (a) Data for the heavier fermion mass. (b) Data for the lighter fermion mass.

general pattern is similar to the one observed in Fig. 2(a). The results are compatible with those obtained in previous works, and finite volume effects become negligible for $L \geq 24$ in this case, as expected given that the masses of the states are lighter almost by a factor 2.

The complete spectrum for the heavier mass is reported in Fig. 3. It is worthwhile to emphasize two important features of the glueball spectrum. The degeneracy observed on small volumes between the A^{++} and the E^{++} states is lifted for $L/a \geq 24$. On the new lattices with $L/a \geq 24$ we are able to extract a signal also for the T^{++} channel, which was not the case on the smaller lattices used in previous investigations. The two states E^{++} and T^{++}, which should have the same mass in the continuum limit, remain degenerate, while the scalar A^{++} becomes lighter. The degeneracy observed on small lattices appears to be a finite volume effect. The temporal and spatial string tensions agree nicely on all the volumes that are analysed in this work. At this value of the fermion mass, all masses can be safely extrapolated to the thermodynamic limit. It should be noted that the glueball states, and the string tension, remain lighter than the mesonic spectrum, and in particular they remain lighter than the mass of the pseudoscalar meson, which is a clear indication that chiral symmetry is not spontaneously broken.

As anticipated in the previous section, we have checked for finite volume effects by comparing the mass spectrum obtained from simulations with different boundary conditions. The results for the pseudoscalar mass are shown in Fig. 4 for the lighter fermion mass. Again we find that finite-volume effects become negligible for $L/a \geq 24$.

Finally the entire spectrum for the lighter fermion mass is reported in Fig. 5. As in the case of the heavier mass discussed above, the glueball spectrum remains

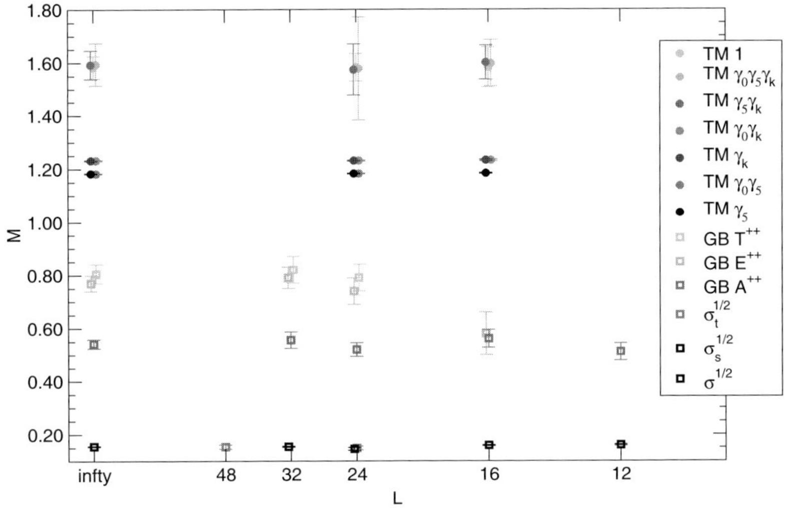

Fig. 3. Masses of all the states measured in this study at the heavier fermion mass.

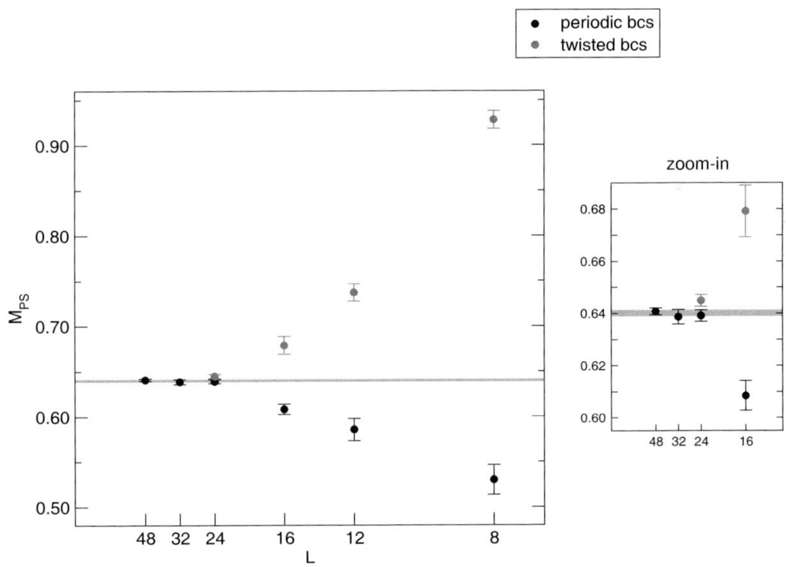

Fig. 4. Mass of the pseudoscalar meson with periodic and twisted boundary conditions. Results obtained for the light fermion mass $am_0 = -1.15$.

lighter than the mesonic one. As the mass is lowered the pseudoscalar meson does not get light compared to the gluonic sector of the theory. As already specified above, this is in sharp contrast with the behavioiur expected in a theory where chiral symmetry is spontaneously broken. However, it is clear from Fig. 5 that our results for the spectrum are still affected by lattice artefacts. For $L/a = 32$ the A^{++} and the E^{++} states are degenerate, and differ from the T^{++} state. This is at odds with the behaviour expected in the thermodynamic limit and close to the continuum limit. On the larger volume, $L/a = 48$, the three states become degenerate. A detailed analysis of the operators in the variational basis is needed in order to understand these results better. Higher statistics may be necessary in order to clarify the pattern of the glueball spectrum.

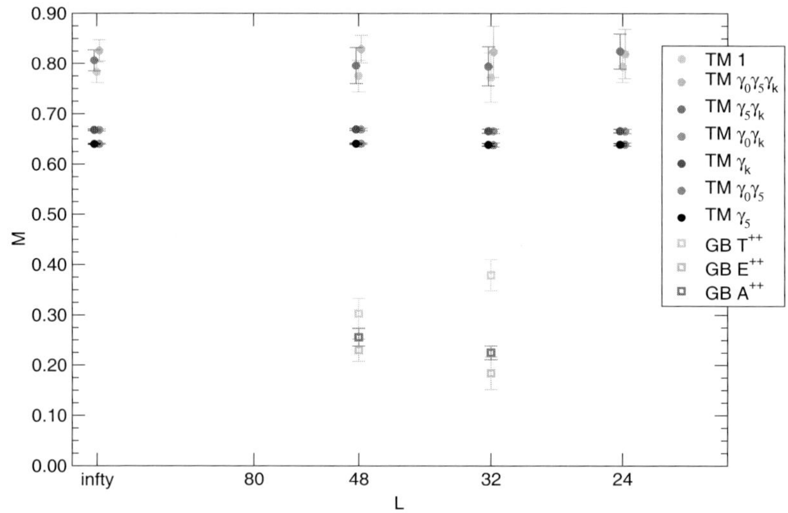

Fig. 5. Spectrum of mesonic and glueball states for the light fermion mass $am_0 = -1.15$.

4. Outlook

Using numerical simulations on large lattices, we have clarified the impact of finite volume effects on the spectrum of MWT. For the lattices we have considered in this study, the mesonic spectrum is free of finite-size effects, and we have a reliable extrapolation of our results to the infinite volume limit. More extensive data are needed in order to reach the same level of accuracy in the gluonic sector. A detailed analysis of the correlators in this sector will appear in a forthcoming publication.

Finding robust evidence for a conformal fixed point is a difficult task, the finite values of the lattice size, and of the fermion mass, introduce corrections to the conformal behaviour that are likely to obscure the sought signals. In this respect, it is important to look for improved methodologies and new observables. A novel proposal for the analysis of two-point correlation functions presented at this conference[19,20] could be tested on the lattices presented in this study.

References

1. W. E. Caswell, *Phys.Rev.Lett.* **33**, p. 244 (1974).
2. T. Banks and A. Zaks, *Nucl.Phys.* **B196**, p. 189 (1982).
3. C. T. Hill and E. H. Simmons, *Phys.Rept.* **381**, 235 (2003).
4. F. Sannino, *Acta Phys.Polon.* **B40**, 3533 (2009).
5. J. Andersen, O. Antipin, G. Azuelos, L. Del Debbio, E. Del Nobile *et al.*, *Eur.Phys.J.Plus* **126**, p. 81 (2011).
6. J. Giedt, *PoS* **LATTICE2012**, p. 006 (2012).
7. T. DeGrand and A. Hasenfratz, *Phys.Rev.* **D80**, p. 034506 (2009).
8. L. Del Debbio and R. Zwicky, *Phys.Rev.* **D82**, p. 014502 (2010).
9. L. Del Debbio and R. Zwicky, *Phys.Lett.* **B700**, 217 (2011).
10. D. D. Dietrich and F. Sannino, *Phys.Rev.* **D75**, p. 085018 (2007).
11. A. Patella, L. Del Debbio, B. Lucini, C. Pica and A. Rago, *PoS* **LATTICE2011**, p. 084 (2011).
12. L. Del Debbio, A. Patella and C. Pica, *Phys.Rev.* **D81**, p. 094503 (2010).
13. L. Del Debbio, B. Lucini, A. Patella, C. Pica and A. Rago, *Phys.Rev.* **D80**, p. 074507 (2009).
14. L. Del Debbio, B. Lucini, A. Patella, C. Pica and A. Rago, *Phys.Rev.* **D82**, p. 014510 (2010).
15. L. Del Debbio, B. Lucini, A. Patella, C. Pica and A. Rago, *Phys.Rev.* **D82**, p. 014509 (2010).
16. F. Bursa, L. Del Debbio, D. Henty, E. Kerrane, B. Lucini *et al.*, *Phys.Rev.* **D84**, p. 034506 (2011).
17. G. 't Hooft, *Nucl.Phys.* **B153**, p. 141 (1979).
18. B. Lucini, A. Rago and E. Rinaldi, *JHEP* **1008**, p. 119 (2010).
19. K.-I. Ishikawa, Y. Iwasaki, Y. Nakayama and T. Yoshie (2013), *Conformal Theories with IR cutoff*, arXiv:1301.4785.
20. K.-I. Ishikawa, Y. Iwasaki, Y. Nakayama and T. Yoshie (2013), *Conformal Behavior in High Temperature QCD*, arXiv:1304.4345.

Reaching the Chiral Limit in Many Flavor Systems

Anna Hasenfratz,* Anqi Cheng, Gregory Petropoulos and David Schaich
Department of Physics, University of Colorado, Boulder, CO 80309, United States
**E-mail: anna@eotvos.colorado.edu*

We present a brief overview of our recent lattice studies of SU(3) gauge theory with $N_f = 8$ and 12 fundamental fermions, including some new and yet-unpublished results. To explore relatively unfamiliar systems beyond lattice QCD, we carry out a wide variety of investigations with the goal of synthesizing the results to better understand the non-perturbative dynamics of these systems. All our findings are consistent with conformal infrared dynamics in the 12-flavor system, but with 8 flavors we observe puzzling behavior that requires further investigation.

Our new Monte Carlo renormalization group technique exploits the Wilson flow to obtain more direct predictions of a 12-flavor IR fixed point. Studies of $N_f = 12$ bulk and finite-temperature transitions also indicate IR conformality, while our current results for the 8-flavor phase diagram do not yet provide clear signs of spontaneous chiral symmetry breaking. From the Dirac eigenvalue spectrum we extract the mass anomalous dimension γ_m, and predict $\gamma_m^\star = 0.32(3)$ at the 12-flavor fixed point. The $N_f = 8$ system again shows interesting behavior, with a large anomalous dimension across a wide range of energy scales. We use the eigenvalue density to predict the chiral condensate, and compare this approach with direct and partially-quenched $\left\langle\overline{\psi}\psi\right\rangle$ measurements.

1. Introduction and overview

The Higgs boson discovered at the Large Hadron Collider in 2012 remains consistent with the minimal standard model.[1] Models of new strong dynamics in which the Higgs is a composite must describe this light scalar in order to remain phenomenologically viable. Such strongly-coupled systems require non-perturbative analysis, which has inspired several recent large-scale lattice studies. Lattice investigations beyond QCD not only need to identify promising models to consider, we also need to determine the best approaches to use; in many cases the most effective methods differ from the familiar techniques of lattice QCD. Our group has recently focused on two models, $N_f = 8$ and 12 fundamental fermions interacting with SU(3) gauge fields.[2–17] We have developed several improved methods to explore these and other models, focusing on systems likely to exhibit conformal or approximately-conformal dynamics in the infrared.

In these proceedings we briefly present some methods we developed and results we obtained for the 8- and 12-flavor models, including some new and preliminary findings that will be published in the future. All our 12-flavor results are consistent with IR conformality, while with $N_f = 8$ we observe some puzzling behavior that we

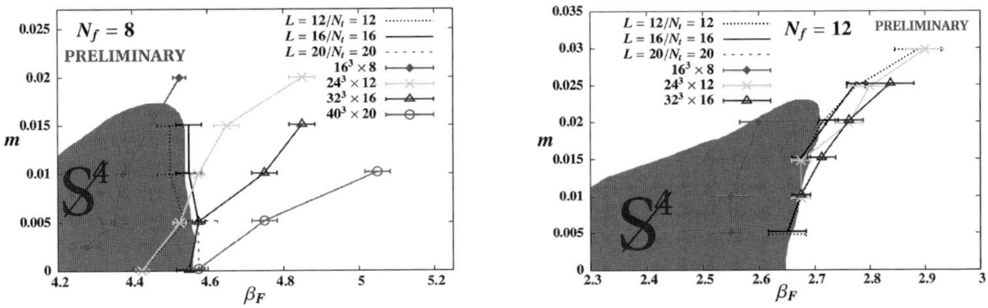

Fig. 1. Bulk and finite-temperature transitions for $N_f = 8$ (left) and $N_f = 12$ (right).[8] The $\mathcal{S}^4$ phase is shaded and the $\mathcal{S}^4$ bulk transitions are shown in black. The colored points indicate finite-temperature transitions determined primarily from the RG-blocked Polyakov loop and eigenvalue density.

continue to investigate. After reviewing the lattice phase diagrams for these models in Section 2, we present our new Wilson-flowed Monte Carlo renormalization group (MCRG) method in Section 3. In Section 4 we summarize our recent studies of the Dirac eigenvalue spectrum, finally considering the chiral condensate in Section 5.

Our studies use a plaquette gauge action that includes an adjoint term with coefficient $\beta_A = -0.25\beta_F$, and nHYP-smeared staggered fermions with smearing parameters $\alpha = (0.5, 0.5, 0.4)$. In earlier work[3,8] we showed that both the 8- and 12-flavor systems possess an unusual phase where the single-site shift symmetry of the staggered action is spontaneously broken ("$\mathcal{S}^4$"). This strong-coupling lattice phase is quite generic and has been observed by all $N_f = 12$ investigations using improved staggered actions.[9,11] First-order transitions separate the $\mathcal{S}^4$ phase from the weak-coupling phase where the continuum limit is defined. This prevents simulations from investigating stronger couplings. For $N_f = 12$ we can still access a large coupling range, but in the 8-flavor case the presence of the $\mathcal{S}^4$ phase prevents, at least for now, resolution of the puzzles we encounter in our studies. In the following we present results up to the $\mathcal{S}^4$ phase, but our focus is on the weak-coupling side.

2. The finite temperature phase structure

We begin by showing in Fig. 1 both finite-temperature transitions (colored points) as well as the first-order bulk transitions (black lines) that separate the $\mathcal{S}^4$ phase from the weak-coupling phase. The bulk transitions are determined from the $\mathcal{S}^4$ order parameters.[3] We also use the Dirac eigenvalue density $\rho(\lambda)$ to identify both bulk and finite-temperature transitions. In addition, we discovered that by measuring observables such as the Polyakov loop on RG-blocked lattices, we can greatly enhance signals of the finite-temperature transition, without distorting the physics.[8]

For $N_f = 12$ in the right panel of Fig. 1, the finite-temperature transitions occur at the same β_F as the $\mathcal{S}^4$ bulk transitions. That is, we move directly from the $\mathcal{S}^4$ phase into a chirally symmetric weak-coupling phase, as expected for an IR-conformal system.

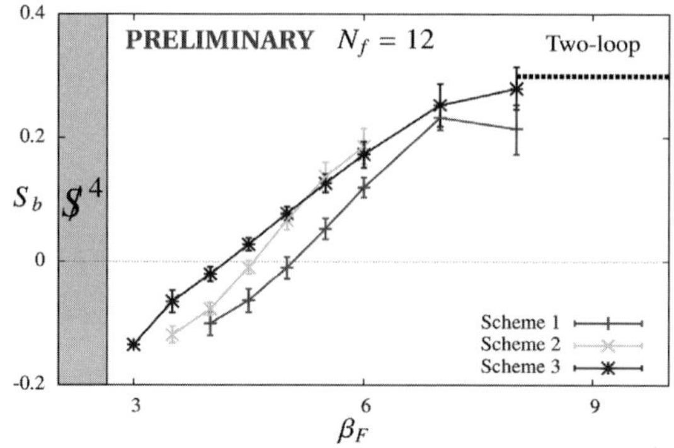

Fig. 2. Preliminary results for the $N_f = 12$ bare step-scaling function s_b from Wilson-flowed MCRG two-lattice matching, illustrating the existence of an IR fixed point. We match $24^3 \times 48$ to $12^3 \times 24$ lattice volumes using three different renormalization schemes corresponding to HYP smearing parameters $\alpha = (0.6, 0.2, 0.2)$, $\alpha = (0.6, 0.3, 0.2)$ and $\alpha = (0.65, 0.3, 0.2)$, respectively. The dashed line is the perturbative prediction for asymptotically weak coupling.

For $N_f = 8$, at sufficiently large fermion mass m there is a region in between the $\not{S}^4$ phase and the chirally symmetric weak-coupling phase, where the lattice systems are confined and chirally broken. For these m, the locations of the finite-temperature transitions change with N_t as expected for QCD-like systems. However, as we approach the chiral limit (which is required to explore IR-conformal or approximately-conformal dynamics), this scaling is lost and all the finite-temperature transitions run into the $\not{S}^4$ phase, with no clear sign of spontaneous chiral symmetry breaking.

At very small mass the finite spatial volume of the lattice can distort the finite-temperature phase diagram. We are currently comparing $24^3 \times 12$ and $36^3 \times 12$ simulations to study these effects, with preliminary results indicating complete consistency for $m \geq 0.005$. It is interesting to note that we are able to simulate directly in the chiral limit on $24^3 \times 48$ volumes everywhere in the weak-coupling and $\not{S}^4$ phases, and on $32^3 \times 64$ volumes for all couplings that we have tested so far ($\beta_F \geq 4.8$). This is very different from the usual QCD-like behavior, which we have checked with $N_f = 4$ simulations. A lattice action where stronger couplings can be reached, before encountering a bulk transition, would help resolve this puzzle.

3. The step scaling function with Wilson-flowed MCRG

Running coupling studies can connect the well-understood perturbative regime to the strong-coupling region where chiral symmetry breaking or IR-conformal dynamics emerges. We use Monte Carlo renormalization group (MCRG) methods to determine the bare step scaling function s_b.[2,12] In the past we used an MCRG two-lattice matching approach that required independent optimization of the RG blocking transformation at every gauge coupling. This predicted a step scaling function that was a composite, probing a different renormalization scheme at every coupling.

Recently we developed a new technique that combines Wilson-flow smearing with

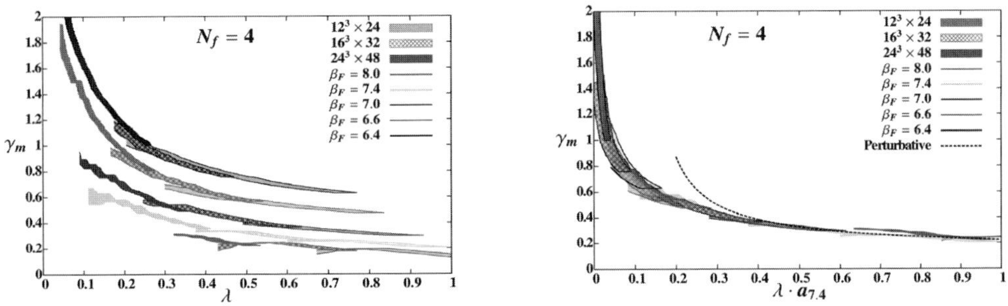

Fig. 3. Eigenmode scaling predictions for the scale-dependent mass anomalous dimension γ_m for $N_f = 4$.[14] In the right panel we rescale λ to a common lattice spacing.

MCRG two-lattice matching.[12] Our new method starts with a short Wilson flow integration before the RG blocking steps. The Wilson flow moves the configurations in the action-space without changing their infrared properties. We use the flow time as the optimization parameter to approach the renormalized trajectory of any fixed RG blocking transformation. This way we determine a step scaling function that corresponds to a unique RG β function.

By changing the blocking transformation we can predict the β function in different renormalization schemes. Fig. 2 shows our preliminary results for the $N_f = 12$ system using three blocking transformations. Each blocking transformation involves two sequential HYP smearings, with different smearing parameters specified in the caption. All three renormalization schemes show an infrared fixed point (IRFP), and we observe the location of the IRFP to depend on the scheme.

The Wilson-flowed MCRG method is fully non-perturbative and can be applied to any other model, even at strong coupling. There is no need for special boundary conditions, allowing configurations to be reused. The ability to compare different renormalization schemes is also valuable, and we can gain additional numerical control by choosing a scheme that moves the IRFP to a convenient coupling.

4. Dirac eigenmodes and the anomalous dimension

Our studies of the Dirac eigenvalue spectrum turned out to be particularly fruitful. We have shown how to extract the energy-dependent mass anomalous dimension $\gamma_m(\lambda)$ from the RG-invariant eigenmode number $\nu(\lambda)$,[7] and found that combining multiple lattice volumes and gauge couplings provides robust predictions across a wide range of energy scales.[14] We tested our procedure with the 4-flavor system, illustrated in Fig. 3. The left panel shows γ_m vs. λ for several values of β_F; in the right panel we rescale λ to a common lattice spacing $a_{7.4}$ corresponding to $\beta_F = 7.4$. We obtain a universal scaling curve that connects the one-loop asymptotic behavior at large λ to the onset of chiral symmetry breaking at small λ. These results give us great confidence in our method.

Fig. 4 illustrates our results for the 12- and 8-flavor systems. The behavior of

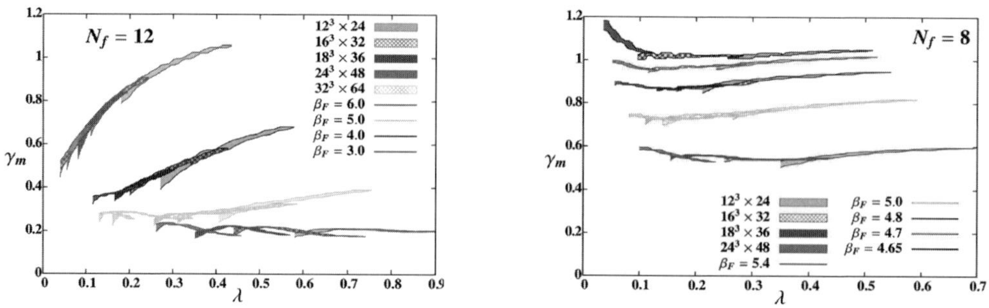

Fig. 4. Eigenmode scaling predictions for the scale-dependent mass anomalous dimension γ_m for $N_f = 12$ (left) and $N_f = 8$ (right).[14]

$N_f = 12$ in the left panel is very different from the 4-flavor case. At the stronger couplings $\beta_F = 3.0$ and 4.0 the anomalous dimension increases with the energy scale, which can be considered a sort of backward flow. We start seeing asymptotically-free dynamics only for $\beta_F > 6.0$. The data are consistent with an IRFP around $\beta_F = 5.0$, and contradict a chirally broken scenario. By extrapolating the predictions at different β_F to the limit $\lambda \to 0$ we predict a common value $\gamma_m^\star = 0.32(3)$ that we identify as the scheme-independent mass anomalous dimension at the IR fixed point. A prominent feature of our 12-flavor results is the strong dependence on the irrelevant, but apparently slowly-running, gauge coupling. With $24^3 \times 48$ and $32^3 \times 64$ lattice volumes, we can only access $\lambda \gtrsim 0.1$ before encountering finite-volume effects. The wide variation $0.2 \lesssim \gamma_m \lesssim 0.6$ predicted by different β_F in this range of λ demonstrates the importance of considering several gauge coupling values, and may be relevant to other analyses.

The right panel of Fig. 4 shows our results for $N_f = 8$. Here the appearance of the $\not{S}^4$ phase around $\beta_F \approx 4.6$ restricts us to a much smaller range of β_F than for $N_f = 12$. Since none of our 8-flavor calculations show spontaneous chiral symmetry breaking (even on volumes as large as $24^3 \times 48$ and $32^3 \times 64$), we carried out this analysis with $m = 0$. For fixed β_F we find hardly any energy dependence in our results for γ_m. At the same time, we observe significant changes as we vary the coupling. These results suggest that the $N_f = 8$ system is very close to the edge of the conformal window.

If our $N_f = 8$ simulations probe the ϵ-regime of a chirally broken system, then the distribution of the lowest eigenmodes could be compared with random matrix theory to predict the chiral condensate Σ. However, our data are not consistent with ϵ-regime scaling; the low-lying eigenmodes do not scale linearly with the volume, but rather with an exponent consistent with the anomalous dimension predicted by the mode number analysis. With our present 8-flavor data we cannot distinguish walking dynamics from IR conformality.

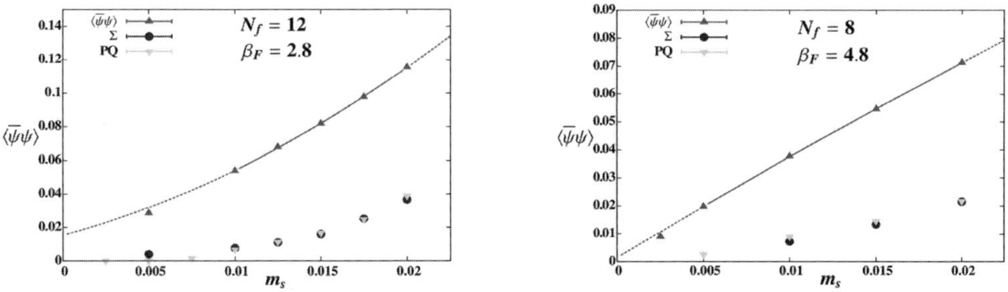

Fig. 5. Direct measurements of $\left\langle\overline{\psi}\psi\right\rangle$ with $m_v = m_s$ (red) compared to $\Sigma \equiv \pi\rho(0)$ (blue) and Σ_{PQ} (green). Left: $N_f = 12$ at $\beta_F = 2.8$, with the line a quadratic fit to $\left\langle\overline{\psi}\psi\right\rangle$ for $0.01 \leq m_s \leq 0.02$. Data with $m \geq 0.01$ have negligible finite-volume effects. Right: $N_f = 8$ at $\beta_F = 4.8$, with the line a quadratic fit to $\left\langle\overline{\psi}\psi\right\rangle$ for $0.005 \leq m_s \leq 0.02$ where finite-volume effects are negligible.

5. Chiral condensate

We typically calculate $\left\langle\overline{\psi}\psi\right\rangle$ with equal sea and valence fermion masses, $m = m_s = m_v$. In the $m = 0$ chiral limit, $\left\langle\overline{\psi}\psi\right\rangle$ is the order parameter of spontaneous chiral symmetry breaking. At non-zero fermion mass, however, $\left\langle\overline{\psi}\psi\right\rangle$ is dominated by a UV-divergent term $\propto m_v/a^2$ (Fig. 5), which complicates the $m \to 0$ chiral extrapolation. We are investigating two alternative ways of reducing this sensitivity to the fermion mass. The first approach is to determine the chiral condensate from the low-lying eigenvalue density of the massless Dirac operator, $\Sigma \equiv \pi\rho(0)$ based on the Banks–Casher relation. Second, we carry out partially-quenched calculations with several $m_v \neq m_s$, and define $\Sigma_{PQ} \equiv \lim_{m_v \to 0} \left\langle\overline{\psi}\psi\right\rangle_{PQ}$ with fixed m_s.

These two approaches give consistent results that are significantly different from direct measurements of $\left\langle\overline{\psi}\psi\right\rangle$ with $m_v = m_s$, as shown in Fig. 5. For the 12-flavor system at the rather strong coupling $\beta_F = 2.8$ (left panel), polynomial extrapolations of $\left\langle\overline{\psi}\psi\right\rangle$ predict a non-zero value in the limit $m_s \to 0$, but this result is inconsistent with the data for Σ. Since Σ is free of ultraviolet divergences and therefore a more reliable observable, we consider this additional evidence that the 12-flavor system is IR conformal, with vanishing chiral condensate.

Our preliminary results for $N_f = 8$ at $\beta_F = 4.8$ are shown in the right panel of Fig. 5. At this coupling, the chiral condensate obtained from extrapolating $\left\langle\overline{\psi}\psi\right\rangle$ is very small, and could be consistent with the prediction from Σ. Even so, we are cautious of such an extrapolation over at least an order of magnitude, which is caused by non-zero valence mass contributions dominating $\left\langle\overline{\psi}\psi\right\rangle$.

6. Conclusion

These proceedings briefly overviewed our ongoing work to synthesize results from wide-ranging lattice investigations of SU(3) gauge theories with $N_f = 8$ and 12. We presented a new Wilson-flowed MCRG method as well as a new technique to extract the mass anomalous dimension γ_m from the Dirac eigenvalue spectrum.

Both of these approaches indicate the presence of an IR fixed point for $N_f = 12$, and the Dirac spectrum predicts the scheme-independent $\gamma_m^\star = 0.32(3)$ at the IRFP. The 12-flavor lattice phase diagram and chiral condensate are also consistent with IR conformality. For $N_f = 8$ our results are more puzzling and require further investigation. The finite-temperature transitions run into the $\not{S}^4$ lattice phase, with no clear signs of spontaneous chiral symmetry breaking, while the Dirac spectrum predicts a large anomalous dimension across a wide range of energy scales.

Acknowledgments: This research was partially supported by U.S. DOE grants DE-FG02-04ER41290 and DE-AC05-06OR23100. Numerical calculations were carried out on the HEP-TH and Janus clusters at the University of Colorado, the latter supported by U.S. NSF grant CNS-0821794; at Fermilab under the auspices of USQCD supported by the DOE SciDAC program; and at the San Diego Computing Center through XSEDE supported by NSF grant OCI-1053575.

References

1. CMS Collaboration, `arXiv:1303.4571` (2013).
2. A. Hasenfratz, *Phys. Rev. Lett.* **108**:061601 (2012) [`arXiv:1106.5293`].
3. A. Cheng, A. Hasenfratz and D. Schaich, *Phys. Rev.* **D85**:094509 (2012) [`arXiv:1111.2317`].
4. X.-Y. Jin and R. D. Mawhinney, *PoS* **Lattice 2011**:066 (2012) [`arXiv:1203.5855`].
5. C.-J. D. Lin, K. Ogawa, H. Ohki and E. Shintani, *JHEP* **1208**:096 (2012) [`arXiv:1205.6076`].
6. Y. Aoki, T. Aoyama, M. Kurachi, T. Maskawa, K.-i. Nagai, H. Ohki, A. Shibata, K. Yamawaki and T. Yamazaki, *Phys. Rev.* **D86**:054506 (2012) [`arXiv:1207.3060`].
7. A. Hasenfratz, A. Cheng, G. Petropoulos and D. Schaich, *PoS* **Lattice 2012**:034 (2012) [`arXiv:1207.7162`].
8. D. Schaich, A. Cheng, A. Hasenfratz and G. Petropoulos, *PoS* **Lattice 2012**:028 (2012) [`arXiv:1207.7164`].
9. A. Deuzeman, M. P. Lombardo, T. Nunes da Silva and E. Pallante, `arXiv:1209.5720` (2012).
10. Z. Fodor, K. Holland, J. Kuti, D. Nogradi, C. Schroeder and C. H. Wong, *PoS* **Lattice 2012**:025 (2012) [`arXiv:1211.3548`].
11. Z. Fodor, K. Holland, J. Kuti, D. Nogradi, C. Schroeder and C. H. Wong, *PoS* **Lattice 2012**:279 (2012) [`arXiv:1211.4238`].
12. G. Petropoulos, A. Cheng, A. Hasenfratz and D. Schaich, *PoS* **Lattice 2012**:051 (2012) [`arXiv:1212.0053`].
13. E. Itou, `arXiv:1212.1353` (2012).
14. A. Cheng, A. Hasenfratz, G. Petropoulos and D. Schaich, submitted to *Phys. Rev. D* (2013) [`arXiv:1301.1355`].
15. Y. Aoki, T. Aoyama, M. Kurachi, T. Maskawa, K.-i. Nagai, H. Ohki, E. Rinaldi, A. Shibata, K. Yamawaki and T. Yamazaki, `arXiv:1302.4577` (2013).
16. Y. Aoki, T. Aoyama, M. Kurachi, T. Maskawa, K.-i. Nagai, H. Ohki, A. Shibata, K. Yamawaki and T. Yamazaki, `arXiv:1302.6859` (2013).
17. J. Giedt, *PoS* **Lattice 2012**:006 (2012).

Phase Structure of Many Flavor Lattice QCD at Finite Temperature

Norikazu Yamada*

KEK Theory Center, Institute of Particle and Nuclear Studies, High Energy Accelerator Research Organization (KEK), Tsukuba 305-0801, Japan
School of High Energy Accelerator Science, The Graduate University for Advanced Studies (Sokendai), Tsukuba 305-0801, Japan

Shinji Ejiri

Graduate School of Science and Technology, Niigata University, Niigata 950-2181, Japan

In realistic technicolor models containing many fermions, the electroweak baryogenesis offers a natural scenario for generating baryon number asymmetry. One of the key ingredients is the occurrence of the first order phase transition at finite temperature. As a first step toward the exploration of this possibility on the lattice, we develop an agile method to identify the critical mass for a given N_f, separating the first order and the crossover transition. We explain the outline of our method and demonstrate it by determining the critical mass of N_f-flavors in the presence of light two-flavors. It is found that the critical mass becomes larger with N_f.

Keywords: Lattice gauge theory, Technicolor models, quark-gluon plasma.

1. Introduction

Technicolor (TC)[1] has been known as one of the most natural candidates for physics beyond the standard model. The classical TC models have been already excluded for many reasons, while those with many fermion flavors in the fundamental representation are expected to escape from various experimental constraints due to special properties called "walking" dynamics.[2] Whether many-flavor TC models really show expected walking behaviors is now actively and rigorously investigated on the lattice.[3]

In this article, we focus on another aspect of TC models, the nature of the thermal phase transition. We consider as a TC model SU(3) gauge theory with $2+N_f$ flavors of techniquarks where the mass of two-flavors are fixed to a small value and the other N_f flavors have arbitrary masses. N_f is also taken to be arbitrary. Based on an analysis of linear sigma model[4a], we infer that the nature of the chiral phase transition at finite temperature changes from crossover to first order at a critical mass as the mass of N_f flavors are decreased from infinity while keeping the two-

*Speaker. E-mail: norikazu.yamada@kek.jp
[a]As for an interesting argument about the phase transition in two-flavor QCD, see Ref. 5.

flavors' mass constant. It should be noted that, if the first order transition is strong enough (or equivalently the masses of N_f-flavors are light enough), the electroweak baryogenesis (EWBG) would become viable. EWBG is, in general, restrictive and attractive in that no tunable parameter exists and its success is solely determined by the dynamics intrinsic to gauge theory. Then, it is interesting to ask what is the upper bound on the masses of N_f-flavors which allows the first order phase transition and successful EWBG. EWBG in the standard model (SM) was studied on the lattice and turned out to fail since $m_H \sim 125$ GeV is too heavy to induce the first order transition.[6] As for TC models, the studies in this direction were carried out in the context of the effective theory and obtained promising results.[7] As a first step toward the rigorous test of this possibility we study the thermal nature of a many-flavor TC model by lattice numerical simulations. To be precise, we aim at putting the upper bound on the mass of N_f-flavors of fermions by requiring the occurrence of first order phase transition. This upper bound can then be translated into that on the technipion mass, which can be directly compared to the results of LHC.

Here let us describe why we consider "2+N_f"-flavors. In many-flavor TC models without the $SU_L(2) \times U_Y(1)$ interactions, two flavors of them have to be exactly massless and the resulting three massless Nambu-Goldstone bosons (NGBs) are absorbed into the longitudinal mode of the weak gauge bosons when one turns on the $SU_L(2) \times U_Y(1)$ interactions. On the other hand, the mass of other N_f flavors must be larger than an appropriate lower bound otherwise SχSB produces too many (light pseudo) NGBs, none of which is observed yet. Furthermore, in the presence of too many massless and almost massless NGBs, S-parameter[8,9] becomes large or even diverges. Thus, N_f-flavors have to have explicit breakings of an appropriate size. In this work, we simply let them have a mass.

Our model based on SU(3) gauge theory is essentially the same as many flavor QCD except for their dynamical scales; ~1 TeV for TC and ~1 GeV for QCD and thus we can simply apply numerical techniques developed in lattice QCD to the study of TC. As discussed below, the critical mass increases with N_f. Hence, from the viewpoint of lattice numerical simulation, the boundary of the first order region can be reached more easily for large N_f.

Another purpose of this study is to understand the real QCD with 2+1 flavors. At the physical masses and zero density, the chiral transition is known to be crossover, and is expected to become first order at a critical density. Toward the determination of the critical density, it is important to find the critical surface in the parameter space spanned by masses and chemical potential.[10,11] However, recent lattice QCD studies suggest that the critical surface at zero density is located in the very light quark mass region and it makes the determination extremely difficult.[12] Fortunately, some of properties are independent of N_f. The study of 2+N_f-flavor QCD is expected to provide important information for 2+1-flavor QCD.

We first describe the method to identify the nature of the phase transition and then present the critical mass separating the first order and crossover regions in 2+N_f-flavor QCD. The work reported here has been already published in Ref. 13.

2. Method

We examine the effective potential defined by the probability distribution function of the gauge action to identify the nature of the phase transition. The first order transition is concluded by the existence of the two peaks in the distribution function.[14,15] We define the distribution function for $2+N_f$-flavor QCD with the quark masses m_f $(f=1,\cdots,2+N_f)$ by

$$w(P;\beta,m_f) = \int \mathcal{D}U\mathcal{D}\psi\mathcal{D}\bar{\psi}\ \delta(P-\hat{P})\ e^{-S_q-S_g}$$
$$= \int \mathcal{D}U\ \delta(P-\hat{P})\ e^{6\beta N_{\rm site}\hat{P}} \prod_{f=1}^{N_f+2} (\det M(m_f)), \qquad (1)$$

where S_g and S_q are the gauge and quark actions, respectively, and M is the quark matrix. $N_{\rm site} \equiv N_{\rm s}^3 \times N_t$ is the number of sites. $\beta = 6/g_0^2$ is the inverse lattice bare coupling, and $\hat{P} = -S_g/(6N_{\rm site}\beta)$. The effective potential is then defined by

$$V_{\rm eff}(P;\beta,m_f) = -\ln w(P;\beta,m_f). \qquad (2)$$

We consider QCD with two degenerate light quarks of the mass $m_{\rm l}$ and N_f quarks of m_h. For later convenience, the potential is separated into two parts; one is the contribution from two-flavor QCD $V_0(P;\beta)$ and the other is the rest,

$$V_{\rm eff}(P;\beta,m_h) = V_0(P;\beta_0) - \ln R(P;\beta,m_h;\beta_0), \qquad (3)$$

with

$$\ln R(P;\beta,m_h;\beta_0) = 6(\beta-\beta_0)N_{\rm site}P + \ln \left\langle \prod_{h=1}^{N_f} \frac{\det M(m_h)}{\det M(\infty)} \right\rangle_{P:{\rm fixed}}, \qquad (4)$$

where $\langle\cdots\rangle_{P:{\rm fixed}} \equiv \langle\delta(P-\hat{P})\cdots\rangle_{\beta_0}/\langle\delta(P-\hat{P})\rangle_{\beta_0}$ and $\langle\cdots\rangle_{\beta_0}$ denotes the ensemble average over two-flavor configurations generated at β_0 and m_l. Since the m_l dependence is not discussed in the following, it is omitted from the arguments. β_0 is the simulation point, which may differ from β. By performing simulations at various β_0, one can obtain the potential in a wide range of P.

Restricting the calculation to the heavy quark region, the determinant for N_f flavors in Eq. (4) is approximated at the leading order as

$$\ln\left[\frac{\det M(\kappa_h)}{\det M(0)}\right] = 288N_{\rm site}\kappa_h^4\hat{P} + 12N_s^3(2\kappa_h)^{N_t}\hat{\Omega} + \cdots \qquad (5)$$

for the standard Wilson quark action and

$$\ln\left[\frac{\det M(m_{\rm h})}{\det M(\infty)}\right] = \frac{36N_{\rm site}}{(2m_{\rm h})^4}\hat{P} + \frac{6N_s^3}{(2m_{\rm h})^{N_t}}\hat{\Omega} + \cdots \qquad (6)$$

for the four-flavor standard staggered quark with $m_{\rm h}$. κ_h in Eq. (5) is the hopping parameter being proportional to $1/m_h$, and $\hat{\Omega}$ is the real part of the Polyakov loop.

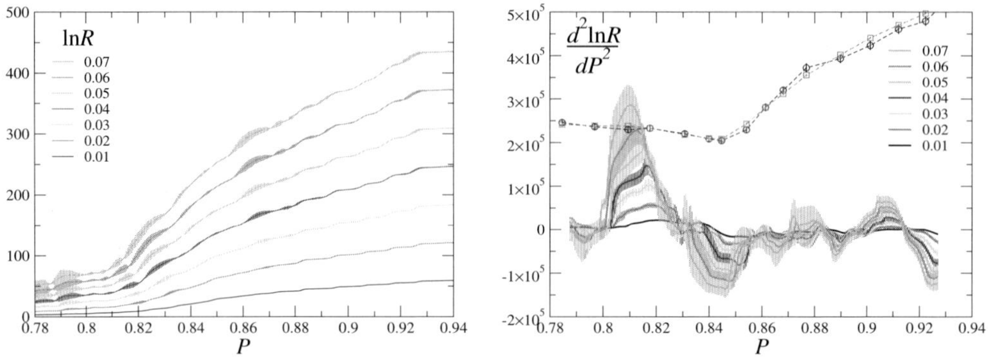

Fig. 1. Left: $\ln \bar{R}(P;h,0)$ as functions of the gauge action. Right: The curvature of $\ln \bar{R}(P;h,0)$ for $h = 0.01 - 0.07$. The circle and square symbols are $d^2V_0/dP^2(P)$.

For improved gauge actions such as $S_g = 6N_{\rm site}\beta[c_0(\text{plaquette}) + c_1(\text{rectangle})]$, additional $c_1 \times O(\kappa^4)$ terms must be contained in Eqs. (5) and (6), where c_1 is the improvement coefficient and $c_0 = 1 - 8c_1$. However, since the improvement term does not affect the physics, we will cancel these terms by a shift of c_1.

Beyond the critical β corresponding to the endpoint of a first order transition, $V_{\rm eff}$ takes a double-well shape as a function of P, and equivalently the curvature of the potential $d^2V_{\rm eff}/d^2P$ will take a negative value. Observing this behavior usually requires a fine-tuning of β. However, $d^2V_{\rm eff}/dP^2$ is independent of β and $d^2V_{\rm eff}/dP^2$ over the wide range of P can be easily obtained by combining data obtained at different β. Thus the fine-tuning of β is not necessary in this case.[14] We focus on the curvature of the effective potential to identify the nature of the phase transition.

Denoting $h = 2N_f(2\kappa_h)^{N_t}$ for N_f degenerate Wilson quarks, or $h = N_f/(4 \times (2m_{\rm h})^{N_t})$ for the staggered quarks, we obtain

$$\ln R(P;\beta,\kappa_h;\beta_0) = \ln \bar{R}(P;h) + (\text{plaquette term}) + O(\kappa_h^{N_t+2}) \tag{7}$$

$$\bar{R}(P;h) = \left\langle \exp[6hN_s^3\hat{\Omega}] \right\rangle_{P:\text{fixed},\beta_0}. \tag{8}$$

Notice that $\bar{R}(P;h)$ is independent of β_0. The plaquette term does not contribute to $d^2V_{\rm eff}/dP^2$ and can be absorbed by shifting $\beta \to \beta^* \equiv \beta + 48N_f\kappa_h^4$ for Wilson quarks. N_f-flavors do not have to be degenerate. The non-degenerate case is realized by redefining $h = 2\sum_{f=1}^{N_f}(2\kappa_f)^{N_t}$ or $h = (1/4)\sum_{f=1}^{N_f}(2m_f)^{-N_t}$. In the following, we discuss the mass dependence of $\bar{R}$ through the parameter h.

3. Numerical results

We explicitly demonstrate the above method. We use the two-flavor QCD configurations generated with p4-improved staggered quark and Symanzik-improved gauge actions in Ref. 16. The lattice size $N_{\rm site}$ is $16^3 \times 4$, and the data are obtained at sixteen values of β from $\beta = 3.52$ to 4.00 keeping the bare quark mass to $ma = 0.1$.

The number of trajectories is 10,000 – 40,000, depending on β. The corresponding temperature normalized by the pseudo-critical temperature is in the range of $T/T_c = 0.76$ to 1.98, and the pseudo-critical point is about $\beta = 3.65$, where the π-ρ ratio is $m_{\rm PS}/m_{\rm V} \approx 0.7$. Further details on the simulation parameters are given in Ref. 16. The same data set is used to study the phase structure of two-flavor QCD at finite density in Ref. 14.

We first calculate the potential in two-flavor QCD, $V_0(P;\beta)$, the first term in Eq. (3). Because the finite temperature transition is crossover for two-flavor QCD at a finite quark mass, the distribution function is always Gaussian type. We thus evaluate the curvature of V_0 using an identity for the Gaussian distribution, $d^2V_0/dP^2 = 6N_{\rm site}/\chi_P$, where χ_P is the gauge action susceptibility,

$$\chi_P \equiv 6N_{\rm site}\langle (P - \langle P \rangle)^2 \rangle. \tag{9}$$

The slope of V_0 in the heavy quark limit can be also measured using an equation derived from Eqs. (3) and (4). When one performs a simulation at β_0, the slope is zero at the minimum of $V_0(P;\beta_0)$, and the minimum is realized at $P \approx \langle \hat{P} \rangle_{\beta_0}$. Hence, we obtain[17]

$$\frac{d\,V_0(\langle \hat{P} \rangle_{\beta_0}, \beta)}{d\,P} = -6(\beta - \beta_0)N_{\rm site}. \tag{10}$$

The result of d^2V_0/dP^2 is plotted in the right panel of Fig. 1. The circles with dashed lines are calculated by χ_P. The squares are computed by the numerical differential of dV_0/dP obtained at the minimum of V_0. dV_0/dP are the squares in Fig. 2. It is seen that two different methods provide the consistent results.

In the calculation of $\bar{R}(P;h)$, we use the delta function approximated by $\delta(x) \approx 1/(\Delta\sqrt{\pi})\exp[-(x/\Delta)^2]$, where $\Delta = 0.0025$ is adopted consulting the resolution and the statistical error. Because $\bar{R}(P;h)$ is independent of β, the data obtained at various β are gathered as is done in Ref. 14. The results for $\ln \bar{R}(P;h)$ are shown by solid curves in the left panel of Fig. 1 for $h = 0.01 - 0.07$. A rapid increase is observed around $P \sim 0.82$. It is also important to note that the gradient becomes larger with h.

The second derivative $d^2 \ln \bar{R}/dP^2$ is calculated by fitting $\ln \bar{R}$ to a quadratic function of P with a range of $P \pm 0.015$ and repeating with various P. The results are plotted in Fig. 1 (right), where d^2V_0/dP^2 is also shown as the circles or the squares with dashed lines. This figure shows that $d^2(\ln \bar{R})/dP^2$ becomes larger with h, and the maximum around $P = 0.81$ exceeds d^2V_0/dP^2 for $h > 0.06$. This indicates that the curvature of the effective potential, $d^2V_{\rm eff}/dP^2 = d^2V_0/dP^2 - d^2(\ln \bar{R})/dP^2$, vanishes at $h \sim 0.06$ and for large h there exists a region of P where the curvature is negative. We estimated the critical value h_c at which the minimum of $d^2V_{\rm eff}/dP^2$ vanishes and obtained $h_c = 0.0614(69)$.

To see the appearance of the first order transition in a different way, we show $dV_{\rm eff}/dP$ at finite h for $\beta^* = 3.65$ in Fig. 2. The shape of the $dV_{\rm eff}/dP$ is independent of β because $d^2V_{\rm eff}/dP^2$ is β-independent. $dV_{\rm eff}/dP$ monotonically increases when

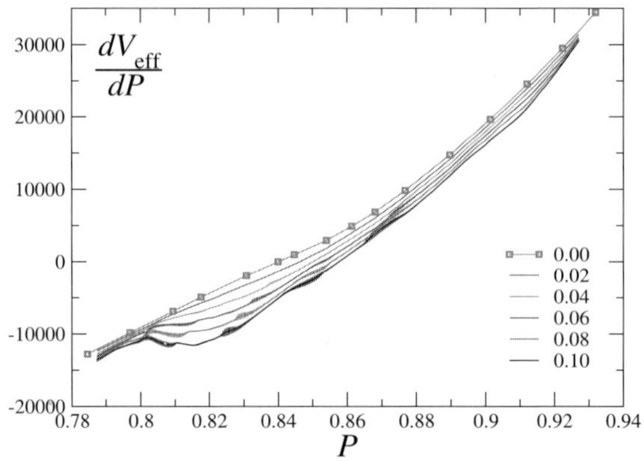

Fig. 2. The slope of $V_{\text{eff}}(P;\beta,h,0)$ normalized at $(\beta,h)=(3.65,0)$ for $h=0.0-0.1$. The squares are dV_0/dP.

h becomes small, indicating that the transition is crossover. However, the shape of dV_{eff}/dP turns into an S-shape at $h\sim 0.06$, corresponding to the double-well potential.

We defined the parameter $h=2N_f\times(2\kappa_h)^{N_t}$ for the Wilson quark. Then, the critical κ_{hc} corresponding h_c decreases as $\kappa_{hc}=[h_c/(2N_f)]^{1/N_t}/2$ with N_f, and the truncation error from the higher order terms in κ_h becomes smaller as N_f increases. The application range of the hopping parameter expansion was examined in quenched QCD simulations with $N_t=4$, by explicitly measuring the size of the next-to-leading order (NLO) terms of the expansion.[18] Then the NLO contribution turned out to be comparable to that in the leading order at $\kappa_h\sim 0.18$. Hence, this method may be applicable up to around $\kappa_h\sim 0.1$. For instance, in the case of $N_f=10$ with $N_t=4$, κ_{hc} is 0.118.

4. Conclusion and outlook

We proposed an agile method to study the thermal nature of many-flavor QCD with EWBG in TC in mind, and applied it to the 2+N_f-flavor QCD. Fixing the mass of two light quarks, we determined the critical mass of the remaining N_f-flavors, which separates the first order and crossover regions. The critical mass is found to become larger with N_f. Further studies using this method are given in Ref. 13, including the investigations at finite density. We find that the critical mass increases with μ in the 2+N_f-flavor QCD.

The next step for the estimation of the baryon number asymmetry in TC scenario is to quantify the strength of the first order phase transition. Another interesting application of our method is to study universal scaling behavior near the tricritical point. If the chiral phase transition in the two flavor massless limit is of second order, the boundary of the first order transition region $m_{\text{l}}^c(m_{\text{h}})$ is expected to behave as $m_{\text{l}}^c\sim|m_h^{\text{tri.}}-m_h|^{5/2}$ in the vicinity of the tricritical point, $(m_{\text{l}},m_{\text{h}},\mu)=(0,m_{\text{h}}^{\text{tri.}},0)$, from the mean field analysis. This power behavior is universal for any N_f. The density dependence is important as well, which is expected to be $m_1^c \sim |\mu|^5$.[19] Starting from large N_f, the systematic study of properties of real QCD phase transition is possible.

Acknowledgments

We would like to thank the organizers of this fruitful workshop. A part of this work was completed at GGI work shop. This work is in part supported by Grants-in-Aid of the Japanese Ministry of Education, Culture, Sports, Science and Technology (No. 22740183, 23540295) and by the Grant-in-Aid for Scientific Research on Innovative Areas (No. 20105002, 20105005, 23105706).

References

1. S. Weinberg, Phys. Rev. D **13**, 974 (1976); L. Susskind, Phys. Rev. D **20**, 2619 (1979).
2. B. Holdom, Phys. Rev. D **24**, 1441 (1981); K. Yamawaki, M. Bando and K. i. Matumoto, Phys. Rev. Lett. **56**, 1335 (1986); T. W. Appelquist, D. Karabali and L. C. R. Wijewardhana, Phys. Rev. Lett. **57**, 957 (1986); T. Akiba and T. Yanagida, Phys. Lett. B **169**, 432 (1986); M. Bando, T. Morozumi, H. So and K. Yamawaki, Phys. Rev. Lett. **59**, 389 (1987).
3. L. Del Debbio, PoS **(LATTICE 2010)** 004 (2010); K. Rummukainen, AIP Conf. Proc. **1343**, 51 (2011) and references therein.
4. R. D. Pisarski and F. Wilczek, Phys. Rev. D **29**, 338 (1984).
5. S. Aoki, H. Fukaya, Y. Taniguchi and , Phys. Rev. D **86**, 114512 (2012).
6. Z. Fodor, Nucl. Phys. Proc. Suppl. **83**, 121 (2000).
7. For semi-quantitative analysis on electroweak baryogenesis in many flavor TC scenario, see T. Appelquist, M. Schwetz and S. B. Selipsky, Phys. Rev. D **52**, 4741 (1995); Y. Kikukawa, M. Kohda and J. Yasuda, Phys. Rev. D **77**, 015014 (2008).
8. M. E. Peskin, T. Takeuchi, Phys. Rev. Lett. **65**, 964 (1990); Phys. Rev. D **46**, 381 (1992).
9. S-parameter can be calculated on the lattice. See, for example, E. Shintani *et al.* [JLQCD Collaboration], Phys. Rev. Lett. **101**, 242001 (2008).
10. P. de Forcrand and O. Philipsen, Nucl. Phys. B **673**, 170 (2003); JHEP **0701**, 077 (2007).
11. Ch. Schmidt *et al.*, Nucl. Phys. B (Proc. Suppl.) **119**, 517 (2003); F. Karsch *et al.*, Nucl. Phys. B (Proc. Suppl.) **129**, 614 (2004); S. Ejiri *et al.*, Prog. Theor. Phys. Suppl. **153**, 118 (2004).
12. S. Ejiri *et al.*, Phys. Rev. D **80**, 094505 (2009).
13. S. Ejiri and N. Yamada, arXiv:1212.5899 [hep-lat], to appear in Physical Review Letter.
14. S. Ejiri, Phys. Rev. D **77**, 014508 (2008).
15. H. Saito *et al.* (WHOT-QCD Collaboration), Phys. Rev. D **84**, 054502 (2011).
16. C. R. Allton *et al.*, Phys. Rev. D **71**, 054508 (2005).
17. S. Ejiri and H. Yoneyama, PoS **(LATTICE 2009)** 173 (2009).
18. WHOT-QCD Collaboration, in preparation.
19. S. Ejiri, PoS **(LATTICE 2008)** 002 (2008).

From Lattice Strong Dynamics to Phenomenology

George T. Fleming[a] and Ethan T. Neil[b*]

(for the Lattice Strong Dynamics (LSD) Collaboration)

[a] *Department of Physics, Sloane Laboratory, Yale University, New Haven, CT 06520, USA*

[b] *Theoretical Physics Group, Fermi National Accelerator Laboratory, Batavia, IL 60510, USA*

We present updated results on the chiral properties of SU(3) gauge theories with $N_f = 2$ and 6 massless Dirac fermions in the fundamental representation. Our focus is on the ratio $\langle\overline{\psi}\psi\rangle/F^3$, where $\langle\overline{\psi}\psi\rangle$ is the chiral condensate and F is the pseudo-Nambu-Goldstone-boson decay constant. This ratio is of interest in the context of fermion mass generation within composite Higgs theories. By re-expanding certain ratios using next-to-leading-order chiral perturbation theory, we confirm our previous result of significant enhancement of this ratio at $N_f = 6$ over $N_f = 2$.

Keywords: SCGT12; chiral condensate; many-fermion theories; lattice simulation.

1. Introduction

Composite Higgs theories, in which a new strongly-coupled sector is responsible for the dynamical breaking of electroweak symmetry, are an attractive possibility for physics beyond the standard model. These extensions of the standard model have a number of attractive theoretical features, in particular resolving the naturalness problem of the fundamental Higgs boson; as a composite bound state, the Higgs mass is generated from the underlying strong dynamics, and is not subject to quadratically divergent quantum corrections.

Lattice simulation plays a crucial role in the study of composite Higgs theories, by allowing rigorous non-perturbative calculations in strongly-coupled gauge theories other than QCD. The properties of these theories can vary greatly depending on the matter content; in particular, with a large number of fermions, the long-distance confinement phase of the theory is lost, giving way to an infrared-conformal phase in which chiral symmetry remains unbroken.[1,2] In recent years, several lattice groups have turned their attention to the study of these gauge theories beyond QCD.[3,4]

In addition to exploring theoretical issues such as the nature and location of the transition from confining to infrared conformal, lattice simulations can be used to calculate phenomenologically relevant quantities for composite Higgs models from

*Speaker. E-mail:eneil@fnal.gov

first principles. One quantity of particular interest is the chiral condensate $\langle\overline{\psi}\psi\rangle$. In the context of a composite Higgs model, the condensate plays the role of the Higgs vacuum expectation value, giving rise to fermion masses through four-fermion couplings $\bar{\psi}\psi\bar{f}f$, in a way analogous to the standard Higgs-Yukawa mechanism. In general, to satisfy experimental constraints based on flavor-changing neutral currents, a large value of $\langle\overline{\psi}\psi\rangle$ relative to the symmetry-breaking scale F is needed, a phenomenon which is conjectured to occur near the critical number of fermions for the infrared-conformal transition.[5–7]

We present here updated results[8] for the quantity $\langle\overline{\psi}\psi\rangle/F^3$ calculated in SU(3) gauge theory, with $N_f = 2$ and 6 Dirac fermions in the fundamental representation. Our main improvement is a new technique for chiral extrapolation, which allows a more precise determination of the ratio of $\langle\overline{\psi}\psi\rangle/F^3$ between the two theories.

2. The chiral condensate

To make the connection to composite Higgs theories, we are interested in studying the chiral limit $m \to 0$, in which the electroweak symmetry is not broken explicitly. Direct extraction of the chiral condensate in the limit $m \to 0$ from lattice calculations is challenging due to the presence of a quadratic divergence in the lattice cutoff contained in the linear term.[9]

We can make use of additional indirect information on the condensate by working in chiral perturbation theory.[10] At leading order, the Gell-Mann-Oakes-Renner (GMOR) relation gives the chiral condensate in terms of the pseudo-Goldstone boson mass M and decay constant F,

$$M_m^2 F_m^2 = 2m\langle\overline{\psi}\psi\rangle_m, \tag{1}$$

where the subscript m denotes evaluation at finite fermion mass m, to distinguish from the chiral-limit values of F and $\langle\overline{\psi}\psi\rangle$.

At higher order, re-expanding in the parameter $z \equiv 2B/(4\pi F)^2 = 2\langle\overline{\psi}\psi\rangle/(4\pi)^2F^4$, we may re-write the mass dependence of M_m[8,9,11]

$$\frac{M_m^2}{2m} = 8\pi^2 F^2 z\left\{1 + zm\left[\alpha_M + \frac{1}{N_f}\log(zm)\right]\right\}, \tag{2}$$

$$F_m = F\left\{1 + zm\left[\alpha_F - \frac{N_f}{2}\log(zm)\right]\right\}, \tag{3}$$

$$\langle\overline{\psi}\psi\rangle_m = 8\pi^2 F^4 z\left\{1 + zm\left[\alpha_c - \frac{N_f^2-1}{N_f}\log(zm)\right]\right\}. \tag{4}$$

From these formulas it is easy to see that chiral perturbation theory is an expansion in $zm = M_P^2/(4\pi F)^2 = M_P^2/\Lambda_\chi^2$.

We are now interested in constructing ratios of observables which will reduce to

$\langle\overline{\psi}\psi\rangle/F^3 = 8\pi^2 Fz$ in the chiral limit. There are three such constructions:

$$X_m^{(FM)} = \frac{M_m^2}{2mF_m}, \tag{5}$$

$$X_m^{(CM)} = \frac{(M_m^2/2m)^{3/2}}{\langle\overline{\psi}\psi\rangle_m^{1/2}}, \tag{6}$$

$$X_m^{(CF)} = \frac{\langle\overline{\psi}\psi\rangle_m}{F_m^3}. \tag{7}$$

Expanding these ratios out using the NLO expressions above, we recover their mass dependence to $\mathcal{O}(zm)$:

$$X_m^{(FM)} = 8\pi^2 Fz\left(1 + zm\left[\alpha_M - \alpha_F + \frac{N_f^2+2}{2N_f}\log(zm)\right]\right) \tag{8}$$

$$X_m^{(CM)} = 8\pi^2 Fz\left(1 + zm\left[\frac{3}{2}\alpha_M - \frac{1}{2}\alpha_C + \frac{N_f^2+2}{2N_f}\log(zm)\right]\right) \tag{9}$$

$$X_m^{(CF)} = 8\pi^2 Fz\left(1 + zm\left[\alpha_C - 3\alpha_F + \frac{N_f^2+2}{2N_f}\log(zm)\right]\right) \tag{10}$$

By construction, all three ratios have the same intercept $8\pi^2 Fz = \langle\overline{\psi}\psi\rangle/F^3$. The coefficient of the chiral logarithm in each expansion is identical, but this is a coincidence, and does not persist for the leading terms at next-to-next-to-leading order (NNLO).[12]

For comparison of results between the $N_f = 2$ and $N_f = 6$ theories, we construct the "ratio of ratios"

$$R_{\tilde{m}}^{(XY)} \equiv \frac{X_m^{(XY)}(N_f = 6)}{X_m^{(XY)}(N_f = 2)}, \tag{11}$$

where (XY) enumerates the three possible constructions $(FM), (CM), (CF)$. The ratio is taken at constant bare mass m_f, but the renormalized masses differ slightly due to the variation of m_{res}; we thus extrapolate in the geometric mean mass, $\tilde{m} \equiv \sqrt{m_{N_f=2}m_{N_f=6}}$. As shown in appendix A, expanding R to NLO in $\tilde{m}$ yields the general functional form

$$R_{\tilde{m}}^{(XY)} = R^{(6)}(1 + \alpha_R^{(XY)}\tilde{m} + \beta_R^{(XY)}\tilde{m}\log\tilde{m} + ...), \tag{12}$$

where the intercept $R^{(6)}$ is independent of the particular ratio XY used in the construction.

Finally, we note that it is possible to construct additional ratios in terms of other quantities related to the PNGB. For example, the $I = 2$ PNGB scattering length a_{PP} is proportional at leading order to $(Ma_{PP}) \propto -M^2/(16\pi F^2)$,[13] so that the appropriate combination is

$$X_m^{(AFM)} = \frac{8\pi F_m M_m a_{PP,m}}{m}. \tag{13}$$

Other constructions are possible. The statistical precision of the scattering length a_{PP} is typically not comparable to the other observables considered here, so we will not include this ratio in our chiral extrapolation to determine $R^{(6)}$. However, agreement between ratios such as this and the more precise set of ratios given above is a useful consistency check.

3. Results

Our lattice calculations are performed using $32^3 \times 64$ domain-wall lattices with the Iwasaki improved gauge action. The gauge coupling is set to $\beta = 2.70$ at $N_f = 2$, and $\beta = 2.10$ at $N_f = 6$. These β values are tuned in order to match the confinement scale of the two theories in the chiral limit. Further details of the simulation are given in a prior reference.[8]

For both $N_f = 2$ and $N_f = 6$, results for six mass points are shown. All chiral extrapolations are considered in terms of the total fermion mass $m \equiv m_f + m_{res}$. We present results for the lightest point $m_f = 0.005$, but do not include them in the analysis due to the likely presence of uncontrolled finite-volume systematic effects. Other simulation results based on these gauge configurations have been presented in previous references.[13–15]

Determination of pseudoscalar masses and decay constants, along with the chiral condensate, is discussed and data tables are presented in a previous work.[8] Data and joint fit results for the three constructed "ratios of ratios" are shown in Fig. 1. The joint fit shown has $\chi^2/\text{d.o.f.} = 5.73$ with 8 degrees of freedom; the relatively large value of χ^2 is likely due to autocorrelation effects in the simulation data, as discussed previously.[8] The intercept of the fit is

$$R^{(6)}(\Lambda) = 1.95(12). \tag{14}$$

This quantity is unrenormalized. Applying one-loop lattice perturbation theory to compute the renormalization factors for the chiral condensate,[16] we find the correction factor[8] $Z_6/Z_2 = 0.8527(97)$, so that

$$R^{(6)}_{\overline{\text{MS}}}(\Lambda) = 1.60(10). \tag{15}$$

The fact that $R^{(6)} > 1$ indicates enhancement of the chiral condensate relative to the $N_f = 2$ case. In particular, this is a substantial and statistically significant increase compared to the perturbative estimate,[8] $R^{(6)}_{\overline{\text{MS}}} \le 1.15$.

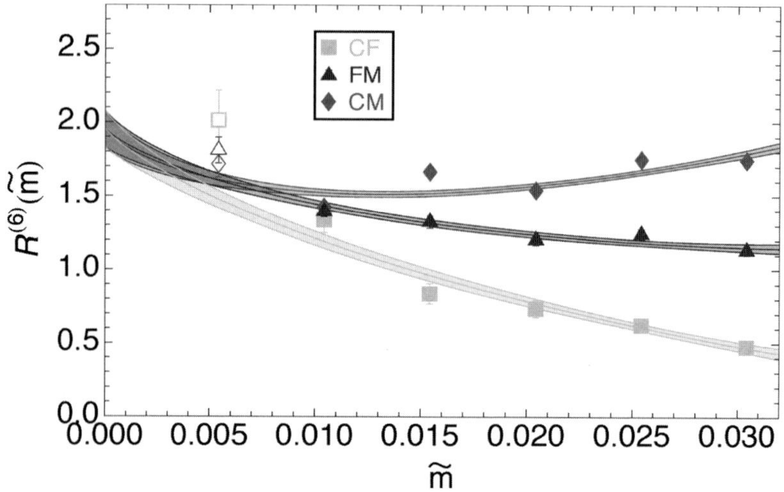

Fig. 1. Extrapolation of $R^{(6)}$ to the chiral limit. "CF", "FM" and "CM" label the three different combinations of observables which all extrapolate to $R^{(6)}$ in the chiral limit, as discussed in the text. The best fit shown is a joint fit to all closed symbols, constrained to have a common intercept value, and includes 1σ error bands. The open symbols at $m = 0.005$ are not included in the analysis, due to possible uncontrolled systematic errors.

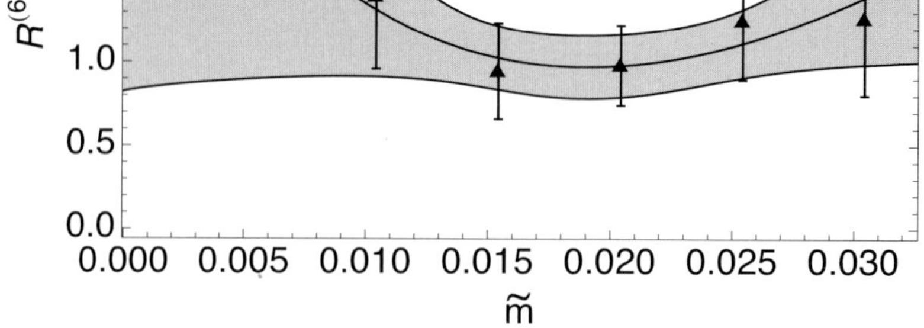

Fig. 2. Alternate determination of $R^{(6)}$ using the pseudoscalar scattering length a_{PP}, as given in eq. 13. A chiral extrapolation with 1σ error bands is also shown. Although the precision of $R^{(6)}$ is much lower in this channel, consistency of the data points with fig. 1 gives a useful cross-check.

Table 1. Simulation results for $N_f = 2$ and 6. The lightest points at $m_f = 0.005$ are excluded from our analysis as discussed in the text.

N_f	m	M_m^2	F_m	$\langle\overline{\psi}\psi\rangle_m$	M_ρ
2	*0.005026*	*0.01183(32)*	*0.02468(55)*	*0.0058293(90)*	*0.2449(60)*
	0.010026	0.01917(18)	0.03080(39)	0.0109544(47)	0.2553(30)
	0.015026	0.02735(30)	0.03294(69)	0.0160535(52)	0.2607(32)
	0.020026	0.04060(45)	0.03586(64)	0.0211960(66)	0.2998(34)
	0.025026	0.04798(54)	0.03709(57)	0.0262204(48)	0.3100(28)
	0.030026	0.05947(54)	0.03795(52)	0.0312388(72)	0.3241(23)
6	*0.005823*	*0.02124(49)*	*0.02113(52)*	*0.007365(15)*	*0.2472(41)*
	0.010823	0.02798(27)	0.02989(53)	0.013395(6)	0.2750(32)
	0.015823	0.04304(44)	0.03718(64)	0.019327(9)	0.3019(37)
	0.020823	0.05959(50)	0.04207(83)	0.025179(9)	0.3429(33)
	0.025823	0.07595(44)	0.04584(90)	0.030964(9)	0.3755(33)
	0.030823	0.09331(56)	0.05120(106)	0.036640(10)	0.4040(35)

4. Conclusion

We have presented updated results for enhancement of the condensate ratio $\langle\overline{\psi}\psi\rangle/F^3$ in SU(3) gauge theories with N_f fundamental fermions, as N_f is increased from 2 to 6. This ratio is important in composite Higgs theories, as the size of the chiral condensate is related to the generation of fermion masses. In general, a large value of the condensate is needed to satisfy phenomenological constraints; the ratio is conjectured to increase dramatically for many-fermion theories near (but below) the transition value N_f^c to infrared-conformal behavior.[5–7]

Our lattice results confirm a significant increase of the enhancement at $N_f = 6$; the "ratio of ratios" comparing $N_f = 6$ to $N_f = 2$ is found to be $R^{(6)}_{\overline{\mathrm{MS}}} = 1.60(10)$. This is large relative to a perturbative estimate, $R^{(6)}_{pert} \lesssim 1.15$. Future work at larger values of N_f closer to the critical value for transition for loss of confinement is in progress, and will provide further information on this interesting trend.

Acknowledgments

The LSD collaboration would like to thank the LLNL Multiprogrammatic and institutional Computing program for Grand Challenge allocations and time on the LLNL BlueGene/L (uBGL) supercomputer. We thank LLNL for funding from LDRD 10-ERD-033 and LDRD 13-ERD-023. This work has been supported by the National Science Foundation under Grant No. NSF PHY11-00905 (G. F.). Fermilab is supported by Fermi Research Alliance, LLC under Contract No. DE-AC02-07CH11359 with the U. S. Department of Energy.

Appendix A. Chiral expansion of ratios with mistuned fermion masses

Comparison of results between two different theories generally requires careful tuning to match physical scales. However, in our simulation results there is a slight mistuning of the fermion masses, due to the residual domain-wall contribution m_{res}. We can correct for mistuning of the fermion mass by using chiral perturbation theory. Suppose that we are working with two theories with matched lattice spacings, $a = a'$, but mismatched fermion masses $m \neq m'$. We first replace the two masses with the geometric mean $\tilde{m}$ and square root of the mass ratio ρ,

$$\tilde{m} \equiv \sqrt{m'm}, \ \rho \equiv \sqrt{\frac{m'}{m}}, \tag{A.1}$$

so that $m = \tilde{m}/\rho$, $m' = \tilde{m}\rho$. Given a quantity X with finite intercept in chiral perturbation theory, we can rewrite its expansion to NLO in the two theories,

$$X'_m = X' \left\{1 + z'\tilde{m}\rho \left[\alpha'_{10,X} + \alpha'_{11,X} \log(z'\tilde{m}\rho)\right]\right\}, \tag{A.2}$$

$$X_m = X \left\{1 + \frac{z\tilde{m}}{\rho} \left[\alpha_{10,X} + \alpha_{11,X} \log\left(\frac{z\tilde{m}}{\rho}\right)\right]\right\}. \tag{A.3}$$

Taking the ratio and discarding terms of higher order in $\tilde{m}$ and ρ, we have

$$\frac{X'_m}{X_m} = \frac{X'}{X} \left[1 + \tilde{m} \left(\beta_{10} + \beta_{11} \log \tilde{m}\right)\right], \tag{A.4}$$

$$\beta_{10} = z'\rho(\alpha'_{10,X} + \alpha'_{11,X} \log(z'\rho) - \frac{z}{\rho}(\alpha_{10,X} + \alpha_{11,X} \log(z/\rho)), \tag{A.5}$$

$$\beta_{11} = z'\rho\alpha'_{11,X} - \frac{z}{\rho}\alpha_{11,X}. \tag{A.6}$$

This makes it clear that in order to recover the correct ratio of intercepts in the chiral limit, we should extrapolate to $\tilde{m} = 0$ while holding the mass ratio ρ fixed.

References

1. W. E. Caswell, *Phys.Rev.Lett.* **33**, p. 244 (1974).
2. T. Banks and A. Zaks, *Nucl.Phys.* **B196**, p. 189 (1982).
3. E. T. Neil, *PoS* **LATTICE2011**, p. 009 (2011).
4. J. Giedt, *PoS* **LATTICE2012**, p. 006 (2012).
5. B. Holdom, *Phys.Rev.* **D24**, p. 1441 (1981).
6. K. Yamawaki, M. Bando and K.-i. Matumoto, *Phys.Rev.Lett.* **56**, p. 1335 (1986).
7. T. W. Appelquist, D. Karabali and L. Wijewardhana, *Phys.Rev.Lett.* **57**, p. 957 (1986).
8. T. Appelquist *et al.*, *Phys.Rev.Lett.* **104**, p. 071601 (2010).
9. E. T. Neil *et al.*, *PoS* **CD09**, p. 088 (2009).
10. J. Gasser and H. Leutwyler, *Nucl.Phys.* **B250**, p. 465 (1985).
11. J. Gasser and H. Leutwyler, *Phys.Lett.* **B184**, p. 83 (1987).
12. J. Bijnens and J. Lu, *JHEP* **0911**, p. 116 (2009).
13. T. Appelquist *et al.*, *Phys.Rev.* **D85**, p. 074505 (2012).
14. T. Appelquist *et al.*, *Phys.Rev.Lett.* **106**, p. 231601 (2011).
15. T. Appelquist *et al.*, *arXiv:1301.1693* (2013).
16. S. Aoki, T. Izubuchi, Y. Kuramashi and Y. Taniguchi, *Phys.Rev.* **D67**, p. 094502 (2003).

Lattice Study of Conformality in Twelve-Flavor QCD

Yasumichi Aoki[a], Tatsumi Aoyama[a], Masafumi Kurachi[a], Toshihide Maskawa[a], Kei-ichi Nagai[a], Hiroshi Ohki[a]*, Akihiro Shibata[b], Koichi Yamawaki[a], Takeshi Yamazaki[a]

(LatKMI Collaboration)

[a] *Kobayashi-Maskawa Institute for the Origin of Particles and the Universe (KMI), Nagoya University, Nagoya, 464-8602, Japan*

[b] *Computing Research Center, High Energy Accelerator Research Organization (KEK), Tsukuba 305-0801, Japan*

We study infrared conformality of the twelve-flavor QCD on the lattice. Utilizing a HISQ type action which is useful to study the continuum physics, we analyze the lattice data of the mass and the decay constant of a pseudoscalar meson and the mass of a vector meson as well at several values of lattice spacing and fermion mass. Our result is consistent with the conformal hypothesis for the mass anomalous dimension $\gamma_m \sim 0.4 - 0.5$.

Keywords: Lattice Gauge Theories, Technicolor, Conformal Window

1. Introduction

There has been a renewed interest in the study of QCD with large number of the massless fermions in the fundamental representation ("large N_f QCD") in the context of walking technicolor having approximate scale invariance in the infrared (IR) region and large anomalous dimension $\gamma_m \sim 1$.[1] Such an approximate scale-invariant dynamics may in fact be realized in the large N_f QCD.[2] As a powerful tool of a nonperturbative study, one may use lattice QCD simulations, which can in principle determine the phase structure of the SU(3) gauge theories with various number of fermions. In addition to the pioneering works,[3] there are many lattice works in the large N_f QCD in the recent years. In particular, the system of the N_f = 12 has been widely investigated by several lattice approaches, such as running coupling, lattice phase diagram, low-energy spectra and so on. (See, for a review, Ref. 4.) We investigate the 12-flavor SU(3) gauge theory using a variant of the highly improved staggered quark (HISQ)[6] action to reduce the discretization error. We study several bound-state masses such as the pseudoscalar meson π and vector meson ρ as well as the decay constant of π, by varying the fermion bare mass m_f. In this work, we introduce a quantity for the scaling test of the conformal hypothesis with finite volume. Using this quantity, we can analyze the data without any assumption of the fitting form in the scaling test. We also discuss possible

*Speaker. E-mail: ohki@kmi.nagoya-u.ac.jp

finite size and mass corrections in the scaling. We find that our results of $N_f = 12$ are consistent with hyperscaling with $\gamma = 0.4 - 0.5$. The details of the study have already been published in Ref. 5.

2. Setup and primary analysis

The lattice gauge configurations are generated by the HMC algorithm using a tree-level Symanzik gauge action and the HISQ[6] action for the dynamical fermions. We simulate with various parameter sets for the fermion mass m_f, volume and the bare coupling $\beta = 6/g^2$. We then calculate several bound-state masses, such as the pseudoscalar meson π and vector meson ρ, and the decay constant of π. The flavor symmetry breaking of this action is highly suppressed in $N_f = 2+1$ QCD[7] and it is almost negligible for the theory in our parameter region.

For a primary analysis, dimensionless ratios composed of these measurements will be plotted against the π mass. If the theory is in the conformal window, the hadron mass M_H and π decay constant F_π obey the conformal hyperscaling

$$M_H \propto m_f^{1/(1+\gamma_*)}, \quad F_\pi \propto m_f^{1/(1+\gamma_*)}, \tag{1}$$

where γ_* denotes the mass anomalous dimension at the IR fixed point. The spectra obtained in our lattice simulation will be tested against this conformal hypothesis in this work.

The left panel of Fig. 1 plots the ratio F_π/M_π for $N_f = 12$ at the largest two volumes and two bare gauge couplings $\beta = 3.7$ and 4. If we look at the results for $\beta = 3.7$ (filled symbol), F_π/M_π tends to be flat for smaller π masses, which shows clear contrast to the ordinary QCD. The behavior is consistent with the hyperscaling Eq. 1. The π mass dependence of the ratio at larger mass can be realized by the correction to the hyperscaling which may be different from one quantity to another. For $\beta = 4$ (open symbol), there is no flat range without volume dependence. Similar observation can be made for the other ratio M_ρ/M_π shown in the right panel of Fig. 1. Here, the flattening is observed for $\beta = 3.7$ again, but the range is wider than F_π/M_π. In this case, $\beta = 4$ shows the flattening, also. The difference of the constant is made possible due to a discretization effect. Existence of the scaling for F_π at $\beta = 3.7$ and absence at $\beta = 4$ at the same aM_π can be made possible if the M_π in the physical unit is larger (thus the correction is no longer negligible) for $\beta = 4$, i.e., the lattice spacing decreases as β increases. Actually a crude analysis of two lattice spacing matching shows the result $a(\beta = 3.7) > a(\beta = 4)$ which is consistent with being in the asymptotically free domain.

3. Finite size hyperscaling analysis

In the conformal window with finite masses and volume, the renormalization group analysis tells us the scaling behavior for low-energy spectra which should obey the

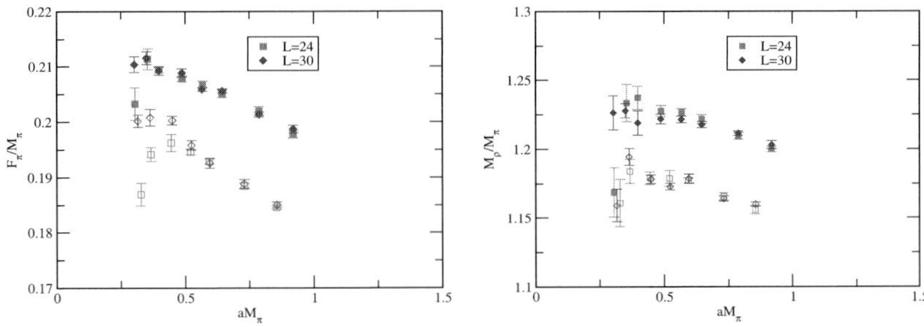

Fig. 1. Dimension-less ratios F_π/M_π and M_ρ/M_π as functions of aM_π for $N_f = 12$ at $\beta = 3.7$ (filled symbol) and 4.0 (open symbol) for two largest volumes.

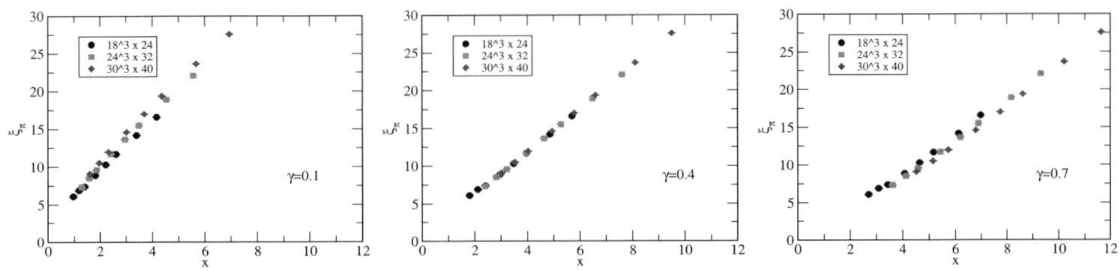

Fig. 2. ξ_π is plotted as a function of the scaling variable x for $\gamma = 0.1$, 0.4, and 0.7 from left to right for $N_f = 12$ at $\beta = 3.7$. An alignment is seen for $\gamma \sim 0.4$.

universal scaling relations [a] as

$$\xi_p \equiv LM_p = f_p(x), \quad \xi_F \equiv LF_\pi = f_F(x), \tag{2}$$

where the subscript p distinguishes the bound state, $p = \pi$ or ρ in this study. The product of bound-state mass or decay constant and linear system size falls into a function of a single scaling variable.

$$x = L \cdot m_f^{\frac{1}{1+\gamma_*}}, \tag{3}$$

where γ_* is the mass anomalous dimension at the IR fixed point. We call these scaling relations "finite-size hyperscaling" (FSHS). While the forms of the scaling functions $f_p(x)$ are unknown in general, the asymptotic form should be $f(x) \sim x$ at large x, since the hyperscaling relation Eq. (1) must be reproduced in large volumes.

Now we examine whether our measurements of bound state mass and decay constant at different m_f and L obey FSHS. To visualize how the scaling works, we show ξ_π as functions of $x = L \cdot m_f^{1/(1+\gamma)}$ for several values of γ in Fig. 2. It is observed that the data points align at around $\gamma = 0.4$, while they become scattered for γ away from that value. This indicates a possible existence of FSHS with $\gamma \sim 0.4$. A similar alignment is observed for ξ_F as well.

[a] For a review of the hyperscaling, see, e.g., Refs. 8,9.

To quantify the "alignment" we introduce an evaluation function $P(\gamma)$ for an observable p as follows. Suppose ξ^j be a data point of measured observable p at $x_j = L_j \cdot m_j^{1/(1+\gamma)}$ and $\delta\xi^j$ is the error of ξ^j. j labels distinction of parameters L and m_f. Let K be a subset of data points $\{(x_k, \xi_k)\}$ from which we construct a function $f^{(K)}(x)$ which represents the subset of data. Then, the evaluation function is defined as

$$P(\gamma) = \frac{1}{\mathcal{N}} \sum_L \sum_{j \notin K_L} \frac{\left|\xi^j - f^{(K_L)}(x_j)\right|^2}{|\delta\xi^j|^2}, \tag{4}$$

where L runs through all the lattice sizes we have, the sum over j is taken for a set of data points which do not belong to K_L which includes all the data obtained on the lattice with size L. $\mathcal{N}$ denotes the total number of summation. Here, we choose for the function $f^{(K_L)}$ a linear or quadratic interpolation of the data points of the fixed lattice size L. In the evaluation function Eq. (4), the data points need to be taken for a range of $x = L \cdot m_f^{1/(1+\gamma)}$ in which there is an overlap of available data for all volumes within the range $[x_{\rm min}, x_{\rm max}]$. We take the value of $x_{\rm min}$ ($x_{\rm max}$) as the smallest (largest) m_f for the largest (smallest) vlume L in our simulation parameters.

The evaluation function for all the quantities is plotted in the left panel of Fig. 3. A clear minimum is observed at which the optimal alignment of the data is achieved and the value of $P(\gamma)$ is $\mathcal{O}(1)$ at the minimum. In order to study systematically how the range of fermion mass affects the result, we consider a window in the x direction which has a span $\Delta x/2$, where $\Delta x = x_{\rm max} - x_{\rm min}$ and slide it with an interval $\Delta x/4$. There are three choices of windows denoted by range 1, 2, and 3, which are $[x_{\rm min}, x_{\rm min} + \Delta x/2]$, $[x_{\rm min} + \Delta x/4, x_{\rm max} - \Delta x/4]$, and $[x_{\rm max} - \Delta x/2, x_{\rm max}]$, respectively. In these windows, the data are taken for all three volumes. For the L direction we have three windows of the data for the combination of two data sets. We repeat this analysis for each observable and at each value of β. The results are tabulated in Table 1.

Let us look at the case of M_π at $\beta = 3.7$ in Table 1. We do not observe clear window dependence. If there is IR conformality, the nonuniversal correction to the hyperscaling is negligible at this precision. In the case of M_ρ, if one restricts the mass range for M_ρ to the smaller side, then the value of $\gamma(M_\rho)$ get closer to that of $\gamma(M_\pi)$, in which $\gamma(M_\rho)$ reduces for smaller mass range (toward range 1) and larger volume (toward $L = 24, 30$). The similar trend for the x-range dependence as for M_ρ at $\beta = 3.7$ is observed for F_π at $\beta = 3.7$, too. The direction of the movement is correct, but it does not get close enough to the value of $\gamma(M_\pi)$. These results may be understood from the fact that in Sec. 2, the scaling is observed only in the very small mass range. In the case of $\beta = 4$ we could not observe such trends for both the data of M_ρ and F_π. As the number of samples have gotten reduced for the fixed range analysis, a statistical instability might have spoiled the result. As we cannot completely resolve these trends in the mass dependence, we regard these variations of γ with respect to the change of the window as the systematic error on the central

Table 1. Summary of the optimal values of γ. See the text for details.

			x			L		
	β	all	range 1	range 2	range 3	(18,24)	(18,30)	(24,30)
M_π	3.7	0.434(4)	0.425(9)	0.436(6)	0.437(4)	0.438(6)	0.433(4)	0.429(8)
M_π	4	0.414(5)	0.420(7)	0.418(6)	0.411(5)	0.397(7)	0.414(4)	0.447(9)
F_π	3.7	0.516(12)	0.481(19)	0.512(19)	0.544(14)	0.526(18)	0.514(11)	0.505(24)
F_π	4	0.580(15)	0.552(21)	0.602(20)	0.605(19)	0.544(27)	0.577(14)	0.645(32)
M_ρ	3.7	0.459(8)	0.411(17)	0.461(10)	0.473(8)	0.491(15)	0.457(8)	0.414(18)
M_ρ	4	0.460(9)	0.458(13)	0.455(14)	0.460(8)	0.457(16)	0.459(8)	0.463(15)

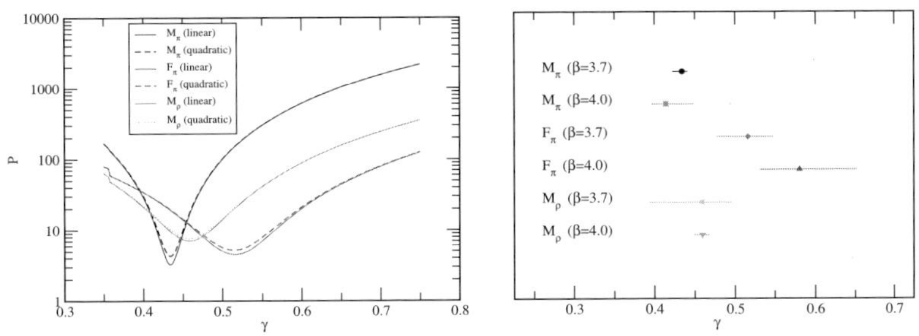

Fig. 3. (Left) The γ dependence of the evaluation function P for M_π, F_π, and M_ρ at $\beta = 3.7$ is plotted. The vertical axis shows the values of P at each of γ where the three volumes and full range of x for the data are considered. The solid and dashed curves show the results of $P(\gamma)$ with the interpolation functions $f(x)$ by the linear and quadratic functions, respectively. (Right) The results of the values of γ for three observables at two β are summarized, where the statistical and systematic errors are added in quadrature.

value of γ. The results with all the errors added in quadrature are summarized in Fig. 3. All the results are consistent with each other within 2σ level except for γ from F_π at $\beta = 4$.

From these analyses, we conclude that our data for the $N_f = 12$ theory by the lattice simulations are reasonably consistent with the FSHS with the IR conformality. The resulting mass anomalous dimensions from different quantities and two lattice spacings are also reasonably consistent. We quote $0.4 \leq \gamma_* \leq 0.5$ for the value of the mass anomalous dimension at the IR fixed point.

4. Fit analyses

We found in the previous section that our data is consistent with the FSHS. The analyses also suggested that the non-universal correction to the hyperscaling could be important. Here we try to test two plausible models for the correction. We use the following formulae for the fit,

$$\xi = c_0 + c_1 L m_f^{1/(1+\gamma)} \quad \cdots \text{fit-A}, \tag{5}$$

$$\xi = c_0 + c_1 L m_f^{1/(1+\gamma)} + c_2 L m_f^{\alpha} \quad \cdots \text{fit-B}. \tag{6}$$

The second formula for fit-B includes a mass correction term. Among various possible choices, we take $\alpha = (3-2\gamma)/(1+\gamma)$, which is inspired by the analytic expression of the solution of the SD equation given in Ref. 10, and also consider $\alpha = 2$ case,

Table 2. The fit results of finite size conformal hypothesis at $\beta = 3.7$(left) and $\beta = 4$(right). The values sandwiched as $[\cdots]$ mean the input in the fit.

	γ	α	χ^2/dof
fit-A	0.449(3)	-	4.52
fit-B1	0.411(9)	$\frac{(3-2\gamma)}{(1+\gamma)}$	1.23
fit-B2	0.423(7)	[2]	1.15

	γ	α	χ^2/dof
fit-A	0.430(3)	-	6.78
fit-B1	0.461(18)	$\frac{(3-2\gamma)}{(1+\gamma)}$	1.86
fit-B2	0.453(11)	[2]	2.00

which is regarded as the small mass correction caused by the explicit chiral symmetry breaking effects or due to the lattice discretization artifact. We denote these fit functions with $\alpha = (3-2\gamma)/(1+\gamma)$ and $\alpha = 2$ as fit-B1 and fit-B2, respectively.

We perform simultaneous fits using M_π, M_ρ and F_π, with common γ for each fit ansatz. Figure 4 shows the fit-B2 as an example. Table 2 shows the resulting fitting parameters. The χ^2/dof is large ≥ 4 for the fit-A, while it is reasonable ($\chi^2/\mathrm{dof} \leq 2$) for the fit-B1 and B2. This implies that the correction is necessary, and both types of mass correction works. Let us look into the fit-B2 at $\beta = 3.7$, for example. The small correction for M_π is consistent with the fact that the value of $\gamma(M_\pi)$ was stable against the x and L range used in the analysis in the previous section. We obtain consistent γ for all the fit-B1 and B2. The resulting values are in the range: $\gamma \sim 0.41 - 0.46$, which is consistent with $0.4 \leq \gamma_* \leq 0.5$ in Sec. 3.

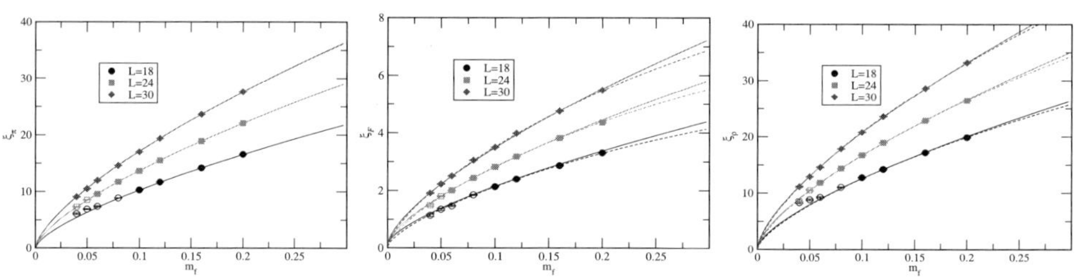

Fig. 4. Spectra ξ_π(left), ξ_F(center) and ξ_ρ(right) as a function of m_f at $\beta = 3.7$. For simplicity, we only show two fit results of fit-A and fit-B2, by the solid and dotted curves, respectively. The data with empty symbols are not used in the fit.

In the test of the chiral broken scenario, it turns out that the natural chiral expansion parameter $\chi = N_f \left(\frac{M_\pi}{4\pi F_\pi/\sqrt{2}}\right)^2$ is very large in the region we simulated, which is evaluated as $\chi \sim 39$ at the smallest M_π using the value of F_π in the chiral limit. With this large χ, we could not consistently analyze the data based on the ChPT. Further efforts would be required to arrive at a decisive conclusion.

5. Summary

We have studied the SU(3) gauge theory with the fundamental 12 fermions using a HISQ type staggered fermion action. Our present data is consistent with the conformal hypothesis. The mass anomalous dimension, γ_* at the infrared fixed point was estimated through the (finite-size) hyperscaling analysis. Our result is $\gamma_* \sim 0.4$–

0.5, which is not as big as $\gamma_* \sim 1$ for the theory to be close to the realistic technicolor model. In the test of chiral broken scenario, the chiral expansion parameter χ is much larger than one even at the smallest M_π. We could not consistently analyze the data based on the chiral expansion. A possibility of the chiral broken phase is not excluded yet. More detailed analyses with more data at larger volume and lighter mass would be required. Another series of the study in our project has been reported for the scalar spectrum from glueball and flavor singlet fermion operators[11] and the test of the walking behavior in $N_f = 8$.[12]

Acknowledgments

Numerical simulation has been carried out on the supercomputer system φ at KMI in Nagoya university. This work is supported by the JSPS Grant-in-Aid for Scientific Research (S) No.22224003, (C) No.23540300 (K.Y.) and (C) No.21540289 (Y.A.), and also by Grants-in-Aid of the Japanese Ministry for Scientific Research on Innovative Areas No. 23105708 (T.Y.).

References

1. K. Yamawaki, M. Bando and K. -i. Matumoto, Phys. Rev. Lett. **56**, 1335 (1986).
2. W. E. Caswell, Phys. Rev. Lett. **33** (1974) 244; T. Banks and A. Zaks, Nucl. Phys. B **196** (1982) 189.
3. Y. Iwasaki, K. Kanaya, S. Sakai and T. Yoshie, Phys. Rev. Lett. **69** (1992) 21; F. R. Brown, H. Chen, N. H. Christ, Z. Dong, R. D. Mawhinney, W. Schaffer and A. Vaccarino, Phys. Rev. D **46** (1992) 5655 [hep-lat/9206001]; P. H. Damgaard, U. M. Heller, A. Krasnitz and P. Olesen, Phys. Lett. B **400** (1997) 169 [hep-lat/9701008].
4. J. Giedt, PoS LATTICE **2012**, 006 (2012); E. T. Neil, PoS LATTICE **2011**, 009 (2011) [arXiv:1205.4706 [hep-lat]], and references therein.
5. Y. Aoki, T. Aoyama, M. Kurachi, T. Maskawa, K. -i. Nagai, H. Ohki, A. Shibata, K. Yamawaki and T. Yamazaki, (LatKMI collaboration), Phys. Rev. D **86**, 054506 (2012) [arXiv:1207.3060 [hep-lat]].
6. E. Follana *et al.* [HPQCD and UKQCD Collaborations], Phys. Rev. D **75** (2007) 054502 [hep-lat/0610092].
7. A. Bazavov, T. Bhattacharya, M. Cheng, C. DeTar, H. T. Ding, S. Gottlieb, R. Gupta and P. Hegde *et al.*, Phys. Rev. D **85**, 054503 (2012) [arXiv:1111.1710 [hep-lat]].
8. T. DeGrand and A. Hasenfratz, Phys. Rev. D **80** (2009) 034506 [arXiv:0906.1976 [hep-lat]].
9. L. Del Debbio and R. Zwicky, Phys. Rev. D **82** (2010) 014502 [arXiv:1005.2371 [hep-ph]].
10. Y. Aoki, T. Aoyama, M. Kurachi, T. Maskawa, K. -i. Nagai, H. Ohki, A. Shibata, K. Yamawaki and T. Yamazaki, (LatKMI collaboration), Phys. Rev. D **85**, 074502 (2012) [arXiv:1201.4157 [hep-lat]].
11. Y. Aoki, T. Aoyama, M. Kurachi, T. Maskawa, K. -i. Nagai, H. Ohki, E. Rinaldi, A. Shibata, K. Yamawaki and T. Yamazaki, (LatKMI collaboration), arXiv:1302.4577 [hep-lat].
12. Y. Aoki, T. Aoyama, M. Kurachi, T. Maskawa, K. -i. Nagai, H. Ohki, A. Shibata, K. Yamawaki and T. Yamazaki, (LatKMI collaboration), arXiv:1302.6859 [hep-lat].

Exploring Walking Behavior in SU(3) Gauge Theory with 8 HISQ Quarks*

Yasumichi Aoki[a], Tatsumi Aoyama[a], Masafumi Kurachi[a], Toshihide Maskawa[a], Kei-ichi Nagai[a†], Hiroshi Ohki[a], Akihiro Shibata[b], Koichi Yamawaki[a] and Takeshi Yamazaki[a]

(LatKMI Collaboration)

[a] *Kobayashi-Maskawa Institute for the Origin of Particles and the Universe, Nagoya University, Nagoya 464-8602, Japan*

[b] *Computing Research Center, High Energy Accelerator Research Organization (KEK), Tsukuba 305-0801, Japan*

We investigate walking signals of $N_f = 8$ QCD through meson spectrum using the highly improved staggered quark (HISQ) action. Our data (the pion decay constant, the pion and ρ meson masses and the chiral condensate) for the $N_f = 8$ QCD are consistent with the spontaneously broken chiral symmetry in the chiral limit extrapolation of the chiral perturbation theory (ChPT). Remarkably enough, while the $N_f = 8$ data near the chiral limit are well described by the ChPT, those for the relatively large fermion bare mass m_f away from the chiral limit actually exhibit a finite-size hyperscaling relation, suggesting a large anomalous dimension $\gamma_m \sim 1$. This implies that there exists a remnant of the infrared conformality, and suggests that a typical technicolor ("one-family model") as modeled by the $N_f = 8$ QCD can be a walking technicolor theory.

Keywords: Lattice Gauge Theory, Walking Technicolor, Conformal Symmetry

1. Introduction

The origin of mass is the most urgent issue of the particle physics today. Although the LHC has discovered a 125 GeV boson roughly consistent with the Standard Model (SM) Higgs boson, there still remain many unsolved problems with the SM, which would require physics beyond the SM. One of the candidates for the theory beyond the SM towards that problem is the Walking Technicolor (WTC)[2] having an approximate scale invariance with a large anomalous dimension $\gamma_m \simeq 1$ due to the "walking" coupling (Fig. 1), which is based on the scale-invariant gauge dynamics (ladder Schwinger-Dyson (SD) equation [3–5]). The walking behavior can in fact be realized in the "large N_f QCD". Actually, the $N_f = 8$ is particularly interesting from the model-building point of view: so-called one-family model (Farhi-Susskind model[6]). Thus if the $N_f = 8$ turns out to be a walking theory, it would be a great message for the phenomenology, which is to be tested by the on-going LHC.

---

*In Ref.[1] the detailed analyses, figures and tables with numerical values are written down.
†Speaker (keiichi.nagai@kmi.nagoya-u.ac.jp)

As in Fig. 1 (the two-loop/ladder analysis), if $N_f = 8$ is a walking theory desired for the WTC, it should be inside the SχSB phase $N_f = 8 < N_f^{\rm cr}$ ($m_D \neq 0$) and at the same time be close to the phase boundary with the conformal window $N_f > N_f^{\rm cr}$ ($m_D = 0$) such that $m_D \ll \Lambda_{\rm QCD}$. Among others the fermion bare mass m_f obviously distorts the ideal behavior of the breaking of the scale symmetry. Then, disregarding the effects of the lattice parameters L and a for the moment[a], we may imagine possible effects of m_f on the walking coupling of our target of study. There is a caveat about the approximate hyperscaling relation to be expected in Case 2 in Fig. 1. There are two infrared mass parameters m_D and m_f which violate the infrared conformality and hence the possible hyperscaling relations for the physical mass quantities should not be universal but do depend on both of them in non-universal ways, in sharp contrast to the hyperscaling relation in the conformal window.[9]

Remarkably enough in our exploration by the numerical simulation, while the $N_f = 8$ data near the chiral limit are well described by the ChPT, those for the relatively large fermion bare mass m_f away from the chiral limit actually exhibit a finite-size hyperscaling relation, suggesting a large anomalous dimension $\gamma_m \sim 1$. It is the first time that the hyperscaling relation is observed in a theory with SχSB.

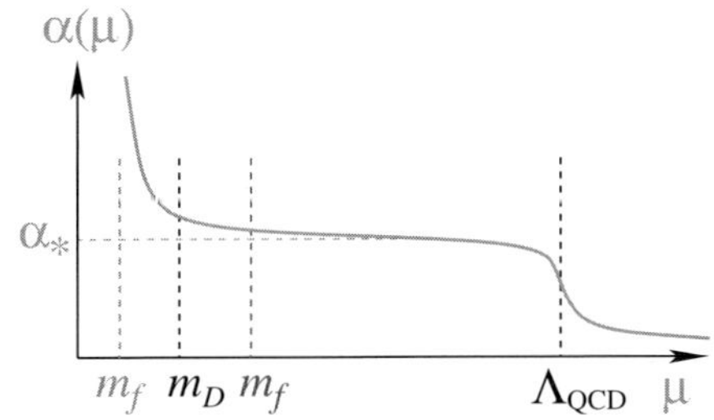

Fig. 1. Schematic two-loop/ladder picture of the gauge coupling of the massless large N_f QCD as a walking gauge theory in the SχSB phase near the conformal window. m_D is the dynamical mass of the fermion generated by the SχSB. The effects of m_f would be qualitatively different depending on the cases: Case 1: $m_f \ll m_D$ (red dotted line) well described by ChPT, and Case 2: $m_f \gg m_D$ (blue dotted line) well described by the hyperscaling. The SχSB order parameter on the lattice is not m_D but would be the decay constant F_π extrapolated to the chiral limit: $F = F_\pi(m_f = 0)$ which would be expected roughly as $m_D = \mathcal{O}(F)$.

2. Lattice simulation and the results

In our simulation, we use the tree-level Symanzik gauge action and HISQ action[10] without the tadpole improvement and the mass correction in the Naik term. It is expected that the flavor symmetry in the staggered fermion and the behavior towards the continuum limit are improved. We carry out the simulation by using

[a]In our simulation we use the parameter region where the effect of the system size is subdominant compared to the mass effect. This strategy is different from the one which is advocated by the authors of Refs.[7,8]

the standard Hybrid Monte-Carlo (HMC) algorithm. We measure the mass of the pion M_π, ρ-meson M_ρ, the decay constant of the pion F_π and the chiral condensate $\langle\bar{\psi}\psi\rangle$ as the basic observable to explore the large-N_f QCD.

The simulation in the preliminary report[11,12] for $N_f = 8$ is carried out at $\beta(= 6/g^2)$=3.6, 3.7, 3.8, 3.9, 4.0 for various quark masses and on various lattices, $L^3 \times T$, where L is the spatial size and T the temporal size with the fixed aspect ratio as $T/L = 4/3$ for $L = 12$, 18, 24, 30 and 36. We need to choose as small value of β as possible to obtain a large enough physical volume to minimize the finite-volume effect. From the pilot study mentioned above, we found that $\beta < 3.8$ is too strong to carry out the HMC simulation with HISQ. Therefore, we choose $\beta = 3.8$ in this article. The boundary condition in the spatial direction is the periodic and the one in the temporal direction is the anti-periodic for fermions in HMC. The well-known technique of combining the valence quark propagators solved with periodic and antiperiodic boundary conditions in the temporal direction enables us to have sufficient range for the fitting due to the effectively doubled temporal size.

We performed the analysis based on ChPT and the finite-size hyperscaling (FSHS); If the system is in the SχSB phase, physical quantities in the spectroscopy, M_H for $H = \pi, \rho, \cdots$ and F_π, are described by the ChPT. The masses and F_π depend on m_f up to chiral log as

$$M_\pi^2 = C_1^\pi m_f + C_2^\pi m_f^2 + \cdots, \quad F_\pi = F + C_1^F m_f + C_2^F m_f^2 + \cdots, \tag{1}$$

where F is the value in the chiral limit.

On the other hand, if the theory is in the conformal window, M_H and F_π on the finite volume are described by the FSHS[13–15]

$$\xi_H = \mathcal{F}_H(L m_f^{\frac{1}{1+\gamma_*}}), \tag{2}$$

where $\xi_p \equiv LM_p$ and $\xi_F \equiv LF_\pi$, for $H = \pi$, ρ or F. The function, $\mathcal{F}_H$, is a some function (unknown *a priori*) of the scaling variable $X = L m_f^{\frac{1}{1+\gamma_*}}$ in which γ_* denotes the mass anomalous dimension γ_m at the infrared fixed point and its value is universal for all channels.

3. Chiral perturbation theory (ChPT) analysis

3.1. *Quadratic fit of F_π (and M_ρ)*

Based on the scenario explained in Fig. 1, we attempt the ChPT analysis in the small m_f. Let us analyze the behavior of F_π in detail, towards the chiral limit. Table 1 is the result of the quadratic fit of F_π. Particularly for the small region, $0.015 \le m_f \le 0.04$, the polynomial fit gives the good χ^2/dof (= 0.46). When we include the data at $m_f = 0.05$, χ^2/dof jumps up. Although this jump might be caused by the instability due to small dof=2, the large χ^2/dof persists for the range with larger masses, thus, with the value of χ^2/dof being more reliable. This suggests that there is a bound, beyond which the ChPT does not describe the data well, and that bound is around $m_f \lesssim 0.05$. With this consideration and the good chiral behavior observed for other quantities (for instance M_ρ and $\langle\bar{\psi}\psi\rangle$) for $m_f = 0.015 - 0.04$, we chose $m_f = 0.015 - 0.04$ for the fitting range of all quantities (roughly corresponding to Case 1 in Fig. 1). The above analysis[b] suggests that our result in $N_f = 8$ is consistent with SχSB phase with $F = 0.0310(13)$ up to chiral log.

Table 1. Results of chiral fit of F_π with $F_\pi = F + C_1^F m_f + C_2^F m_f^2$ and the the expansion parameter in ChPT, $\mathcal{X} = N_f \left(\frac{M_\pi}{4\pi F/\sqrt{2}}\right)^2$, for various fit ranges.

fit range (m_f)	F	$\mathcal{X}(m_f^{\min} = 0.015)$	$\mathcal{X}(m_f = m_{\max})$	χ^2/dof	dof
0.015–0.04	0.0310(13)	3.74	11.80	0.46	1
0.015–0.05	0.0278(8)	4.64	19.28	5.56	2
0.015–0.06	0.0284(6)	4.44	23.2	4.09	3
0.015–0.10	0.0311(3)	3.70	37.0	7.85	6
0.015–0.16	0.0349((2)	2.94	54.0	34.2	9

As shown in Fig. 2, the linear fit in the small region of m_f is in good agreement with ChPT analysis. Therefore $N_f = 8$ QCD is consistent with that of SχSB from the analysis of F_π (and M_ρ). As shown in Fig. 2, the linear fit in the small region

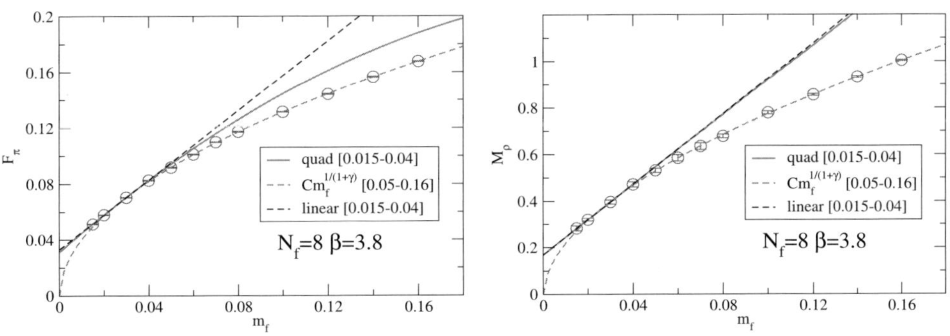

Fig. 2. Fitting results of F_π (left) and M_ρ (right). Linear and quadratic fits in $0.015 \le m_f \le 0.04$, Power fit ($y = Cm_f^\alpha$) in $0.05 \le m_f \le 0.16$. In $0.015 \le m_f \le 0.04$, $F_\pi = 0.0310(13) + 1.4(1) \cdot m_f - 2.6(1.7) \cdot m_f^2$ (χ^2/dof $= 0.46$), $F_\pi = 0.0329(3) + 1.24(1) \cdot m_f$ (χ^2/dof $= 1.4$), and $M_\rho = 0.168(32) + 7.6(4.1) \cdot m_f - 1.2(73.4) \cdot m_f^2$ (χ^2/dof $= 0.0017$), $M_\rho = 0.169(13) + 7.6(5) \cdot m_f$ (χ^2/dof $= 0.0010$). In $0.05 \le m_f \le 0.16$, $F_\pi = 0.431(2) \cdot m_f^{0.515(2)}$ (χ^2/dof $= 0.66$) and $M_\rho = 2.75(5) \cdot m_f^{0.550(9)}$ (χ^2/dof $= 0.21$).

of m_f gives the good fitting result as well as the quadratic fit. Therefore $N_f = 8$ QCD is consistent with that of SχSB from the analysis of F_π (and M_ρ).

[b]The ChPT analysis in this case is self-consistent. The expansion parameter, $\mathcal{X} = N_f \left(\frac{M_\pi}{4\pi F/\sqrt{2}}\right)^2$ in Table 1, is $\mathcal{X} = O(1)$ in our smallest m_f. Thus the ChPT is barely self-consistent in contrast to the $N_f = 12$ case[9] where $\mathcal{X} \simeq 40$ at $m_f^{\min}$.

3.2. *Chiral condensate*

In this subsection, we analyze the chiral condensate whether this quantity shows the SχSB behavior. We perform a direct measurement $\langle\bar{\psi}\psi\rangle = \mathrm{Tr}[D^{-1}_{HISQ}(x,x)]/4$ and compare it with the quantity $\Sigma \equiv \frac{F_\pi^2 M_\pi^2}{4m_f}$, through GMOR relation. As seen in Fig. 3, both the results in the chiral limit are non-zero, and are consistent with each other in the small m_f region $0.015 \le m_f \le 0.04$: $\langle\bar{\psi}\psi\rangle\big|_{m_f\to 0} = 0.00052(5)$, $\Sigma|_{m_f\to 0} = 0.00059(13)$. We also estimate the chiral condensate in the chiral limit by multiplying F with the extrapolated value of M_π^2/m_f, as $F^2 \cdot \left(\frac{M_\pi^2}{4m_f}\right)\Big|_{m_f\to 0} = 0.00050(3)$, which is consistent with $\langle\bar{\psi}\psi\rangle$ and Σ in the small m_f defined by the analysis of F_π above.

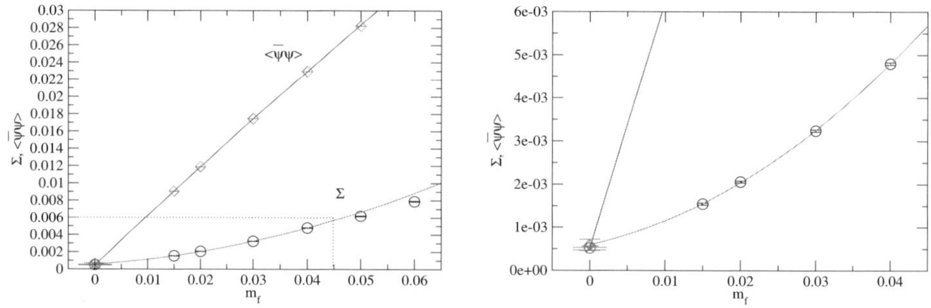

Fig. 3. Σ and $\langle\bar{\psi}\psi\rangle$ (left panel) as a function of m_f. The small region, $0 \le m_f \le 0.045$ and $0 \le \Sigma, \langle\bar{\psi}\psi\rangle \le 0.0006$, in the left panel is enlarged to the right panel. The quadratic fit curves by using the data in $m_f \le 0.04$ are shown. The green symbol is the value of $F^2 \cdot (\frac{M_\pi^2}{4m_f})|_{m_f\to 0}$.

From the analyses up to chiral log of all the observables, F_π, M_π, M_ρ and the chiral condensate, the chiral property of $N_f = 8$ QCD is consistent with that of SχSB in the small m_f region ($0.015 \le m_f \le 0.04$).

4. Study of remnants of conformality

We find the two regions of m_f having qualitatively different properties: $m_f = 0.015$–0.04 and $m_f = 0.05 - 0.16$ by the analysis of ChPT shown in the previous section.

Furthermore, based on the scenario of WTC described in Fig. 1, if this theory is near the conformal phase boundary, it is expected that some remnants of the conformal symmetry appear in physical quantities.

Figure 2 shows the results of ChPT behavior in $0.015 \le m_f \le 0.04$. On the other hand, it is remarkable that in Fig. 2 the fit results in the mass range, $m_f \gtrsim 0.05$, are consistent with the power behavior, $F_\pi(M_\rho) = C_1 m_f^{1/(1+\gamma)}$. This suggests that, although $N_f = 8$ QCD is in the SχSB phase, there exists a remnant of the conformality in the m_f region away from the chiral limit. Therefore, we will carry out further in depth analysis, which employs the hyperscaling test on the finite volume, to investigate whether the remnant of the conformality really persists.

4.1. *FSHS fits with the correction term*

If the system is in the conformal window, the data on a finite volume is in good agreement with the finite size hyperscaling (FSHS) having a universal value of $\gamma = \gamma_*$ at IRFP for observables as given in Eq. (2). In general our data of $N_f = 8$ cannot satisfy the FSHS with universal γ in the whole range of m_f as shown in Fig. 4, because we showed that the theory is in the SχSB phase. However, because of the power behavior in the middle range of the fermion mass as mentioned in Fig. 2, we carry out the FSHS test in our data to find a remnant of the conformality.

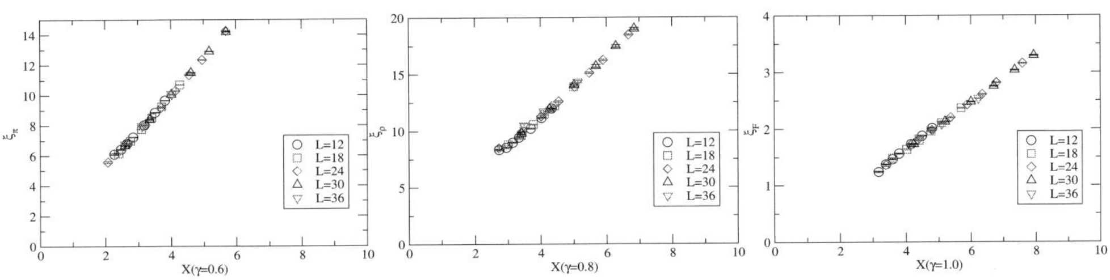

Fig. 4. ξ_π plotted as functions of X with $\gamma = 0.6$ (left), ξ_ρ with $\gamma = 0.8$ (center), ξ_F with $\gamma = 1.0$ (right) for the FSHS test. The data of $N_f = 8$ cannot satisfy the FSHS with universal γ in the whole range of m_f, in contrast to the conformal case ($N_f = 12$)[9] showing the existence of alignment and the QCD case ($N_f = 4$)[1] showing no alignment.

Since $N_f = 8$ theory is in SχSB phase, FSHS cannot become accurate by approaching to the chiral limit, which is in contrast to the $N_f = 12$ case.[9] Since it might be possible to obtain a common value of the γ, FSHS is only expected for larger mass region, where mass corrections may not be negligible,[16]

$$\xi_H = C_0^H + C_1^H X + C_2^H L m_f^\alpha , \quad \text{for} \quad m_f \geq 0.05 \quad \text{and} \quad \xi_\pi \geq 8 . \tag{3}$$

We perform simultaneous fit[1] using all the observables M_π, F_π, and M_ρ with a common γ. Under the assumption that all the observables give a universal γ, we estimate $\gamma = 0.78$–0.93 with $\chi^2/\text{dof} = O(0.1)$. (The example with $\alpha = 1$ of this analysis is shown in Fig. 5 as a typical result of the simultaneous fit.) It is noted that a simultaneous fit including the lighter mass with $m_f \geq 0.015$ in $\xi_\pi \geq 6.8$ fails with a large $\chi^2/\text{dof} = 3.5$ even if the mass correction is included. This is because the chiral property is dictated by SχSB and should not be consistent with universal hyperscaling near the chiral limit.

To summarize, using the fits with the correction term, we estimated the value of γ from the three observables, and obtain $\gamma = 0.62$–0.97 which depends on the observables and the correction term in the fit form. Furthermore we carry out simultaneous FSHS fits with the correction term, since a universal γ would be expected if the theory is very close to the conformal phase boundary even in the SχSB phase. The resulting γ in the simultaneous fits reads 0.78–0.93. These estimated γ's would be identified as the mass anomalous dimension in the walking regime.

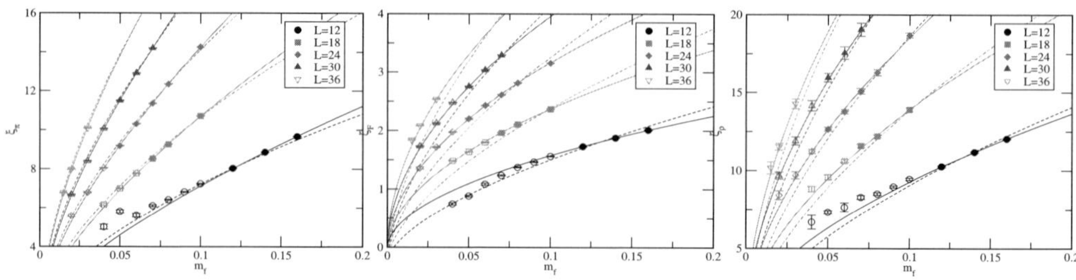

Fig. 5. Simultaneous FSHS fit in ξ_π(left), ξ_F(center) and ξ_ρ(right) with $\alpha = 1$. The filled symbols are included in the fit, but the open symbols are omitted. The fitted region is $m_f \geq 0.05$ and $\xi_\pi \geq 8$. The solid curve is the fit result. For a comparison, the simultaneous fit result without correction terms is also plotted by the dashed curve, whose $\chi^2/dof = 83$.

5. Summary and discussion

In search for a candidate for the Walking Technicolor, we have investigated meson spectrum of $N_f = 8$ QCD by the lattice simulations based on the HISQ action for $\beta = 6/g^2 = 3.8$, and for the fermion bare mass range $m_f = 0.015 - 0.16$ depending on the size $L^3 \times T$ with $(L,T) = (12,16), (18,24), (24,32), (30,40), (36,48)$.

We found that the data of F_π, M_π, M_ρ and $\langle\overline{\psi}\psi\rangle$ are consistent with the SχSB well described by the ChPT in the small m_f region, $0.015 \leq m_f \leq 0.04$, (Case 1 in Fig. 1 for a schematic view), suggesting that $F = 0.031(1)(^{+2}_{-10})$, $\langle\overline{\psi}\psi\rangle\big|_{m_f\to 0} = 0.00052(5)(^{+8}_{-29})$, and $\frac{M_\rho}{F/\sqrt{2}} = 7.7(1.5)(^{+3.8}_{-0.4})$ in the chiral limit extrapolation with the systematic error. Remarkably enough, in contrast to the data near the chiral limit ($m_f \leq 0.04$) indicating the SχSB, those for the relatively large fermion bare mass $m_f \geq 0.05$ away from the chiral limit actually exhibited a FSHS with the scaling exponent $\gamma(M_\pi) \simeq 0.57$, $\gamma(F_\pi) \simeq 0.93$ and $\gamma(M_\rho) \simeq 0.80$. This, non unversal value of γ, implies that there exists a remnant of the infrared conformality where the SχSB effects are negligible (Case 2 in Fig. 1). Therefore, there could exist large mass corrections[16] on the FSHS. If we include possible mass corrections on FSHS, we obtained $0.78 \lesssim \gamma \lesssim 0.93$ in simultaneous FSHS fits. Summarizing all our analyses we may infer that a typical technicolor ("one-family model") as modeled by the $N_f = 8$ QCD can be a walking technicolor theory having an approximate scale invariance with large anomalous dimension $\gamma_m \sim 1$.

Finally, we should comment on the possible light flavor-singlet scalar meson in $N_f = 8$ QCD. The walking technicolor predicts a light composite Higgs-like scalar boson (techni-dilaton) as a pseudo NG boson of the approximate scale invariance. We made the studies of both flavor-singlet scalar and scalar glueballs for $N_f = 12$, in which we found a hint of a flavor-singlet scalar bound state lighter than π for $N_f = 12$.[17,18] Then, if the $N_f = 8$ QCD behaves as a walking theory with approximate scale invariance, it would be expected that a light flavor-singlet scalar composite does exist. These studies are currently under way and details will be reported elsewhere.

Acknowledgments

We would like to thank Anna Hasenfratz, Yoichi Iwasaki and Julius Kuti for fruitful discussions, and acknowledge Katsuya Hasebe for encouragements. We also thank Enrico Rinaldi for helpful discussions. Numerical simulation has been carried out on the supercomputer system φ at KMI, Nagoya University and on the computer facilities of the Research Institute for Information Technology at Kyushu University. This work is supported by the JSPS Grant-in-Aid for Scientific Research (S) No.22224003, (C) No.23540300 (K.Y.) and (C) No.21540289 (Y.A.), and also by Grants-in-Aid of the Japanese Ministry for Scientific Research on Innovative Areas No. 23105708 (T.Y.).

References

1. Y. Aoki, T. Aoyama, M. Kurachi, T. Maskawa, K.-i. Nagai, H. Ohki, A. Shibata, K. Yamawaki and T. Yamazaki, *Phys. Rev. D* **87**, p. 094511 (2013).
2. K. Yamawaki, M. Bando and K.-i. Matumoto, *Phys. Rev. Lett.* **56**, p. 1335 (1986).
3. T. Maskawa and H. Nakajima, *Prog. Theor. Phys.* **52**, p. 1326 (1974).
4. R. Fukuda and T. Kugo, *Nucl.Phys.* **B117**, p. 250 (1976).
5. V. Miransky, *Nuovo Cim.* **A90**, 149 (1985).
6. E. Farhi and L. Susskind, *Phys.Rev.* **D20**, 3404 (1979).
7. Y. Iwasaki, arXiv:1212.4343 [hep-lat] (2012).
8. K.-I. Ishikawa, Y. Iwasaki, Y. Nakayama and T. Yoshie, *Phys. Rev. D* **87**, p. 071503(R) (2013).
9. Y. Aoki, T. Aoyama, M. Kurachi, T. Maskawa, K.-i. Nagai, H. Ohki, A. Shibata, K. Yamawaki and T. Yamazaki, *Phys. Rev. D* **86**, p. 054506 (2012)
10. E. Follana, Q. Mason, C. Davies, K. Hornbostel, G. P. Lepage, J. Shigemitsu, H. Trottier and K. Wong, *Phys. Rev. D* **75**, p. 054502 (2007).
11. Y. Aoki, T. Aoyama, M. Kurachi, T. Maskawa, K.-i. Nagai, H. Ohki, A. Shibata, K. Yamawaki and T. Yamazaki, *PoS* **LATTICE2011**, p. 080 (2011).
12. Y. Aoki, T. Aoyama, M. Kurachi, T. Maskawa, K.-i. Nagai, H. Ohki, A. Shibata, K. Yamawaki and T. Yamazaki, *PoS* **LATTICE2012**, p. 035 (2012).
13. T. DeGrand, *Phys. Rev. D* **80**, p. 114507 (2009).
14. L. Del Debbio and R. Zwicky, *Phys. Rev. D* **82**, p. 014502 (2010).
15. L. Del Debbio, B. Lucini, A. Patella, C. Pica and A. Rago, *Phys. Rev. D* **82**, p. 014510 (2010).
16. Y. Aoki, T. Aoyama, M. Kurachi, T. Maskawa, K.-i. Nagai, H. Ohki, A. Shibata, K. Yamawaki and T. Yamazaki, *Phys. Rev. D* **85**, p. 074502 (2012).
17. Y. Aoki, T. Aoyama, M. Kurachi, T. Maskawa, K.-i. Nagai, H. Ohki, E. Rinaldi, A. Shibata, K. Yamawaki and T. Yamazaki, arXiv:1302.4577 [hep-lat] (2013).
18. Y. Aoki, T. Aoyama, M. Kurachi, T. Maskawa, K.-i. Nagai, H. Ohki, E. Rinaldi, A. Shibata, K. Yamawaki and T. Yamazaki, arXiv:1305.6006 [hep-lat] (2013).

The Scalar Spectrum of Many-Flavour QCD

Yasumichi Aoki[a], Tatsumi Aoyama[a], Masafumi Kurachi[a], Toshihide Maskawa[a], Kei-ichi Nagai[a], Hiroshi Ohki[a], Enrico Rinaldi[a,b*], Akihiro Shibata[c], Koichi Yamawaki[a], Takeshi Yamazaki[a]

(LatKMI Collaboration)

[a]*Kobayashi-Maskawa Institute for the Origin of Particles and the Universe (KMI), Nagoya University, Nagoya, 464-8602, Japan*
[b]*SUPA and School of Physics and Astronomy, University of Edinburgh, Edinburgh, EH9 3JZ, UK*
[c]*Computing Research Center, High Energy Accelerator Research Organization (KEK), Tsukuba 305-0801, Japan*

The LatKMI collaboration is studying systematically the dynamical properties of $N_f = 4, 8, 12, 16$ SU(3) gauge theories using lattice simulations with (HISQ) staggered fermions. Exploring the spectrum of many-flavour QCD, and its scaling near the chiral limit, is mandatory in order to establish if one of these models realises the Walking Technicolor scenario. Although lattice technologies to study the mesonic spectrum are well developed, scalar flavour-singlet states still require extra effort to be determined. In addition, gluonic observables usually require large-statistic simulations and powerful noise-reduction techniques. In the following, we present useful spectroscopic methods to investigate scalar glueballs and scalar flavour-singlet mesons, together with the current status of the scalar spectrum in $N_f = 12$ QCD from the LatKMI collaboration.

Keywords: Lattice Gauge Theories, Walking Technicolour, Glueballs

1. Introduction

It has been known for more than 30 years that the electroweak symmetry breaking (EWSB) mechanism could be due to a new strongly-interacting sector at energies of $\mathcal{O}$(TeV). Since then, this idea has evolved to become one of the most promising scenarios for physics beyond the Standard Model, the so-called Walking Technicolour with a large anomalous dimension $\gamma_m \sim 1$ and an approximate scale invariance[1] (See also similar works[2–4] without notion of anomalous dimension and scale invariance).

Recently, the discovery of a new particle of mass $\sim$ 125 GeV at the Large Hadron Collider[5,6] has attracted a lot of attention and it is now a pressing issue to determine its properties and its nature. The new particle's branching ratios are consistent with the ones of the Standard Model Higgs boson, but the very interesting possibility of this particle being of composite nature, a techni-dilaton,[1] as predicted by Walking Technicolour models is also viable.[7,8] The LHC experiments will be working very

*Speaker. E-mail: e.rinaldi@sms.ed.ac.uk

hard in the following years, trying to assess the validity of their analysis for the different decay channels of the new particle. At the same time, the effort of the theory community will go in a similar direction; new increasingly precise lattice simulations might help addressing the question whether or not a more fundamental strongly-interacting gauge sector is responsible for the origin of mass (a review of recent lattice studies can be found in Refs. 9,10).

The LatKMI collaboration is focusing on a systematic study of SU(3) gauge theories with different numbers N_f of fundamental massless fermions using lattice Monte Carlo simulations (see Refs. 11–13 for recent updates). The main purpose of such on-going project is to find a near-conformal theory with a large anomalous dimension $\gamma_m \sim 1$ by investigating the infrared dynamics of the theory which is in the non-perturbative domain. If such a theory will be found, the important question will be to identify the composite scalar state, techni-dilaton, mimicking the role of the *Higgs* boson of the Standard Model and to study its mass as it arises from a purely non-perturbative dynamics.

The study of scalar single-particle resonances on the lattice is complicated due to the poor quality of the signal-to-noise ratio in Monte Carlo simulation. Even in state-of-the-art lattice QCD calculations, where an extreme degree of accuracy has been reached in recent years for many physical observables, the spectrum of scalar mesons is still an open issue. Moreover, scalar excitations of gluonic degrees of freedom (glueballs) introduce an additional factor of complexity since they can mix with fermionic states: a careful study is necessary to distinguish the nature of states in the scalar channel (for an up to date review see Ref. 14). In the following, we describe the status of LatKMI investigation of the scalar spectrum from lattice simulations of $N_f = 12$ QCD. The data shown in the following are preliminary.

2. Lattice simulations

The lattice gauge configurations for the SU(3) theory with $N_f = 12$ fundamental flavours are generated by the HMC algorithm using a tree-level Symanzik gauge action and the HISQ (Highly Improved Staggered Quark[15]) action for the dynamical fermions. The flavour symmetry breaking of this action (measured as taste splitting) is highly suppressed in QCD[16] and it is almost negligible for the theory we investigate. At the couplings used in this study, the masses of bound states in the same taste multiplet can not be distinguished from one another as shown in Fig. 1. More details on the calculated mesonic spectrum and decay constants can be found in Ref. 11.

At fixed lattice spacing defined by the bare coupling constant $\beta = 6/g^2 = 4.0$, we simulate two physical volumes $V = (aL)^3$ with $L = 18, 24$ and aspect ratio $T/L = 4/3$. On the smaller volume we investigate the scalar spectrum at five different bare quark masses $0.06 < am_f < 0.16$; the bare mass changes by almost a factor of 3 and this allows us to identify a possible quark mass dependence in the measured observables. Another important feature of our simulation is the large

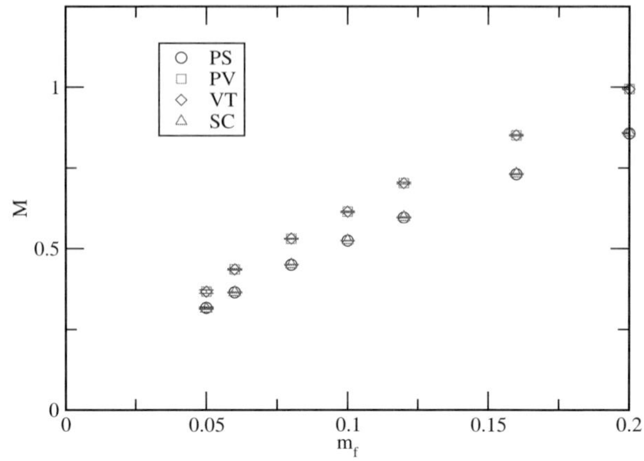

Fig. 1. An example of the small taste splitting in the spectrum of $N_f = 12$ QCD with HISQ fermions at $\beta = 4.0$. The figure is taken from Ref. 11. M and m_f are the meson mass and the bare quark mass respectively, both in units of the common fixed lattice spacing a. PS and SC refer to pseudoscalar mesons coming from operators with different internal taste structure, whereas PV and VT refer to vector mesons.

number of Monte Carlo trajectories obtained from uninterrupted Markov chains after thermalization. For any set of parameters explored, we collect more than 15000 trajectories (with MD step set to unity), which is necessary to contrast the rapid degradation of signal to noise ratio of the scalar correlators. A summary of the simulated parameters is found in Tab. 1.

Table 1. Parameters of lattice simulations for $N_f = 12$ QCD at fixed $\beta = 4.0$. $N_{\rm traj}$ is the number of Monte Carlo trajectories after thermalisation, $N_{\rm cfgs}$ is the number of saved gauge configurations. For different ensembles, configurations are saved every 2 or 5 trajectories.

$L \times T$	am_f	$N_{\rm traj}$	$N_{\rm cfgs}$
18×24	0.06	15445	3090
18×24	0.08	22200	4440
18×24	0.10	15300	3060
18×24	0.12	21500	4300
18×24	0.16	18450	3690
24×32	0.06	28000	14000
24×32	0.08	19280	9640
24×32	0.10	18960	9480

3. Glueball measurement

It has been pointed out[17] that signs of infrared conformality can be found by comparing the mesonic spectrum with the gluonic one. Such a comparison was done in the SU(2) gauge theory with two flavours of adjoint fermions[18] and the results supported the hypothesis of an infrared fixed point being responsible for the dynamics of the theory on the critical massless surface.

Addressing a measurement of the glueball spectrum is intrinsically difficult due to the poor signal over noise ratio in gluonic correlators. The theory behind the measurement of glueball masses in lattice spectroscopy has been known for a long time.[19] Recently, a powerful framework to measure gluonic bound states has been developed[20] that improves upon previous results. The method employs a large number of different operators built from gauge-invariant combinations of gauge links in such a way that a robust basis for a variational ansatz can be created. The variational approach, with such a large basis, has been shown to produce excellent results for different gauge theories.[20–22] In the following we briefly review the technology involved in such a variational method and we show the preliminary results obtained by the LatKMI collaboration using the gauge configurations in Tab. 1. This is the first time that glueball states are investigated in many-flavour SU(3) gauge theory.

Eigenstates of the lattice Hamiltonian fall into irreducible representations of the symmetry group characterising the four-dimensional lattice system. In particular, there are 20 symmetry channels for glueballs on a three-dimensional spatial lattice, corresponding to the 5 irreducible representations of the cubic group and including both parity and charge quantum numbers. In the following, we focus on the scalar 0^{++} channel that is expected to feature the lightest state. On each spatial point of the lattice we build gauge-invariant interpolating operators $\mathcal{O}_\alpha(x,t)$ with well-defined rotational quantum number by using a prescribed linear combination of traced spatial Wilson loops. Zero-momentum operators consist in an average over the whole volume V. By using differently shaped Wilson loops we construct 32 different basis operators for the scalar glueball (each of these operators come in different blocking sizes and they are all included in the variational basis).

Let us now give an example of a typical glueball measurement. For every configuration we measure a matrix of correlation functions, whose elements are

$$\tilde{C}_{\alpha\beta}(t) \;=\; \sum_{\tau} \langle 0|\, \mathcal{O}^\dagger_\alpha(t+\tau)\mathcal{O}_\beta(\tau)\,|0\rangle \;. \tag{1}$$

By solving the generalised eigenvalue problem for the matrix above, optimal operators (i.e. those that create almost pure states $|i\rangle$) can be found that are a linear sum of the basis vectors

$$\tilde{\mathcal{O}}_i(t) \;=\; \sum_\alpha v^i_\alpha \mathcal{O}_\alpha(t)\,; \qquad \tilde{\mathcal{O}}_i(t)\,|0\rangle \approx |i\rangle \;, \tag{2}$$

where v^i_α are the components of the ith eigenvector. Different eigenvectors v^i correspond to different states. The mass m_i of the ith state is then extracted by fitting correlators of optimal operators using

$$\bar{C}_{ii}(t) = |c_i|^2\left(e^{-m_i t} + e^{-m_i(T-t)}\right)\,, \tag{3}$$

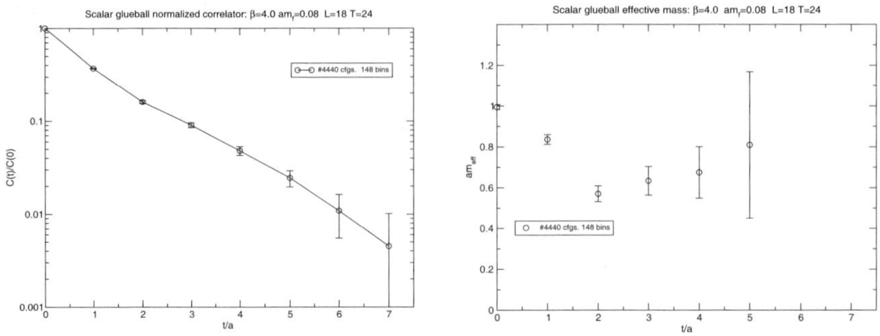

(a) (b)

Fig. 2. Analysis of 4440 configurations for $\beta = 4.0$ and $am_f = 0.08$ on the smaller volume. (a) The time dependence of the normalized scalar correlator in logarithmic scale; the linear behaviour corresponds to the propagation of a single state (b) The effective mass of the corresponding correlator; the signal is lost after 6 timeslices, but the plateaux is already present at short temporal distances.

where T is the length of the lattice in the time direction and the functional form is a consequence of the usual exponential decay in a lattice with periodic boundary condition in the time direction.

In general, glueball correlators are very noisy and this limits the usefulness of numerical correlators to short time separations. However, although Eq. (3) is only valid at large t, if the overlap with an Hamiltonian state is almost perfect, it is possible to extract a reliable value for the mass at short time separation, since the decay is largely dominated by a single state. For this to be true, a careful construction of the variational basis is paramount. Whether the state created by an optimal operator $\tilde{\mathcal{O}}_i(t)$ is a good approximation of the Hamiltonian eigenstates can be checked by looking at the value of the overlap $|c_i|^2$: the closer this number is to one, the better is the variational calculation.

We measure the correlators of scalar gluonic operators on the configurations of Tab. 1. The measurements are taken for every configurations and statistical errors are estimated using jack-knife. We check for autocorrelations by changing the size of the jack-knife bins.

For each set of simulated parameters, we perform the variational analysis described above and we obtain the correlator corresponding to the optimal operator that projects on the ground state ($i = 1$). In some cases, the correlator for the first excited state ($i = 2$) is studied as well in order to estimate the higher states contamination in the ground state estimate. We calculate the effective mass and we look for a plateaux where Eq. (3) can be applied. An example of the scalar ground state correlator and its effective mass is shown in Fig. 2.

The high number of analysed configurations is very helpful when determining the effective mass plateaux. For example Fig. 3 shows how the effective mass changes when the statistic is increased.

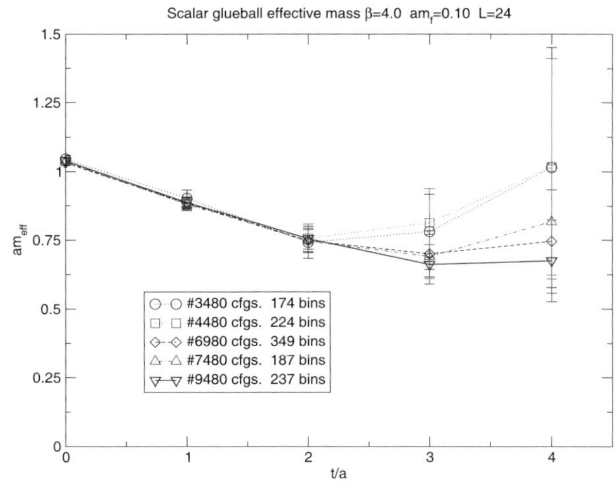

Fig. 3. An example of the scalar effective mass at $\beta = 4.0$ and $am_f = 0.10$ on the larger volume $L \times T = 24 \times 32$ as the number of configuration is increased. A stable plateaux becomes more and more visible as the statistics increases.

4. Flavour-singlet scalar meson

In the staggered discretization, fermionic fields defined on the lattice sites have an additional *taste* structure. The scalar 0^+ fermionic interpolating operator that we used in the construction of our propagators is the simplest one and has a (1×1) spin–taste content. As is generally the case, this operator couples also to a negative parity 0^- state whose contribution must be accounted for in the analysis of the full connected correlator $C(t)$. However, for the scalar flavour-singlet case (referred to as σ in the following), we need to compute disconnected diagrams and the vacuum expectation value of the mesonic operator.

Measuring disconnected contributions to mesonic observables is very challenging and ad-hoc noise reduction techniques have been developed and used successfully in lattice QCD (LQCD). We used the stochastic source method[23,24] with Gaussian noise and local source and sink insertions. The number of noise vectors for each configuration has been chosen to the minimal value that yields the asymptotic value of the disconnected correlator $D(t)$ on the available statistics.

Thanks to the relative high number of configurations analysed at the lightest bare fermion mass (cfr. Tab. 1), we are able to extract a clean signal for the disconnected correlator $D(t)$ at large temporal distances and including the vacuum expectation value subtraction, as it is shown in Fig. 4(a). The scalar singlet correlator $C_\sigma(t)$ is extracted from combinations of $C(t)$ and $D(t)$ that properly account for the effects of the parity partner meson and of the vacuum fluctuations:

$$\tilde{C}(2t) = 2C(2t) + C(2t+1) + C(2t-1)\,, \tag{4}$$

$$\tilde{D}(2t) = 2D(2t) + D(2t+1) + D(2t-1)\,, \tag{5}$$

$$C_\sigma(2t) = -\tilde{C}(2t) + 3\tilde{D}(2t)\,. \tag{6}$$

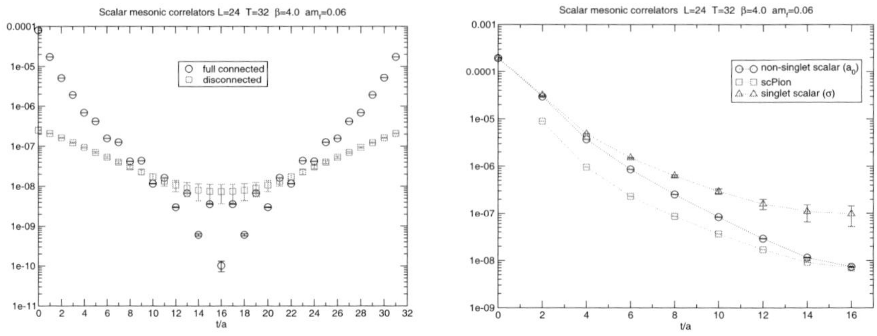

(a) (b)

Fig. 4. Results for the analysis of scalar mesonic interpolating operators (a) The full connected and disconnected correlators where the vev as been subtracted in the latter. The connected correlator still includes the parity partner contributions. (b) The correlators of the different channels extracted from $C(t)$, where disconnected terms have been included for the singlet case only.

The factor of 3 in the last equation comes from the number of actual fermions in the simulation (3 staggered fermions $\times$ 4 tastes $=$ 12 flavours). From the full connected correlator, the non-singlet (a_0) and the pseudoscalar (scPion) states can be isolated as well. The three different channels are compared in Fig. 4(b) and the results strongly indicates that the σ meson mass is smaller than the pseudoscalar one. Moreover, although the scalar effective mass plateaux appears only at very large temporal separation and the statistical error are large, the corresponding mass is compatible with the one obtained using only purely gluonic operators. The comparison between the gluonic scalar spectrum and the fermionic scalar singlet is reported in Fig. 5.

5. Conclusions

We performed a study of the scalar glueball and scalar flavour-singlet fermionic state. Advanced lattice spectroscopy techniques that have proven successful in LQCD studies have been applied for the first time to explore the scalar spectrum of many-flavour QCD.

The results shown in the previous sections clearly show that the variational method used to extract masses of gluonic bound states is very efficient and allowed us to study the qualitative quark mass dependence of scalar glueballs. Although the results are preliminary and the statistical errors of the order $\sim 10\%$, it is very encouraging that a signal for an effective mass plateaux could be obtained even for relatively small time separation. As for the scalar fermionic correlator including the disconnected term, although an effective mass plateaux is observed at somewhat larger time due to the lack of improved operators, it is remarkable that the effective mass is consistent with the mass obtained from the gluonic operators.

The most striking feature of the measured scalar spectrum is the appearance of a state lighter than the pion for low bare quark masses as it is shown in Fig. 5. Such a state appears both in gluonic and fermionic correlators, indicating that mixing between gluons and fermions plays a relevant role. A future variational analysis including both kind of operators described in the previous sections will help disentangle mixed contributions. Moreover, a careful finite size study is under completion and possible systematic effects of the lattice regularization are being investigated.

Despite it being studied by several groups using different approaches, $N_f = 12$ QCD has not yet been identified as an infrared-conformal or near-conformal theory. The majority of studies suggest the presence of an infrared fixed point with a small anomalous dimension. If this turns out to be the case, such a theory would not be a viable candidate for a phenomenologically interesting Walking Technicolour model. A walking theory, which may be realised near the edge of the conformal window, should have its property influenced by the near-conformal theory. The light scalar state observed for $N_f = 12$ in this study is a promising signal for the search of a successful walking theory, where the existence of a light scalar is now required by LHC experiment. A possible future direction is to look at more viable candidates for Walking Technicolour models; for example, it will be interesting to investigate the scalar spectrum of the $N_f = 8$ SU(3) theory where, if the theory has a walking regime, the scalar state could be identified with the pseudo-NG boson coming from the breaking of conformal symmetry.

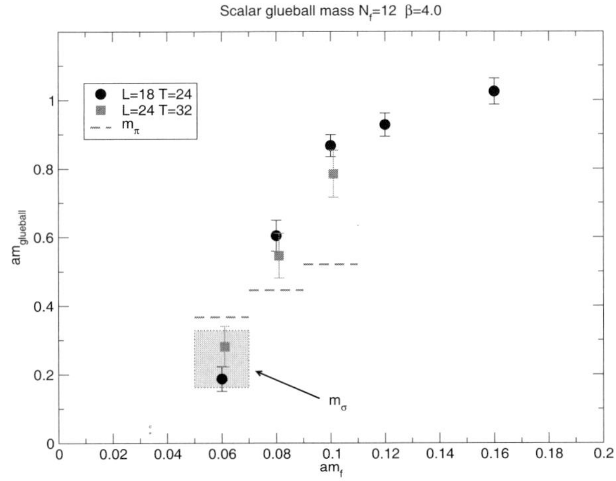

Fig. 5. Summary of the fitted scalar glueball masses for $N_f = 12$ QCD at $\beta = 4.0$ and for several bare fermion masses. Some values of the pion mass are shown for comparison. The point on different volumes have been slightly displaced for clarity. The grey box indicates the location of the scalar flavour-singlet mass m_σ on the ensemble with the lightest bare quark mass.

Acknowledgments

Numerical simulation has been carried out on the supercomputer system φ at KMI in Nagoya university, and on the computer facilities of the Research Institute for Information Technology at Kyushu University.
This work is supported by the JSPS Grant-in-Aid for Scientific Research (S) No.22224003, (C) No.23540300 (K.Y.) and (C) No.21540289 (Y.A.), and also by Grants-in-Aid of the Japanese Ministry for Scientific Research on Innovative Areas No. 23105708 (T.Y.). E.R. is currently supported by a SUPA Prize Studentship and was supported by a FY2012 JSPS Postdoctoral Fellowship for Foreign Researchers (short-term), ID PE12518.

References

1. K. Yamawaki, M. Bando and K.-i. Matumoto, *Phys.Rev.Lett.* **56**, p. 1335 (1986).
2. B. Holdom, *Phys.Lett.* **B150**, p. 301 (1985).
3. T. W. Appelquist, D. Karabali and L. Wijewardhana, *Phys.Rev.Lett.* **57**, p. 957 (1986).
4. T. Akiba and T. Yanagida, *Phys.Lett.* **B169**, p. 432 (1986).
5. CMS, *Phys.Lett.* **B716**, 30 (2012).
6. ATLAS, *Phys.Lett.* **B716**, 1 (2012).
7. S. Matsuzaki and K. Yamawaki, *Phys.Rev.* **D86**, p. 035025 (2012).
8. S. Matsuzaki and K. Yamawaki, *Phys.Rev.* **D86**, p. 115004 (2012).
9. E. T. Neil, *PoS* **LATTICE2011**, p. 009 (2011).
10. J. Giedt, *PoS* **LATTICE2012**, p. 006 (2012).
11. Y. Aoki, T. Aoyama, M. Kurachi, T. Maskawa, K.-i. Nagai, H. Ohki, A. Shibata, K. Yamawaki and T. Yamazaki, *Phys.Rev.* **D86**, p. 054506 (2012).
12. Y. Aoki, T. Aoyama, M. Kurachi, T. Maskawa, K.-i. Nagai, H. Ohki, A. Shibata, K. Yamawaki and T. Yamazaki, *PoS* **LATTICE2012**, p. 035 (2012).
13. Y. Aoki, T. Aoyama, M. Kurachi, T. Maskawa, K.-i. Nagai, H. Ohki, A. Shibata, K. Yamawaki and T. Yamazaki, *PoS* **LATTICE2012**, p. 029 (2012).
14. W. Ochs, *arxiv:1301.5183* (2013).
15. E. Follana *et al.*, *Phys.Rev.* **D75**, p. 054502 (2007).
16. A. Bazavov, T. Bhattacharya, M. Cheng, C. DeTar, H. Ding *et al.*, *Phys.Rev.* **D85**, p. 054503 (2012).
17. L. Del Debbio, B. Lucini, A. Patella, C. Pica and A. Rago, *Phys.Rev.* **D80**, p. 074507 (2009).
18. L. Del Debbio, B. Lucini, A. Patella, C. Pica and A. Rago, *Phys.Rev.* **D82**, p. 014510 (2010).
19. B. Berg and A. Billoire, *Nucl. Phys.* **B221**, p. 109 (1983).
20. B. Lucini, A. Rago and E. Rinaldi, *JHEP* **1008**, p. 119 (2010).
21. L. Del Debbio, A. Hart and E. Rinaldi, *JHEP* **1207**, p. 178 (2012).
22. E. Gregory, A. Irving, B. Lucini, C. McNeile, A. Rago *et al.*, *JHEP* **1210**, p. 170 (2012).
23. L. Venkataraman and G. Kilcup, *Phys.Rev.D* (1997).
24. E. B. Gregory, A. C. Irving, C. M. Richards and C. McNeile, *Phys.Rev.* **D77**, p. 065019 (2008).

Conformal Dynamics and Thermal/Bulk Phase Transition in Lattice Gauge Theory with Many Fermion Species

Kohtaroh Miura* and Maria Paola Lombardo
INFN Laboratori Nazionali di Frascati,
I-00044, Frascati (RM), Italy
**E-mail: Kohtaroh.Miura@lnf.infn.it*

Elisabetta Pallante and Tiago Nunes da Silva
Centre for Theoretical Physics, University of Groningen,
9747 AG, Netherlands

Albert Deuzeman
Albert Einstein Center for Fundamental Physics, University of Bern,
CH-3012 Bern, Switzerland

We summarize our recent results[1,2] which investigate the conformal dynamics associated with the infra-red fixed point in colour SU($N_c = 3$) gauge theory with a variable number (N_f) of fundamental fermions by using lattice Monte-Carlo simulations. We evaluate the thermal chiral phase transition as a function of N_f, and investigate the vanishing of the chiral dynamics in the large N_f region. Then, three independent analyses consistently indicate the onset of the conformal phase for $N_f \gtrsim N_f^* \sim 9.2 - 14.1$. Furthermore, we study the bulk transition in $N_f = 12$ system with a focus on the interplay between the transition property and the improvements for the lattice action. We show that the bulk transition observed in the case of the naive action evolves into two rapid crossovers due to the improvement for the fermion sector.

Keywords: Lattice Gauge Theory; Conformal Symmetry; Chiral Symmetry; Finite Temperature.

1. Introduction

The infra-red fixed point (IRFP) is anticipated to appear in asymptotically free non-Abelian gauge theories when the number of fermion species, N_f, exceeds a critical value $N_f = {N_f}^*$.[3] The IRFP leads to a new class of a gauge theory in which a conformal phase dynamically emerges due to the non-perturbative interaction. The approach to the conformal phase from below ($N_f \nearrow N_f^*$) probably accompanies with the pre-conformal gauge theory with a slowly running (walking) coupling, which has been advocated as a basis for strongly interacting mechanisms of electroweak symmetry breaking.[4,5] Lattice Monte-Carlo simulations are now expected to provide a solid theoretical base for the IRFP associated (pre-)conformal gauge theory.[6]

In this proceedings, we investigate two subjects closely related to each other based on our recent lattice results: The first topic is the estimate of N_f^* with respect for the vanishing of the thermal chiral transition with increasing N_f.[1] The second topic is the bulk phase transition in the $N_f = 12$ system with a focus on the interplay between the transition property and the improvements for the lattice action.[2]

2. Estimate of N_f^*

We investigate the thermal chiral phase transition for $N_f = (4, 6, 8)$ and the deconfinement transition for the quenched system $N_f = 0$. Then we study the vanishing nature of the chiral phase transition for larger N_f, which signals the emergence of the conformal phase.

We use the Asqtad action with a one-loop Symanzik and tadpole improved gauge action. The tadpole factor u_0 is determined by performing zero temperature simulations, and used as an input for finite temperature simulations. We have used the rational hybrid Monte Carlo algorithm. The quenched ($N_f = 0$) system has been realised by using massive bare fermion mass $ma = 1.0$ in the four flavour system. The six flavour system has been realised by using two pseudo-fermions in the rational approximation with a quarter root technique, $N_f = 4 \cdot 2 \cdot 3/4 = 6$. We assume the rooting does not affect the results within the accuracy of our simulation.

We evaluate the critical lattice coupling ${\beta_{\rm L}}^{\rm c}$ associated with the chiral (deconfinement) crossover by locating the peak position of the chiral (Polyakov loop) susceptibilities for each finite N_f (quenched). In table 1, we summarize the obtained ${\beta_{\rm L}}^{\rm c}$ as a function of (N_f, N_t). As we have shown in Ref. 1, the (approximate) asymptotic scaling for the normalized critical temperature $T_c/\Lambda_{\rm L}$ varying N_t holds at each N_f, where $\Lambda_{\rm L}$ is so-called lattice Lambda. This indicates that our ${\beta_{\rm L}}^{\rm c}$ have been determined near the continuum limit.

Table 1. The (pseudo) critical lattice couplings ${\beta_{\rm L}}^{\rm c}(N_f, N_t)$.[1,7]

$N_f \backslash N_t$	4	6	8	12
0	7.35 ± 0.05	7.97 ± 0.07	8.26 ± 0.06	—
4	5.65 ± 0.05	6.00 ± 0.05	6.15 ± 0.15	—
6	4.675 ± 0.05	5.025 ± 0.05	5.20 ± 0.05	5.55 ± 0.1
8	—	4.1125 ± 0.0125	4.275 ± 0.05	4.34 ± 0.04

By using obtained ${\beta_{\rm L}}^{\rm c}$, we investigate the onset of the conformal window N_f^* by performing three independent analyses as follows. The first method is based on the well-known step-scaling function: $\Delta {g_{\rm L}}^s = g_{\rm L} - {g_{\rm L}}'$, where the lattice coupling $g_{\rm L}$ and $g_{\rm L}'$ give a same physical scale ξ: $\xi = a(g_{\rm L})\hat{\xi} = a({g_{\rm L}}')\hat{\xi}'$, with $\hat{\xi}$ being the dimensionless lattice correlation length, and $\hat{\xi}/\hat{\xi}' = s$. As discussed in the previous study Ref. 8, $\Delta g_{\rm L}^s = 0$ holds at the IRFP regardless the scale factor s. We here identify $(\xi, \hat{\xi}, \hat{\xi}')$ to be (T_c^{-1}, N_t, N_t'). Then, the vanishing of the step scaling function reads $T_c^{-1} = N_t\ a({g_{\rm L}}^{\rm c}(N_f^*, N_t)) = N_t'\ a({g_{\rm L}}^{\rm c}(N_f^*, N_t'))$, which defines N_f^*. Here, ${g_{\rm L}}^{\rm c}(N_f, N_t) = \sqrt{10/{\beta_{\rm L}}^{\rm c}(N_f, N_t)}$ (see Table 1).

Consider the pseudo-critical line which connects ${g_{\rm L}}^{\rm c}$ with a constant N_t, ${g_{\rm L}}^{\rm c}(N_f)|_{N_t=\rm fixed}$ in the space spanned by $g_{\rm L}$ and N_f (Fig. 1). The resultant phase diagram may be seen as an extension of the well-known Miransky-Yamawaki (MY) phase diagram[9] to finite temperature. The increasing trend of ${g_{\rm L}}^{\rm c}$ for larger N_f would be a consequence of enhanced fermionic screening due to the fermion multiplicity.[10] The crossing point of the pseudo-critical lines in Fig. 1 indicates the vanishing of the step scalings, *i.e.*, the IRFP. Using two lines ${g_{\rm L}}^{\rm c}(N_f)|_{N_t=6}$ and ${g_{\rm L}}^{\rm c}(N_f)|_{N_t=12}$, the IRFP is estimated to be $({g_{\rm L}}^{\rm c}, N_f^*) = (1.79 \pm 0.12, 11.1 \pm 1.6)$.

Now we are going to the second estimate of N_f^*: Consider the two-loop renormalisation flow,

$$R(g_{\rm L}, {g_{\rm L}}^{\rm ref}) \equiv \frac{\Lambda_{\rm ref}({g_{\rm L}}^{\rm ref})}{a^{-1}(g_{\rm L})} = \exp\left[\int_{g_{\rm L}}^{{g_{\rm L}}^{\rm ref}} \frac{dg}{\beta_{\rm 2loop}(g)}\right] . \tag{1}$$

We set the initial condition of the flow by using the critical lattice coupling: $g_{\rm L} \to {g_{\rm L}}^{\rm c} = \sqrt{10/{\beta_{\rm L}}^{\rm c}}$. Here we particularly pick up those obtained by using the largest temporal extension for each N_f from Table 1. The reference scale $\Lambda_{\rm ref}$ is specified to be the critical temperature itself: $\Lambda_{\rm ref}({g_{\rm L}}^{\rm ref}) = 1/(N_t\ a({g_{\rm L}}^{\rm c}))$. Thus we have introduced the thermal critical coupling $g_{\rm T}^{\rm c}(N_f)$ via $R({g_{\rm L}}^{\rm c}, g_{\rm T}^{\rm c}) = 1/N_t$.

In the right panel of Fig. 1, we plot $g_{\rm T}^{\rm c}$ as a function of N_f (red (□) symbol) in comparisons with zero temperature critical couplings $g_{\rm SD}^{\rm c}$ obtained by the two-loop Schwinger-Dyson equation[11] and the four-loop IRFP $g_{41}^{\rm IRFP}$[12] where the error bars for $g_{41}^{\rm IRFP}$ implements a possible scheme dependence. The emergence of the conformal phase would be seen as a coincidence of them: $g_{\rm T}^{\rm c}(N_f^*) = g_{\rm SD}^{\rm c}(N_f^*)$ or $g_{\rm T}^{\rm c}(N_f^*) = g_{41}^{\rm IRFP}(N_f^*)$. These two conditions are consistent within the largish errors, and we quote the average of the two estimates: $N_f^* \sim 12.5 \pm 1.6$. In addition, on a slightly different aspect, the non-Abelian plasma gets to be more strongly interacting for larger N_f[13] as we see from the increasing $g_{\rm T}^{\rm c}$.

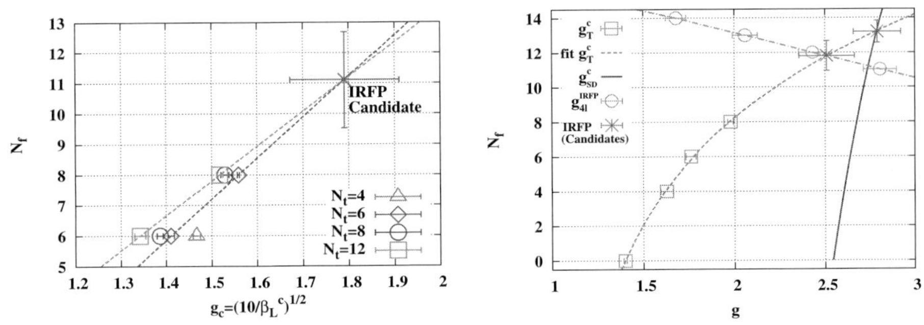

Fig. 1. Left: (Pseudo) critical values of the lattice coupling ${g_{\rm L}}^{\rm c}(N_f, N_t) = \sqrt{10/{\beta_{\rm L}}^{\rm c}(N_f, N_t)}$ for in the parameter space spanned by $(N_f, g_{\rm L})$. The IRFP is extracted from the intersection of two pseudo-critical lines ${g_{\rm L}}^{\rm c}(N_f)|_{N_t=6}$ and ${g_{\rm L}}^{\rm c}(N_f)|_{N_t=12}$. Right: The thermal critical coupling (red □) as a function of N_f and the fit for them (dashed red line) in comparisons with $g_{\rm SD}^{\rm c}$ and $g_{41}^{\rm IRFP}$: See text for the details. Both figures are quoted from Ref. 1.

Now we explain the final method to evaluate N_f^* with respect for the vanishing of the critical temperature T_c in the conformal phase. We utilize Eq. (1), where we use again the thermal critical coupling explained above as the reference coupling: ${g_{\rm L}}^{\rm ref} = g_{\rm T}^{\rm c}$. In turn, the initial condition is determined as follows: We pick up a tadpole factor $u_0 = \langle P\rangle^{1/4}$ from the lattice measurements at $T = 0$ in N_f independent way, for instance $u_0 = 0.8$, and read off the corresponding bare lattice couplings at each N_f. The obtained $g_{\rm UV}(N_f) = g_{\rm L}(N_f)$ defines the UV initial condition $M(N_f) = a^{-1}(g_{\rm UV}(N_f))$ in Eq. (1).

At the scale $M \gg T_c$, the physics does not depend on N_f in the sense that the normalized force which is nothing but $\langle P\rangle = u_0^4$ at M has no N_f independence. Starting from such a UV initial condition, the physics at the scale of T_c gets to be N_f dependent due to the growing IRFP dynamics for larger N_f. This would be exposed by the vanishing of the normalized critical temperature $R(g_{\rm UV}(N_f), g_{\rm T}^{\rm c}(N_f)) = (T_c/M)(N_f) \to 0$. Thus, N_f^* is defined by the condition $(T_c/M)(N_f^*) = 0$.

Note that the coupling at the lattice cutoff $a^{-1}({g_{\rm L}}^{\rm c})$ is $N_t \gg 1$ times larger than T_c. The scale hierarchy $T_c(N_f) \ll a^{-1}({g_{\rm L}}^{\rm c}(N_f))$ admits the condition $T_c(N_f) \ll M(g_{\rm UV}) < a^{-1}({g_{\rm L}}^{\rm c}(N_f))$ by picking up u_0 satisfying $g_{\rm UV} \ll g_{\rm T}^{\rm c}$ for ${}^\forall N_f$. We find that $u_0 \sim 0.8$ meets this requirement.

The left panel of Fig. 2 displays $(T_c/M)(N_f)$ for $u_0 = 0.79$, 0.80, and 0.81. Here, we fit the data points for $N_f \geq 4$ by using the FRG motivated ansatz,[14] $T_c = K|N_f^* - N_f|^{(-2b_0^2/b_1)(N_f^*)}$, where $b_{0,1}$ represents the first two coefficient of the two-loop beta-function $\beta_{\rm 2loop}$. We extract the average among the results obtained for $u_0 = (0.79, 0.80, 0.81)$: $N_f^* = 10.4 \pm 1.2$. The use of so-called Miransky scaling[9] for the fit ansatz should be investigated in future.

As shown in the right panel of Fig. 2, N_f^* is relatively stable within the range $0.79 \leq u_0 \leq 0.94$. The scale cannot be pushed further towards the UV because of discritization errors. On the other hand, a small $u_0 \lesssim 0.7$ leads to $M \sim T_c$ or smaller. In such a case, the scale M is apparently affected by IR physics, and cannot be used to study the vanishing of T_c.

3. Bulk transition and lattice action improvement for $N_f = 12$

We here extend our study to $N_f = 12$ system, for which a part of the authors have recently examined the chiral transition in finite T setup, and observed the bulk transition rather than the thermal.[15] This has implied the absense of the physical finite T chiral transition, and thereby, the onset of the conformal phase within the available data set. The detailed discussions on the chiral and conformal dynamics for $N_f = 12$ are found in Refs. 6.

The study of the bulk transition then gives an important step for better understandings of the IRFP conformality: If the transition is second order, a new continuum limit can emerge at a second non-trivial UVFP. If instead the transition is first order, no continuum limit exists, but the UVFP can still emerge at the end-point of a line of first order bulk transitions when we fine-tune a bare fermion

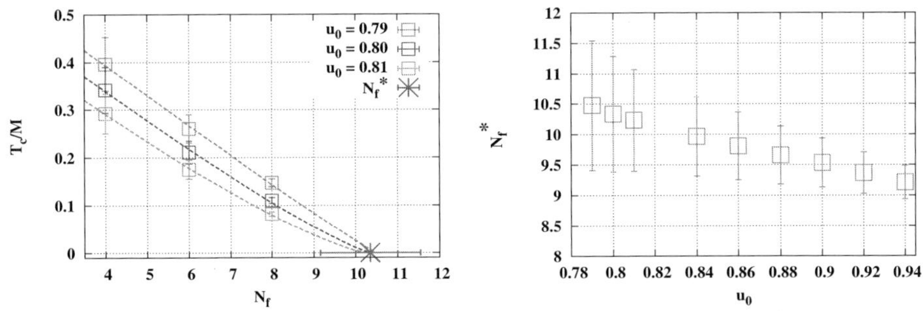

Fig. 2. Left: The N_f dependence of T_c/M where M is determined to be a UV scale corresponding to $u_0 = 0.79$ (red box), 0.80 (blue ○), and 0.81 (magenta triangle) for each fixed N_f. The dashed lines represent fits for data (see text). Right: The u_0 dependence of N_f^*. The three data in the left side are determined within the condition $M \lesssim a^{-1}(g_L{}^c)$. Both figures are quoted from Ref. 1.

mass. Then, the conformal phase transition ($N_f \searrow N_f^*$) may be observed as the annihilation of a pair of IR and UV fixed points.

In Ref. 2, the study of the bulk transition has been further elaborated with a particular emphasis on the interplay between the transition pattern and the lattice action improvements by a part of the authors: Let us consider four different combinations of (un)improved gauge and fermion

A without any improvements (*i.e.*, naive action),
B with a tree level Symanzik improvement for the gauge sector,
C with three-, five- and seven-link (Naik) improvements for the fermion sector,
D with both improvements adopted in B and C.

In the left (right) panel in Fig. 3, we compare the ensemble average of the chiral condensate obtained in A and D (B and C). The most notable is the presence of two rapid crossovers in both C and D cases, which indicates that a novel intermediate phase comes out due to the Naik improvement for the fermion action. Importantly, the rapid crossover at weaker coupling diminishes with a decreasing bare mass. Hence the crossover at stronger coupling is interpreted as a *true* chiral crossover.

Inside the exotic intermediate phase, the pseudoscalar corrector develops an oscillatory component, as reported in Ref. 2. This is explained by the competition between the effects of local and less local interactions to the baryon number conservation. In the present system, such a competition arises from the Naik improved fermion action. This characterization of the intermediate phase would be consistent with the $\mathbb{S}_4$ shift symmetry breaking discussed in Ref. 16. We note that the Naik improved fermion action either the Symanzik improved gauge action leads to (pairwise) complex eigen values in the transfer matrix in the strong coupling regime. It is interesting as a future work to investigate quantitatively the connection between such a non-Hermiticity of the transfer matrix and the oscillatory behaviour of the correlation functions in the intermediate phase.

More generally speaking, the emergence of some complicated phase structure and/or the oscillatory behavior of observables may be a general feature resulting from the competition between the nearest and non-nearest-neighbor interactions. Actually, so-called ANNNI model is known as a typical example showing such a competition leading to new phases. In the improved lattice action, the non-nearest-neighbor effects become dominant in the strong coupling region, and the resultant competition could originate the intermediate phases.

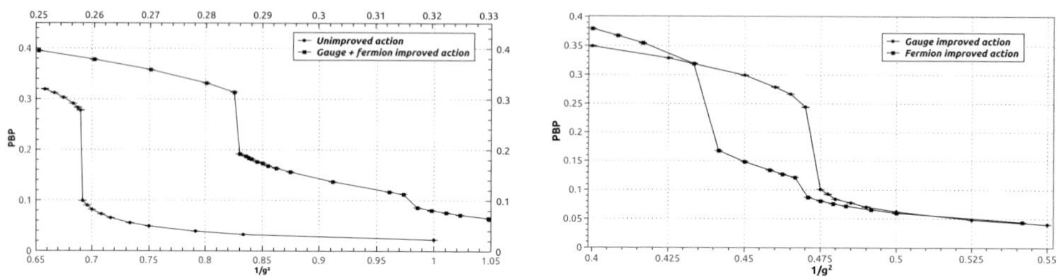

Fig. 3. The comparison of the chiral condensate (PBP) for $N_f = 12$ as a function of $1/g^2$ with g begin the lattice bare coupling for $am = 0.025$ and volume $16^3 \times 24$ (Ref. 2). Left: action A (black) versus D (red). Right: action B (black) versus C (red). See text for the details of action A - D.

4. Summary

We have reported our recent lattice studies[1,2] on thermal and/or bulk phase transitions in colour SU($N_c = 3$) QCD with a variable number (N_f) of fundamental fermions by using lattice Monte-Carlo simulations. One of the main goals was to estimate the number of flavour N_f^{*1} where the thermal chiral transition vanishes and the conformal phase sets in, by investigating the vanishing nature of the thermal chiral transition for larger N_f with three independent theoretical ansatz: $N_f^* \sim 11.1 \pm 1.6$ from the vanishing of the thermal scaling (*i.e.*, typical scale of the theory), $\sim 12.5 \pm 1.6$ from the coincidence between the coupling at $T = T_c$ and $T = 0$ or the 4-loop IRFP, $\sim 10.4 \pm 1.2$ from the vanishing of critical temperature T_c. Three methods have given consistent results within the errors.

Another main subject investigated in this proceedings was the bulk transition in $N_f = 12$ system with a particular focus on the interplay between the transition property and the improvements for the lattice action.[2] We have shown that a novel intermediate phase appears due to the Naik improvement for the fermion action, and this would be understood by the competition between the effects of local and less local interactions to the baryon number conservation. Then we have argued that the emergence of new phases would be a general feature resulting from the competition between the nearest and non-nearest-neighbor interactions.

As a future perspective, we plan to perform more rigorous scale settings, by exploiting state-of-art measurements of the heavy fermion potential. This is now work in progress for $N_f = 6$ and 8 systems based on the successes reported in this proceedings. It is also mandatory to investigate the chiral and thermodynamic limits. This, together with a more extended set of flavour numbers, will allow a quantitative analysis of the critical behaviour in the vicinity of the conformal IRFP. We also expect that our thermodynamic lattice study plays an important role as a new connection between the lattice and the Gauge/Gravity duality.[17]

Acknowledgements

This work was in part based on the MILC Collaboration's public lattice gauge theory code.[18] The numerical calculations for $N_f = (0, 4, 6, 8)$ were carried out on the IBM-SP6 and BG/P at CINECA, Italian-Grid-Infrastructures in Italy, and the Hitachi SR-16000 at YITP, Kyoto University in Japan. The numerical calculations for $N_f = 12$ were performed in Dutch National Computing Foundation (NCF) and the University of Groningen.

References

1. K. Miura, M. P. Lombardo and E. Pallante, *Phys. Lett. B* **710**, 676 (2012); K. Miura and M. P. Lombardo, *Nucl. Phys. B* **871**, pp. 52-81 (2013).
2. A. Deuzeman, M. P. Lombardo, T. N. da Silva and E. Pallante, *Phys. Lett. B* in Press, arXiv:1209.5720 [hep-lat]; T. N. da Silva and E. Pallante, *PoS* **LAT2012**, 052 (2012).
3. W. E. Caswell, *Phys. Rev. Lett.* **33**, 244 (1974); T. Banks and A. Zaks, *Nucl. Phys. B* **196**, 189 (1982).
4. For a historical review of the walking dynamics, see K. Yamawaki, arXiv:hep-ph/9603293.
5. For a review of LHC physics in the light of walking technicolor model, see R. S. Chivukula, P. Ittisamai, J. Ren, and E. H. Simmons, arXiv:1202.1505 [hep-ph].
6. For recent lattice reviews, see J. Giedt, *PoS* **LAT2012**, 006 (2012); E. T. Neil, *PoS* **LAT2011**, 009 (2011); L. Del Debbio, *PoS* **LAT2010**, 004 (2010); E. Pallante, *PoS* **LAT2009**, 015 (2009).
7. A. Deuzeman, M. P. Lombardo and E. Pallante, *Phys. Lett. B* **670**, 41 (2008).
8. A. Hasenfratz, MCRG matching," *Phys. Rev. Lett.* **108** 061601 (2012).
9. V. A. Miransky and K. Yamawaki, *Phys. Rev. D* **55** 5051 (1997) [Erratum-ibid. *D* **56** 3768 (1997)].
10. J. B. Kogut, J. Polonyi, H. W. Wyld, and D. K. Sinclair, *Phys. Rev. Lett.* **54**, 1475 (1985).
11. T. Appelquist, A. Ratnaweera, J. Terning, and L. C. R. Wijewardhana, *Phys. Rev. D* **58**, 105017 (1998).
12. T. A. Ryttov and R. Shrock, *Phys. Rev. D* **86**, 085005 (2012).
13. J. Liao and E. Shuryak, *Phys. Rev. Lett.* **109**, 152001 (2012).
14. J. Braun, C. S. Fisher, and H. Gies, *Phys. Rev. D* **84** 034045 (2011); J. Braun and H. Gies, *JHEP* **1005** 060 (2010); **0606** (2006) 024.
15. A. Deuzeman, M. P. Lombardo and E. Pallante, *Phys. Rev. D* **82** 074503 (2010).
16. A. Cheng, A. Hasenfratz, and D. Schaich, *Phys. Rev. D* **85** 094509 (2012).
17. U. Gursoy, E. Kiritsis, L. Mazzanti, G. Michalogiorgakis, and F. Nitti, *Lect. Notes Phys.* **828**, 79 (2011); T. Alho, M. Jarvinen, K. Kajantie, E. Kiritsis, and K. Tuominen, *JHEP* **1301**, 093 (2013).
18. MILC Collaboration, http://www.physics.indiana.edu/~sg/milc.html

Lattice QCD with 12 Quark Flavors: A Careful Scrutiny

Xiao-Yong Jin

RIKEN Advanced Institute for Computational Science,
Kobe, Hyogo 650-0047, Japan
E-mail: xjin@riken.jp

Robert D. Mawhinney

Department of Physics, Columbia University,
New York, NY 10027, USA
E-mail: rdm@physics.columbia.edu

With a substantial amount of simulations, we have explored the system across a wide range of lattice scales. We have located a lattice artifact, first order bulk transition, have studied its properties, and found that the flavor-singlet scalar meson mass vanishes at the critical endpoint. We will discuss the lattice phase diagrams and the continuum limits for both a SχSB phase and an IR conformal phase, and compare results with other groups.

Keywords: Lattice QCD; Many-flavor-QCD; Walking theory; Conformal theory

1. Introduction

It is both phenomenologically and theoretically interesting to study strongly coupled gauge theories with the number of massless fermions, N_f, tuned close to the lower limit of the conformal window, where the theory remains asymptotically free at high energies while developing a non-trivial infrared (IR) fixed point. While the system with N_f close to the upper limit of the conformal window can be studied perturbatively,[1] for systems with N_f close to the lower limit, the behavior is less clear due to the non-perturbative nature of gauge theories.

We have been studying the SU(3) gauge theory with 8 and 12 fundamental fermions on the lattice, using the naive staggered fermion action and the DBW2 gauge action. We have explored the system through basic low-energy QCD observables with multiple volumes. In this paper, we will concentrate on our recent findings of the lattice artifact bulk transition with 12 flavors, discuss the continuum limit, and compare results with other groups. We quote results from largest volume available at each set of parameters, with finite volume effects of about a few percent.

2. Lattice critical point

We have located a line of first order bulk transitions[2] in the bare lattice parameter space: quark mass, m_q, and coupling, $\beta = 6/g^2$. This line of first order transitions appears at relatively strong lattice couplings $\beta \sim 0.46$ and small input quark masses $m_q \lesssim 0.008$. Discontinuities in hadronic observables at the first order transition become smaller, as the quark mass, m_q, is increased. The system shows a quick cross-over region,[3] at $m_q \geq 0.01$, with a marked change in lattice scales, from strong couplings to weaker couplings.

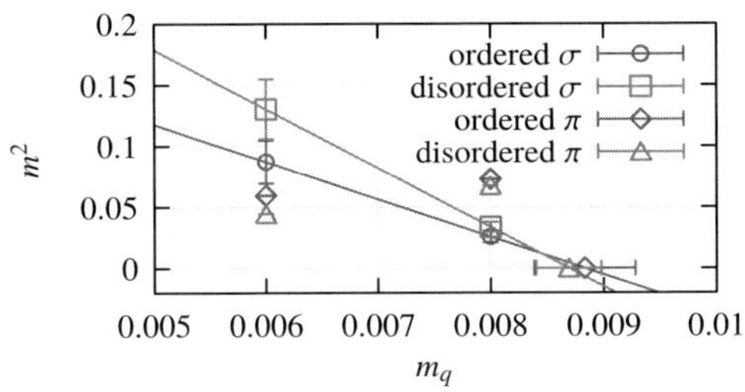

Fig. 1. The critical endpoint from extrapolating the mass of the flavor-singlet scalar meson

The system is approaching a second order transition, or a critical endpoint, as the quark mass, m_q, is increased while the lattice coupling, β, is tuned toward the end of the first order bulk transition. We have measured the mass of the flavor-singlet scalar meson, σ, using metastable ensembles at two sets of input parameters, detailed in Ref. 2. The mass of the flavor-singlet scalar meson, m_σ, depicted in Fig. 1, decreases on the line of the first order bulk transition, as m_q increases, and becomes even smaller than pion. To fit the critical behavior, we have used the mean-field scaling law, $m_\sigma \propto (m_q - m_q^C)^{1/2}$, which happens to describe our data quite well. The fitted location of the endpoint from ordered-start configurations (weak coupling) agrees with the fitted result from disordered-start configurations (strong coupling).

This critical behavior of m_σ, resembles the behavior found in finite temperature transitions.[4] However, our findings with 12 quark flavors at zero temperature occur at finite m_q and β. While the continuum physics dictated by asymptotic freedom, can only be extracted at the limit of $m_q \to 0$ and $\beta \to \infty$, this critical point certainly does not have such a conventional continuum limit, and is irrelevant to the continuum physics we are after. Nonetheless, this critical point does possess a continuum limit, since the σ correlation length diverges at this point, while all other hadronic scales decouple from the theory. This second order endpoint thus gives a scalar field theory, and is likely non-interacting.[a]

The first order bulk transition and critical endpoint[2] almost surely depends on the lattice action, as a few other groups have found bulk behaviors[5–7] with different lattice actions and bare parameters. In the parameter space we have simulated, however, we did not observe the same characteristics of the bulk transition those

[a]We don't have hard proof, but it is hard to construct a nontrivial scalar field theory.

groups advertised. We acknowledge that certain kind of lattice artifacts are highly likely to exist in different lattice actions under different disguise and with different degrees of severity. Our findings of the critical behavior of the flavor-singlet scalar meson suggest that caution must be exercised in interpreting results from the scalar channel on the lattice, especially when asymptotic scaling is not well established.

3. Continuum limit

The continuum limit of lattice gauge theories can be analyzed by the renormalization group equation of any physical observable, $P(a, g, \hat{m})$, which depends on lattice spacing, a, bare lattice coupling,[b] g, and bare lattice input quark mass, $\hat{m} = ma$. Near the continuum limit, P should mildly depend on a,

$$a\frac{\mathrm{d}}{\mathrm{d}a}P = \left\{ a\frac{\partial}{\partial a} - \beta(g,\hat{m})\frac{\partial}{\partial g} - \gamma(g,\hat{m})\frac{\partial}{\partial \hat{m}} \right\} P(a,g,\hat{m}) = \mathcal{O}(a)\,, \tag{1}$$

where $\beta(g, \hat{m})$ and $\gamma(g, \hat{m})$ are lattice version of the renormalization group beta functions, and the right hand side of the equation is the usual scaling error.[c] In an asymptotically free theory, to find the lattice beta functions, one would perturbatively compute some observable P, and plug the perturbative form back to Eq. (1). See Ref. 8 for a massless derivation using the quark potential. One will find the beta functions coincide with the continuum ones up to the lowest universal orders.[d] In an asymptotically free theory, this procedure is self-consistent, because $g \to 0$, together with a fine-tuned $\hat{m} \to 0$, leads to $a \to 0$ according to the beta function calculated from this procedure.

Away from the physical continuum limit, one can supposedly use a form of the observable P that contains lattice higher order operators, say $P \to P + P_l$, where P_l is irrelevant in the usual continuum limit of physical QCD. With this additional term P_l, solving Eq. (1) has the possibility to lead to lattice beta functions, $\beta_l(g, \hat{m})$ and $\gamma_l(g, \hat{m})$, that are completely different from the usual continuum QCD, and contain lattice artifacts. One can solve these new lattice beta functions for $a \to 0$, and obtain a new continuum limit at $g \to g_C$ and $\hat{m} \to \hat{m}^C$, which is different from the usual continuum limit of physical QCD at $g \to 0$ and $\hat{m} \to 0$. As long as the result of observable $P + P_l$ is consistent with calculations in the limit of $g \to g_C$ and $\hat{m} \to \hat{m}^C$, such that Eq. (1) is self-consistent, we arrive at a lattice artifact continuum limit.

The second order critical endpoint we found with 12 flavors at finite lattice coupling β and input quark mass m_q, presented in the previous section, is such a lattice artifact continuum limit. The empirical lattice phase diagram is sketched in Fig. 2(a), where the axes are lattice bare parameters. We have two data points, which

---

[b]Not to be confused with renormalization beta functions, we use β and g interchangeably as the lattice bare coupling for ease of discussion.
[c]The size of the scaling violation depends on the lattice action and observable.
[d]Minus signs in Eq. (1) ensure this agreement.

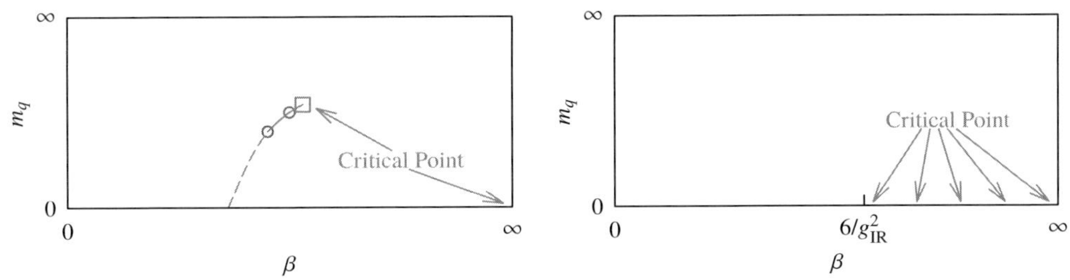

(a) Empirical diagram with 12 flavors (b) Infinite volume diagram with IR conformal

Fig. 2. Lattice phase diagrams

are located on the transition line and extended to smaller masses with a dashed line. We note two critical points that correspond to the lattice artifact continuum limit and the physical QCD continuum limit, respectively.

To verify whether the 12-flavor theory is in the conformal window or remains in the spontaneously chiral symmetry breaking phase, one needs to know how the above picture of continuum limits would change if the underlying continuum theory has a non-trivial IR fixed point. In the following discussion, we assume the lattice is infinitely large, $V \to \infty$. Since the weak coupling limit of the theory is asymptotically free, the limit with $\beta \to \infty$ and $m_q \to 0$ is a valid continuum limit representing the true IR conformal continuum theory. It is conceivable that any non-zero quark mass term would break the conformal symmetry and the renormalized theory would lose the IR fixed point. If we keep $m_q = 0$, and vary β away from the weak coupling limit, since the renormalized coupling remains the fixed-point value, g_{IR}, the physics governed by the IR fixed point does not change, and neither do the lattice beta functions. Thus the theory remains IR conformal with a vanishing lattice spacing, $a = 0$. In terms of Eq. (1), any low-energy observable P remains under control of the IR fixed point, and is independent of the bare coupling, g, such that increasing g from 0 does not need to be compensated by increasing lattice spacing, a. In this case, Eq. (1) is self-consistently satisfied with $a = 0$ for $\hat{m} = 0$ and $0 \leq g \leq g_{\mathrm{IR}}$. In conclusion, shown in Fig. 2(b), in the infinite volume limit, any point along the line with $m_q = 0$ and $\beta \geq 6/g_{\mathrm{IR}}^2$ is a lattice critical point, and shares the same continuum limit as the weak coupling limit, which describes the same conformal theory at low energies. On the other hand, if $m_q \neq 0$, the theory is no longer conformal, and lattice artifacts are present with $a \neq 0$, which can be controlled by the bare coupling, β. This proposed scenario appears consistent with the data and we are investigating it further.

4. Mass ratios and universality

The dimensionless quantity, m_π/f_π, is shown in Fig. 3 versus the bare input quark mass, for both 8 flavors (left panel) and 12 flavors (right panel), respectively. We have drawn simple polynomial fit curves on top of our data. The discontinuity at

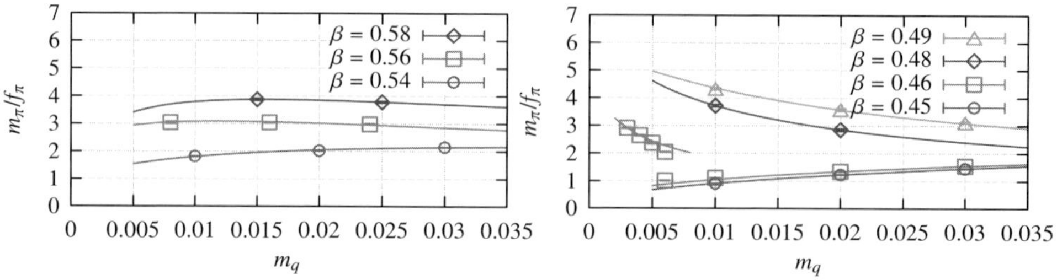

(a) 8 flavors (b) 12 flavors

Fig. 3. Scaling of constant physics by m_π/f_π

$m_q = 0.006$ and $\beta = 0.46$ with 12 flavors represents the bulk transition. No bulk transition has been observed within our simulation parameters with 8 flavors.

We observed that, at strong couplings, $\beta = 0.54$ and $\beta = 0.45$ for 8 and 12 flavors respectively, the value of m_π/f_π decreases as m_q decreases. This behavior is expected in a physical QCD simulation, where chiral symmetry is spontaneously broken at $m_q \to 0$, and m_π vanishes while f_π remains finite in the limit. On the contrary, at weaker couplings, $\beta \geq 0.56$ for 8 flavors, and $\beta \geq 0.46$ after the bulk transition for 12 flavors, the m_π/f_π increases as m_q decreases. With 8 flavors, m_π/f_π is almost constant, and the fitted curves start to bend towards zero at small m_q, while the fitted curves with 12 flavors continues to grow in the parameter range we have simulated. Despite this difference, at weak couplings, m_π/f_π with 12 flavors resembles that with 8 flavors at relatively large $m_q \gtrsim 0.015$.

If we look at the constant physics naively with fixed m_π/f_π at weak couplings of the 12-flavor simulations, in order to reach the massless quark limit, we have to decrease β and go to stronger couplings. This is in contrast with the continuum limit discussed in the last section, and thus we cannot proceed with this procedure. The reason for this could be: a) lattice artifacts; b) physical behaviors.

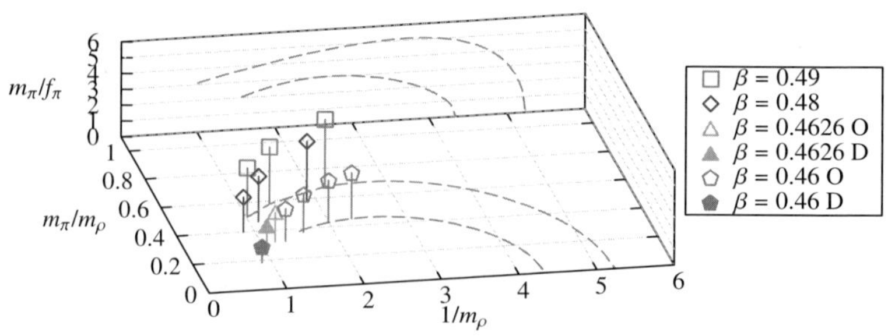

Fig. 4. Diagram of dimensionless ratios versus typical lattice correlation length

We have assembled values of both m_π/f_π and m_π/m_ρ from our 12-flavor simulations and plotted them against a typical lattice correlation length $1/m_\rho$ in Fig. 4.

In the key of the figure, for ensembles with fixed β across both sides of the first order bulk transition, weak coupling side (ordered-started) and strong coupling side (disordered-started) are labeled with O and D, respectively. The critical endpoint is located near the data points with $\beta = 0.4626$. In the figure, the length of the vertical line attached to each symbol represents m_π/f_π, and its position on the bottom layer shows the value of m_π/m_ρ versus the corresponding $1/m_\rho$. Dashed lines are polynomial extrapolations inspired by spontaneous-chiral-symmetry-breaking (SχSB) behaviors,

$$m_\pi^2 = c_0 m_q (1 + c_1 m_q)\,, \tag{2}$$

$$m_\rho \text{ or } f_\pi = b_0 + b_1 m_q + b_2 m_q^2\,, \tag{3}$$

acted on each individual channel separately. Assuming SχSB behaviors, $1/m_\rho$ needs to be increasingly larger, with the coupling weakening, to observe the ratios decrease as a result of vanishing m_π. On the other hand, if the theory does indeed have an IR fixed point, ratios in the infinite volume limit should approach constant values with different couplings, as $1/m_\rho \to \infty$, while all the dimensionful quantities vanish in the conformal limit. Therefore, smaller quark masses would have stronger differentiating power to tell the true behavior of the system.

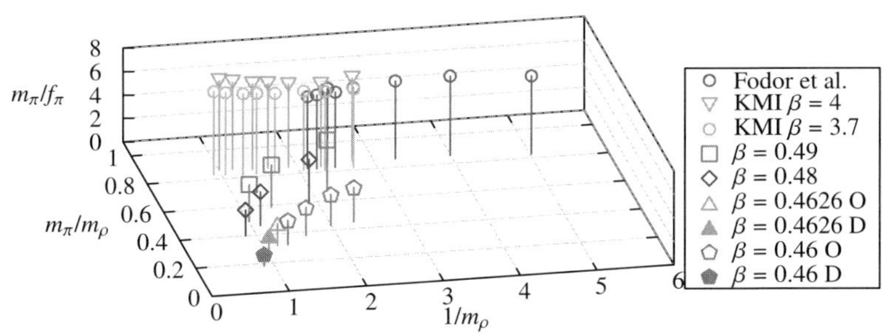

Fig. 5. Comparing dimensionless ratios with results from Fodor et al.[9] and LatKMI[10]

To compare our results with results from other groups, we have included recent 12-flavor results[e] from Fodor et al.[9] and LatKMI,[10f] in Fig. 5. Although results of different groups are from different lattice actions, there is the universal behavior to be learned from the similar behaviors within the parameter space that has been explored. Apparently, both ratios, m_π/f_π and m_π/m_ρ are not changing noticeably as $1/m_\rho$ increases with fixed β within the parameter space limited by the computing resources. However, this behavior is not different from the behavior of m_π/f_π, shown in Fig. 3(a), for a system with 8 flavors. On the other hand, the ratios do seem to approach a universal continuum limit at $1/m_\rho \to \infty$, if the tendency continues without alteration in the region with small quark masses.

[e]Results from the largest volume available are taken, and the finite volume effects are small.
[f]Values of f_π from LatKMI are normalized by dividing $\sqrt{2}$.

5. Summary

We have located a lattice artifact, critical endpoint, where the scalar correlation length diverges, for 12-flavor QCD with naive staggered fermions and the DBW2 gauge action. In the weak coupling side of the lattice artifact bulk transition, we see confined, massive particle states for all simulations done to date. As the input quark mass decreases at a fixed lattice coupling, the hadronic scales (m_ρ and f_π) decrease much more rapidly than the pion mass, such that m_π/f_π or m_π/m_ρ increases on the weak coupling side of the bulk transition. The lattice input quark mass at the critical endpoint of the lattice artifact transition gives us a hint of the size of quark masses, at that lattice coupling, where the vanishingly small scalar mass can wildly distort continuum physics.

We have argued that an exact same continuum limit of an asymptotically free, IR conformal theory can be approached at the massless quark limit and the infinite volume limit, with any lattice bare coupling that is weaker than the renormalized coupling at the IR fixed point.

It may not be surprising that much lighter quark masses are required in lattice simulations for the SχSB behavior to be numerically indisputable. Nevertheless, how to effectively scale towards the continuum limit within our current computational resources remains a question for both SχSB and IR conformal scenarios.

Acknowledgments

X.-Y. Jin would like to thank the organizers of SCGT12 for the invitation to this wonderful workshop. Our calculations were done on the QCDOC and NY Blue at BNL. X.-Y. Jin wants to express the sincere gratitude especially towards all the brilliant minds behind the QCDOC project. This research utilized resources at the New York Center for Computational Sciences at Stony Brook University/Brookhaven National Laboratory which is supported by the U.S. Department of Energy under Contract No. DE-FG02-92ER40699 and by the State of New York.

References

1. T. Banks and A. Zaks, *Nucl. Phys.* **B196**, p. 189 (1982).
2. X.-Y. Jin and R. D. Mawhinney, *PoS* **LATTICE2011**, p. 066 (2011).
3. X.-Y. Jin and R. D. Mawhinney, *PoS* **LAT2009**, p. 049 (2009).
4. X. Liao, *Nucl.Phys.Proc.Suppl.* **106**, 426 (2002).
5. A. Cheng, A. Hasenfratz and D. Schaich, *Phys.Rev.* **D85**, p. 094509 (2012).
6. A. Deuzeman, M. P. Lombardo, T. N. da Silva and E. Pallante, *PoS* **LATTICE2011**, p. 321 (2011).
7. Z. Fodor, K. Holland, J. Kuti, D. Nogradi, C. Schroeder *et al.*, *PoS* **LATTICE2012**, p. 279 (2012).
8. J. B. Kogut, *Rev.Mod.Phys.* **55**, p. 775 (1983).
9. Z. Fodor, K. Holland, J. Kuti, D. Nogradi, C. Schroeder *et al.*, *Phys.Lett.* **B703**, 348 (2011).
10. Y. Aoki, T. Aoyama, M. Kurachi, T. Maskawa, K.-i. Nagai *et al.*, *Phys.Rev.* **D86**, p. 054506 (2012).

On Gauge Theory Phase Diagrams at Zero and Finite Temperature

Kimmo Tuominen*

Department of Physics, University of Jyväskylä, P.O.Box 35, FIN-40014 Jyväskylä, Finland
Helsinki Institute of Physics, P.O.Box 64, FIN-00014 University of Helsinki, Finland

We consider gauge theory phase diagrams at zero and finite temperature. In the former case, we discuss some recent progress in the lattice computations using improved Wilson fermions. In the latter case we describe the general features of the phase diagrams and also discuss the directions they imply for lattice simulations.

1. Introduction

The phase diagrams of massless gauge theories, as a function of the number of colours, N, flavours N_f and fermion representations have recently been studied numerically.[1–12] The motivation arises on one hand from applications of strong dynamics in elementary particle phenomenology, including unparticles[13] and (extended) technicolor scenarios,[14,15] and on the other hand from the purely theoretical aim of understanding the nonperturbative gauge theory dynamics from first principles.

A traditional method to estimate the vacuum phase diagram of a gauge theory is to compare the critical coupling for chiral symmetry breaking $\alpha_c = \pi/(3C_2(R)$, obtained from the Schwinger-Dyson equation for the fermion propagator in the ladder approximation, with the value of the fixed point coupling α^* of the two-loop beta-function $\beta(\lambda) = -b_0\lambda^2 - b_1\lambda^3$. This allows one to estimate the location of the boundary between conformal and confining phases as a function of N, N_f and fermion representations.[16,17] The phenomenologically most interesting is the region directly below the lower boundary of the conformal window where the theory may exhibit the walking behavior.

2. Improved actions for lattice strong dynamics

On the lattice, the computations with Wilson fermions are subject to lattice artifacts proportional to the lattice spacing a. To obtain high precision results for QCD like theories these lattice artefacts are cancelled by the methods developed in.[18,19] The basic idea in this approach is to introduce local counterterms whose coefficients will be finetuned to cancel all $\mathcal{O}(a)$ contributions. On lattices with periodic boundaries,

*E-mail: kimmo.i.tuominen@jyu.fi

the $\mathcal{O}(a)$ discretization errors in the basic Wilson lattice action can be fully removed by introducing the Sheikholeslami-Wohlert-term[18] and tuning its coefficient.

For the measurement of the coupling, one usually employs the Schrödinger functional scheme, where further contributions to linear lattice artefacts arise due to the fixed spatially constant boundary conditions at times $t = 0$ and $t = L$ (here L^4 is the volume of the lattice):

$$U_k(x)|_{(x_0=0)} = \exp(aC_k), \quad U_k(x)|_{(x_0=L)} = \exp(aC_k'), \tag{1}$$

where a is the lattice spacing and

$$C_k = i\mathrm{diag}(\phi_1, \ldots, \phi_{N_c})/L, \quad C_k' = i\mathrm{diag}(\phi_1', \ldots, \phi_{N_c}')/L, \tag{2}$$

with the constraint $\sum_{i=1}^{N_c} \phi_i = \sum_{i=1}^{N_c} \phi_i' = 0$. For SU(2) and SU(3) the boundary fields are parametrized as[20,21]

$$(\phi_1, \phi_2) = (\eta, -\eta), \quad (\phi_1', \phi_2') = (\rho - \eta, -\rho + \eta) \quad \mathrm{SU}(2) \tag{3}$$

$$\begin{array}{l} (\phi_1, \phi_2, \phi_3) = (\eta - \rho, -\eta/2, -\eta/2 + \rho) \\ (\phi_1', \phi_2', \phi_3') = (-\eta - 3\rho, \eta/2 + \rho, \eta/2 + 2\rho) \end{array} \quad \mathrm{SU}(3). \tag{4}$$

The conventional choice of parameters is $(\rho, \eta) = (\pi, \pi/4)$ for SU(2) and $(\rho, \eta) = (\pi/3, 0)$ for SU(3). The spatial boundary conditions are taken to be periodic, and the fermion fields are set to vanish at the $t = 0$ and $t = L$ boundaries. For complete analysis of the necessary counterterms introduced by these boundary conditions, see[22] and for the perturbative evaluation of the coefficients relevant for higher fermion representations, see.[23] Classically, the nontrivial boundary conditions generate a spatially constant chromoelectric field and the derivative of the action with respect to η is proportional to the inverse of the bare coupling g_0^2. The coupling is then defined by[21]

$$\frac{g_0^2}{g^2} = \langle \frac{\partial S}{\partial \eta} \rangle \bigg/ \frac{\partial S^{\mathrm{cl.}}}{\partial \eta} \tag{5}$$

The running of the coupling is quantified using the step scaling function:

$$\Sigma(u, s, L/a) = g^2(g_0, sL/a)|_{g^2(g_0, L/a)=u} = u + (\Sigma_{1,0} + \Sigma_{1,1} N_f) u^2. \tag{6}$$

The factor $\Sigma_{1,1}$ ($\Sigma_{1,0}$) denotes the fermion (pure gauge) contribution at one loop. The continuum limit of $\Sigma_{1,1}$ ($\Sigma_{1,0}$) is given by the fermionic (bosonic) contribution to the one loop coefficient $b_0 = \beta_0/(16\pi^2)$ of the beta function, where $\beta_0 = 11/3N_c - 4/3T(R)N_f$. In other words,

$$\delta_i = \lim_{L/a \to 0} \Sigma_{1,i}/(2b_{0,i} \ln 2) = 1, \; i = 0, 1 \tag{7}$$

where $b_{0,0} = 11N_c/(48\pi^2)$ and $b_{0,1} = N_f T(R)/(12\pi^2)$, and we chose $s = 2$.

Using the perturbation theory results,[23] we highlight two issues here: First, improvement should be carried out in order to obtain reasonably accurate results on lattices of reasonable size. Second, the improvement program has to be carried out consistently and correctly.

To illustrate that improvement is essential for taming the $\mathcal{O}(a)$ artefacts, in the left panelf of Fig. 1 we show δ_1 defined in (7) for SU(N) gauge theory with fundamental fermions for $N = 2$, 3 and 4.

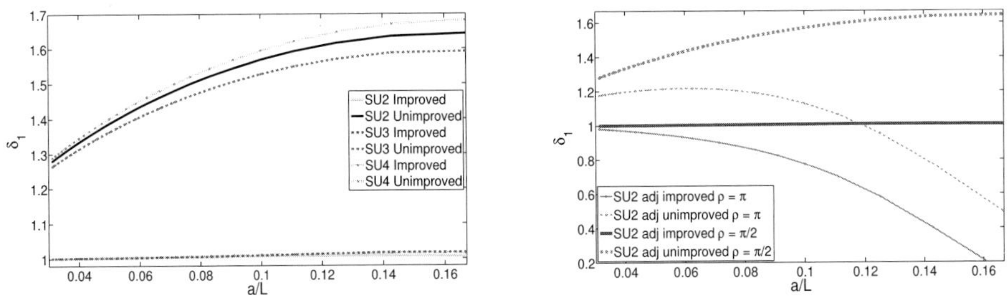

Fig. 1. Left: Comparison between unimproved and improved results for the fermionic contribution to the perturbative step scaling function for SU(2), SU(3) and SU(4) gauge theories with fundamental representation matter fields. Right: Comparison between two different boundary fields in SU(2) gauge theory with adjoint fermions. The thin solid and dashed curves correspond, respectively, to improved and unimproved results with $\rho = \pi$. The thick solid (dashed) curves correspond to improved (unimproved) results for $\rho = \pi/2$

Also, the precise values of the boundary fields must be chosen carefully to guarantee rapid convergence of the results. In the right panel of Fig. 1 we show the results for $\Sigma_{1,1}$ for SU(2) gauge theory and adjoint fermions. To guarantee that the $\mathcal{O}(a^2)$ corrections remain small the boundary values must be optimized. The optimal choice is $\rho = \pi/2$ and $0 < \eta < \pi/2$ excluding the value $\eta = \pi/4$. To understand this intuitively, we note that the boundary field is generically a diagonal matrix with eigenvalues ϕ and $-\phi$, but the adjoint fermions see these eigenvalues as 2ϕ, -2ϕ and 0. Hence halving the background field for the adjoint fermions leads to similar convergence as in the case of fundamental fermions.

The analysis is straightforward also for SU(3) with higher representation fermions. As in the case of SU(2), when translating the boundary matrix to the adjoint or sextet representation of SU(3), one sees that the background field is effectively doubled. For the adjoint representation, shown in the left panel of Fig. 2, the best convergence of δ_1 is obtained for $\rho = \pi/6$ and $-4\pi/9 < \eta < \pi/9$ excluding the value $\eta = -\pi/6$. For the sextet representation we find that the convergence is optimal for $\rho = 67\pi/150 \approx 0.45\pi$ and $-166\pi/225 < -7\pi/45$ excluding $\eta = -67\pi/150$. The convergence is dominantly controlled by the value of the ρ while η can be chosen from a wide interval. For the figure we used $L/a = 10$, but the results remain quantitatively similar for other values of L/a as well.

Finally, as shown in the right panel of Fig. 2, the improvement coefficients must be included consistently: only when the Sheikholeslami-Wohlert coefficient $c_{\rm sw}$ and the boundary improvement coefficients are consistently taken into account, optimal convergence towards the continuum limit is obtained.[23]

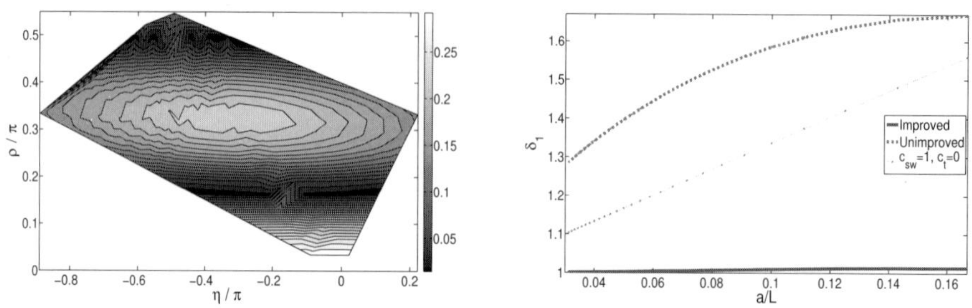

Fig. 2. Left: Convergence of $|\delta_1 - 1|$ as a function of ρ and η for fermions in the adjoint representation of SU(3). Right: The thick solid and dashed curves correspond, respectively, to improved and unimproved results. The thin dotted curves correspond to the results obtained by setting $c_{\rm sw} = 1$ and $c_t = 0$. The curves are for fermions in the adjoint representation of SU(3).

3. Finite temperature gauge theories

A next logical direction, to further understand strong coupling gauge theory dynamics, is to consider these theories at finite temperature.[24] The phenomenological applications include phase transitions in early universe.[25,26] So far less analytic and lattice efforts (see[27–29]), have been devoted to expose the finite temperature phase diagrams. There are several interesting qualitative structures shown in Fig. 3, which are open for further more precise studies. The figure shows the possible phases of SU(N) gauge theory as a function of temperature and $x_f \equiv N_f/N_c$. Fig. 3 corresponds to theories which exhibit walking behavior[†]. There are two scales, the ultraviolet scale $\Lambda_{\rm UV}$, at which asymptotic freedom sets in, and the infrared scale $\Lambda_{\rm IR}$ where the coupling diverges leading to confinement. For QCD like theory $\Lambda_{\rm IR} \approx \Lambda_{\rm UV}$, and for walking theory $\Lambda_{\rm IR} \ll \Lambda_{\rm UV}$. Inside the conformal window there is no confinement and no infrared scale $\Lambda_{\rm IR}$.

The phase boundaries can correspond to first or higher order phase transition or to a continuous crossover; the order depends on the underlying theory. Below the conformal window, for chiral symmetry restoring transition the order parameter is provided by the chiral condensate while for the phase boundaries between chirally symmetric phases the order parameters is provided by the free energy itself. The latter provides a direct measure for a (substantial) change in the number of effective degrees of freedom. We will now discuss different regions of the phase diagram in more detail.

An essential feature arises when approaching the conformal window from below: At $x_f = x_c$ and $T = 0$ there is a second order quantum phase transition. Directly below the conformal window at some fixed x_f such that $x^* < x_f < x_c$ the theory exhibits walking. This is reflected at finite temperature as an existence of a quasi conformal intermediate phase where $p/T^4 \sim$ const. over a temperature range whose

[†]This requires that the zero of the beta function at the lower boundary of the conformal window is at least of second order. If the zero is first order, there is no walking[30]

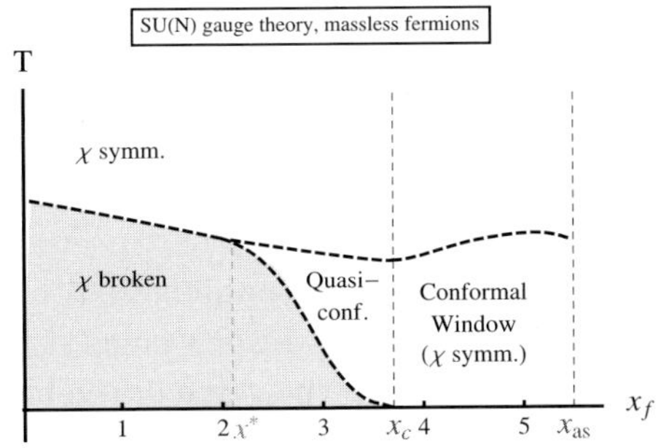

Fig. 3. A schematic (x_f, T) phase diagram of gauge theory with massless fermions where the dynamics of the theory provides walking behavior. The shaded region corresponds to the phase where chiral symmetry is broken.

width increases rapidly as the conformal window is approached; see Fig. 3.

To see how this structure arises, consider the theory first at high temperatures. We know that at high temperatures due to asymptotic freedom, the system is in the deconfined phase. Then, as temperature is lowered below $T_{c,\mathrm{UV}} \sim \Lambda_{\mathrm{UV}}$, the system moves out of this partonic plasma phase. On the other hand, at low temperature the system is in the confined hadronic phase. The zero temperature spectrum directly displays the defining feature of walking, i.e. Miransky scaling: $m \propto \Lambda_{\mathrm{IR}} \propto e^{-1/\sqrt{x_c - x_f}} \Lambda_{\mathrm{UV}}$ as $x_f \to x_c$ from below. At finite temperature these bound states melt, and this defines the critical temperature $m \sim 2\pi T_{c,\mathrm{IR}}$. Quasi-conformality at $T_{c,\mathrm{IR}} < T < T_{c,\mathrm{UV}}$ arises due to the fact that the trace of the energy momentum tensor is vanishingly small between Λ_{IR} and Λ_{UV} implying $p \sim T^4$. Note that x^* can in principle be determined by studying the behavior of the pressure at different values of x_f at finite temperature using e.g. lattice simulations.

Then, let us consider further interesting features of this phase diagram: First, at small values of x_f the theory displays QCD-like behavior; hadronic phase and spontaneous breaking of chiral symmetry at low temperatures, and chirally symmetric partonic plasma at high temperature. These behaviors are well understood. For fundamental representation matter there is a single phase transition driven by restoration of chiral symmetry, but leading also to observed discontinuities or rapid variations in observables associated with deconfinement, like the Polyakov loop.[31] With higher representations the possibilities are richer.[29,31–33]

Second, inside the conformal window, $x_f > x_c$ there is a transition between two phases whose free energies behave approximately conformally: from a low temperature IR phase of "unparticles", $\sim g_{\mathrm{IR}} T^4$ to deconfined partonic plasma at high temperatures, $\sim g_{\mathrm{UV}} T^4$. ‡

‡Note that only the $T = 0$ phase is strictly conformal in the sense that its Green functions are scale invariant. Finite temperature provides a scale, and physical quantities like screening lengths scale with T. This is very different from theories like QCD where the infrared properties are dictated by the intrinsic scale Λ_{QCD}

4. Conclusions

We have described developments in analyses of gauge theories with higher representation matter fields on the lattice. First we demonstrated the sensitivity of the results on the correct values of the improvement coefficients and boundary conditions when using Wilson fermions. Using the perturbative analysis of the counterterm coefficients required for the step scaling, we emphasized the choice of the boundary conditions; for a related study, see.[34] Our results suggest that in studies of higher representations there is need to optimize the value of the background field carefully. This need is driven by the fermionic contribution; we have explicitly checked that the effect on the pure gauge contribution is negligible, on the level of few percents, for the cases we have considered.

Then we considered generic features of gauge theory phase diagrams at finite temperature. Outside the conformal window, an obvious goal for lattice simulations would be to determine the value of x^* above which the quasiconformal behavior should appear and phase structure changes from QCD-like theory with a single finite temperature phase transition. The critical temperature of the chiral symmetry restoring phase transition will exhibit Miransky scaling. However, studying the scaling properties with x_f is difficult and also mapping out the temperature dependence of the quasiconformal intermediate phase at larger values of x_f might be numerically challenging problem due to the large scale separation, implying the need for large lattices.

Hence, it might be more fruitful to start inside the conformal window, where it should be possible to determine the phase boundary between the low temperature unparticle phase and high temperature partonic plasma phase. The lattice results, in combination with analytic insights obtained so far should help in drawing a coherent picture. For example there is currently a consensus that the simple SU(2) gauge theory with two Dirac fermions in the adjoint representation of SU(2) has an IRFP controlling the vacuum properties[2,4,6,11] . Heating up this theory would provide a testbed for studying the existence of the two distinct conformal phases at finite temperature.

References

1. S. Catterall and F. Sannino, Phys. Rev. D **76**, 034504 (2007) [arXiv:0705.1664 [hep-lat]].
2. A. J. Hietanen, J. Rantaharju, K. Rummukainen and K. Tuominen, JHEP **0905**, 025 (2009) [arXiv:0812.1467 [hep-lat]]
3. L. Del Debbio, A. Patella and C. Pica, Phys. Rev. D **81** (2010) 094503 [arXiv:0805.2058 [hep-lat]].
4. S. Catterall, J. Giedt, F. Sannino and J. Schneible, arXiv:0807.0792 [hep-lat].
5. A. J. Hietanen, K. Rummukainen and K. Tuominen, Phys. Rev. D **80**, 094504 (2009) [arXiv:0904.0864 [hep-lat]].
6. T. DeGrand, Y. Shamir, B. Svetitsky, [arXiv:1102.2843 [hep-lat]].
7. T. Appelquist, G. T. Fleming and E. T. Neil, Phys. Rev. Lett. **100**, 171607 (2008) [arXiv:0712.0609 [hep-ph]].

8. Z. Fodor, K. Holland, J. Kuti, D. Nogradi and C. Schroeder, Phys. Lett. B **681**, 353 (2009) [arXiv:0907.4562 [hep-lat]];
9. A. Deuzeman, M. P. Lombardo and E. Pallante, Phys. Lett. B **670**, 41 (2008) [arXiv:0804.2905 [hep-lat]];
10. Y. Shamir, B. Svetitsky and T. DeGrand, Phys. Rev. D **78**, 031502 (2008) [arXiv:0803.1707 [hep-lat]].
11. F. Bursa, L. Del Debbio, L. Keegan, C. Pica and T. Pickup, arXiv:1010.0901 [hep-ph]
12. T. Karavirta, J. Rantaharju, K. Rummukainen and K. Tuominen, arXiv:1111.4104 [hep-lat].
13. H. Georgi, Phys. Rev. Lett. **98**, 221601 (2007)
14. S. Weinberg, Phys. Rev. D **19**, 1277 (1979); L. Susskind, Phys. Rev. D **20**, 2619 (1979).
15. F. Sannino, arXiv:0804.0182 [hep-ph].
16. T. Appelquist, A. Ratnaweera, J. Terning and L. C. R. Wijewardhana, Phys. Rev. D **58**, 105017 (1998) [arXiv:hep-ph/9806472].
17. F. Sannino and K. Tuominen, Phys. Rev. D **71**, 051901 (2005) [arXiv:hep-ph/0405209];
18. B. Sheikholeslami and R. Wohlert, Nucl. Phys. B **259** (1985) 572.
19. M. Luscher, R. Narayanan, P. Weisz and U. Wolff, Nucl. Phys. B **384**, 168 (1992) [arXiv:hep-lat/9207009].
20. M. Luscher, R. Sommer, U. Wolff and P. Weisz, Nucl. Phys. B **389**, 247 (1993) [hep-lat/9207010].
21. M. Luscher, R. Sommer, P. Weisz and U. Wolff, Nucl. Phys. B **413**, 481 (1994) [arXiv:hep-lat/9309005].
22. M. Luscher, S. Sint, R. Sommer and P. Weisz, Nucl. Phys. B **478**, 365 (1996) [arXiv:hep-lat/9605038].
23. T. Karavirta, K. Tuominen, K. Rummukainen and , Phys. Rev. D **85**, 054506 (2012) [arXiv:1201.1883 [hep-lat]].
24. K. Tuominen, arXiv:1206.5772 [hep-lat].
25. J. M. Cline, M. Jarvinen and F. Sannino, Phys. Rev. D **78**, 075027 (2008) [arXiv:0808.1512 [hep-ph]].
26. M. Jarvinen and F. Sannino, JHEP **1102**, 081 (2011) [arXiv:1009.5380 [hep-ph]].
27. J. Braun and H. Gies, JHEP **1005**, 060 (2010) [arXiv:0912.4168 [hep-ph]].
28. K. Miura, M. P. Lombardo and E. Pallante, "Chiral phase transition at finite temperature and conformal dynamics in large Nf QCD," Phys. Lett. B **710**, 676 (2012) [arXiv:1110.3152 [hep-lat]].
29. J. B. Kogut and D. K. Sinclair, Phys. Rev. D **85**, 054505 (2012) [arXiv:1111.3353 [hep-lat]].
30. F. Sannino, arXiv:1205.4246 [hep-ph].
31. A. Mocsy, F. Sannino and K. Tuominen, Phys. Rev. Lett. **92**, 182302 (2004) [hep-ph/0308135].
32. F. Sannino, Phys. Rev. D **66**, 034013 (2002) [hep-ph/0204174].
33. T. Kahara, M. Ruggieri and K. Tuominen, Phys. Rev. D **85**, 094020 (2012) [arXiv:1202.1769 [hep-ph]].
34. S. Sint, P. Vilaseca and , PoS LATTICE **2012**, 031 (2012) [arXiv:1211.0411 [hep-lat]].

Latest ATLAS Results on Higgs and BSM Physics

Osamu Jinnouchi

Tokyo Institute of Technology,
Meguro-ku, Tokyo 152-8551, Japan
E-mail: jinnouchi@phys.titech.ac.jp

This article reviews the latest results from the energy frontier experiment, ATLAS at the Large Hadron Collider (LHC), using the data collected during 2011 and 2012 corresponding to the recorded integrated luminosity of proton proton collision about 4.9 fb^{-1} at $\sqrt{s} = 7\,\text{TeV}$ and 20.8 fb^{-1} at $\sqrt{s} = 8\,\text{TeV}$ * respectively.

Keywords: LHC, SM, Higgs, Supersymmetry, Extra dimensions, Technicolor

1. Introduction

It has been three years since the last LHC talk was given in this workshop in 2009, when we barely had the solid prospects for the coming LHC data taking. Nevertheless, the speaker gave a good prediction[1] out of the limited circumstantial information before the start of LHC. He predicted there be a clear answer for distinguishing between models (Light Higgs + SUSY, Low mass TC model, or mixed scenarios of TC + SUSY) by the year 2012. The prediction turned out to be quite right, we could bring some messages out of the successful LHC runs during the past three years.

Concerning the status of the 2012 runs, the accelerator condition was largely stable through the year, the collision energy was increased from 7 to 8 TeV, and as of the beginning of December, total recorded LHC data at ATLAS was 20.8 fb^{-1}[a]. As widely known among the HEP community, the LHC stops its operation for two years and will resume the *pp* runs from 2015. Therefore the results shown here are presumably considered as the preliminary. The final results would be obtained after the refinement of the analysis over the full luminosity data by 2013 summer. The data taking efficiency of the ATLAS experiment has been continuously high ($> 93\%$) throughout the year, among them more than 93% are flagged as good data for the physics analysis. The mean number of interactions per crossing is about 20. This amount is still within the tolerance of the original detector design. In

*as of the day when this talk was given

[a]2012 *pp* run ended at 2013/12/16 and 21.7$\text{fb}^{-1}$was recorded in ATLAS.

addition, various techniques were invented to suppress the effect from these multiply interacting events, hence although the analyses are challenging, they are mostly robust.

This article would not review the detail of recent well known news about the discovery of the Higgs like boson, nor about the lacking of any hint for traditional SUSY or other BSM searches. Instead, this article outlines the latest results on the property measurements of 'Higgs like particle', the search strategies in SUSY signatures with natural or more complicated scenarios, a status of the low scale Technicolor searches, and finally about the experimental sensitivities to find new physics signals in the coming years of LHC runnings.

2. Properties of Higgs like particle

The famous 'July papers',[2,3] joint articles by the two large collaborations ATLAS and CMS reported the large significance in confirming the existence of Higgs like particle, e.g. ATLAS reported 4.5σ, 3.4σ, and 2.8σ for the $\gamma\gamma$, $ZZ \to 4\ell$, and $WW \to \ell\nu\ell\nu$ decay modes respectively, and 5.8σ as a combined significance. The increase in significance would provide further confidence, but after 5σ, the significance value itself is not so interesting anymore. A fact that we have found a new particle would not change. The current experimental priority/stress is more on the property measurements of this new particle to identify it. From the trivial considerations, it is already known to have zero electric charge, and ought to have even number spin (spin=1 is disfavored from the observed $\gamma\gamma$ decay signals), its mass is around 126 GeV. As of 2012 December, ATLAS released the mass and coupling study results, while CMS has additionally shown the study of the CP state. In followings, these new results are discussed. **The mass measurement**: Figure 1(a) shows the contour plot of the fit likelihood ratio as a function of its mass and the signal strength (μ). Two highest resolution channels ($\gamma\gamma$, 4ℓ decays) are used to deter-

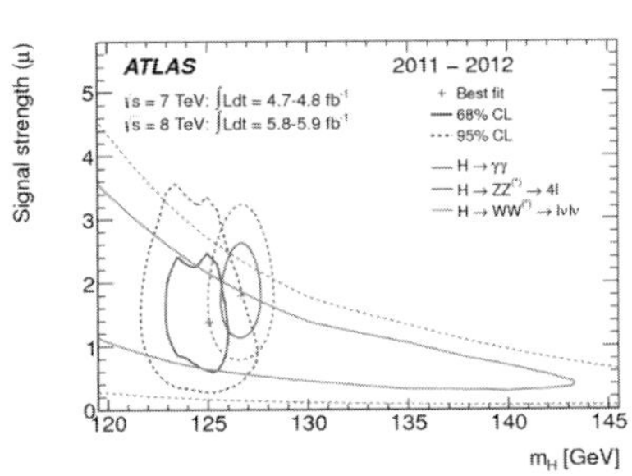

(a)

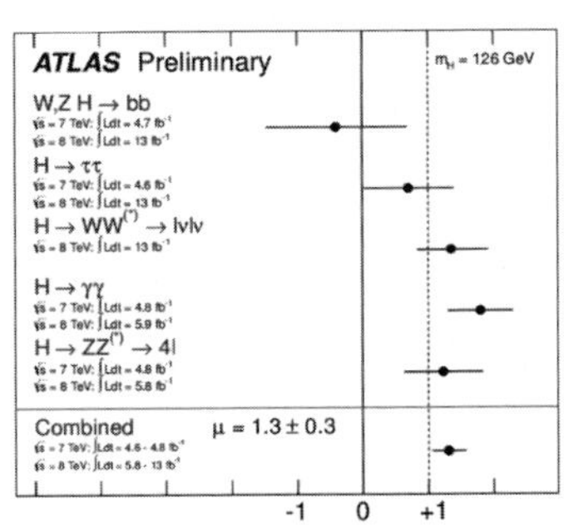

(b)

Fig. 1. (a) The contour of the likelihood fit of Higgs mass m_H and the signal strength (μ). (b) The signal strength (μ) for each process and the combined value.

mine the mass and the signal strength, while the $WW(\to \ell\nu\ell\nu)$ decay mode is used to confirm the consistency between the channels.The best fit value is found to be $m = 126.0 \pm 0.4(stat.) \pm 0.4(syst.)$ A remaining systematic error mainly comes from the uncertainties of the energy scale and the energy resolution. It is about 0.45% uncertainty, and could already be considered as a precision measurement. **The signal strength**: After the famous July paper,[2,3] the new/update results are obtained for $H \to WW^* \to \ell\nu\ell\nu$,[4] $H \to \tau\tau$[5] and $H \to bb$.[6] From the channel combination result, the updated overall signal strength is obtained as, $\mu = \sigma/\sigma_{SM} = 1.3 \pm 0.3$. Figure 1(b) shows the μ values for each decay mode. A decay mode for $\gamma\gamma$[7] may possibly be enhanced. Since $H \to \gamma\gamma$ process occurs via loop diagrams, this channel is sensitive to BSM physics. The overall consistency with SM ($\mu = 1$) is 23%. **The signal strength between the production modes**: The analysis is based on the profiled likelihood fit, with the parameterization of signal strength in each channel. During this fit, m_H is fixed to 126 GeV. $\gamma\gamma$ and WW channels are used to separate the two Higgs production modes (N.B. $H \to ZZ \to 4\ell$ is a inclusive analysis). ggF and ttH production modes correspond to fermionic couplings. On the other hand VBF and VH production have a vector boson couplings. The measurements are consistent with the SM within $1 \sim 2\%$.[8] **The coupling measurements**: Various methods to measure the couplings were proposed with the benchmark models by LHC Higgs XS working group.[9] Any deviation from the SM coupling strength is considered as a sign of BSM physics, hence the series of parameters κ which represents the coupling strength with respect to SM are investigated. Following seven symmetries with or without fixing Higgs total width, are probed. 1) $\kappa = \sqrt{\mu}$ checking the universal scaling of the couplings. 2) κ_V vs. κ_F to see if the coupling depends on the spin, i.e. vector bosons or fermions. 3) κ_W vs. κ_Z to check the Custodial symmetry. 4) κ_q vs. κ_ℓ flavor of fermions 5) κ_u vs. κ_d fermion type, i.e. up or down quark 6) κ_g vs. κ_γ to see the effects from loop couplings made by the heavy BSM particles. The last one 7) κ_H vs. $\mathrm{BR}_{\mathrm{inv}}$ takes into account the light invisible BSM searches. All of these efforts are summarized in the note.[8] For instance, a possible deviation from the SM ggH coupling is expressed as, $K_g^2 = \sigma_{ggH}/\sigma_{ggH}^{SM}$, and will be extracted from the observa-

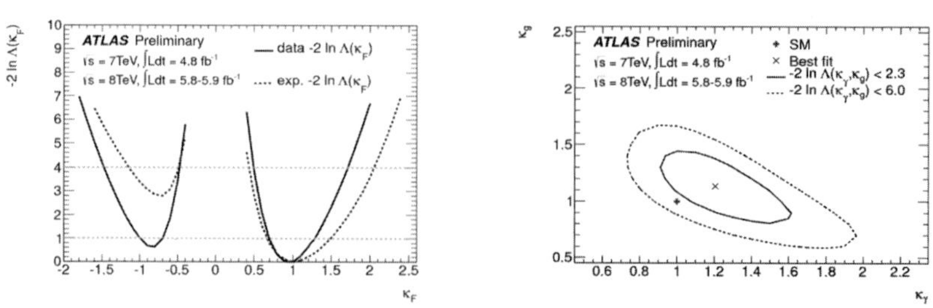

(a) (b)

Fig. 2. Fits for benchmark models probing for (a) different coupling strength scale factor for fermions κ_F. (b) contributions from non-SM particles, probing the $gg \to H$ and $H \to \gamma\gamma$ loops.

tion as, $(\sigma \cdot BR)_{obs}(gg \to H \to \gamma\gamma) = \sigma_{SM}(gg \to H) \cdot BR_{SM}(H \to \gamma\gamma) \cdot \frac{K_g^2 \cdot K_\gamma^2}{K_H^2}$. For the sake of space, I would not go into the detail of each measurement, and would suggest interested readers to check the reference[8] for detail. However one of the studies comparing vector bosons and fermions showed an interesting result, which is worth repeated here. The couplings κ_V, κ_F are the independent strengths and common to all the bosons and fermions, and used for the likelihood fits. Figure 2(a) shows the fit likelihood as a function of κ_F. The Fermiophobic Higgs scenario (i.e. κ_F=0) is strongly disfavored. The double asymmetric log likelihood in figure is due to the interference between the W and Top loops in $H \to \gamma\gamma$ diagrams. From the other comparisons in the list above, it was shown that 3) Custodial symmetry is consistent with SM, 6) the κ_g and κ_γ are consistent with SM with the compatibility estimated with p-value is 18% as seen in Fig. 2(b). The rest of the studies in the list did not show a clear result, i.e. 4) 5) do not have enough statistics to distinguish SM from BSM yet. 6) only showed course agreement with SM.

Property measurements other than the couplings are also carried out. CMS has shown the results on J^{CP} of the new particle, based on the angular correlation in $H \to ZZ^* \to 4\ell$, excluded the possibility of 0^- at more than 97% CL.[10] ATLAS had not shown this result yet[b]. In order to determine the spin, the large statistics and good control over the systematic error is required. No results were obtained from both experiments yet.

3. Searching for supersymmetry (SUSY)

In spite of the thorough searches, the SUSY (or physics beyond the SM) has not been found at LHC so far. This fact suggests there should generally be three directions in which we take for the next step in SUSY searches. The first one bases on the orthodox scenario, the one with the strong production with R-parity conservation (RPC). This channel looks for the signatures from the gluino, or 1st and 2nd generation squark production. The mass region up to $m_{\tilde{q}} \sim m_{\tilde{g}} = 1.5\,\text{TeV}$ is so far excluded. This channel would keep exploring for even higher mass region.[12–14] This channel is skipped in this article. The second direction is to look for the natural spectrum in RPC scenario. This channel assumes the gluino or light squarks are very heavy, and only the 3rd generation squarks or weak gauginos are light just to satisfy the hierarchy problem of the quantum loop correction over the Higgs mass. The results from these channels are shown in this article later on. The third direction is the R-parity violating scenarios or the long-lived SUSY particle searches, which heavily relies on the characteristics and the performance of the detector components. This topic is not covered in this article. The interested readers are recommended to check the following references.[15,16] **The sbottom direct production:** The results from the 7 TeV 4.7fb^{-1}data are available. The sbottoms decay into b-jets are searched for with the 0-lepton, 2-bjets and missing E_T signature.[17] $m(\tilde{b}) < 500\,\text{GeV}$ is excluded.

[b]later on, ATLAS also released the results on the same subject[11]

The sbottoms decay directly into charginos are also investigated, where the events with 3-leptons, more than 4 jets, and the missing E_T are looked for. $m(\tilde{b}) < 430\,\text{GeV}$ is excluded.[18] **The stop direct production:** The analyses look for the scalar top direct production events. The search strategies are prepared for various stop mass vs. lightest neutralino mass regions,[19–23] As shown in Fig.3(a), significant fraction of the phase space in stop mass vs. lightest neutralino mass is already excluded. **The gluino mediated stop decay:** There is a significant update in this channel out of 8 TeV results. The search looks for the final states with 4 top quarks. The main background is the ttbar production. 0-lepton plus 3 b-jets mode is the highest sensitivity channel,[24] and the 0-lepton plus 6 to 9 jets mode is the next.[12] There are yet other modes. One looks for the same sign lepton pairs[25] and the other looks for 3 leptons, jets and missing E_T.[26] The 3 lepton channel is complementary to the other channel near the $\tilde{g} \to t\bar{t}\tilde{\chi}_1^0$ border line. $m_{\tilde{g}} < 1240\,\text{GeV}$, $m_{\tilde{\chi}} < 500\,\text{GeV}$ are excluded with 95% CL from these searches. These results are summarized in Fig. 3(b). **The direct gaugino production:** Another candidate for the naturalness scenarios is the direct production of the gauginos. The analysis looks for the events with low jet activities and multi-lepton final states. The main backgrounds are, *ttbar*, *Z+jets* and *WW* for 2 lepton + missing E_T channel (7 TeV),[27] and $W\gamma$ or $Z\gamma$ events for 3 lepton + missing E_T channel (8 TeV).[28] These results are interpret in simplified model, where gluino and squark masses are set very heavy, and masses of $\tilde{\chi}_2^0$ and $\tilde{\chi}_1^{\pm}$ are set the same.

4. Current status of the LSTC searches at LHC

Two experiments, ATLAS and CMS, have searched for large number of exotic signatures for three years. Therefore indeed there are huge number of results. This article skips all of them, but instead only reviews about the results related to the Low Scale TechniColor (LSTC) scenarios which is one of the main subjects of this

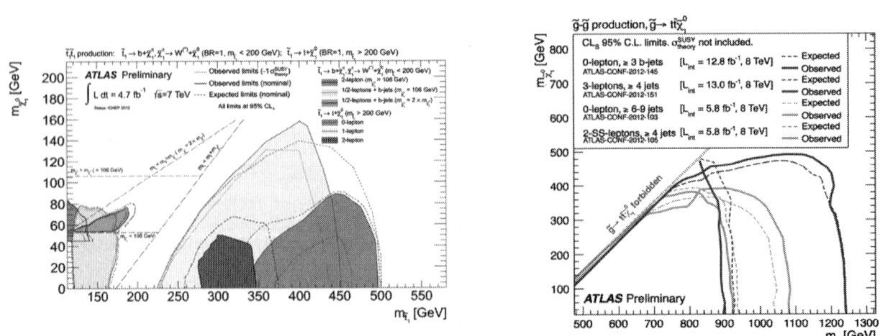

(a) (b)

Fig. 3. (a) Summary of the dedicated ATLAS searches for top squark pair production. Exclusion limits at 95% CL are shown in the $\tilde{t}_1 - \tilde{\chi}_1^0$ mass plane. (b) Exclusion limits at 95% CL for 8 TeV analyses in the ($m_{\tilde{g}}$, $m_{\tilde{\chi}_1^0}$) plane for the *Gtt* simplified model.

conference. The typical LSTC signatures expected in LHC experiments are di-lepton and di-boson resonances. In some published papers, low mass vector techni-mesons such as ρ_{TC}, ω_{TC} and a_{TC} are looked for in the resonance searches based on specific LSTC models, the others base on generic models as SSM,[29] then the spectrum or the model independent cross section limit becomes important. Here, two signatures are considered, one with the vector mesons decaying into di-leptons, and the other is the resonance decay into di-boson e.g. $WZ, Z\gamma$ and $W\gamma$. The searches look for the QCD-like spectrum in the scale $O(\Lambda_{TC}) \sim 100\,\mathrm{GeV}$.

In the region $(m(\rho_T) > 2{\times}m(\pi_T))$, very narrow resonances are expected from the suppression of $V \to n{\times}\pi_T$ decays due to the walking couplings, and ATLAS look for the narrow resonances via $\rho_T^0 \to \ell\ell$ and $\rho_T^{\pm} \to WZ$. Previous limits from Tevatron experiments were obtained in the region $(m(\rho_T) > m(W) + m(\pi_T)$ and $m(\rho_T) < 2 \times m(\pi_T))$ as $m(\rho_T) < 250\,\mathrm{GeV}$,[30] and in the region $(m(\rho_T) < m(W) + m(\pi_T))$ as $m(\rho_T) < 400\,\mathrm{GeV}$[31]

In the generic di-lepton resonance channel, the primary focus of the analysis lies at the high mass tail region ($m_{\ell\ell} > 1\,\mathrm{TeV}$) for BSM physics models such as Z'_{SSM}. A smaller cross section is expected for LSTC model than the generic BSM models, thus the di-lepton invariant mass distribution (see Fig. 4(a) for m_{ee} and (b) for $m_{\mu\mu}$) in modest scale (200-600 GeV) is reinterpreted in terms of LSTC. As seen in Fig. 5(a), with the mass degenerate assumption the observed limit was set as $m(\omega_T) = m(\rho_T) > 0.85\,\mathrm{TeV}$ by the center of mass energy 7 TeV with $5\,\mathrm{fb}^{-1}$. This is a significant improvement over Tevatron and previous ATLAS result.[32]

Di-bosons final state searches are carried out in various BSM scenarios, where new heavy state decays into a pair of vector bosons. The various channels are categorized in terms of the number of leptons ($ZZ \to \ell\ell\ell\ell$, [7 TeV, $1\,\mathrm{fb}^{-1}$[33]]; $WZ \to \ell\nu\ell\ell$, [7 TeV, $1\,\mathrm{fb}^{-1}$[34]]; $WW \to \ell\nu\ell\nu$, [7 TeV, $1\,\mathrm{fb}^{-1}$[35]]; $Z + W/Z \to \ell\ell qq$, [8 TeV, $1\,\mathrm{fb}^{-1}$[36]]). Despite of the enthusiastic searches all these yielded null results so far. Among these, the WZ channel is re-interpreted with a LSTC channel. For instance, the final state with 3 charged leptons (two opposite sign and the same flavor, ex-

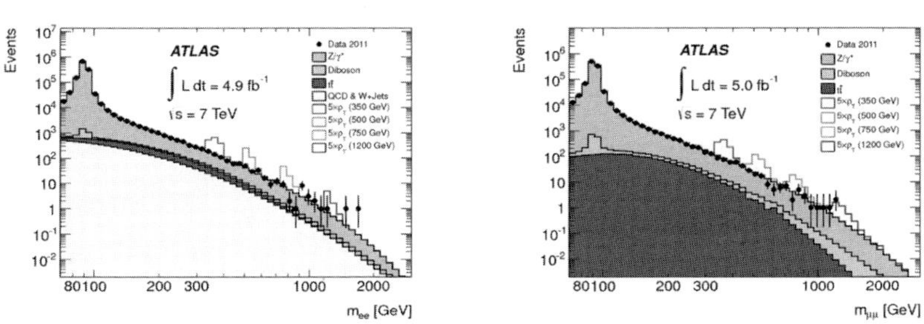

(a) (b)

Fig. 4. Dilepton invariant mass ($m_{\ell\ell}$) distribution after final selection, compared to the stacked sum of expected backgrounds, with four selected ρ_T signals overlaid (scaled by 5). (a) dielectron channel. (b) dimuon channel.

cluding the region $|m_{\ell\ell} - m_Z| < 20\,\mathrm{GeV}$) and missing $E_T > 25\,\mathrm{GeV}$ is looked for. No significant excess was observed against the SM prediction, and the limit on ρ_T mass was set as 467 (506) GeV for observed (expected)[34] as seen in Fig. 5(b).

The other related topic to the TechniColor scenario is the high energy WW scatterings which is important subject to be checked even if we confirm the existing of the SM Higgs boson. However due to the very small cross section, this measurement requires a large amount of luminosity and need to wait for the future high-luminosity runs. Hence this topic is skipped despite of its importance.

5. Prospects for the future searches

Extensive efforts have been spent for the Higgs property measurements aiming for the winter conferences. The analysis for 2011/2012 full data would allow for the spin and parity measurements. After the first long shutdown (2013-2014), it is envisaged that total amount of around $400\,\mathrm{fb}^{-1}$integrated luminosity is expected at 14 TeV by the year 2021, then the high-luminosity LHC (HL-LHC) starts from 2022, where we expect order of $3000\,\mathrm{fb}^{-1}$at 14 TeV during the 10 years. In coming years, the precise Higgs coupling measurements will be the most important task for LHC community, on average 3% and 9% accuracies are expected for vector boson, and fermion couplings respectively.[37] The search for BSM is even more important in future LHC program. The discovery reach will be extended with the higher CM energy and integrated luminosity. The standard example is the di-lepton resonance search for new Z' in the Sequential SM(SSM). In di-electron (di-muon) channel, 6.5 (6.4) TeV 95% CL limit is expected for $300\,\mathrm{fb}^{-1}$, and similarly 7.8 (7.6) TeV limit is expected for $3000\,\mathrm{fb}^{-1}$.

6. Summary

For Higgs like particle: there is no question concerning its existence anymore, and various measurements have started. The signal strength measurements so far agree

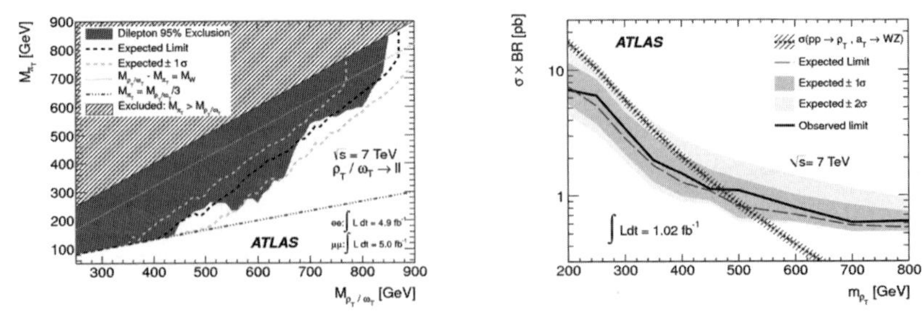

(a) (b)

Fig. 5. (a) The 95% CL excluded region (in red) in the plane π_T mass vs. ρ_T/ω_T mass, assuming $m_{a_T} = 1.1 \times m_{\rho_T/\omega_T}$, for the combination of dielectron and dimuon channels. (b) The observed and expected limits on $\sigma \times \mathrm{BR}$ for $pp \to \rho_T, a_T \to WZ$. The green and yellow bands represent respectively the $1\,\sigma$ and $2\,\sigma$ uncertainties on the expected limit.

with SM predictions within the current experimental uncertainties. The coupling measurements so far suggest they generally favor SM, some channels already have sensitivities in judging the agreements. For SUSY: no hint of the new phenomena observed yet, hence the search strategies have evolved to look for various naturalness scenarios to further explore unveiled regions. For LSTC: ATLAS has set severe constraints to the models.

References

1. S. Asai, (2009), A talk at the Strong Coupling Gauge Theories in the LHC Perspective (SCGT09).
2. ATLAS Collaboration, *Phys.Lett.* **B716**, 1 (2012), arXiv:1207.7214.
3. CMS Collaboration, *Phys.Lett.* **B716**, 30 (2012), arXiv:1207.7235.
4. ATLAS Collaboration, (2012), ATL-CONF-2012-158.
5. ATLAS Collaboration, (2012), ATL-CONF-2012-160.
6. ATLAS Collaboration, (2012), ATL-CONF-2012-161.
7. ATLAS Collaboration, (2012), ATL-CONF-2012-162.
8. ATLAS Collaboration, (2012), ATL-CONF-2012-127.
9. LHC Higgs Cross Section Working Group, and David, A. and Denner, A. and Duehrssen, M. and Grazzini, M. and others, (2012), arXiv:1209.0040.
10. CMS Collaboration, (2012), CMS-PAS HIG-12-045.
11. ATLAS Collaboration, (2012), ATL-CONF-2012-168, ATL-CONF-2012-169.
12. ATLAS Collaboration, (2012), ATL-CONF-2012-103.
13. ATLAS Collaboration, (2012), ATL-CONF-2012-104.
14. ATLAS Collaboration, (2012), ATL-CONF-2012-109.
15. ATLAS Collaboration, *Phys.Lett.* **B719**, 280 (2013).
16. ATLAS Collaboration, *JHEP* **1301**, p. 131 (2013).
17. ATLAS Collaboration, (2012), ATL-CONF-2012-106.
18. ATLAS Collaboration, (2012), ATL-CONF-2012-108.
19. ATLAS Collaboration, (2012), ATL-CONF-2012-059.
20. ATLAS Collaboration, (2012), ATL-CONF-2012-070.
21. ATLAS Collaboration, (2012), ATL-CONF-2012-071.
22. ATLAS Collaboration, (2012), ATL-CONF-2012-073.
23. ATLAS Collaboration, (2012), ATL-CONF-2012-074.
24. ATLAS Collaboration, (2012), ATL-CONF-2012-145.
25. ATLAS Collaboration, (2012), ATL-CONF-2012-105.
26. ATLAS Collaboration, (2012), ATL-CONF-2012-151.
27. ATLAS Collaboration, *Phys.Lett.* **B718**, 879 (2013).
28. ATLAS Collaboration, (2012), ATL-CONF-2012-154.
29. P. Langacker, *Rev.Mod.Phys.* **81** (2009), The Physics of Heavy Z′ Gauge Bosons.
30. CDF Collaboration, *Phys.Rev.Lett.* **104**, p. 111802 (2010).
31. D0 Collaboration, *Phys.Rev.Lett.* **104**, p. 061801 (2010).
32. ATLAS Collaboration, (2011), ATL-CONF-2011-125.
33. ATLAS Collaboration, *Phys.Lett.* **B712**, 331 (2012).
34. ATLAS Collaboration, *Phys.Rev.* **D85**, p. 112012 (2012).
35. ATLAS Collaboration, *Phys.Lett.* **B718**, 860 (2013).
36. ATLAS Collaboration, (2012), ATL-CONF-2012-150.
37. ATLAS Collaboration, (2012), ATL-PHYS-PUB-2012-001, ATL-PHYS-PUB-2012-004.

Coloron Models and LHC Phenomenology

Elizabeth H Simmons*, Anupama Atre, R. Sekhar Chivukula,
Pawin Ittisamai and Natascia Vignaroli

Department of Physics and Astronomy, Michigan State University
East Lansing, MI 48824, USA
**E-mail: esimmons@msu.edu*
www.msu.edu

Arsham Farzinnia

Institute of Modern Physics, Center for High Energy Physics Research,
Tsinghua University, Beijing 100084, China

Roshan Foadi

Service de Physique Theorique, Universite Libre de Bruxelles, Brussels, Belgium
and Centre for Cosmology, Particle Physics and Cosmology (CP3),
Universite catholique de Louvain, B-1348 Louvain-la-Neuve, Belgium

This talk discusses the possibility of new physics within the strong gauge interactions, specifically the idea of an extended color gauge group that is spontaneously broken to QCD. After a brief review of the literature, three of our recent pieces of work on coloron phenomenology are summarized. First, some key results on coloron production to NLO at hadron colliders are described. Next, a method of using associated production of colorons and weak vector bosons to better determine coloron couplings is discussed. Finally, a new model that naturally realizes flavor physics is reviewed.

Keywords: BSM physics, colorons, LHC phenomenology

1. Introduction

The LHC Era is proving to be extremely exciting: not only have the experiments decisively rediscovered all of the familiar particles of the Standard Model, confirming that the operations of the accelerator and detectors are well understood, but both ATLAS[1] and CMS[2] have also found a new scalar particle that appears to be the long-awaited Higgs Boson. Since the Standard Model with one Higgs doublet is not natural up to arbitrarily high energies, and since it leaves many questions (including the origin of flavor) unanswered, we anticipate that some physics beyond the Standard Model remains to be discovered.

An intriguing possibility is that an extended color gauge sector may exist. In particular, new colored states beyond the familiar quarks and gluons could be awaiting discovery at the LHC. These could reflect a variety of kinds of theories beyond the standard model. One class of theories are those in which the strong interactions are extended from the standard $SU(3)_{QCD}$ to a larger $SU(3)_1 \times SU(3)_2$ group and in which spontaneous symmetry breaking reduces the larger group to its diagonal subgroup which is identified with $SU(3)_{QCD}$. These models include topcolor,[3] the flavor-universal topcolor,[4] classic chiral color,[5] chiral color with unequal gauge couplings[6] and a newer flavor non-universal chiral color model.[7] Each of these models includes new heavy colored gauge bosons (colorons, topgluons, or axigluons) transforming as a color octet. Other theories with new color octet states include theories of new extra spacetime dimensions that incorporate Kaluza-Klein partners for the gluons, as in Refs.[8–10] or technicolor models with colored technifermions that bind into color-octet techni-rho mesons.[11] An entire catalog of possible new colored states including color sextet fermions, colored scalars, and low-scale string resonances[12] has also been reviewed in.[13]

If an extended color gauge sector does exist, then there are indications that it could be likely to couple more strongly to the third generation than the light quarks. For instance, if the new scalar state with a mass of 125 GeV is actually a composite, rather than a fundamental, scalar, it could potentially be a bound state of top quarks,[14–18] as realized in topcolor,[3] topcolor-assisted technicolor[19] and top seesaw[20–22] models, and as analyzed phenomenologically in.[23,24] There is also the puzzling question of how to explain the forward-backward asymmetry observed by the Tevatron experiments[25,26] in the production of top-quark pairs. Models involving flavor non-universal axigluons[7] have been cited as a possible explanation, and there has been discussion in the literature[27,28] of the degree to which the properties of those axigluons would be constrained by data on flavor-changing neutral currents.

This talk discusses recent work on several different aspects of coloron theory and phenomenology. We begin by reviewing the basic gauge structure of coloron models. We then present results on coloron production at NLO, indicating how this impacts colliders searches for new colored resonances. In the following section, we discuss a method for using production of a coloron in association with a W or Z boson to determine the chiral couplings of the coloron. Finally, we present a new model for a coloron model that incorporates a realistic theory of flavor physics. Each of these topics is discussed in greater detail elsewhere, and we provide the reader with the appropriate references.

2. Gauge Structure

The coloron models discussed in this talk are simple, renormalizable, models with the gauge structure $SU(3)_1 \times SU(3)_2 \times SU(2)_W \times U(1)_Y$. We name the $SU(3)_1 \times SU(3)_2$ gauge bosons $A^a_{1\mu}$ and $A^a_{2\mu}$, respectively, and call the corresponding gauge couplings g_1, g_2. The two $SU(3)$ gauge couplings are related to the QCD coupling g_S through

$$g_S = g_1 \sin\omega = g_2 \cos\omega, \tag{1}$$

where ω is a new gauge mixing angle.

Gauge symmetry breaking occurs in two steps:

- $SU(3)_1 \times SU(3)_2 \to SU(3)_C$ due to the (diagonal) expectation value $\langle\Phi\rangle \propto u\cdot\mathcal{I}$, where the fundamental or composite scalar, Φ, transforms as a $(3,\bar{3})$ under $SU(3)_1 \times SU(3)_2$ and $\mathcal{I}$ is the identity matrix,
- $SU(2)_W \times U(1)_Y \to U(1)_{em}$ in the usual way due to a Higgs doublet ϕ transforming as a $2_{1/2}$ of the electroweak group, and with the usual vacuum expectation value given by $v \approx 246$ GeV.

We will assume that the color-group symmetry breaking occurs at a scale large compared to the weak scale, $u \gg v$.[a]

The mass-squared matrix for the colored gauge bosons is given by

$$-\frac{1}{2}u^2 \begin{pmatrix} g_1^2 & -g_1 g_2 \\ -g_1 g_2 & g_2^2 \end{pmatrix} . \tag{2}$$

Diagonalizing this matrix reveals mass eigenstates G^a and C^a

$$G^a_\mu = \sin\omega A^a_{1\mu} + \cos\omega A^a_{2\mu} \tag{3}$$
$$C^a_\mu = \cos\omega A^a_{1\mu} - \sin\omega A^a_{2\mu} \tag{4}$$

with masses

$$M_G = 0 \qquad M_C = u\sqrt{g_1^2 + g_2^2} = \frac{g_S\, u}{\sin\omega\cos\omega} \tag{5}$$

If we name the color current associated with $SU(3)_i$ by the symbol $J_i^{a\mu}$, then the gluon and coloron, respectively couple to the following currents:

$$g_S J_G^{a\mu} = g_S(J_1^{a\mu} + J_2^{a\mu}) \tag{6}$$
$$g_S J_C^{a\mu} = g_S(\cot\omega J_1^{a\mu} - \tan\omega J_2^{a\mu}) \tag{7}$$

From this, we calculate the decay width of the coloron into massless color-triplet fermions to be

$$\Gamma_C = \frac{g_S^2 M_C}{24\pi}\left(n_1\cot^2\omega + n_2\tan^2\omega\right) \tag{8}$$

where n_1 and n_2 correspond to the number of Dirac fermion states charged under $SU(3)_1$ and $SU(3)_2$ respectively. Finally, we note that at energy scales well below the coloron mass, coloron exchange may be approximated by the current-current interaction:

$$-\frac{g_S^2}{2M_C^2} J_C^{a\mu} J^a_{C\mu} . \tag{9}$$

[a]The vacuum expectation values for ϕ and Φ occur for a choice of parameters in the most general, renormalizable, potential for these fields, and the vacuum is unique up to an arbitrary global gauge transformation. We will assume that the additional physical singlet and color-octet fields in Φ are heavy, and neglect them in what follows.

3. NLO Coloron Production

In Ref.[29] we report the first complete calculation of QCD corrections to the production of a massive color-octet vector boson. We treat the coloron as an asymptotic state in our calculations, employing the narrow width approximation. Our next-to-leading-order (NLO) calculation includes both virtual corrections as well as corrections arising from the emission of gluons and light quarks, and we demonstrate the reduction in factorization-scale dependence relative to the leading-order (LO) approximation used in previous hadron collider studies.

The QCD NLO calculation of coloron production reported here differs substantially from the classic computation of the QCD NLO corrections to Drell-Yan production,[30] because the final state is colored. In particular, Drell-Yan production involves the coupling of the light quarks to a conserved (or, in the case of W- or Z-mediated processes, conserved up to quark masses) current. Hence, in computing the NLO corrections to Drell-Yan processes, the current conservation Ward identity insures a cancellation between the UV divergences arising from virtual quark wave function and vertex corrections. These cancellations do not occur in the calculation of the NLO corrections to coloron production, because of vertex corrections involving the 3-point non-Abelian colored-boson vertices. We use the "pinch technique"[31] to divide the problematic non-Abelian vertex corrections into two pieces – a "pinched" piece whose UV divergence contributes to the renormalization of the coloron wavefunction (and, ultimately, a renormalization of the coloron coupling) and an "unpinched" part whose UV divergence (when combined with an Abelian vertex correction) cancels against the UV divergences in quark wavefunction renormalization. Once the UV divergences are properly accounted for, the IR divergences cancel in the usual way: the IR divergences arising from real quark or gluon emission cancel against the IR divergences in the virtual corrections, and the IR divergences arising from collinear quarks or gluons in the initial state are absorbed in the properly defined parton distribution functions (PDFs).

We compute the gauge-, quark-, and self-couplings of the coloron assuming the extended color structure described in Sec. 2. The calculation yields the minimal coupling of gluons to colorons, and allows for the most general couplings of quarks to colorons. The cancellation of UV divergences described above, however, occurs only when the 3-coloron coupling has the strength that arises from the dimension-four gauge-kinetic energy terms of the extended color gauge structure. Our computation applies directly to any theory with this structure, *i.e.* to massive color-octet vector bosons in axigluon, topcolor, and coloron models. In general, the triple coupling of KK gluons in extra-dimensional models, or of colored technivector mesons in technicolor models, will not follow this pattern. However our results apply approximately to these cases as well, to the extent that the $SU(3)_1 \times SU(3)_2$ model is a good low-energy effective theory for the extra dimensional model (a "two-site" approximation in the language of deconstruction[32,33]) or for the technicolor theory (a hidden local symmetry approximation for the effective technivector meson sector[34,35]).

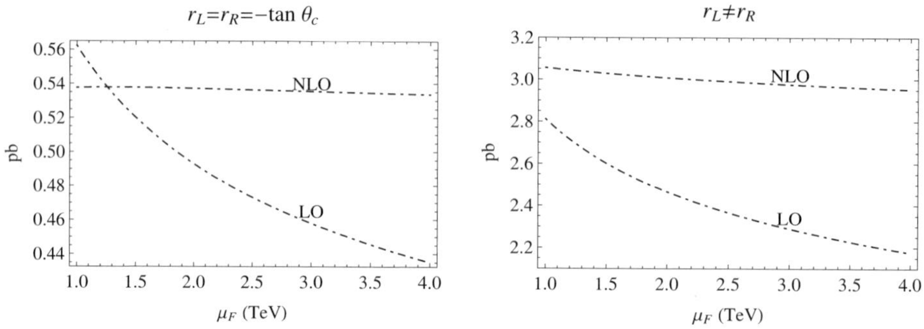

Fig. 1. Dependence of LO and NLO cross sections at the LHC ($\sqrt{s} = 7$ TeV), as a function of factorization scale μ_F for $M_C = 2.0$ TeV, $\sin^2\theta_c|_{\mu=2.0\text{ TeV}} = 0.25$, and two quark charge assignments. The NLO cross section has a much weaker (formally, two-loop) residual scale-dependence.

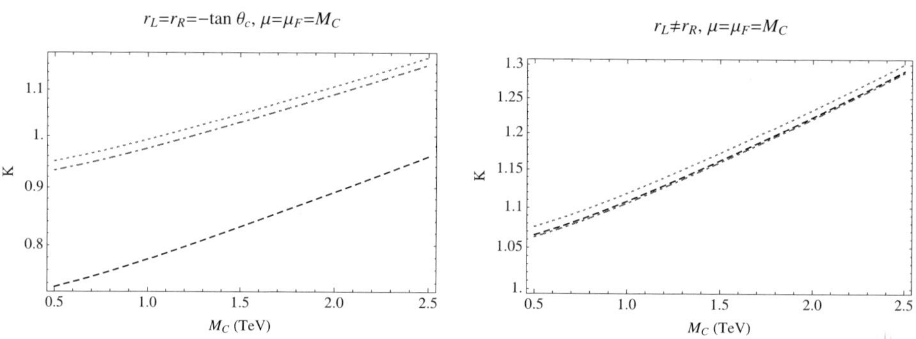

Fig. 2. "K-factor", the ratio of the NLO to LO cross section for coloron production at the LHC ($\sqrt{s} = 7$ TeV), plotted as a function of M_C for $\sin^2\theta_c = 0.05$ (dashed), 0.25 (dot-dashed) and 0.50 (dotted), $\mu_F = M_C$, and the two quark charge assignments.

We find that incorporating NLO corrections makes a substantial difference to the production cross-section and kinematic distributions of the colorons. Depending on the coloron mass, the production rate can differ significantly from tree-level predictions. As searches for new resonances decaying to dijets are routinely undertaken by the LHC experiments, taking these results into account is important for setting accurate limits. Moreover, at NLO, one sees that singly-produced colorons can have substantial transverse momentum (while at tree-level they should have nearly none).

The figures illustrating our results use the following notation: We denote the gauge boson mixing angle ω by θ_c (as was done in Ref.[29]) and we call the coupling of a left(right)-handed quark to colorons, divided by the quark's coupling to gluons g_S, by the name r_L (r_R). Plots are shown for two of the three possible flavor-universal charge assignments for the quarks: all quarks charged under $SU(3)_1$ [$r_L = r_R = -\tan\theta_c$] and left(right)-handed quarks charged under $SU(3)_{1,(2)}$ [$r_L \neq r_R$]; results for the third case, in which all quarks charged under $SU(3)_2$ [$r_L = r_R = \cot\theta_c$], are substantially similar to those for the first case.

Fig. 1 shows that the scale-dependence of the LO cross section is of order 30% while at NLO this is reduced to about 2% percent. Fig. 2 shows the "K-factor" for coloron production.

$$K(M_C, \sin\theta_c|_{\mu=M_C}, \mu_F = M_C) \equiv \frac{\sigma^{NLO}(M_C, \sin\theta_c|_{\mu=M_C}, \mu_F = M_C)}{\sigma^{LO}(M_C, \sin\theta_c|_{\mu=M_C}, \mu_F = M_C)}\ , \tag{10}$$

for $\sin^2\theta_c = 0.05$ (dashed), 0.25 (dot-dashed) and 0.50 (dotted). Again, we see that the NLO corrections are of order 30%. We have provided detailed tables of our results for the coloron K-factors for the LHC and Tevatron at a variety of collision energies in Refs.[29,36]

Finally, we found that of order 30% of the colorons in this generic model with masses in the TeV range are produced with $p_T \geq 200$ GeV, once NLO corrections are taken into account. In other words, the tree-level picture of a new resonance produced at rest and decaying to back-to-back jets receives significant modification at next-to-leading order. Once a coloron is discovered, studies of the angular distributions of its decay products will need to take this into account.

4. Associated Coloron Production

The high production rate of a colored resonance (due to the strong coupling value) and the simple topology of the final state (decay into two jets) makes the search for di-jet resonances one of the early signatures that are studied at hadron colliders. Once a new colored resonance is discovered, measuring its properties will be the next important task. The di-jet invariant mass m_{jj} and the angular distributions of energetic jets relative to the beam axis are sensitive observables to determine the properties, such as mass and spin of the resonance. Although one can constrain the coupling strength of the colored resonance to the Standard Model (SM) quarks using the total cross section, this is not sufficient to determine the chiral structure of the couplings.

In Ref.,[37] we propose a new channel for studying coloron couplings: the associated production of a W or Z gauge boson with the color-octet at the LHC. The chiral couplings of the weak gauge bosons to the fermions in the associated production channel provides additional information about the chiral structure of the new strong dynamics. Combining the associated production channel with the di-jet channel makes it possible to extract the chiral couplings of the colored resonance because the cross-sections of each channel have a different dependence on the coloron's couplings to fermions. The functional form of the dependance of these measurements on the chiral couplings in the di-jet channel is $g_L^2 + g_R^2$; in the Wjj channel it is g_L^2 and in the Zjj channel it is $ag_L^2 + bg_R^2$. A cartoon illustration of these three measurements along with the di-jet measurement is shown in the left-hand pane of Fig. 3. Notice that while combining the different channels will narrow the allowed range of couplings, there remains an ambiguity in extracting the sign of the couplings. This method of using the associated production of a weak gauge boson

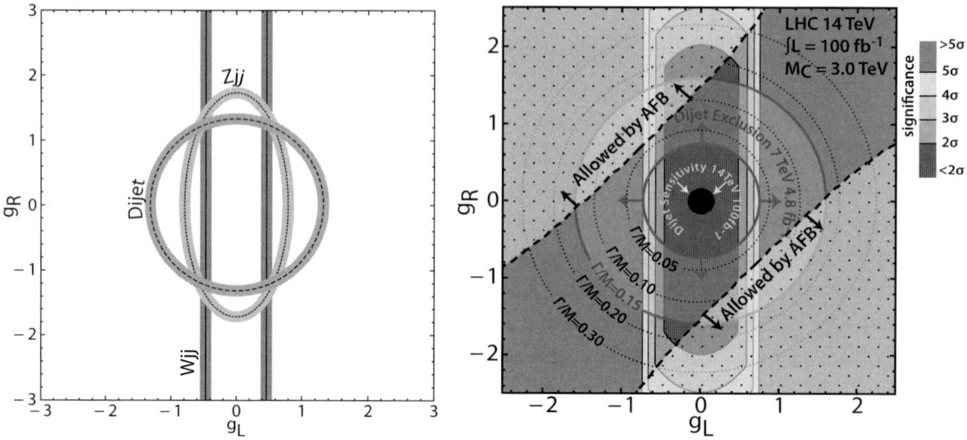

Fig. 3. Left: A cartoon illustration of the form of prospective constraints on chiral couplings from the di-jet channel (dashed black circle with red band), the channel with associated production of a W boson (solid black parallel lines with green band) and the channel with associated production of a Z boson (dotted black ellipse with blue band). Combining the constraints from different channels will narrow the range of allowed couplings. Right: same as Fig. 4(a) but for $M_C = 3$ TeV and 100 fb^{-1} of data. Here we have combined the sensitivity in the associated production channel from both W and Z bosons. The couplings allowed by the current measurement of A_{FB}^{tt} (see text) are shown by the translucent yellow dotted region.

to illuminate the properties of a new resonance was studied earlier in the context of the measurement of Z' couplings.[38,39]

The color-octet states (C) are produced and decay to two jets via the process $pp \xrightarrow{C} j\ j$. They can also be produced in association with a weak gauge boson via the processes $pp \xrightarrow{C} j\ j\ W^{\pm}$ and $pp \xrightarrow{C} j\ j\ Z$ where $j = u, d, s, c, b$. The associated production modes include s and t channel diagrams with the emission of the gauge bosons in either the initial or final state. The final state channels that we study are $pp \to \ell^{\pm} E_T^{miss}\ 2j, \quad \ell^+\ell^-\ 2j$, coming from $W^{\pm}(\to \ell^{\pm}\nu)$ or $Z(\to \ell^+\ell^-)$, respectively and $\ell = e, \mu$. Although the inclusion of the τ lepton in the final state could increase signal statistics, for simplicity we ignore this experimentally more challenging channel. Our simulations included the relevant backgrounds

- $W+$ jets, $Z+$ jets with W, Z leptonic decays;
- top pair production with fully leptonic, semi-leptonic and hadronic decays (where some final state particles may be missed or mis-identified);
- single top production leading to a $W^{\pm}b\ q$ final state;
- W^+W^-, $W^{\pm}Z$ and ZZ with all possible decay combinations leading to the final state of interest.

For details on the simulations, including cuts and acceptances, see Ref.[37]

We present the results of our analysis for the LHC in the plane of the couplings g_L^q, g_R^q for different masses, M_C, of the color-octet. The sensitivity for the channel with associated production of a $W(Z)$ gauge boson is presented in the upper (lower) panels of Fig. 4, while the left (right) panels are for integrated luminosity of 10 (100) fb^{-1}. The different colored bands represent varying significance of signal observation from 2σ to greater than 5σ. The black dotted curves are contours of constant widths and we show the curves for several values of $\Gamma_C/M_C = 0.05,\ 0.10,\ 0.20$ and 0.30. The small black region in the center lies outside the projected dijet sensitivity (i.e. couplings within the black region result in a dijet production rate too small to be observed) at the LHC with 100 fb^{-1} of data citeGumus:2006mxa.

The results in Fig. 4 illustrate several features. In the channel with associated production of a W boson, there is no sensitivity in the region near $g_L^q = 0$ due to the left-handed couplings of the W boson, and the sensitivity improves as we move away from the $g_L^q = 0$ axis. The channel with the associated production of a Z boson on the other hand is sensitive to both left and right-handed couplings and sensitivity in the region close to $g_L^q = 0$ is non-zero. The smaller production cross section for this channel along with the small leptonic branching fractions for Z decays limit the gain in sensitivity. Nonetheless this channel provides an additional measurement and hence useful information in untangling the couplings.

To summarize our results for a variety of coloron masses and couplings with either 10 fb^{-1} or 100 fb^{-1} of data: As expected, the longer run with more data has better sensitivity and can probe masses up to 4.5 (4.0) TeV in the channel with associated production of a $W(Z)$ gauge boson. The projected limit of sensitivity in the dijet channel for the LHC with 100 fb^{-1} data is shown in the figures as the small black circle in the center with small couplings. It is safe to assume that the final acceptance will be at least equal to or even better than the current one and the small black region in the center could shrink even further. If the LHC were to discover a resonance with such small couplings, one would have to find other novel ways of understanding the chiral structure of couplings as the associated production channel does not have sensitivity in those regions of parameter space.

Separately, we note that the Tevatron has made a measurement of the top-pair forward-backward asymmetry (A_{FB}^{tt}).[46,47] The authors of Ref.[48] have translated this measurement into constraints on the couplings of color-octet resonances. We show this additional constraint for the case of a color-octet with mass $M_C = 3$ TeV in the right-hand pane of Fig. 3. The region consistent with the A_{FB}^{tt} measurement is shown in translucent yellow with small dots. Note that for clarity of presentation we have combined the sensitivity from coloron plus W and coloron plus Z channels in this figure.

The results are encouraging: The LHC will be able to provide information about the chiral structure for a wide range of couplings and masses and hence point us in the direction of the underlying theoretical structure.

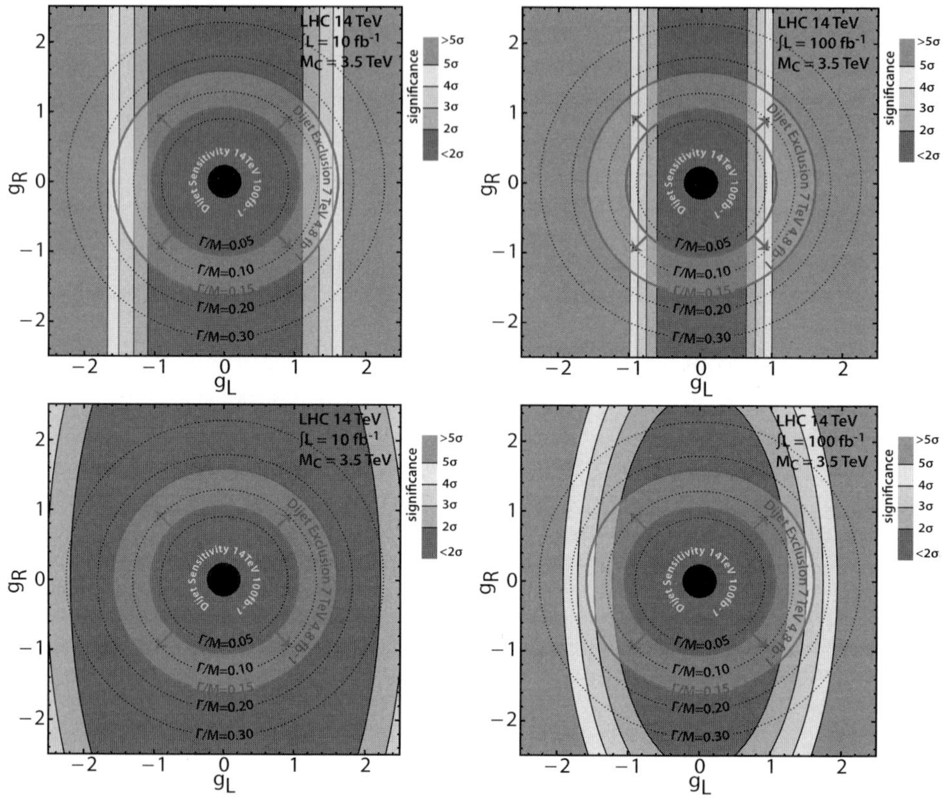

Fig. 4. (a) Top left: sensitivity plot for a color-octet of mass $M_C = 3.5$ TeV produced in association with a W gauge boson, in the plane of the couplings g_L^q, g_R^q at the LHC with 10 fb^{-1} of data and $\sqrt{s} = 14$ TeV; (b) top right: same for 100 fb^{-1} of data; (c) bottom left: same as (a) but for a color-octet produced in association with a Z boson; (d) bottom right: same as (c) but for 100 fb^{-1} of data. The colored bands show regions with significance from 2σ to $> 5\sigma$. The inner solid green circle is the limit from current direct searches for narrow dijet resonances at the LHC.[40–44] The outer green circle corresponds to a contour with $\Gamma_C/M_C = 0.15$. The faded grey region between the two circles is excluded for narrow resonances while the region beyond the outer green circle is allowed for broad resonances. The small black region in the center lies beyond the projected dijet sensitivity at the LHC with 100 fb^{-1} of data citeGumus:2006mxa. Black dotted contours indicate the combinations of couplings that give rise to varying widths $\Gamma_C/M_C = 0.05$, 0.10, 0.20 and 0.30.

5. A Flavorful Coloron Model

In Ref.,[49] we introduced a model in which the strong interactions are extended to an $SU(3)_1 \times SU(3)_2$ structure in a way that causes the new heavy coloron states to couple differently to the third generation quarks than to the lighter generations. What is novel here is that the model also naturally addresses the experimental observation that the third family of quarks have only a small mixing with the lighter families. This is implemented through the presence of heavy weak-vector quarks that transform in the same way under the extended color group as the third-generation quarks. Mixing between the ordinary quark generations occurs only because all three generations mix with the vector quarks; the different gauge charges of the third and lighter generations of quarks thus leads to naturally small mixing between those generations. Effectively, this model realizes next-to-minimal flavor violation.[50]

The color gauge sector of the model remains as described above in Sec. 2. The matter fields of this model are summarized[b] below in Table 1. Those coupled to $SU(3)_1$ include one chiral quark generation (q_L, t_R, and b_R), which will be associated primarily with the third generation quarks, and one vectorial quark generation ($Q_{L,R}$). The two remaining (chiral) quark generations ($\vec{\psi}_L$, $\vec{u}_R$, and $\vec{d}_R$) are coupled to $SU(3)_2$ and will be associated primarily with the two light quark generations. Noting that Q_L and q_L transform in the same way under the gauge symmetries, we define $\vec{\mathcal{Q}}_L \equiv (q_L, Q_L)$ and observe that the flavor symmetries (ignoring gauge anomalies) of the quark kinetic energy terms in this model are

$$U(2)_{\vec{\psi}_L} \times U(2)_{\vec{u}_R} \times U(2)_{\vec{d}_R} \times U(2)_{\vec{\mathcal{Q}}_L} \times U(1)_{t_R} \times U(1)_{b_R} \times U(1)_{Q_R} \,. \tag{11}$$

As discussed in Ref.[49] the flavor symmetries may be used to simplify our analysis of the fermion masses and Yukawa couplings.

Particle	$SU(3)_1$	$SU(3)_2$	$SU(2)$	$U(1)$
$\vec{\mathcal{Q}}_L = \begin{pmatrix} q_L \\ Q_L \end{pmatrix}$	3	1	2	+1/6
t_R	3	1	1	+2/3
b_R	3	1	1	-1/3
Q_R	3	1	2	+1/6
$\vec{\psi}_L = \begin{pmatrix} \psi_L^1 \\ \psi_L^2 \end{pmatrix}$	1	3	2	+1/6
$\vec{u}_R = \begin{pmatrix} u_R^1 \\ u_R^2 \end{pmatrix}$	1	3	1	+2/3
$\vec{d}_R = \begin{pmatrix} d_R^1 \\ d_R^2 \end{pmatrix}$	1	3	1	-1/3
ϕ	1	1	2	+1/2
Φ	3	$\bar{3}$	1	0

Let us take a closer look at those mass and Yukawa terms. The existence of the right-handed weak doublet state Q_R permits the Dirac mass term

$$\bar{\vec{\mathcal{Q}}}_L \cdot \vec{\mathcal{M}}\, Q_R + h.c.\,, \tag{12}$$

where $\vec{\mathcal{M}}$ is an arbitrary two-component complex mass matrix. Using the $U(2)_{\vec{\mathcal{Q}}_L}$ symmetry of the quark kinetic terms, we will choose to work in a basis in which

[b]The lepton fields are assigned to $SU(2) \times U(1)$ just as in the standard model. We normalize hypercharge such that $Q = T_3 + Y$.

$\vec{\mathcal{M}}^T = (0\ M)$ where M is real and positive. This defines what we will mean by q_L and Q_L in Table 1 from here on.

The third-generation quark Yukawa couplings are given by

$$\frac{\sqrt{2}M}{v} \cdot \left(\beta_b \bar{q}_L \phi b_R + \beta_t \bar{q}_L \tilde{\phi} t_R\right) + h.c.\ , \tag{13}$$

where the $\beta_{t,b}$, can be chosen to be real, using the $U(1)_{t_R} \times U(1)_{b_R}$ symmetries. The Yukawa couplings for the light two generations are given by

$$\frac{\sqrt{2}M}{v} \cdot \left(\vec{\bar{\psi}}_L \phi \lambda_d \vec{d}_R + \vec{\bar{\psi}}_L \tilde{\phi} \lambda_u \vec{u}_R\right) + h.c.\ . \tag{14}$$

where $\lambda_{u,d}$ are 2×2 complex matrix Yukawa couplings. Neglecting the (small) mixing of the third-generation quarks with the first two generations, the parameters $\beta_{t,b}$ and $\lambda_{u,d}$ are just equal to the corresponding parameters in the standard model, up to the factor of $\sqrt{2}M/v$ which is included for later convenience.[c]

Mixing of the third quark generation with the first two occurs only because all three generations mix with the (heavy) vector quarks. Mixing between the third-generation quarks and the vector quarks occur through

$$\frac{\sqrt{2}M}{v} \cdot \left(\lambda'_b \bar{Q}_L \phi b_R + \lambda'_t \bar{Q}_L \tilde{\phi} t_R\right) + h.c.\ , \tag{15}$$

and mixing between the first- and second-generation quarks and the vector quarks occurs through the Yukawa couplings to the color-octet scalar

$$\frac{M}{u} \cdot \left(\vec{\bar{\psi}}_L \cdot \vec{\alpha}\, \Phi Q_R\right) + h.c.\ . \tag{16}$$

Here the $\lambda'_{b,t}$ are complex numbers, while $\vec{\alpha}$ is a two-component complex vector, whose phases and orientations can be simplified using the residual flavor symmetries. Note that in the limit that either $\lambda'_{b,t} \to 0$ or $\vec{\alpha} \to 0$, third-generation quark number is conserved separately from quark number for the light quarks,[d] and mixing between the third generation and the first two vanishes.

As a result, having the mixing between the heavy and light quark generations be small is natural in this model. In fact, we find that the pattern of quark masses and CKM mixings is reproduced naturally in this model. Specifically, the CKM matrix is correctly reproduced by:

$$\begin{aligned} V_{ub} &= \alpha_1 d = \alpha_1 \left(\frac{\lambda'_t}{\beta_t} - \frac{\lambda'_b}{\beta_b}\right) = A\lambda^3(\rho - i\eta) = 0.00131 - i0.00334 \\ V_{cb} &= \alpha_2 d = \alpha_2 \left(\frac{\lambda'_t}{\beta_t} - \frac{\lambda'_b}{\beta_b}\right) = A\lambda^2 = 0.0415\ , \end{aligned} \tag{17}$$

[c]While incorporating M into the Yukawa couplings is convenient for subsequent calcualtions, its presence obscures the *decoupling* properties[51] of the theory in the limit $M \to \infty$.

[d]In the limit $\lambda'_{t,b} \to 0$, top- and bottom-quark number is conserved separately, while in the limit $\vec{\alpha} \to 0$ it is conserved in combination with vector-quark number.

where both d and α_2 are real, and where α_2 is $\mathcal{O}(\lambda^2)$ while α_1 is $\mathcal{O}(\lambda^3)$. Here, the parameters A, λ, ρ, and η belong to the conventional Wolfenstein parameterization of the CKM matrix.

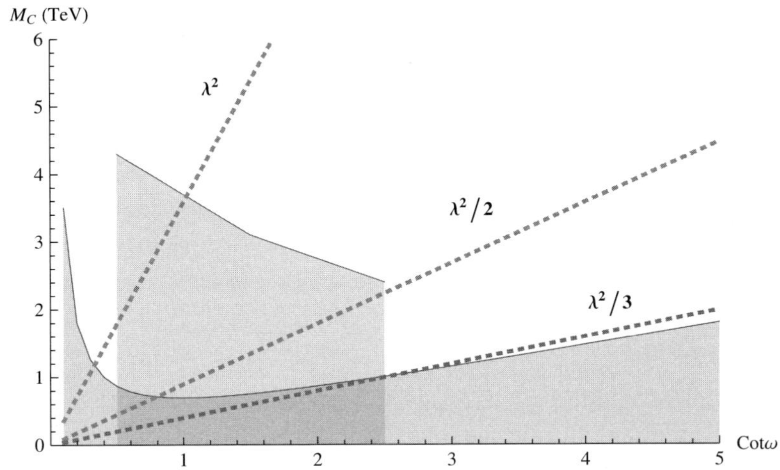

Fig. 5. Left: Exclusion regions on the plane $(\cot\omega, M_C)$ from the ATLAS search for dijet resonances (pink region, beneath the short upper curve) and from B_s mixing, assuming that Yukawa couplings take on the maximum value allowed by $b \to s\gamma$, (blue region, beneath the long lower curve). In addition, a bound from Kaon mixing excludes the region below the dashed line whose label matches the value of the coupling $|\alpha_2|$; a larger value yields a stronger bound. See Ref.[49] for details.

We are now prepared to examine the phenomenology of the model in more detail.

New contributions to flavor-changing neutral currents (FCNC) from the mixing of the ordinary fermions with the new weak vector femion states, and through the couplings of the colorons to fermions. We find that data on $b \to s\gamma$ and $\Delta F = 2$ meson mixing processes place bounds on the model parameters but leave substantial regions of allowed parameter space, as indicated in Fig. 5.

We have also compared the predictions of the model to data from searches for new resonances decaying to dijets performed by the ATLAS[52] and CMS[53] experiments at the LHC. We find that the lower bound on the coloron mass ranges from $M_C \geq 2.4$ TeV for $\cot\omega \approx 2.5$, when the coloron couples mainly to third-generation quarks, all the way to $M_C \geq 4.3$ TeV for $\cos\omega \approx 0.5$, when the coloron couples more strongly to first and second generation quarks. In contrast, the ATLAS[54] and CMS[55] searches for resonances decaying to $t\bar{t}$ put weaker bounds on the coloron mass than the searches for resonances in dijets. Limits from the dijet searches are also summarized in Fig. 5.

Finally, it is worth noting that LHC data provides a lower bound on the masses of the heavy quark states that are mostly composed of the vector fermions. The heavy B, T states (partners of b and t) can be produced at the LHC in pairs via gluon-gluon fusion[56,57] or singly, through their interactions with W, Z, or h.[57–59] The ATLAS search in these channels[60] for a 4th generation down-type quark, which decays predominantly into Wt, puts a limit on the 4th generation quark mass that we can directly apply to the B vector fermion mass: $M_{\mathsf{B}} \gtrsim 0.67$ TeV. The analogous limit on the heavy up-type vector quark is weaker.

6. Conclusions

The LHC is poised to seek signs of new physics beyond the standard model in the strong-interaction gauge sector. This paper has reviewed new developments in theory, model-building, and phenomenology that offer promising avenues for further exploration in the LHC's upcoming high-energy run. We look forward to seeing what the experiments will discover.

Acknowledgments

We thank John M. Campbell, Wayne W. Repko, Carl Schmidt, and Devin G.E. Walker for their input and discussions. The authors were supported, in part, by the U.S. National Science Foundation under Grants No. PHY-0854889 and PHY-0855561. PI is supported by Development and Promotion of Science and Technology Talents Project (DPST), Thailand. We also acknowledge the support of the Michigan State University High Performance Computing Center and the Institute for Cyber Enabled Research. R.S.C. and E.H.S. acknowledge the hospitality of the Galileo Galilei Institute for Theoretical Physics, where part of this work was accomplished. A.F. acknowledges the hospitality of TASI 2011, where part of this work was completed.

References

1. G. Aad *et al.* [ATLAS Collaboration], Phys. Lett. B **716**, 1 (2012)
2. S. Chatrchyan *et al.* [CMS Collaboration], Phys. Lett. B **716**, 30 (2012)
3. C. T. Hill, Phys. Lett. B **266**, 419 (1991).
4. R. S. Chivukula, A. G. Cohen and E. H. Simmons, Phys. Lett. B **380**, 92 (1996)
5. P. H. Frampton and S. L. Glashow, Phys. Lett. B **190**, 157 (1987).
6. M. V. Martynov and A. D. Smirnov, Mod. Phys. Lett. A **24**, 1897 (2009)
7. P. H. Frampton, J. Shu and K. Wang, Phys. Lett. B **683**, 294 (2010)
8. D. A. Dicus, C. D. McMullen and S. Nandi, Phys. Rev. D **65**, 076007 (2002)
9. H. Davoudiasl, J. L. Hewett and T. G. Rizzo, Phys. Rev. D **63** (2001) 075004
10. B. Lillie, L. Randall and L. -T. Wang, JHEP **0709**, 074 (2007)
11. E. Farhi and L. Susskind, Phys. Rept. **74**, 277 (1981).
12. I. Antoniadis, Phys. Lett. B **246**, 377 (1990).
13. T. Han, I. Lewis and Z. Liu, JHEP **1012**, 085 (2010)
14. V. A. Miransky, M. Tanabashi and K. Yamawaki, Phys. Lett. B **221**, 177 (1989).
15. V. A. Miransky, M. Tanabashi and K. Yamawaki, Mod. Phys. Lett. A **4**, 1043 (1989).
16. Y. Nambu, EFI-89-08.
17. W. J. Marciano, Phys. Rev. Lett. **62**, 2793 (1989).
18. W. A. Bardeen, C. T. Hill and M. Lindner, Phys. Rev. D **41**, 1647 (1990).

19. C. T. Hill, Phys. Lett. B **345**, 483 (1995)
20. B. A. Dobrescu and C. T. Hill, Phys. Rev. Lett. **81**, 2634 (1998)
21. R. S. Chivukula, B. A. Dobrescu, H. Georgi and C. T. Hill, Phys. Rev. D **59**, 075003 (1999)
22. H. -J. He, C. T. Hill and T. M. P. Tait, Phys. Rev. D **65**, 055006 (2002)
23. R. S. Chivukula, P. Ittisamai, E. H. Simmons, B. Coleppa, H. E. Logan, A. Martin and J. Ren, Phys. Rev. D **86**, 095017 (2012)
24. R. Barbieri, D. Buttazzo, F. Sala, D. M. Straub and A. Tesi, arXiv:1211.5085 [hep-ph].
25. T. Aaltonen *et al.* [CDF Collaboration], Phys. Rev. D **83**, 112003 (2011)
26. V. M. Abazov *et al.* [D0 Collaboration], Phys. Rev. D **84**, 112005 (2011)
27. R. S. Chivukula, E. H. Simmons and C. -P. Yuan, Phys. Rev. D **82**, 094009 (2010)
28. Y. Bai, J. L. Hewett, J. Kaplan and T. G. Rizzo, JHEP **1103**, 003 (2011)
29. R. S. Chivukula, A. Farzinnia, E. H. Simmons, R. Foadi and , Phys. Rev. D **85**, 054005 (2012) [arXiv:1111.7261 [hep-ph]].
30. G. Altarelli, R. K. Ellis, G. Martinelli, Nucl. Phys. **B157**, 461 (1979).
31. D. Binosi, J. Papavassiliou, Phys. Rept. **479**, 1-152 (2009). [arXiv:0909.2536 [hep-ph]].
32. N. Arkani-Hamed, A. G. Cohen, H. Georgi, Phys. Rev. Lett. **86**, 4757-4761 (2001). [hep-th/0104005].
33. C. T. Hill, S. Pokorski, J. Wang, Phys. Rev. **D64**, 105005 (2001). [hep-th/0104035].
34. M. Bando, T. Kugo, S. Uehara, K. Yamawaki, T. Yanagida, Phys. Rev. Lett. **54**, 1215 (1985).
35. M. Bando, T. Kugo, K. Yamawaki, Nucl. Phys. **B259**, 493 (1985).
36. R. S. Chivukula, A. Farzinnia, J. Ren, E. H. Simmons and , arXiv:1303.1120 [hep-ph].
37. A. Atre, R. S. Chivukula, P. Ittisamai, E. H. Simmons, J. -H. Yu and , Phys. Rev. D **86**, 054003 (2012) [arXiv:1206.1661 [hep-ph]].
38. M. Cvetic, P. Langacker and , Phys. Rev. D **46**, 4943 (1992) [Erratum-ibid. D **48**, 4484 (1993)] [hep-ph/9207216].
39. F. del Aguila, M. Cvetic, P. Langacker and , Phys. Rev. D **48**, 969 (1993) [hep-ph/9303299].
40. G. Aad *et al.* [ATLAS Collaboration], Phys. Rev. Lett. **105**, 161801 (2010) [arXiv:1008.2461 [hep-ex]].
41. V. Khachatryan *et al.* [CMS Collaboration], Phys. Rev. Lett. **105**, 211801 (2010) [arXiv:1010.0203 [hep-ex]].
42. S. Chatrchyan *et al.* [CMS Collaboration], Phys. Lett. B **704**, 123 (2011) [arXiv:1107.4771 [hep-ex]].
43. G. Aad *et al.* [ATLAS Collaboration], Phys. Lett. B **708**, 37 (2012) [arXiv:1108.6311 [hep-ex]].
44. Search for new phenomena in dijet mass and angular distributions using 4.8 fb-1 of pp collisions at sqrt(s) = 7 tev collected by the atlas detector. ATL-CONF-2012-038
45. K. Gumus, N. Akchurin, S. Esen and R. MHarris, CMS-NOTE-2006-070.
46. Measurement of the inclusive forward-backward asymmetry and its rapidity dependence afb(delta y) of tt production in 5.3 fb-1 of tevatron data, CDF-note-10185.
47. Measurement of the forward-backward production asymmetry of t and tbar quarks in ppbar to ttbar events, D0-Note 6062-CONF.
48. G. Rodrigo, P. Ferrario and , Nuovo Cim. C **033N4**, 221 (2010) [arXiv:1007.4328 [hep-ph]].
49. R. S. Chivukula, E. H. Simmons, N. Vignaroli and , arXiv:1302.1069 [hep-ph].
50. K. Agashe, M. Papucci, G. Perez and D. Pirjol, hep-ph/0509117.
51. T. Appelquist and J. Carazzone, Phys. Rev. D **11**, 2856 (1975).
52. ATLAS Collaboration, ATLAS-CONF-2012-148.

53. CMS Collaboration, CMS PAS EXO-12-016.
54. G. Aad *et al.* [ATLAS Collaboration], arXiv:1207.2409.
55. [CMS Collaboration], arXiv:1209.4397.
56. R. Contino, G. Servant, JHEP **0806** (2008) 026, arXiv:0801.1679 [hep-ph].
57. J. A. Aguilar Saavedra, JHEP **0911** (2009) 030, arXiv:0907.3155 [hep-ph].
58. J. Mrazek, A. Wulzer, Phys. Rev. D **81** (2010) 075006, arXiv:0909.3977 [hep-ph].
59. N. Vignaroli, JHEP **1207** (2012) 158, arXiv:1204.0468 [hep-ph].
60. ATLAS Collaboration, ATLAS-CONF-2012-130.

Studies on the UV to IR Evolution of Gauge Theories and Quasiconformal Behavior

R. Shrock*

C. N. Yang Institute for Theoretical Physics, Stony Brook University, Stony Brook, NY 11794, USA
and Department of Physics, Sloane Laboratory
*Yale University, New Haven, CT 06520, USA *E-mail: robert.shrock@stonybrook.edu*

We describe recent results from our studies of the UV to IR evolution of asymptotically free vectorial gauge theories and quasiconformal behavior. These include higher-loop calculations of the IR zero of the beta function and of the anomalous dimension of the fermion bilinear. Effects of scheme-dependence of higher-loop results are assessed in detail. Applications to models with dynamical electroweak symmetry breaking are discussed.

Keywords: UV to IR evolution, IR fixed point, quasiconformal behavior

1. Introduction

The evolution of an asymptotically free gauge theory from the ultraviolet (UV) to the infrared (IR) is of fundamental field-theoretic interest. In this SCGT12 talk we report on our calculations in[1–3] with T. A. Ryttov of higher-loop corrections to this UV to IR evolution and on some new results (submitted shortly after the SCGT12 workshop in[4,5]). The dependence of the gauge coupling $g(\mu)$ on the Euclidean momentum scale, μ, is determined by the β function $\beta \equiv \beta_\alpha \equiv d\alpha/dt$, where $t = \ln \mu$ and $\alpha(\mu) = g(\mu)^2/(4\pi)$ (and μ will often be suppressed in the notation). This function has the series expansion

$$\beta = -2\alpha \sum_{\ell=1}^{\infty} b_\ell \, a^\ell = -2\alpha \sum_{\ell=1}^{\infty} \bar{b}_\ell \, \alpha^\ell \ , \tag{1}$$

where $a = \alpha/(4\pi)$, ℓ denotes the number of loops involved in the calculation of the coefficient b_ℓ, and $\bar{b}_\ell = b_\ell/(4\pi)^\ell$. The first two coefficients in β, b_1 and b_2, are scheme-independent and were calculated in.[6,7] The b_ℓ have been calculated up to 4-loop order in the $\overline{MS}$ scheme in.[8,9] The n-loop ($n\ell$) beta function, denoted $\beta_{n\ell}$, is given by the RHS of Eq. (1) with the upper bound on the sum set equal to n instead of ∞.

We consider the UV to IR evolution of an asymptotically free vectorial gauge theory with gauge group G and N_f massless fermions transforming according to a representation R of G. There are two possibiities for this evolution: (i) there may not be any IR zero in β, so that as μ decreases, $\alpha(\mu)$ increases, eventually beyond the perturbatively calculable region (which is the case for QCD); (ii) β may have an IR zero at a certain value denoted α_{IR}, so that as μ decreases, $\alpha(\mu)$ increases from 0 toward α_{IR}. In this class of theories, there are two further generic possibilities, namely $\alpha_{IR} < \alpha_{cr}$ or $\alpha_{IR} > \alpha_{cr}$, where α_{cr} is the critical minimal value of α (depending on R) for spontaneous chiral symmetry breaking (SχSB). If $\alpha_{IR} < \alpha_{cr}$, then the zero of β at α_{IR} is an exact IR fixed point (IRFP) of the renormalization group; as $\mu \to 0$ and $\alpha \to \alpha_{IR}$, $\beta \to \beta(\alpha_{IR}) = 0$, and the theory becomes exactly scale-invariant with nontrivial anomalous dimensions.[7,10]

If β has no IR zero, or an IR zero at $\alpha_{IR} > \alpha_{cr}$, then as μ decreases through a scale denoted Λ, $\alpha(\mu)$ exceeds α_{cr} and spontaneous chiral symmetry breaking occurs, so that the fermions gain dynamical masses $\sim \Lambda$. In this case, in the low-energy effective field theory applicable for $\mu < \Lambda$, one integrates these fermions out, and the β function becomes that of a pure gauge theory, which has no (perturbative) IR zero. Hence, if β has a zero at $\alpha_{IR} > \alpha_{cr}$, this is only an approximate IRFP. If $\alpha_{IR} > \alpha_{cr}$, the effect of the approximate IRFP at α_{IR} depends on how close it is to α_{cr}. The desire to understand better both quantum chromodynamics (QCD) and the properties of this IR zero have motivated calculations of higher-loop terms in β[8,9] and higher-loop corrections to the 2-loop result for the IR zero.[1,11,12] We denote the IR zero of the n-loop beta function, $\beta_{n\ell}$, as $\alpha_{IR,n\ell}$. The need to go beyond the 2-loop result for α_{IR} in studies of quasiconformal theories is evident from the fact that $\alpha_{cr} \sim O(1)$ and one is interested in values of α_{IR} near to α_{cr}. Although the coefficients in β at $\ell \geq 3$ loop order are scheme-dependent, the results give a measure of accuracy of the 2-loop calculation of the IR zero, and similarly with the value of the anomalous dimension of the fermion bilinear (discussed below).

If α_{IR} is only slightly greater than α_{cr}, then, as $\alpha(\mu)$ approaches α_{IR}, since $\beta = d\alpha/dt \to 0$, $\alpha(\mu)$ varies very slowly as a function of the scale μ, i.e., there is approximately scale-invariant (equivalently, dilatation-invariant, slow-running, or "walking") behavior.[13] For these theories, this is equivalent to quasiconformal behavior.[14] The spontaneous chiral symmetry breaking and attendant fermion mass generation at Λ spontaneously break the approximate dilatation symmetry, plausibly leading to a resultant light Nambu-Goldstone boson, the dilaton.[15] The dilaton is not massless, because β is not exactly zero for $\alpha(\mu) \neq \alpha_{IR}$. Studies of gauge-singlet bound states in confining quasiconformal gauge theories, using Schwinger-Dyson and Bethe-Salpeter equations, find a significant reduction in the scalar mass divided by the vector meson mass.[16] Eventually, lattice gauge measurements may be able to determine the mass of a dilaton in a theory of this type. The quasiconformal behavior associated with an approximate IRFP has been an ingredient of DEWSB models since the 1980s, as a means of enhancing Standard-Model (SM) fermion masses while keeping neutral flavor-changing current (FCNC) processes sufficiently suppressed.[13]

2. Basic Properties of β

Since $b_1 = (1/3)(11C_A - 4N_fT_f)$,[6,19] the asymptotic freedom of the theory requires $N_f < N_{f,b1z}$, where $N_{f,b1z} = 11C_A/(4T_f)$. (The subscript $b\ell z$ stands for "b_ℓ zero".) Since $\beta_{2\ell} = -[\alpha^2/(2\pi)](b_1 + b_2 a)$, this function has an IR zero at

$$\alpha_{IR,2\ell} = -\frac{4\pi b_1}{b_2} , \tag{2}$$

which is physical for $b_2 < 0$. Now b_2 decreases linearly as a function of N_f; for small N_f, $b_2 > 0$, but as N_f increases through the value

$$N_{f,b2z} = \frac{34C_A^2}{4T_f(5C_A + 3C_f)} , \tag{3}$$

b_2 reverses its sign and becomes negative. Since $N_{f,b2z} < N_{f,b1z}$, there is always an interval of $N_f < N_{f,b1z}$ for which β has an IR zero, namely the interval

$$I: \quad N_{f,b2z} < N_f < N_{f,b1z} \ . \tag{4}$$

If R is the fundamental (fund.) representation of SU(N), then

$$I: \quad \frac{34N^3}{13N^2 - 3} < N_f < \frac{11N}{2} , \qquad R = \text{fund.} \tag{5}$$

For example, for $N = 2$ and $N = 3$, these intervals are $5.55 < N_f < 11$ and $8.05 < N_f < 16.5$, respectively. As $N \to \infty$, the interval I is $2.62N < N_f < 5.5N$. Here and below, when an expression is given for N_f that formally evaluates to a non-integral real value ν, it is understood implicitly that one infers an appropriate integral value of N_f from it.

If $N_f \in I$ is near to $N_{f,b1z}$, so that $\alpha_{IR,2\ell}$ is small, then the theory evolves from the UV to the IR without any spontaneous chiral symmetry breaking. As N_f decreases in I, $\alpha_{IR,2\ell}$ increases and, as N_f decreases through a critical value denoted $N_{f,cr}$, the IR zero of β increases through α_{cr}, so that the UV to IR evolution leads to SχSB. For N_f near lower end of I, $b_2 \to 0$ and $\alpha_{IR,2\ell}$ is too large for the 2-loop calculation to be reliable.

3. Calculations of Higher-Loop Corrections to UV to IR Evolution

In this section we discuss our calculations of higher-loop corrections to the UV to IR evolution of a gauge theory and, in particular, higher-loop calculations of the IR zero of the beta function. We first recall our 3-loop calculation of the IR zero of β in the $\overline{MS}$ scheme. The 3-loop coefficient in β, b_3, is a quadratic function of N_f and vanishes, with sign reversal, at two values of N_f, denoted $N_{f,b3z,1}$ and $N_{f,b3z,2}$. Now b_3 is positive for small N_f and vanishes first at $N_{f,b3z,1}$, which is smaller than $N_{f,b2z}$, the left endpoint of the interval I. Furthermore, $N_{f,b3z,2} > N_{f,b1z}$, the right endpoint of I. For example, for $N = 2$, $N_{f,b3z,1} = 3.99 < N_{f,b2z} = 5.55$ and $N_{f,b3z,2} = 27.6 > N_{f,b1z} = 11$, while for $N = 3$, $N_{f,b3z,1} = 5.84 < N_{f,b2z} = 8.05$ and $N_{f,b3z,2} = 40.6 > N_{f,b1z} = 16.5$. Hence, $b_3 < 0$ for $N_f \in I$, the interval of interest for the IR zero of β. At this 3-loop level, $\beta_{3\ell} = -[\alpha^2/(2\pi)](b_1 + b_2 a + b_3 a^2)$, so β vanishes away from $\alpha = 0$ at two values. In terms of α, these are

$$\alpha = \frac{2\pi}{b_3}\Big(- b_2 \pm \sqrt{b_2^2 - 4b_1 b_3}\ \Big) \tag{6}$$

Since $b_2 < 0$ in I and the $\overline{MS}$ $b_3 < 0$ in I also, this can be expressed in terms of positive quantities as

$$\alpha = \frac{2\pi}{|b_3|}\Big(- |b_2| \mp \sqrt{b_2^2 + 4b_1|b_3|}\ \Big) \ . \tag{7}$$

One of these solutions is negative and hence is unphysical; the other is manifestly positive, and is $\alpha_{IR,3\ell}$. Using this result, we proved in[1] that in the $\overline{MS}$ scheme, $\alpha_{IR,3\ell} < \alpha_{IR,2\ell}$.

A natural question that arises from the analysis in[1] is how general this inequality is and, specifically, whether it also holds for other schemes. We address and answer this question here.[4] To do this, we observe that if a scheme had $b_3 > 0$ in I, then, since $b_2 \to 0$ at the lower end of I, it would follow that $b_2^2 - 4b_1 b_3 < 0$, so this scheme would not have a physical (real, positive) $\alpha_{IR,3\ell}$ in this region. Since the existence of the IR zero in $\beta_{2\ell}$ is a scheme-independent property, one may require that an acceptable scheme should preserve this to higher-loop order. To satisfy this requirement, by the argument above, it should have $b_3 < 0$ for $N_f \in I$, as is the case in the $\overline{MS}$ scheme. We now prove that in all such schemes,

$$\alpha_{IR,3\ell} < \alpha_{IR,2\ell} \ . \tag{8}$$

To prove this, we consider the difference

$$\alpha_{IR,2\ell} - \alpha_{IR,3\ell} = \frac{2\pi}{|b_2 b_3|}\bigg[2b_1|b_3| + b_2^2 - |b_2|\sqrt{b_2^2 + 4b_1|b_3|}\ \bigg] \ . \tag{9}$$

The expression in square brackets is positive if and only if

$$(2b_1|b_3| + b_2^2)^2 - b_2^2(b_2^2 + 4b_1|b_3|) > 0 \ . \tag{10}$$

This difference is equal to the nonnegative quantity $(2b_1 b_3)^2$, which proves the inequality. Note that, since $b_1 > 0$ from the asymptotic freedom property, this difference vanishes if and only if $b_3 = 0$, in which case $\alpha_{IR,3\ell} = \alpha_{IR,2\ell}$. This proof holds for general G, R, and $N_f \in I$.

As noted above, $\alpha_{IR,2\ell}$ is a monotonically decreasing function of $N_f \in I$. With $b_3 < 0$ for $N_f \in I$, this monotonicity property is also true of $\alpha_{IR,3\ell}$. As N_f increases from $N_{f,b2z}$ to $N_{f,b1z}$ in the interval I, $\alpha_{IR,3\ell}$ decreases from

$$\alpha_{IR,3\ell} = 4\pi\sqrt{\frac{b_1}{|b_3|}} \quad \text{at } N_f = N_{f,b2z} \tag{11}$$

to zero at the upper end of the interval I, vanishing like

$$\alpha_{IR,3\ell} = \frac{4\pi b_1}{|b_2|}\bigg[1 - \frac{|b_3|b_1}{|b_2|^2} + O(b_1^2) \bigg] \quad \text{as } N_f \nearrow N_{f,b1z} \ . \tag{12}$$

For the 4-loop analysis of β, we use b_4, which is a cubic polynomial in N_f. This coefficient is positive for $N_f \in I$ for $N = 2, 3$ but is negative in part of I for higher N. The 4-loop β function is $\beta_{4\ell} = -[\alpha^2/(2\pi)](b_1 + b_2 a + b_3 a^2 + b_4 a^3)$, so β has three zeros away from the origin. We determine the smallest positive real zero as $\alpha_{IR,4\ell}$.

In addition to the inequality (8), we find the following general results: (i) going from the 3-loop to 4-loop level, there is a slight change in the value of the IR zero, but this change is smaller than the decrease from the 2-loop to 3-loop level, so $\alpha_{IR,4\ell} < \alpha_{IR,2\ell}$; and (ii) fractional changes in the value of the IR zero of β decrease in magnitude as N_f increases toward $N_{f,b1z}$, and all of the values of $\alpha_{IR,n\ell} \to 0$. Our finding that the fractional change in the location of the IR zero of β is reduced at higher-loop order agrees with the general expectation that calculating a quantity to higher order in perturbation theory should give a more stable and accurate result. Since $\alpha_{cr} \sim O(1)$, the decrease in α_{IR} at higher-loop order, together with the property that α_{IR} increases as N_f decreases, means that one must go to smaller N_f for $\alpha_{IR,n\ell}$ to grow to a given size for the $n = 3$ and $n = 4$ loop level as compared with the $n = 2$ loop level. This suggests that the actual lower boundary of the IR-conformal phase could lie somewhat below an estimate obtained setting $\alpha_{IR,2\ell} = \alpha_{cr}$.

Some numerical values of $\alpha_{IR,n\ell}$ at the 2-loop, 3-loop, and 4-loop level for fermions in the fundamental representation, $N_f \in I$, and the illustrative groups $G = \mathrm{SU}(2)$ and $G = \mathrm{SU}(3)$ from[1] are given in Table 1.

Table 1. Values of $\alpha_{IR,n\ell}$ for the beta function of an SU(N) gauge theory with N_f fermions in the fundamental representation, for $N = 2$, 3 and $N_f \in I$.

N	N_f	$\alpha_{IR,2\ell}$	$\alpha_{IR,3\ell}$	$\alpha_{IR,4\ell}$
2	7	2.83	1.05	1.21
2	8	1.26	0.688	0.760
2	9	0.595	0.418	0.444
2	10	0.231	0.196	0.200
3	10	2.21	0.764	0.815
3	11	1.23	0.578	0.626
3	12	0.754	0.435	0.470
3	13	0.468	0.317	0.337
3	14	0.278	0.215	0.224
3	15	0.143	0.123	0.126
3	16	0.0416	0.0397	0.0398

Corresponding higher-loop calculations were carried out in[1] for SU(N) gauge theories with N_f fermions in the adjoint, symmetric and antisymmetric rank-2 tensor representations. The general result $\alpha_{IR,3\ell} < \alpha_{IR,2\ell}$ continues to apply. The difference $\alpha_{IR,4\ell} - \alpha_{IR,3\ell}$ tends to be relatively small, but can have either sign. For example, for R = adjoint, $N_{f,b1z} = 11/4$ and $N_{f,b2z} = 17/16$ (independent of N), so the interval I where β has an IR zero, viz., $N_{f,b2z} < N_f < N_{f,b1z}$, is

$1.06 < N_f < 2.75$. This includes only one physical, integral value, $N_f = 2$. For this value of N_f and $N = 2$, one has $\alpha_{IR,2\ell} = 0.628$, $\alpha_{IR,3\ell} = 0.459$, $\alpha_{IR,4\ell} = 0.450$, while for $N = 3$, $\alpha_{IR,2\ell} = 0.419$, $\alpha_{IR,3\ell} = 0.306$, $\alpha_{IR,4\ell} = 0.308$.

4. Anomalous Dimension of Fermion Bilinear

The anomalous dimension $\gamma_m \equiv \gamma$ for the fermion bilinear operator has the series expansion

$$\gamma = \sum_{\ell=1}^{\infty} c_\ell a^\ell = \sum_{\ell=1}^{\infty} \bar{c}_\ell \alpha^\ell \, , \tag{13}$$

where $\bar{c}_\ell = c_\ell/(4\pi)^\ell$ is the ℓ-loop coefficient. The one-loop coefficient c_1 is scheme-independent, while the c_ℓ with $\ell \geq 2$ are scheme-dependent. The c_ℓ have been calculated up to 4-loop level in the $\overline{MS}$ scheme.[20] The first two coefficients are $c_1 = 6C_f$ and $c_2 = 2C_f[(3/2)C_f + (97/6)C_A - (10/3)T_f N_f]$.

It is of interest to calculate γ at the exact IRFP in the IR-conformal phase and at the approximate IRFP in the phase with spontaneous chiral symmetry breaking. We denote γ calculated to n-loop ($n\ell$) level as $\gamma_{n\ell}$ and, evaluated at the IR zero of $\beta_{n\ell}$, as $\gamma_{IR,n\ell} \equiv \gamma_{n\ell}(\alpha = \alpha_{IR,n\ell})$. In the IR-conformal phase, an exact calculation of γ evaluated at the IRFP would be an exact (scheme-independent) property of the theory, but in the broken phase, just as the IR zero of β is only an approximate IRFP, so also, γ is only approximate, describing the running of $\bar{\psi}\psi$ and the dynamically generated fermion mass near the zero of β, according to

$$\Sigma(k) \sim \Lambda\Big(\frac{\Lambda}{k}\Big)^{2-\gamma} \tag{14}$$

for Euclidean momenta $k >> \Lambda$. This, in turn, affects SM fermion masses in DEWSB theories.[13] In both the conformal and the chirally broken IR phases, the upper bound $\gamma < 2$ holds.

At the 2-loop level we calculate

$$\gamma_{IR,2\ell} = \frac{C_f(11C_A - 4T_f N_f)[455C_A^2 + 99C_A C_f + (180C_f - 248C_A)T_f N_f + 80(T_f N_f)^2]}{12[-17C_A^2 + 2(5C_A + 3C_f)T_f N_f]^2} \tag{15}$$

Our analytic expressions for $\gamma_{IR,n\ell}$ at the 3-loop and 4-loop level are more complicated.[1] Illustrative numerical values of $\gamma_{IR,n\ell}$ at the 2-, 3-, and 4-loop level are given below in Table 2 and Figs. 1 and 2 for $R = fund.$, and the illustrative values $N = 2,\ 3$. (Values in parentheses violate the upper bound $\gamma < 2$ and reflect the inadequacy of perturbation theory if α is too large.)

We have also performed these calculations for $G = \mathrm{SU}(N)$ and higher fermion representations R. In general, we find that, for a given N, R, and N_f, the values of $\gamma_{IR,n\ell}$ calculated to 3-loop and 4-loop order are smaller than the 2-loop value. The value of these higher-loop calculations is evident from the figures. A necessary condition for a perturbative calculation to be reliable is that higher-order contributions do not modify the result too much. One sees from the tables and figures that,

Table 2. Values of $\gamma_{IR,n\ell}$ for an SU(N) gauge theory with N_f fermions in the fundamental representation, for $N = 2, 3$ and $N_f \in I$.

N	N_f	$\gamma_{IR,2\ell}$	$\gamma_{IR,3\ell}$	$\gamma_{IR,4\ell}$
2	7	(2.67)	0.457	0.0325
2	8	0.752	0.272	0.204
2	9	0.275	0.161	0.157
2	10	0.0910	0.0738	0.0748
3	10	(4.19)	0.647	0.156
3	11	1.61	0.439	0.250
3	12	0.773	0.312	0.253
3	13	0.404	0.220	0.210
3	14	0.212	0.146	0.147
3	15	0.0997	0.0826	0.0836
3	16	0.0272	0.0258	0.0259

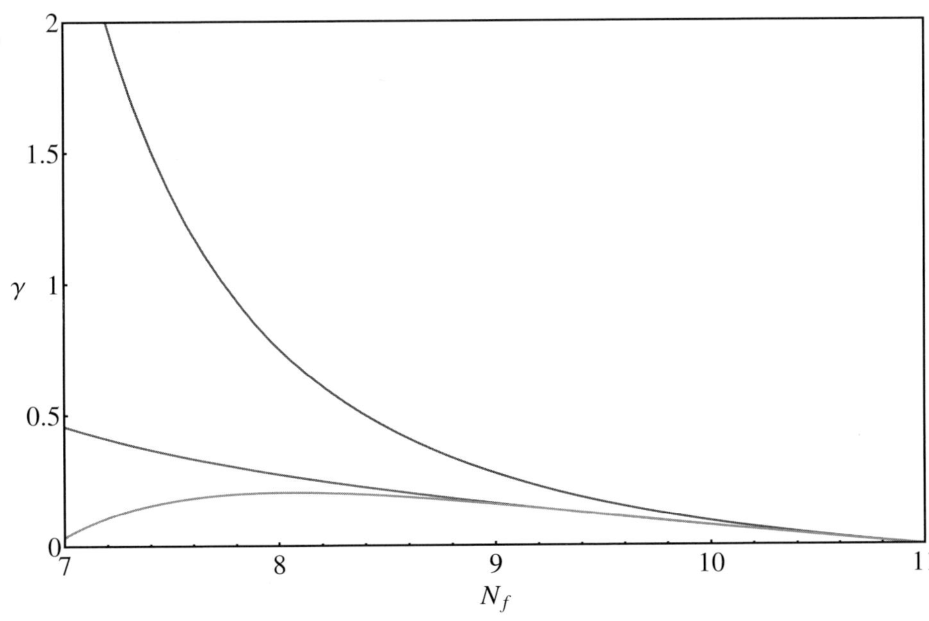

Fig. 1. γ_{IR} for SU(2) and fermions in the fundamental representation. From top to bottom, the curves show $\gamma_{IR,2\ell}$, $\gamma_{IR,3\ell}$, and $\gamma_{IR,4\ell}$.

especially for smaller N_f, there is a substantial decrease in $\alpha_{IR,n\ell}$ and $\gamma_{IR,n\ell}$ when one goes from 2-loop to 3-loop order, but for a reasonable range of N_f, the 3-loop and 4-loop results are close to each other. Thus, our higher-loop calculations of α_{IR} and γ allow us to probe the theory reliably down to smaller values of N_f and thus stronger couplings.

It is useful to give a comparison of our calculations with lattice measurements.

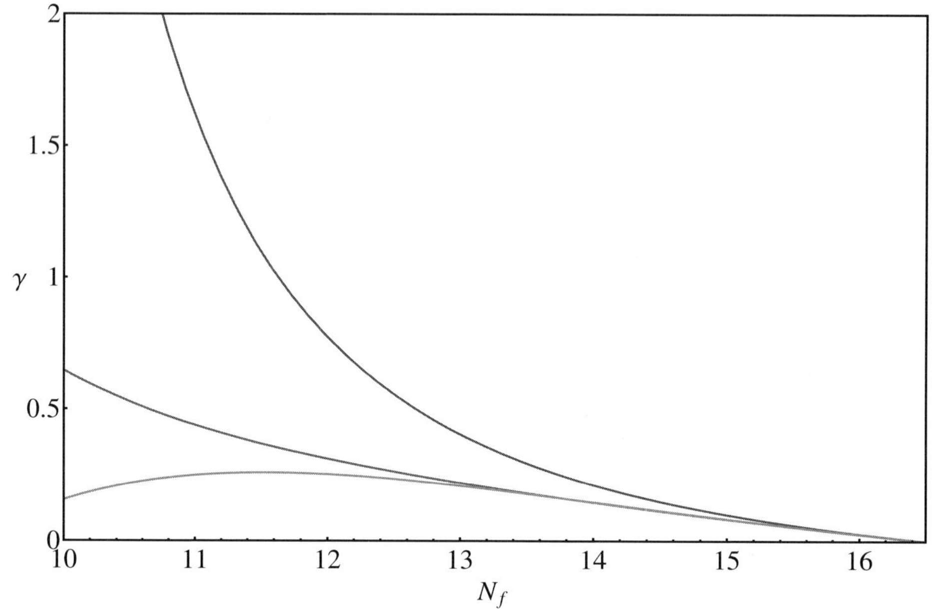

Fig. 2. γ_{IR} for SU(3) and fermions in the fundamental representation. From top to bottom, the curves show $\gamma_{IR,2\ell}$, $\gamma_{IR,3\ell}$, and $\gamma_{IR,4\ell}$.

The theory with SU(3), $R = fund.$, and $N_f = 12$ has been the subject of intensive lattice study, so we focus on this for the comparison. For this theory we calculate (see Table 2) $\gamma_{IR,2\ell} = 0.77$, $\gamma_{IR,3\ell} = 0.31$, and $\gamma_{IR,4\ell} = 0.25$ (to two significant figures). Lattice results include $\gamma = 0.414 \pm 0.016$;[21] $\gamma \sim 0.35$;[22] $0.2 \leq \gamma \leq 0.4$;[23] $\gamma = 0.4 - 0.5$;[24] and $\gamma = 0.27 \pm 0.03$.[25] Here the 2-loop value of γ is larger than, and the 3-loop and 4-loop values are closer to, these lattice measurements. This illustrates how higher-loop calculations of γ can improve agreement with lattice measurements. Note that there is not yet a consensus among lattice groups as to whether this theory has an IR phase with chiral symmetry or spontaneous chiral symmetry breaking. Similar comparisons can be given for other values of G, R and N_f. In particular, for SU(3), $R = fund.$, $N_f = 10$, one group obtains $\gamma_{IR} \sim 1$,[26] consistent with the idea that $\gamma_{IR} \sim 1$ at the lower end of the IR-conformal phase. This region is difficult to probe perturbatively because of the strongly coupled nature of the physics.

An interesting property of the values of $\alpha_{IR,n\ell}$ and $\gamma_{IR,n\ell}$ in the case where $R = fund.$ is that these are similar in theories with different values of N and N_f, provided that these theories have the same or similar values of the ratio N_f/N. This can be understood as a result of a rapid approach to the 't Hooft-Veneziano limit $N \to \infty$, $N_f \to \infty$ with N_f/N fixed.[1,4,5]

5. Supersymmetric Gauge Theory

It is valuable to carry out a similar analysis in an asymptotically free $\mathcal{N} = 1$ supersymmetric gauge theory with vectorial chiral superfield content Φ_i, $\tilde{\Phi}_i$, $i = 1, ..., N_f$ in the R, $\bar{R}$ reps., respectively.[2] An appeal of this analysis is that exact results on the IR properties of the theory are known.[27] Thus, one can compare results from higher-loop perturbative calculations with exact results, in particular, for $N_{f,cr}$. The coefficients of β and γ have been calculated up to 3-loop order.[28] The one-loop coefficient is $b_{1,s} = 3C_A - 2T_f N_f$. (Here and below, the subscript s, for "supersymmetric", is appended to distinguish these quantities from those for the nonsupersymmetric theory). Asymptotic freedom requires that $N_f < N_{f,b1z,s}$, where $N_{f,b1z,s} = 3C_A/(2T_f)$. The 2-loop coefficient of β_s is $b_{2,s} = 6C_A^2 - 4T_f N_f(C_A + 2C_f)$, which is positive for small N_f, but vanishes with sign reversal as N_f increases through the value

$$N_{f,b2z,s} = \frac{3C_A^2}{2T_f(C_A + 2C_f)} , \tag{16}$$

which is smaller than $N_{f,b1z,s}$. Thus, for this theory, there is again an interval I_s in which $\beta_{2\ell}$ has an IR zero, namely

$$I_s : \quad N_{f,b2z,s} < N_f < N_{f,b1z,s} . \tag{17}$$

For $R = fund.$ this interval I_s is $3N^3/(2N^2 - 1) < N_f < 3N$. In general, this IR zero of $\beta_{2\ell,s}$ occurs at

$$\alpha_{IR,2\ell,s} = \frac{2\pi(3C_A - 2T_f N_f)}{2T_f N_f(C_A + 2C_f) - 3C_A^2} . \tag{18}$$

The 3-loop coefficient $b_{3,s}$ is positive for small N_f and vanishes at two values of N_f, denoted $N_{f,b3z,1,s}$ and $N_{f,b3z,2,s}$. As before, we find that $N_{f,b3z,1,s} < N_{f,b2z,s}$ and $N_{f,b3z,2,s} > N_{f,b1z,s}$, so $b_{3,s} < 0$ for $N_f \in I_s$. Since $b_{3,s} < 0$ for $N_f \in I$, we find, by the same type of proof as given above, that for any G, R, and $N_f \in I_s$

$$\alpha_{IR,3\ell,s} < \alpha_{IR,2\ell,s} . \tag{19}$$

For fixed N, we find that $\alpha_{IR,n\ell}$ increases monotonically with decreasing N_f at both the 2-loop and 3-loop level.

Next, we analyze the anomalous dimension γ of the superfield operator product $\Phi\tilde{\Phi}$ containing the term $\theta\theta\psi\tilde{\psi}$. In a conformally invariant d-dimensional field theory (whether supersymmetric or not), unitarity yields a lower bound on the dimension $D_\mathcal{O}$ of a spin-0 operator $\mathcal{O}$ (other than the identity), namely, $D_\mathcal{O} \geq (d-2)/2$, where d = spacetime dim.; so $D_\mathcal{O} \geq 1$ here.[29] In the nonsupersymmetric theory, with $\dim(\bar{\psi}\psi) = 3 - \gamma$, this constraint is $D_{\bar{\psi}\psi} = 3 - \gamma > 1$, so $\gamma < 2$. In the supersymmetric theory, with $\dim(\theta) = -1/2$ and $\dim(\psi\tilde{\psi}) = 3 - \gamma$, the constraint is $D_{\Phi\tilde{\Phi}} = -1 + 3 - \gamma > 1$, so $\gamma < 1$.

At the 2-loop level, we find

$$\gamma_{IR,2\ell,s} = \frac{C_f(3C_A - 2T_fN_f)(2T_fN_f - C_A)(2T_fN_f - 3C_A + 6C_f)}{[2(C_A + 2C_f)T_fN_f - 3C_A^2]^2} \ . \tag{20}$$

We have also calculated $\gamma_{IR,3\ell}$ and find that, as in the nonsupersymmetric case, $\gamma_{IR,3\ell} < \gamma_{IR,2\ell}$. Let us focus on the case $R = fund.$, for which $N_{f,b1z,s} = 3N$ and $N_{f,cr,s} = (3/2)N$. One perturbative estimate of $N_{f,cr}$ can be obtained by assuming that the upper bound $\gamma \le 1$ is saturated as $N_f \searrow N_{f,cr,s}$. Solving this, we obtain that for $N = 2$, $N_{f,cr,s,est.} = 4.24$, a factor of 1.41 larger than exact $N_{f,cr,s} = 3$, while for $N = 3$ $N_{f,cr,s,est.} = 6.15$ a factor of 1.37 larger than exact $N_{f,cr,s} = 4.5$. As $N \to \infty$, this procedure yields $N_{f,cr,s,est.} \to 2N$, a factor of (4/3) larger than the exact result, $N_{f,cr,s} = (3/2)N$.

This comparison for the $\mathcal{N} = 1$ supersymmetric gauge theory suggests that the perturbative calculation slightly overestimates the value of $N_{f,cr}$, i.e., slightly underestimates the size of the IR-conformal phase, similar to what we found for the nonsupersymmetric theory.

6. Scheme-Dependence in Calculation of IR Fixed Point

Since the coefficients in β at the level of $n \ge 3$ loops are scheme-dependent, so is the resultant value of $\alpha_{IR,n\ell}$. It is important to assess quantitatively the uncertainty in the analysis of the UV to IR evolution due to this scheme dependence. A way to do this is to perform scheme transformations and determine how much of a change there is in $\alpha_{IR,n\ell}$.[3]

A scheme transformation (ST) is a map between α and α' or equivalently, a and a', where $a = \alpha/(4\pi)$, which can be expressed as

$$a = a'f(a') \tag{21}$$

with $f(0) = 1$ to keep the UV properties unchanged. We write

$$f(a') = 1 + \sum_{s=1}^{s_{max}} k_s(a')^s = 1 + \sum_{s=1}^{s_{max}} \bar{k}_s(\alpha')^s \ , \tag{22}$$

where the k_s are constants, $\bar{k}_s = k_s/(4\pi)^s$, and s_{max} may be finite or infinite. Hence, the Jacobian

$$J = \frac{da}{da'} = \frac{d\alpha}{d\alpha'} \tag{23}$$

satisfies $J = 1$ at $a = a' = 0$. We have

$$\beta_{\alpha'} \equiv \frac{d\alpha'}{dt} = \frac{d\alpha'}{d\alpha}\frac{d\alpha}{dt} = J^{-1}\beta_\alpha \ . \tag{24}$$

This has the expansion

$$\beta_{\alpha'} = -2\alpha' \sum_{\ell=1}^{\infty} b'_\ell(a')^\ell = -2\alpha' \sum_{\ell=1}^{\infty} \bar{b}'_\ell(\alpha')^\ell \ , \tag{25}$$

where $\bar{b}'_\ell = b'_\ell/(4\pi)^\ell$.

Using these two equivalent expressions for $\beta_{\alpha'}$, one can solve for the b'_ℓ in terms of the b_ℓ and k_s. This yields the well-known result that $b'_1 = b_1$ and $b'_2 = b_2$. To assess the scheme-dependence of an IRFP, we have calculated the relations between the b'_ℓ and b_ℓ for higher ℓ values.[3] For example, for $\ell = 3,\ 4,\ 5$, we obtain

$$b'_3 = b_3 + k_1 b_2 + (k_1^2 - k_2)b_1 \ , \tag{26}$$

$$b'_4 = b_4 + 2k_1 b_3 + k_1^2 b_2 + (-2k_1^3 + 4k_1 k_2 - 2k_3)b_1 \tag{27}$$

$$b'_5 = b_5 + 3k_1 b_4 + (2k_1^2 + k_2)b_3 + (-k_1^3 + 3k_1 k_2 - k_3)b_2$$
$$+ (4k_1^4 - 11k_1^2 k_2 + 6k_1 k_3 + 4k_2^2 - 3k_4)b_1 \ . \tag{28}$$

To be physically acceptable, a ST must satisfy several conditions, C_i:

- C_1: the ST must map a real positive α to a real positive α', since a map taking $\alpha > 0$ to $\alpha' = 0$ would be singular, and a map taking $\alpha > 0$ to a negative or complex α' would violate the unitarity of the theory.
- C_2: the ST should not map a moderate value of α, for which perturbative calculations may be reliable, to an excessively large value of α' where perturbative calculations are inapplicable.
- C_3: J should not vanish in the region of α and α' of interest, or else there would be a pole in the relation between β_α and $\beta_{\alpha'}$.
- C_4: The existence of an IR zero of β is a scheme-independent property, depending (in an asymptotically free theory) only on the condition that $b_2 < 0$. Hence, a ST should satisfy the condition that β_α has an IR zero if and only if $\beta_{\alpha'}$ has an IR zero.

These four conditions can always be satisfied by scheme transformations near a UV fixed point, and hence in applications to perturbative QCD calculation, since α is small, and one can choose the k_s to be small also, so $\alpha' \simeq \alpha$. However, these conditions C1-C4 are not automatically satisfied, and are a significant constraint, on a scheme transformation applied in the vicinity of an IRFP, where α may be O(1). For example, consider the scheme transformation

$$\alpha = \tanh(\alpha') \tag{29}$$

with inverse

$$\alpha' = \frac{1}{2}\ln\left(\frac{1+\alpha}{1-\alpha}\right) \ . \tag{30}$$

If $\alpha << 1$, as at a UVFP, this is acceptable, but as α approaches 1 from below it maps a moderate value of α to an arbitrarily large α' and hence fails condition C2, and if α exceeds 1, even if by a small amount, then it fails conditions C1 and C4, since it maps a real positive α to a complex α'.

We have studied the scheme dependence of the IR zero of β in[3] using several scheme transformations; e.g., the ST (depending on a parameter r)

$$S_{sh,r}: \quad a = \frac{\sinh(ra')}{r} \tag{31}$$

Since $\sinh(ra')/r$ is an even fn. of r, we take $r > 0$ with no loss of generality. This transformation has the inverse

$$a' = \frac{1}{r} \ln \left[ra + \sqrt{1+(ra)^2} \right] \tag{32}$$

and the Jacobian $J = \cosh(ra')$. For this ST,

$$f(a') = \frac{\sinh(ra')}{ra'} \ . \tag{33}$$

This $f(a')$ has a series expansion with $k_s = 0$ for odd s and, for even s,

$$k_2 = \frac{r^2}{6} \ , \quad k_4 = \frac{r^4}{120} \ , \quad k_6 = \frac{r^6}{5040} \ , \quad k_8 = \frac{r^8}{362880} \ , \tag{34}$$

etc. for higher s. Substituting these results for k_s into the equations for the b'_ℓ, we obtain

$$b'_3 = b_3 - \frac{r^2 b_1}{6} \ , \quad b'_4 = b_4 \ , \tag{35}$$

$$b'_5 = b_5 + \frac{r^2 b_3}{6} + \frac{31 r^4 b_1}{360} \ , \tag{36}$$

and so forth for higher ℓ.

We apply this S_{sh_r} ST to the β function in the $\overline{MS}$ scheme, calculated up to $\ell = 4$ loop level. For $N_f \in I$ where $\beta_{2\ell}$ has an IR zero, we then calculate the resultant IR zeros in $\beta_{\alpha'}$ at the 3-loop and 4-loop order and compare the values with those in the $\overline{MS}$ scheme. We list some numerical results for illustrative values of r and for $N = 2, \ 3$ below. We denote the IR zero of $\beta_{\alpha'}$ at the n-loop level as $\alpha'_{IR,n\ell} \equiv \alpha'_{IR,n\ell,r}$. For example, for $N = 3$, $N_f = 10$, $\alpha_{IR,2\ell} = 2.21$, and:

$$\alpha_{IR,3\ell,\overline{MS}} = 0.764, \quad \alpha'_{IR,3\ell,r=3} = 0.762, \quad \alpha'_{IR,3\ell,r=6} = 0.754, \tag{37}$$

$$\alpha'_{IR,3\ell,r=9} = 0.742, \quad \alpha'_{IR,3\ell,r=4\pi} = 0.723 \tag{38}$$

$$\alpha_{IR,4\ell,\overline{MS}} = 0.815, \quad \alpha'_{IR,4\ell,r=3} = 0.812, \quad \alpha'_{IR,4\ell,r=6} = 0.802, \tag{39}$$

$$\alpha'_{IR,4\ell,r=9} = 0.786, \quad \alpha'_{IR,4\ell,r=4\pi} = 0.762 \ . \tag{40}$$

In general, the effect of scheme dependence tends to be reduced (i) for a given N and N_f, as one calculates to higher-loop order, and (ii) for a given N, as $N_f \to N_{f,b1z}$, so that the value of $\alpha_{IR} \to 0$. These results provide a quantitative measure of scheme dependence of the location of an IR zero of β.

Since the β function coefficients b_ℓ with $\ell \geq 3$ are scheme-dependent, there should exist a ST that renders these coefficients equal to zero (i.e., maps to the 't

Hooft scheme). We constructed an explicit ST that can does this at a UVFP. This necessarily has $s_{max} = \infty$. For simplicity, we set $k_1 = 0$. Our solutions for the first few k_s are

$$k_2 = \frac{b_3}{b_1} , \quad k_3 = \frac{b_4}{2b_1} , \tag{41}$$

$$k_4 = \frac{b_5}{3b_1} - \frac{b_2 b_4}{6b_1^2} + \frac{5b_3^2}{3b_1^2} , \tag{42}$$

$$k_5 = \frac{b_6}{4b_1} - \frac{b_2 b_5}{6b_1^2} + \frac{2b_3 b_4}{b_1^2} + \frac{b_2^2 b_4}{12b_1^3} - \frac{b_2 b_3^2}{12b_1^3} , \tag{43}$$

and so forth for higher s.

7. Application to Models of Dynamical Electroweak Symmetry Breaking

Models with dynamical EWSB have been of interest as one way to avoid the hierarchy (fine-tuning) problem with the Standard Model, while supersymmetry provides another way to do this. DEWSB models use an asymptotically free vectorial gauge interaction with a certain set of fermions subject to this gauge interaction, which becomes strong at the TeV scale, producing bilinear condensates of these fermions. These models involve further ingredients to give masses to SM fermions. The running mass $m_{f_i}(p)$ of a SM fermion of generation i is constant up to the scale where it is dynamically generated, and has a power-law decay above this scale.[30] Quasiconformal behavior in these models resulting from an approximate IRFP[13] is important as a means to produce a substantial γ_m and hence enhance SM fermion masses. Since the quasiconformal property depends on having an approximate IRFP α_{IR} slightly greater than α_{cr}, the higher-loop calculations of the UV to IR evolution described here are valuable not only for their intrinsic field-theoretic interest, but also as applied to the construction of such models. Studies of reasonably UV-complete DEWSB models showed that approximate residual generational symmetries suppress FCNC effects,[31,32] but it is challenging to reproduce all features of SM fermion masses, such as $m_t >> m_b$, etc.

Concerning dilatons resulting from quasiconformal DEWSB models, we note that such models need not have any color-nonsinglet fermions subject to the new gauge interaction. The minimal SM-nonsinglet fermion content consists of one $SU(2)_L$ doublet with corresponding right-handed $SU(2)_L$ singlets, all of which are color-singlets. DEWSB models of this type can exhibit quasiconformal behavior.[33] Hence, in contrast to one-family DEWSB models, in these minimal models, the resultant dilaton would be comprised purely of color-singlet constituents. This difference is recognized in some of the phenomenological papers on dilatons. The boson with mass $\simeq$ 125 GeV discovered at the LHC by ATLAS and CMS is consistent with being the SM Higgs. An important question is to determine how well it can, alternatively, be modelled as a dilaton resulting from a quasiconformal DEWSB model. Further experimental and theoretical work should settle this question.

8. Conclusions

In conclusion, understanding the UV to IR evolution of an asymptotically free gauge theory and the nature of the IR behavior is of fundamental field-theoretic interest. Our higher-loop calculations give new information on this UV to IR flow and on the determination of $\alpha_{IR,n\ell}$ and $\gamma_{IR,n\ell}$. It is valuable to compare our higher-order calculations of $\gamma_{IR,n\ell}$ with lattice measurements. The higher-loop study of the UV to IR flow for supersymmetric gauge theories yields further insights. We have carried out a study of scheme-dependence in higher-loop calculations, noting that scheme transformations are subject to constraints that are easily satisfied at a UVFP but are quite restrictive at IRFP. In addition to their intrinsic interest, quasiconformal gauge theories are relevant to models of dynamical EWSB, plausibly yielding a light pseudo-Nambu-Goldstone boson, the dilaton.

Acknowledgments

I thank K. Yamawaki for the invitation to speak at SCGT12 and the excellent organization of the conference. I thank T. Ryttov for collaboration on.[1–3] This talk was given on Dec. 5, 2012 at Nagoya University, during a sabbatical at Yale University, and I thank T. Appelquist and the theory group at Yale for warm hospitality during the sabbatical. This research was partially supported by the grant NSF-PHY-09-69739.

References

1. T. A. Ryttov and R. Shrock, Phys. Rev. D **83**, 056011 (2011), arXiv:1011.4542.
2. T. A. Ryttov and R. Shrock, Phys. Rev. D **85**, 076009 (2012), arXiv:1202.1297.
3. T. A. Ryttov and R. Shrock, Phys. Rev. D **86**, 065032 (2012), arXiv:1206.2366; Phys. Rev. D **86**, 085005 (2012), arXiv:1206.6895.
4. R. Shrock, Phys. Rev. D, in press, arXiv:1301.3209.
5. R. Shrock, arXiv:1302.5434.
6. D. J. Gross and F. Wilczek, Phys. Rev. Lett. **30**, 1343 (1973); H. D. Politzer, Phys. Rev. Lett. **30**, 1346 (1973); G. 't Hooft, unpublished.
7. W. E. Caswell, Phys. Rev. Lett. **33**, 244 (1974); D. R. T. Jones, Nucl. Phys. B **75**, 531 (1974).
8. O. V. Tarasov, A. A. Vladimirov, and A. Yu. Zharkov, Phys. Lett. B **93**, 429 (1980); S. A. Larin and J. A. M. Vermaseren, Phys. Lett. B **303**, 334 (1993).
9. T. van Ritbergen, J. A. M. Vermaseren, S. A. Larin, Phys. Lett. B **400**, 379 (1997).
10. T. Banks and A. Zaks, Nucl. Phys. B **196**, 189 (1982).
11. E. Gardi and M. Karliner, Nucl. Phys. B **529**, 383 (1998); E. Gardi and G. Grunberg, JHEP 03, 024 (1999).
12. C. Pica and R. Sannino, Phys. Rev. D **83**, 035013 (2011), arXiv:1011.5917.
13. K. Yamawaki, M. Bando, and K. Matumoto, Phys. Rev. Lett. **56**, 1335 (1986); T. Appelquist, D. Karabali, and L. C. R. Wijewardhana, Phys. Rev. Lett. **57**, 957 (1986);

T. Appelquist and L. C. R. Wijewardhana, Phys. Rev. D **35**, 774 (1987); B. Holdom, Phys. Rev. Lett. **60**, 1233 (1988).
14. J.-F. Fortin, B. Grinstein, and A. Stergiou, arXiv:1208.3674v2.
15. There is a large literature on dilatons. A few of the many papers giving estimates of the mass of a dilaton in a quasi-scale-invariant gauge theory are K. Yamawaki, M. Bando, and K. Matumoto, Phys. Rev. Lett. **56**, 1335 (1986); W. A. Bardeen, C. N. Leung, and S. T. Love, Phys. Rev. Lett. **56**, 1230 (1986); B. Holdom and J. Terning, Phys. Lett. B **200**, 338 (1988); D. D. Dietrich, F. Sannino, and K. Tuominen, Phys. Rev. D **72** , 055001 (2005); W. D. Goldberger, B. Grinstein, and W. Skiba, Phys. Rev. Lett. **100**, 111802 (2008); T. Appelquist and Y. Bai, Phys. Rev. D **82**, 071701 (2010); D. Elander, C. Núñez, and M. Piai, Phys. Lett. B **686**, 64 (2010); S. Matsuzaki and K. Yamawaki, Phys. Rev. D **85**, 095020 (2012); Phys. Rev. D **86**, 035025, 115004 (2012); R. Lawrance and M. Piai, arXiv:1207.0427; D. Elander and M. Piai, arXiv:1208.0546; Z. Chacko, R. Franceschini, and R. Mishra, arXiv:1209.3259; B. Bellazzini, Csáki, Hubisz, Serra and Terning, arXiv:1209.3299. Other papers report fits of a dilaton to LHC data using various assumptions concerning dilaton properties.
16. M. Harada, M. Kurachi, and K. Yamawaki, Phys. Rev. D **70**, 033009 (2004); M. Kurachi and R. Shrock, JHEP **12**, 034 (2006).
17. ATLAS Collab., Phys. Lett. B **716**, 1 (2012), arXiv:1207.7214.
18. CMS Collab., Phys. Lett. B **716**, 30 (2012), arXiv:1207.7235.
19. Here $C_A \equiv C_2(G)$ is the quadratic Casimir invariant for the adjoint representation, $C_f \equiv C_2(R)$, and T_f is the trace invariant.
20. J. A. M. Vermaseren, S. A. Larin, T. van Ritbergen, Phys. Lett. B **405**, 327 (1997).
21. T. Appelquist, G. Fleming, M. Lin, E. Neil, D. Schaich, Phys. Rev. D **84**, 054501 (2011).
22. T. DeGrand, Phys. Rev. D **84**, 116901 (2011).
23. Z. Fodor, K. Holland, J. Kuti, D. Nogradi, C. Schroeder, and C.-H. Wong, Phys. Lett. B **703**, 348 (2011); Phys. Lett. B **718**, 657 (2012).
24. Y. Aoki et al., Phys. Rev. D **86**, 054506 (2012).
25. A. Hasenfratz, A. Cheng, G. Petropoulos, and D. Schaich, arXiv:1207.7162.
26. T. Appelquist et al., arXiv:1204.6000.
27. N. Seiberg, Nucl. Phys. B **435**, 129 (1995); K. A. Intriligator and N. Seiberg, Nucl. Phys. B **444**, 125 (1995); see also V. A. Novikov, M. A. Shifman, A. I. Vainshtein, and V. I. Zakharov, Nucl. Phys. B **229**, 381 (1983).
28. I. Jack, D. R. T. Jones, and C. G. North, Phys. Lett. B **386**, 138 (1996); Nucl. Phys. B **473**, 308 (1996); R. V. Harlander et al., JHEP 0612, 024 (2006); R. Harlander et al., Eur. Phys. J. C **63**, 383 (2009) and references therein.
29. G. Mack, Commun. Math. Phys. **55**, 1 (1977); B. Grinstein, K. Intriligator, I. Rothstein, Phys. Lett. B **662**, 367 (2008).
30. N. D. Christensen and R. Shrock, Phys. Rev. Lett. **94**, 241801 (2005).
31. T. Appelquist, M. Piai, and R. Shrock, Phys. Rev. D **69**, 015002, (2004).
32. T. Appelquist, M. Piai, R. Shrock, Phys. Lett. B **593**, 175 (2004); Phys. Lett. B **595**, 442 (2004); T. Appelquist, N. D. Christensen, M. Piai, R. Shrock, Phys. Rev. D **70**, 093010 (2004); T. A. Ryttov and R. Shrock, Phys. Rev. D **81**, 115013 (2010).
33. D. D. Dietrich, F. Sannino, and K. Tuominen, Phys. Rev. **72**, 055001 (2005); N. D. Christensen and R. Shrock, Phys. Lett. B **632**, 92 (2006); F. Sannino, Acta Phys. Polon. B **40**, 3533 (2009).

Vector Bosons Signals of Electroweak Symmetry Breaking*

R. Sekhar Chivukula*, Elizabeth H. Simmons

Department of Physics and Astronomy, Michigan State University
East Lansing, Michigan 48824, USA
** Speaker at Conference*

Chun Du, Hong-Jian He, Yu-Ping Kuang, Bin Zhang

Center for High Energy Physics, Tsinghua University
Beijing 100084, China

Neil D. Christensen

Pittsburgh Particle Physics, Astrophysics and Cosmology Center
Department of Physics and Astronomy
University of Pittsburgh, Pittsburgh, PA 15260, USA

We study the physics potential of the 8 TeV LHC (LHC-8) to discover signals of extended gauge models or extra dimensional models whose low energy behavior is well represented by an $SU(2)^2 \otimes U(1)$ electroweak gauge structure. We find that with a combined integrated luminosity of 40 fb^{-1}, the first new Kaluza-Klein mode of the W gauge boson can be discovered up to a mass of about 400 GeV, when produced in association with a Z boson.

1. Introduction

The ATLAS and CMS experiments at the LHC have now each collected over 20 fb^{-1} of data at an 8 TeV collision energy. These data will enable the LHC to make incisive tests of the predictions of many competing models of the origin of electroweak symmetry breaking (EWSB), from the Standard Model (SM) with a single Higgs boson, to models with multiple Higgs bosons, and to so-called Higgsless models of the EWSB. The Higgsless models[2] contain new spin-1 gauge bosons which play a key role in EWSB by delaying unitarity violation of longitudinal weak boson scattering up to a higher ultraviolet (UV) scale.[3] Very recently, the effective UV completion of the minimal three-site Higgsless model[4] was presented and studied in[5] which showed that the latest LHC signals of a Higgs-like state with mass around $125-126\,\text{GeV}$[6] can be readily explained, in addition to the signals of new spin-1 gauge bosons studied in the present paper.

*This contribution is an abbrievated version of ref.[1]

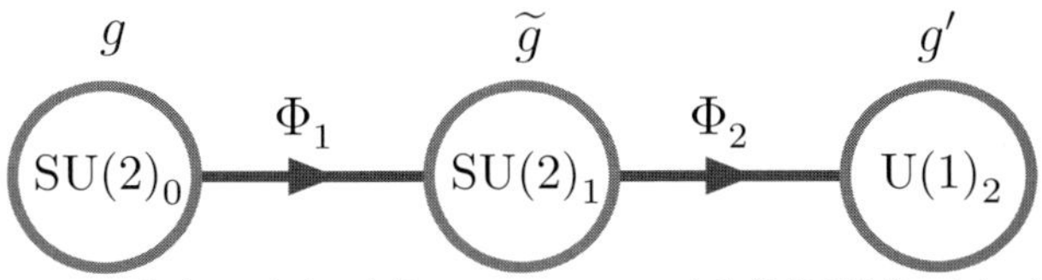

Fig. 1. Moose diagram of the minimal linear moose model (MLMM) with the gauge structure $SU(2)_0 \times SU(2)_1 \times U(1)_2$ as well as two independent link fields Φ_1 and Φ_2 for spontaneous symmetry breaking. The relevant parameter space of phenomenological interest is where the gauge couplings obey $g, g' \ll \tilde{g}$.

In this talk, we explore the physics potential of the LHC-8 to discover a relatively light fermiophobic electroweak gauge boson W_1 with mass $250 - 400\,\text{GeV}$, as predicted by the minimal three-site moose model[4] and its UV completion.[5] Being fermiophobic or nearly so, the W_1 state is allowed to be fairly light. More specifically, the 5d models that incorporate ideally[7] delocalized fermions,[8,9] in which the ordinary fermions propagate appropriately in the compactified extra dimension (or in deconstructed language, derive their weak properties from more than one $SU(2)$ group in the extended electroweak sector[10,11]), yield phenomenologically acceptable values for all Z-pole observables.[4] In this case, the leading deviations from the SM appear in multi-gauge-boson couplings, rather than the oblique parameters S and T. Ref. [12] demonstrates that the LEP-II constraints on the strength of the coupling of the Z_0-W_0-W_0 vertex allow a W_1 mass as light as $250\,\text{GeV}$, where W_0 and Z_0 refer to the usual electroweak gauge bosons.

2. The Model

We study the minimal deconstructed moose model at LHC-8 in a limit where its gauge sector is equivalent to the "three site model"[4] or its UV completed "minimal linear moose model" (MLMM).[5] Both the three site model and the MLMM are based on the gauge group $SU(2)_0 \otimes SU(2)_1 \otimes U(1)_2$, as depicted by Fig. 1 and its gauge sector is the same as that of the BESS models[14,15] or the hidden local symmetry model.[16–20] The extended electroweak symmetry spontaneously breaks to electromagnetism when the distinct Higgs link-fields Φ_1 connecting $SU(2)_0$ to $SU(2)_1$ and Φ_2 connecting $SU(2)_1$ to $U(1)_2$ acquire vacuum expectation values (VEVs) f_1 and f_2. The weak scale $v \simeq 246\,\text{GeV}$ is related to those VEVs via $v^{-2} = f_1^{-2} + f_2^{-2}$ and, for illustration, we take $f_1 = f_2 = \sqrt{2}v$. Below the symmetry breaking scale, the gauge boson spectrum includes an extra set of weak bosons (W_1, Z_1), in addition to the standard-model-like weak bosons (W_0, Z_0) and the photon. Furthermore, the scalar sector of the MLMM[5] contains two neutral physical Higgs bosons (h^0, H^0), as well as the six would-be Goldstones eaten by the corresponding gauge bosons (W_0, Z_0) and (W_1, Z_1).

One distinctive feature of the MLMM is that the unitarity of high-energy longitudinal weak boson scattering is maintained jointly by the exchange of both the new spin-1 weak bosons and the spin-0 Higgs bosons.[5] This differs from either the

SM (in which unitarity of longitudinal weak boson scattering is ensured by the exchange of the Higgs boson alone) [21] or the conventional Higgsless models (in which unitarity of longitudinal weak boson scattering is ensured by the exchange of spin-1 new gauge bosons alone) .[3] It has been shown[12] that the scattering amplitudes in such highly deconstructed models with only three sites can accurately reproduce many aspects of the low-energy behavior of 5d continuum theories.

The unitarity of the generic longitudinal scattering amplitude of $W_0^L W_0^L \to W_0^L W_0^L$, in the presence of any numbers of spin-1 new gauge bosons $V_k\,(= W_k, Z_k)$ and spin-0 Higgs bosons h_k, was recently studied in Ref. [5] For the MLMM, tree-level unitarity implies sum rule,[5]

$$G_{4W_0} - \frac{3M_{Z_0}^2}{4M_{W_0}^2} G_{W_0 W_0 Z_0}^2 = \frac{3M_{Z_1}^2}{4M_{W_0}^2} G_{W_0 W_0 Z_1}^2 + \frac{G_{W_0 W_0 h}^2 + G_{W_0 W_0 H}^2}{4M_{W_0}^2}, \tag{1}$$

where the symbols $(h,\,H)$ denote the two mass-eigenstate Higgs bosons. The sum rule illustrates how exchanging both the new spin-1 weak bosons W_1/Z_1 and the spin-0 Higgs bosons h/H is required to ensure the unitarity of longitudinal weak boson scattering in the MLMM.[5a]

3. Analysis of $\mathbf{W_1^\pm}$ Detection at the LHC-8

Extrapolating from our previous work[13] at a 14 TeV LHC, we have found that the best process for detecting W_1 at LHC-8 is associated production, $pp \to W_1 Z_0 \to W_0 Z_0 Z_0 \to jj\ell^+\ell^-\ell^+\ell^-$, where we select the W_0 decays into dijets and the Z_0 decays into electron or muon pairs.

We have systematically computed all the major SM backgrounds for the $jj4\ell$ final state, including the irreducible backgrounds $pp \to W_0 Z_0 Z_0 \to jj4\ell$ $(jj = qq')$ without the contribution of W_1, as well as the reducible backgrounds $pp \to gg Z_0 Z_0 \to jj4\ell$, $pp \to Z_0 Z_0 Z_0 \to jj4\ell$, and the SM $pp \to jj4\ell$ other than the above reducible backgrounds.

We performed parton level calculations at tree-level using two different methods and two different gauges to check the consistency. In one calculation, we used the helicity amplitude approach[24] to generate the signal and backgrounds. We also calculated both the signal and background using CalcHEP.[27,28] For the signal calculation in CalcHEP, we used FeynRules[22] to implement the minimal Higgsless model.[23] We found satisfactory agreement between these two approaches and between both unitary and 't Hooft-Feynman gauge. We used a scale of $\sqrt{\hat{s}}$ for the strong coupling in the backgrounds and $\sqrt{\hat{s}}/2$ for the CTEQ6L[25] parton distribution functions. We included both the first and second generation quarks in the protons and jets, and both electrons and muons in the final-state leptons.

---

[a]We also note that the hWW and hZZ couplings are generally suppressed [5] relative to the SM values because of the VEV ratio $f_2/f_1 = O(1)$ and the $h - H$ mixing.

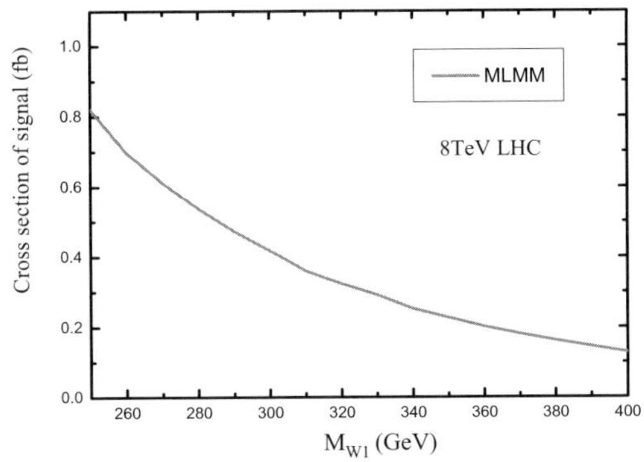

Fig. 2. Predicted signal cross section for $pp \to W_1 Z_0 \to W_0 Z_0 Z_0 \to jj4\ell$ as a function of the W_1 mass in the MLMM after all cuts at the LHC-8.

In our calculations, we impose basic acceptance cuts,

$$\begin{aligned} p_{T\ell} &> 10\,\mathrm{GeV}, \qquad |\eta_\ell| < 2.5\,, \\ p_{Tj} &> 15\,\mathrm{GeV}, \qquad |\eta_j| < 4.5\,, \end{aligned} \tag{2}$$

and also a reconstruction cut for identifying W_0 bosons that decay to dijets,

$$M_{jj} = 80 \pm 15\,\mathrm{GeV}\,. \tag{3}$$

We further analyzed the distributions of the dijet opening-angle $\Delta R(jj)$ in the decays of $W_0 \to jj$ for both the signal and SM background events. We find that the signal events are peaked in the small opening-angle region around $\Delta R(jj) \sim 0.6$, while the SM backgrounds tend to populate the range of larger opening angles, with a broad bump around $\Delta R(jj) = 1.5 - 3.3$. In order to sufficiently suppress the SM backgrounds, we find the following opening-angle cut[b] to be very effective,[26]

$$\Delta R(jj) \ < \ 1.6\,. \tag{4}$$

At the LHC-8, we note that the above cut reduces the signal events by only 10−15%, but removes about 72 − 80% of the SM backgrounds.

In Fig. 2, we display the predicted total signal cross section for the process $pp \to W_0 Z_0 Z_0 \to jj4\ell$ after all cuts at the LHC-8 have been imposed; this is shown as a function of the W_1 mass for the range 250 − 400 GeV.[c]

In Fig. 3, we display the required integrated luminosities for detecting the $W_1^\pm$ signal at the 3σ and 5σ levels as a function of the $W_1^\pm$ mass M_{W_1}. We see the LHC-8 should be able to observe the $W_1^\pm$ gauge bosons of the minimal linear moose model studied up to masses of order 400 GeV. We look forward to seeing the results.

[b]These are somewhat weaker than the cut of $\Delta R(jj) < 1.5$ imposed in.[13]

[c]Here, we define the signal region to include all events satisfying the condition, $M(Z_0 jj) \ = \ M_{W_1} \pm 20\,\mathrm{GeV}$.

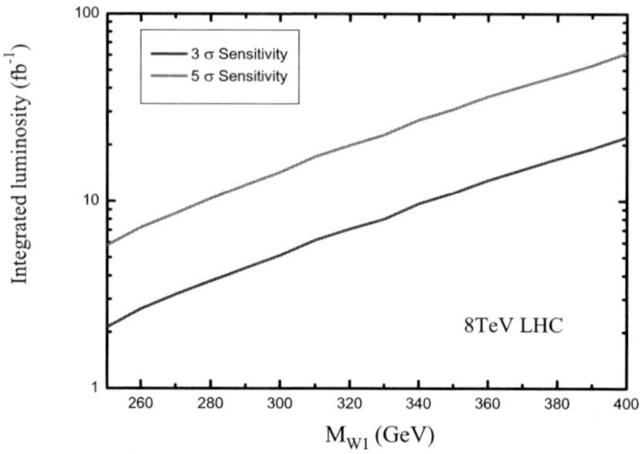

Fig. 3. Integrated luminosities required for detection of new $W_1^{\pm}$ gauge bosons at the 3σ level in the MLMM (lower blue curve), and at the 5σ level (upper red curve) as a function of the W_1 mass, at the LHC-8.

Acknowledgments

This research was supported by the NSF of China (grants 11275101, 10625522, 10635030, 11135003, 11075086) and the National Basic Research Program of China (grant 2010CB833000); by the U.S. NSF under Grants PHY-0854889 and PHY-0705682; and by the University of Pittsburgh Particle Physics, Astrophysics, and Cosmology Center. HJH thanks CERN Theory Division for hospitality.

References

1. C. Du, H. -J. He, Y. -P. Kuang, B. Zhang, N. D. Christensen, R. S. Chivukula, E. H. Simmons and , Phys. Rev. D **86**, 095011 (2012) [arXiv:1206.6022 [hep-ph]].
2. C. Csaki, C. Grojean, H. Murayama, L. Pilo and J. Terning, Phys. Rev. D **69**, 055006 (2004) [hep-ph/0305237]; C. Csaki, C. Grojean, L. Pilo and J. Terning, Phys. Rev. Lett. **92**, 101802 (2004) [hep-ph/0308038].
3. R. S. Chivukula, D. A. Dicus, H. J. He, Phys. Lett. B **525**, 175 (2002) [hep-ph/0111016]; R. S. Chivukula and H. J. He, Phys. Lett. B **532**, 121 (2002) [hep-ph/0201164]; R. S. Chivukula, D. A. Dicus, H. J. He, S. Nandi, Phys. Lett. B **562**, 109 (2003) [hep-ph/0302263]; H. J. He, Int. J. Mod. Phys. A **20**, 3362 (2005) [hep-ph/0412113] R. S. Chivukula, H. J. He, M. Kurachi, E. H. Simmons, M. Tanabashi, Phys. Rev. D **78**, 095003 (2008) [arXiv:0808.1682]
4. R. S. Chivukula, B. Coleppa, S. Di Chiara, E. H. Simmons, H. J. He, M. Kurachi, M. Tanabashi, Phys. Rev. D **74**, 075011 (2006) [arXiv:hep-ph/0607124].

5. T. Abe, N. Chen, H. J. He, JHEP (2012), [arXiv:1207.4103 [hep-ph]].

6. G. Aad *et al.*, [ATLAS Collaboration], Phys. Lett. B **716**, 1 (2012) [arXiv:1207.7214 [hep-ex]];

S. Chatrchyan *et al.*, [CMS Collaboration], Phys. Lett. B **716**, 30 (2012) [arXiv:1207.7235 [hep-ex]].

7. R. S. Chivukula, E. H. Simmons, H. J. He, M. Kurachi and M. Tanabashi, Phys. Rev. D **72**, 015008 (2005) [arXiv:hep-ph/0504114].

8. G. Cacciapaglia, C. Csaki, C. Grojean and J. Terning, Phys. Rev. D **71**, 035015 (2005) [arXiv:hep-ph/0409126].

9. R. Foadi, S. Gopalakrishna and C. Schmidt, Phys. Lett. B **606**, 157 (2005) [arXiv:hep-ph/0409266].

10. R. S. Chivukula, E. H. Simmons, H. J. He, M. Kurachi and M. Tanabashi, Phys. Rev. D **71**, 115001 (2005) [arXiv:hep-ph/0502162].

11. R. Casalbuoni, S. De Curtis, D. Dolce and D. Dominici, Phys. Rev. D **71**, 075015 (2005) [arXiv:hep-ph/0502209].

12. A. Belyaev, R. S. Chivukula, N. D. Christensen, H. J. He, M. Kurachi, E. H. Simmons, M. Tanabashi, Phys. Rev. D **80**, 055022 (2009) [arXiv:0907.2662]; and presentation in the Proceedings of International Workshop on "Strong Coupling Gauge Theories in LHC Era" (SCGT-2009), arXiv:1003.1786 [hep-ph].

13. H. J. He, Y. P. Kuang, Y. Qi, B. Zhang, A. Belyaev, R. S. Chivukula, N. D. Christensen, A. Pukhov, E. H. Simmons, Phys. Rev. D **78**, 031701 (2008) [arXiv:0708.2588].

14. R. Casalbuoni, S. De Curtis, D. Dominici and R. Gatto, Phys. Lett. B **155**, 95 (1985).

15. R. Casalbuoni *et al.*, Phys. Rev. **D53**, 5201 (1996) [hep-ph/9510431].

16. M. Bando, T. Kugo, S. Uehara, K. Yamawaki and T. Yanagida, Phys. Rev. Lett. **54**, 1215 (1985).

17. M. Bando, T. Kugo and K. Yamawaki, Nucl. Phys. B **259**, 493 (1985).

18. M. Bando, T. Fujiwara, and K. Yamawaki, Prog. Theor. Phys. **79**, 1140 (1988).

19. M. Bando, T. Kugo, and K. Yamawaki, Phys. Rept. **164**, 217 (1988).

20. M. Harada and K. Yamawaki, Phys. Rept. **381**, 1 (2003) [hep-ph/0302103].

21. J. M. Cornwall, D. N. Levin, and G. Tiktopoulos, Phys. Rev. Lett. **30** (1973) 1268; Phys. Rev. D **10** (1974) 1145; C. H. Llewellyn Smith, Phys. Lett. **46**B (1973) 233. D.

A. Dicus and V. S. Mathur, Phys. Rev. D **7** (1973) 3111; B. W. Lee, C. Quigg, and H. B. Thacker, Phys. Rev. Lett. **38** (1977) 883; Phys. Rev. D **16** (1977) 1519; M. S. Chanowitz and M. K. Gaillard, Nucl. Phys. B **261** (1985) 379.

22. N. D. Christensen and C. Duhr, Comput. Phys. Commun. **180**, 1614 (2009) [arXiv:0806.4194 [hep-ph]].
23. N. D. Christensen, P. de Aquino, C. Degrande, C. Duhr, B. Fuks, M. Herquet, F. Maltoni and S. Schumann, Eur. Phys. J. C **71**, 1541 (2011) [arXiv:0906.2474 [hep-ph]].
24. K. Hagiwara and D. Zeppenfeld, DESY-85-133; K. Hagiwara, R. D. Peccei, D. Zeppenfeld and K. Hikasa, DESY-86-058 (1986); K. Hagiwara and D. Zeppenfeld, KEK Preprint 87-158 (1988).
25. J. Pumplin, D. R. Stump, J. Huston, H. L. Lai, P. Nadolsky and W. K. Tung, JHEP **07**, 012 (2002).
26. This kind of cut should be applied to separated jets. Experimentally, two jets are separable if $\Delta R_{jj} > 0.5$ [see for instance, S. Ask (ATLAS Collaboration), arXiv:1106.2061], and two jet-cones do not overlap at all if $\Delta R_{jj} > 1$. Hence the cut (4) can be realized.
27. A. Pukhov, E. Boos, M. Dubinin, V. Edneral, V. Ilyin, D. Kovalenko, A. Kryukov and V. Savrin, *et al.*, arXiv:hep-ph/9908288.
28. A. Pukhov, arXiv:hep-ph/0412191.

Higgs Discovery: Impact on Composite Dynamics Technicolor & eXtreme Compositeness *Thinking Fast and Slow*

Francesco Sannino*

CP^3-Origins & the Danish Institute for Advanced Study DIAS,
*University of Southern Denmark, Campusvej 55, DK-5230 Odense M, Denmark. *E-mail: sannino@cp3.dias.sdu.dk*

I discuss the impact of the discovery of a Higgs-like state on composite dynamics starting by critically examining the reasons in favour of either an elementary or composite nature of this state. Accepting the standard model interpretation I re-address the standard model vacuum stability within a Weyl-consistent computation.

I will carefully examine the fundamental reasons why what has been discovered might not be the standard model Higgs. Dynamical electroweak breaking naturally addresses a number of the fundamental issues unsolved by the standard model interpretation. However this paradigm has been challenged by the discovery of a not-so-heavy Higgs-like state. I will therefore review the recent discovery[1] that the standard model top-induced radiative corrections naturally reduce the intrinsic non-perturbative mass of the composite Higgs state towards the desired experimental value. Not only we have a natural and testable working framework but we have also suggested specific gauge theories that can realise, at the fundamental level, these minimal models of dynamical electroweak symmetry breaking. These strongly coupled gauge theories are now being heavily investigated via first principle lattice simulations with encouraging results.

The new findings show that the recent naive claims made about new strong dynamics at the electroweak scale being disfavoured by the discovery of a not-so-heavy composite Higgs are unwarranted.

I will then introduce the more speculative idea of extreme compositeness according to which not only the Higgs sector of the standard model is composite but also quarks and leptons, and provide a toy example in the form of gauge-gauge duality.

Keywords: Technicolor and eXtreme compositeness

1. Thinking fast and slow

Excitingly, both the ATLAS and CMS collaborations at the Large Hadron Collider (LHC) have independently reported the discovery of a new particle[2,3] with properties consistent with the standard model (SM) Higgs. The current status is summarised in Fig. 1. The burning question is whether the new particle state is indeed the SM Higgs, given that the current experimental status does not allow for a precise determination of all its properties.

1.1. *Thinking fast*

It is tempting to accept the simplest paradigm, i.e. the SM Higgs, for the following reasons:

- It is the most minimal renormalizable model one can write able to break the electroweak symmetry preserving the $SU(2)_c$ custodial symmetry and compatible with all the experimental data.
- It gives masses to the SM fermions, including the SM neutrinos [a], while enacting the largest number of symmetries protecting against unobserved flavour changing neutral currents.

[a]Although the SM Higgs mechanism does not require the presence of right-handed neutrinos, at the same time, it does not forbid their existence. Henceforth a Dirac mass term is easy to add in the SM. A direct Majorana mass term for the SM left-handed neutrinos requires the existence of new physics since there are no renormalizable operators within the SM able to accommodate it. However, to date, there is no experimental evidence favouring a Majorana mass term. Furthermore, once one has accepted the fine-tuning in the Higgs sector the quark and lepton Yukawa couplings can be fine-tuned too.

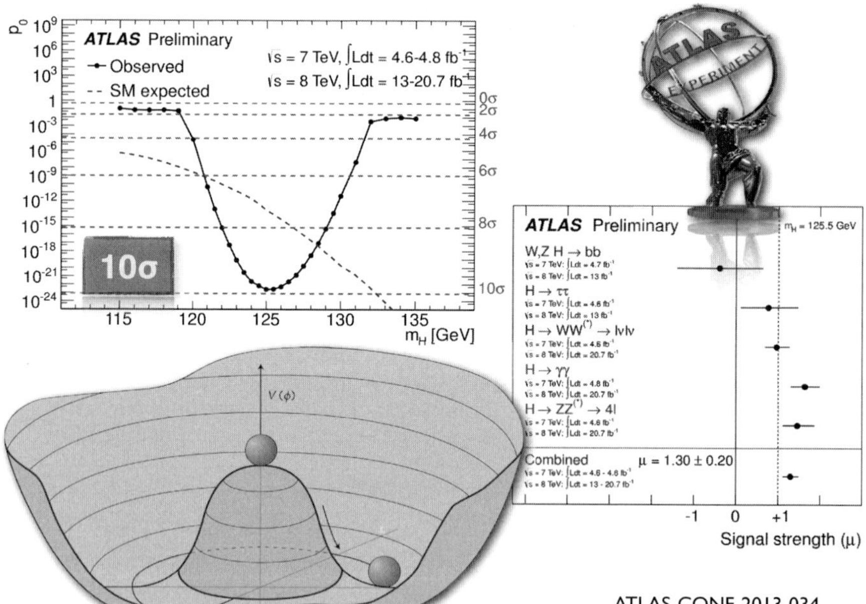

Fig. 1. Top left plot is the local probability p0 for a background-only experiment to be more signal-like than the observation as a function of the mass of the Higgs or the combination of all channels. The dashed curve shows the median expected local p0 under the hypothesis of a standard model Higgs boson production at that mass. The horizontal dashed lines indicate the p values corresponding to significances from 0 to 10. Right bottom plot reports the measurements of the signal strength parameter for a mass of the Higgs of 125.5 GeV for the individual channels and their combination. The results are from.[4] The mexican hat potential represents the standard model Higgs potential.

If one accepts the SM Higgs paradigm to be valid it becomes imperative to investigate its vacuum stability[5,6] against quantum corrections. Because at high energies the SM displays classical conformality, the SM quantum corrections must satisfy the Weyl consistency conditions taken into account for the first time in.[6,7] These conditions help organise the SM perturbative series and the resulting stability plot is shown in Fig. 2. The line of absolute stability, in the top-mass versus Higgs mass plane, is within two sigma from the SM point with the largest uncertainty coming from the measurement of the top mass.[6] The next-to-leading Weyl consistent analysis is therefore relevant and should be performed as suggested in.[6] However according to this analysis the SM might very well be in a metastable state to tunnel to the true groundstate located at much higher values of the Higgs field. The unity probability that this occurs within the age of the Universe defines the boundary between the stable and metastable region.

1.2. *Thinking slow*

Accepting the SM Higgs paradigm does not come cheap and following Daniel Kahneman[b] this is the time to think slow and examine the reasons why what has been discovered might not be the SM Higgs.

[b]Winner of the Nobel Prize in Economics and the author of the book *Thinking Fast and Slow*.

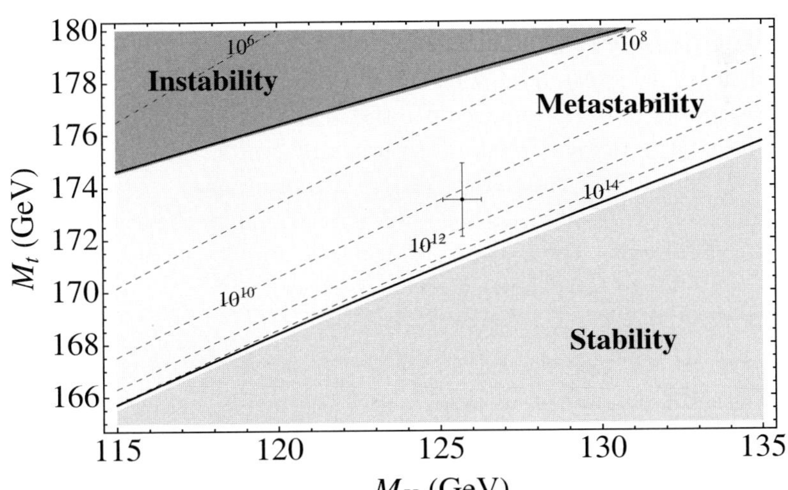

Fig. 2. Standard model stability analysis based on the effective standard model Higgs quartic coupling and respecting the Weyl consistency conditions.[6] The red region indicates instability, the yellow metastability and the green absolute stability following the analysis presented in.[6] For comparison, the black lines indicate the bounds from.[5] The point with error bars shows the experimental values of the top[8] and Higgs[9] masses. The red dashed lines show the value in GeV at which the Higgs effective self-coupling crosses zero.

- The SM Higgs *does not explain* spontaneous symmetry breaking, at best parametrizes the phenomenon.
- Scalars are not fundamental representations of the Lorentz group, spin one-half fermions are. No fundamental (pseudo)scalar has ever been discovered so far in Nature. It would be the most important discovery made at the LHC [c].
- The masses of the SM fermions are not explained but fitted to their value. The same is true for any other extension of the SM with fundamental scalars.
- The SM Higgs sector alone cannot explain why we observe at least three matter generations.
- The dark matter problem is unexplained in the SM, although it is straightforward to construct extensions of the SM featuring perturbative or non-perturbative dark matter components.
- There is no explanation for the strong CP problem.
- There is no working mechanism for baryogenesis and dark matter genesis, if asymmetric.

I have not considered the grand unification problem of the SM gauge couplings since, although interesting, requires a number of ad hoc theoretical assumptions such as, for example, single step unification. There is no reason to believe that all the couplings must unify simultaneously. In fact, we already assume, given that we do not know how to quantise gravity, that gravity unifies at a higher energy scale.

A composite Higgs and associate composite sector represent a natural solution to several of the above mentioned issues [d]. The time-honoured example of composite dynamics is Technicolor (TC).[13,14] In TC the Higgs sector of the Standard Model is replaced by a new gauge dynamics featuring fermionic matter.

Supersymmetry and TC require both the introduction of yet another sector. For supersymmetry it is the supersymmetry breaking sector and for TC one needs the new sector to account for the masses of the SM fermions.

Additional exotic possibilities have been envisioned by theorists and are actively pursued in experiments. For a complete overview and status of different extensions of the SM, we refer to the following recent reviews.[15–17]

---

[c] By linking scalars degrees of freedom to fermions, which are fundamental representations of the Lorentz group, supersymmetry[10,11] apparently gives a natural way to understand why scalars (and by supersymmetry also pseudoscalars) might also be natural, However, supersymmetry itself might be an emergent phenomenon.[12] *Why would Nature possess such a large symmetry and then hide it so well?*

[d] By composite, I mean composite by four-dimensional fermionic matter in the form of a strongly coupled gauge theory. One can, of course, enlarge the space of theories or the idea of compositeness, but should, at the same time, declare the SM problems is set to solve. For example, simply writing an effective Lagrangian to replace the Higgs sector does not imply that the new Higgs sector is made by more fundamental matter. In fact, underlying purely bosonic nonsupersymmetric theories admit low energy effective theories describing the associated Goldstone excitations. If this happens the first three problems of the list are not addressed.

2. Light scalar from a heavy composite state: The power of radiative corrections

If new strong dynamics [18–22] underlies the Higgs mechanism, is it possible for the corresponding spectrum to feature a 125 GeV composite scalar? To answer this question one must disentangle the SM radiative corrections from the dynamical mass M_H^{TC}, *i.e.* the mass stemming purely from composite dynamics. Because of the large and negative radiative corrections from the top loop, we argued in[1] that, for a SM-like top-Yukawa coupling, the dynamical mass of the scalar required to match the observations is of the order of $M_H^{TC} \sim 600/\sqrt{N_{\rm TD}}$ GeV, where $N_{\rm TD}$ is the number of weak technidoublets. This is a significant increase in the value of the dynamical mass compared to the observed 125 GeV value, often incorrectly identified with the dynamical mass. Additionally, if the dominant decay channels are into SM states, then for a physical mass of 125 GeV the TC Higgs is narrow simply because of kinematics. A famous example in strong dynamics is the $f_0(980)$ resonance, which is extremely narrow because the decay mode into $K\bar{K}$ is below threshold.[23,24]

To make the point above explicit consider TC theories (or any other new similar strong dynamics) featuring, at scales below the mass of the technirho M_ρ, only the eaten Goldstone bosons and the TC Higgs H. We assume the TC dynamics to respect $SU(2)_c$ custodial isospin symmetry, and adopt a nonlinear realization for the composite states. The latter are thus classified according to linear multiplets of $SU(2)_c$: the electroweak Goldstone bosons π^a, with $a = 1, 2, 3$, form an $SU(2)_c$ triplet, whereas the TC Higgs is an $SU(2)_c$ singlet. The elementary SM fields are multiplets of the electroweak group. The Yukawa interactions of the TC Higgs with SM fermions are induced by interactions beyond the TC theory itself, *e.g.* extended technicolor (ETC).[19,20]

Assuming that the only non-negligible sources of custodial isospin violation are due to the Yukawa interactions, and retaining only the leading order terms in a momentum expansions, leads to the effective Lagrangian

$$\begin{aligned}\mathcal{L} = \mathcal{L}_{\overline{\rm SM}} &+ \left(1 + \frac{2r_\pi}{v} H + \frac{s_\pi}{v^2} H^2\right) \frac{v^2}{4} {\rm Tr}\, D_\mu U^\dagger D^\mu U + \frac{1}{2}\, \partial_\mu H\, \partial^\mu H - V[H] \\ &- m_t \left(1 + \frac{r_t}{v} H\right) \left[\overline{q}_L\, U \left(\frac{1}{2} + T^3\right) q_R + {\rm h.c.}\right] - m_b \left(1 + \frac{r_b}{v} H\right) \\ &\times \left[\overline{q}_L\, U \left(\frac{1}{2} - T^3\right) q_R + {\rm h.c.}\right] + \cdots \\ &- \Delta S\; W^a_{\mu\nu} B^{\mu\nu}\, {\rm Tr}\, T^a U T^3 U^\dagger + \mathcal{O}\left(\frac{1}{M_\rho}\right) \end{aligned} \tag{1}$$

where $\mathcal{L}_{\overline{\rm SM}}$ is the SM Lagrangian without Higgs and Yukawa sectors, the ellipses denote Yukawa interactions for SM fermions other than the top-bottom doublet $q \equiv (t, b)$, and $\mathcal{O}(1/M_\rho)$ includes higher-dimensional operators, which are suppressed by powers of $1/M_\rho$. In this Lagrangian $v \simeq 246$ GeV is the electroweak vev, U is the usual exponential map of the Goldstone bosons produced by the breaking of the electroweak symmetry, $U = \exp\left(i2\pi^a T^a/v\right)$, with covariant derivative $D_\mu U \equiv \partial_\mu U - igW_\mu^a T^a U + ig' U B_\mu T^3$, $2T^a$ are the Pauli matrices, with $a = 1, 2, 3$, and $V[H]$ is the TC Higgs potential. ΔS is the contribution to the S parameter from the physics at the cutoff scale, and is assumed to vanish in the $M_\rho \to \infty$ limit. The interactions contributing to the Higgs self-energy are

$$\mathcal{L}_H \supset \frac{2\, m_W^2\, r_\pi}{v}\; H\, W_\mu^+\, W^{-\mu} + \frac{m_Z^2\, r_\pi}{v}\; H\, Z_\mu\, Z^\mu - \frac{m_t\, r_t}{v}\; H\, \bar{t}\, t$$
$$+ \frac{m_W^2\, s_\pi}{v^2}\; H^2\, W_\mu^+\, W^{-\mu} + \frac{m_Z^2\, s_\pi}{2\, v^2}\; H^2\, Z_\mu\, Z^\mu\,. \qquad (2)$$

The tree-level SM is recovered for $r_\pi = s_\pi = r_t = r_b = 1$. We divide the ra-

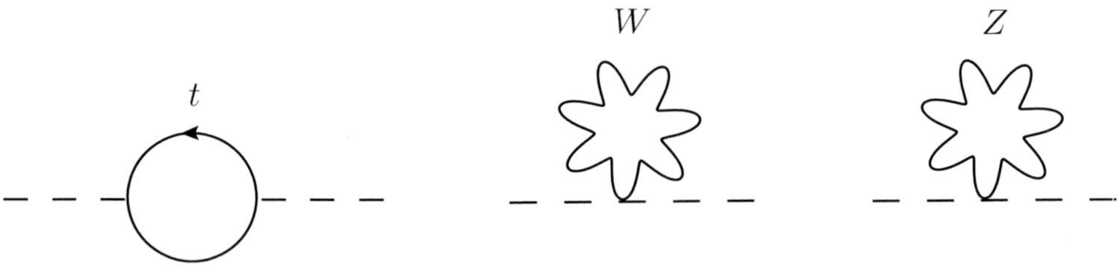

Fig. 3. Quadratically divergent diagrams contributing to the Higgs mass. The gauge boson exchanges are computed in Landau gauge: then the seagull diagrams, with a single W and Z exchange, are the only quadratically divergent one-loop diagrams with gauge boson exchanges.

diative corrections to the TC Higgs mass into two classes: external contributions, corresponding to loop corrections involving elementary SM fields, and TC contributions, corresponding to loop corrections involving TC composites only. The latter contribute to the dynamical mass M_H^{TC} which has been estimated in.[1] In order to isolate the SM contributions we work in Landau gauge. Here transversely polarized gauge boson propagators correspond to elementary fields, and massless Goldstone boson propagators correspond to TC composites.[25] The only SM contributions to the TC Higgs mass which are quadratically divergent in the cutoff come from the diagrams of Fig. 3. Retaining only the quadratically divergent terms leads to a physical mass M_H given by

$$M_H^2 = (M_H^{TC})^2 + \frac{3(4\pi\kappa F_\Pi)^2}{16\pi^2 v^2}\left[-4r_t^2 m_t^2 + 2s_\pi\left(m_W^2 + \frac{m_Z^2}{2}\right)\right] + \Delta_{M_H^2}(4\pi\kappa F_\Pi)\,, \qquad (3)$$

where F_Π is the TC-pion decay constant, and $\Delta_{M_H^2}(4\pi\kappa F_\Pi)$ is the counterterm. The cutoff is estimated to be $4\pi\kappa F_\pi$, where κ is a number of order one. The latter scales like $1/\sqrt{d(R_{\rm TC})}$ if the cutoff is identified with the technirho mass, or is a constant if the cutoff is of the order of $4\pi F_\Pi$. Provided r_t is also of order one, the dominant radiative correction is due to the top quark. For instance, if $F_\Pi = v$, which is appropriate for a TC theory with one weak technidoublet, then

$\delta M_H^2 \sim -12\kappa^2 r_t^2 m_t^2 \sim -\kappa^2 r_t^2 (600\,\text{GeV})^2$. In Fig. 4 we plot the mass of the TC Higgs as a function of the product $\kappa\, r_t$ using the formula $M_H^{TC} = \sqrt{M_H^2 + 12\kappa^2 r_t^2 m_t^2}$. This is obtained in the simple approximation of neglecting the weak gauge boson contributions, and having set to zero the counterterm in (3). This shows that the dynamical mass of the TC Higgs can be substantially heavier than the physical mass, $M_H \simeq 125$ GeV.

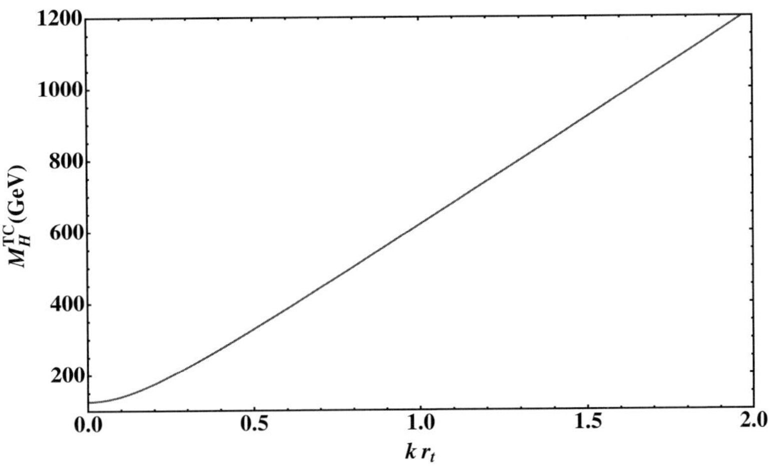

Fig. 4. Mass of the TC Higgs as a function of the product κr_t, using the formula $M_H^{TC} = \sqrt{M_H^2 + 12\kappa^2 r_t^2 m_t^2}$.

We have also argued that there exist a large number of underling fundamental theories available[26–30] to construct minimal models of dynamical electroweak symmetry breaking. Minimal refers to the fact that we have only a doublet of QCD neutral techni-fermions gauged under the electroweak symmetry featuring a gauge dynamics drastically different from a scaled up version of QCD.

Furthermore some of these theories can feature a scalar singlet with dynamical mass $M_H^{TC} \sim 600/\sqrt{N_{\text{TD}}}$ GeV[1,21e]. These theories are being investigated via first principle lattice simulations with promising results[41–49] with interesting results for the physical spectrum.[50–52] More recently we have also been investigating on the lattice new classes of models featuring interesting patterns of chiral symmetry breaking[51,52] leading to highly interesting models of both light and heavy composite dark matter[53–57] as well as asymmetric dark matter genesis.[53,54,56,58] The

[e]Furthermore, reduction of the TC-Higgs dynamical mass may originate from walking (or near-conformal) dynamics,[31] both for the fundamental or for higher-dimensional representations. Walking dynamics is useful to alleviate the tension with flavor changing neutral currents, and to reduce the value of the S-parameter.[33,59] The latter, however, is not expected to vanish in conformal field theories,[34–36] as well as in TC theories featuring near-conformal dynamics.[25] In the literature a light TC Higgs originating from walking-type dynamics is also known as *technidilaton*.[37–40]

phenomenology associated to minimal models of dynamical electroweak symmetry breaking is summarised in.[59–63]

Furthermore the couplings to WW and ZZ of the composite Higgs are naturally of the order of the SM Higgs in minimal models of composite dynamics as explained in.[1]

The results[1] reviewed above clearly demonstrate that several recent and highly naive claims made about new strong dynamics at the electroweak scale being disfavoured by the discovery of a not so heavy composite Higgs are unwarranted.

3. eXtreme compositeness: An alternative paradigm & a toy model

Often in TC-inspired models, or any other composite Higgs theory, only the Higgs sector, and perhaps part of the flavour sector, are imagined to be made of something else. This theory is expected to be strongly coupled at the electroweak theory but cannot be directly used to give masses to the SM fermions. In fact in these frameworks the flavour sector requires extra strong dynamics.[64–68] On the other hand it is quite natural to expect that the flavour and the Higgs sector are intimately related[69] and that a better understanding would require an unified description of these sectors.

Extreme compositeness aims, indeed, to a deeper understanding of the interplay between these two sectors. There might be many ways in which this relation can manifest itself, here we explore the possibility that extreme compositeness manifests itself in the form of electro-magnetic duality at the SM model level.[70]

According to this idea of extreme compositeness the SM is currently written in terms of *magnetic* fields and the dual *electric* description contains, besides gauge bosons, features only fermionic degrees of freedom.[70] We will elucidate our idea using the investigations on gauge-gauge duality presented in[35,71–73] and already used for relevant phenomenological predictions in.[35] The gauge-gauge duality framework was pioneered in a series of ground breaking papers by Seiberg[74,75] for supersymmetric gauge theories.

Quarks, leptons and the Higgs itself are degrees of freedom which are to be considered elementary in the magnetic description, but arise as composite states in the dual electric variables. The reason why it has been natural to introduce these states first when introducing the SM is that the magnetic description is perturbative at the electroweak energy scale while the electric one is strongly coupled rendering harder the identification with the electrical variables. We will therefore make use of the logical possibility that asymptotically free gauge theories have magnetic duals.

Experimentally three generations of quarks and leptons have been discovered with the left handed fields transforming according to the doublet representation of the weak interactions and the right handed fields are singlets with respect to the weak interactions. Electromagnetic interactions are felt by left and right transforming electrically charged states. The Higgs sector of the SM is being experimentally tested. To illustrate how the basic idea works I start with a toy model featuring a more symmetric looking toy SM matter content summarized in Table 1.

Fields	$[SU(3)]$	$SU(N_f)_L$	$SU(N_f)_R$
q	$\square$	$\overline{\square}$	1
$\widetilde{q}$	$\overline{\square}$	1	$\square$
l	1	$\overline{\square}$	1
$\widetilde{l}$	1	1	$\square$
H	1	$\square$	$\overline{\square}$

With $q^i_{\alpha,c}$ I indicate the two component left spinor where $\alpha = 1, 2$ is the spin index, $c = 1, ..., 3$ is the color index while $i = 1, ..., N_f$ represents the flavor. $\widetilde{q}^{\alpha,c}_i$ is the two component conjugated right spinor. Similarly the leptonic fields are summarized in the table with l^i_α the two component left spinor and $\widetilde{l}^{\alpha,c}_i$ the two component conjugated right spinor. Here $N_f = 2n_g$ with n_g the number of SM generations. The weak interactions are naturally embedded within the flavor group $SU(2n_g)_L \times SU(2n_g)_R$ by opportunely gauging n_g times the $SU(2)_L \times U(1)_Y$ subgroup. The generalized Higgs field has been chosen to transform according to the bifundamental representation of $SU(2n_g)_L \times SU(2n_g)_R$ and therefore is not the minimal choice but it is the most natural one here. We set $SU(3)$ in between square brackets in the table to indicate that this is the gauge group we are concentrating on to discuss how our gauge-gauge duality may work. We therefore switch off the weak interactions for the time being. From Table 1 one is naturally led to consider the leptons as the forth color of an extended color group $[SU(4)]$. This is the renowned Pati-Salam[76–78] extension of the SM generalized to $N_f/2 = n_g$ generations. In Table 2 the spectrum of the SM is summarized with respect to the Pati-Salam $SU(4)$ gauge group. We

Fields	$[SU(4)]$	$SU(N_f)_L$	$SU(N_f)_R$
p	$\square$	$\overline{\square}$	1
$\widetilde{p}$	$\overline{\square}$	1	$\square$
H	1	$\square$	$\overline{\square}$

have now $p^i_{\alpha,C}$ ($\widetilde{p}^i_{\alpha,C}$) with $C = 1, 2, 3$ representing the ordinary left-handed quarks (conjugated right spinors) while $C = 4$ are the leptons, and therefor C is the vector index of the Pati-Salam group $SU(4)$. The $B - L$ symmetry is automatically embedded as one of the generators of $SU(4)$.[79] This description of the SM fields is, in practice, of book-keeping nature. To upgrade this model to a more realistic one Pati and Salam introduced several new scalar degrees of freedom with the hope that one day there might be a more profound understanding of the origin behind the scalar sector. Here, we will not duel with the specific details of the scalar potential and the pattern of chiral symmetry breaking. Therefore, we add only the minimum number of matter fields allowing for such a possibility to manifest itself. We start by introducing the new complex scalars Φ_p ($\widetilde{\Phi}_{\widetilde{p}}$) transforming according to the fundamental (antifunamental) representation of the $[SU(4)]$ gauge group and fundamental (antifundamental) representation of the first (second) flavor group.

At this point the spectrum of the theory is close to the *magnetic* gauge dual envisioned in.[73] The states to add are a magnetic Weyl fermion λ_m transforming according to the adjoint representation of $[SU(4)]$ and a Weyl fermion M transforming as the Higgs with respect to the non-abelian flavor group. Adding these states leads to the spectrum reported in Table 3. We also make explicit the global

Fields	$[SU(4)]$	$SU(N_f)_L$	$SU(N_f)_R$	$U(1)_p$	$U(1)_{AF}$
λ_m	Adj	1	1	0	1
p	$\Box$	$\overline{\Box}$	1	$\frac{2n_g-4}{4}$	$-\frac{4}{2n_g}$
$\widetilde{p}$	$\overline{\Box}$	1	$\Box$	$-\frac{2n_g-4}{4}$	$-\frac{4}{2n_g}$
Φ_p	$\Box$	$\overline{\Box}$	1	$\frac{2n_g-4}{4}$	$-\frac{2n_g-4}{2n_g}$
$\widetilde{\Phi}_{\widetilde{p}}$	$\overline{\Box}$	1	$\Box$	$-\frac{2n_g-4}{4}$	$-\frac{2n_g-4}{2n_g}$
M	1	$\Box$	$\overline{\Box}$	0	$-1+\frac{8}{2n_g}$
H	1	$\Box$	$\overline{\Box}$	0	$\frac{8}{2n_g}$

symmetries of the new theory which are constituted by a new vector-like $U(1)_p$ and an axial one $U(1)_{AF}$ which is anomaly free. These global symmetries play a fundamental role via the 't Hooft anomaly conditions in order to identify the correct electric theory. By determining the most general set of solutions to these conditions, together with requiring consistent flavor decoupling and involution at the level of the electric and magnetic gauge groups in[73] it was argued that the natural non-supersymmetric electric dual theory is the one summarized in Table 4. In[73] it was further shown that it is possible to construct all the singlet states of the magnetic theory as composites of the electric ones. The only state we need to add to the table of[73] is H which corresponds naturally to the electric composite gauge singlet $P\lambda\lambda\widetilde{P}$. It is remarkable that the proposed electric dual theory does not contain any scalar degrees of freedom.

Since the magnetic extension of the SM mimics the Pati-Salam one I expect that part of the phenomenological analysis will resemble to the one already present in the literature (see for example[80]) . However, it is also clear that as we go higher in energy the magnetic description must deform itself into the electric one. In the same way the hadronic description dissolves itself into the quark and gluons one. This will require the discovery of copies of the SM fields at higher energies interacting via momentum dependent form factors.

Following[73] the dual *electric* gauge group is therefore $SU(2n_g-4) = SU(N_f-4)$.

Fields	$[SU(2n_g-4)]$	$SU(2n_g)_L$	$SU(2n_g)_R$	$U(1)_p$	$U(1)_{AF}$
λ	Adj	1	1	0	1
P	$\square$	$\square$	1	1	$-\frac{2n_g-4}{2n_g}$
$\widetilde{P}$	$\overline{\square}$	1	$\overline{\square}$	-1	$-\frac{2n_g-4}{2n_g}$

In order for the magnetic theory to be nonabelian we must have $2n_g - 4 \geq 2$ yielding the phenomenologically intriguing result that $n_g \geq 3$. Of course, if $n_g = 3$ the electric gauge group is $SU(2)$ and we expect an enhanced accidental global symmetry to occur i.e. $SU(2n_g)_L \times SU(2n_g)_R \times U(1)_p \subset SU(4n_g)$, however if $n_g > 3$ the electric theory has the same global symmetries of the magnetic one. Requiring the magnetic theory to remain asymptotically free we deduce the upper bound on the number of generations to be 6 and therefore:

$$3 \leq n_g \leq 6 \ . \tag{4}$$

Interestingly duality, via a Pati-Salam extension of the SM, hints to an elegant solution to the mystery of the phenomenological existence of, at least, three generations of quarks and leptons.

Our construction predicts the existence of few more matter fields around the energy scale where the Pati-Salam extended color gauge group $SU(4)$ appears. We expect this scale to be above the TeV scale. The reader will recognize that our magnetic spectrum resembles a supersymmetric one, however, the magnetic theory is not supersymmetric since we do not invoke supersymmetric relations among its spectrum and couplings.[73]

The dual electric theory is expected to be strongly coupled at the energy scale where the magnetic one is weakly coupled explaining why the quarks, the leptons and the Higgs seem elementary. The Pati-Salam like extension of the SM and our electric dual might not be unique and, in principle, an even more minimal magnetic extension of the SM with associated electric description could exist. A similar supersymmetric construction was attempted in.[81]

Some caveats and further explanations are in order. Our results rely on the potential existence of gauge duals for nonsupersymmetric gauge theories featuring only fermionic matter whose dual involves both fermions and scalars. This type of duality has not yet been fully established for the theory presented here. There is, however, a known example in nature. This is ordinary QCD. The dual theory of QCD is the theory of hadrons which features composite fermions and scalars. The hadronic picture is weakly coupled at large N. Therefore nonsupersymmetric gauge theories can have dual gauge theories in a fashion similar to QCD of supersymmetric examples. Arguably the first test for the possible existence of gauge duals is the demonstration that the dual passes the 't Hooft anomaly matching conditions. For the theory used here in[73] it has been exhibit the infinite set of solutions of the 't Hooft anomaly conditions for any number of flavors and colors. We then required a number of universal (i.e. susy independent) constraints.[73]

The introduced scalars in the magnetic theory are composite states of the electric degrees of freedom in the same way that Seiberg's postulated dual quarks are composite states of the electric quarks. A famous example is QCD itself where many spin zero states emerge and are instrumental to explain the low energy physics of QCD.

4. What will be discovered next?

According to the minimal models of dynamical electroweak breaking we expect to see at the LHC the emergence of heavy technnirho and techniaxial states in the one to three TeV region with very small couplings to the SM fields.[60–62] These states will be discovered via Drell-Yan production and/or studied indirectly by investigating the composite Higgs production in association with a SM gauge boson.[60–62]

If, on the other hand, composite dynamics emerges in the form of extreme compositeness we will observe a new zoo of composite particles some of which may carry colour. For this kind of extreme dynamics we do not yet have a firm prediction for the new scale of the theory although it is natural to expect it in the multi TeV range as well.

Acknowledgments

I thank Claudio Pica, Robert Shrock and Kimmo Tuominen for useful comments on the manuscript. CP^3-Origins is partially supported by the Danish National Research Foundation, grant number DNFR 90.

References

1. R. Foadi, M. T. Frandsen and F. Sannino, Phys. Rev. D **87**, 095001 (2013) [arXiv:1211.1083 [hep-ph]].
2. G. Aad *et al.* [ATLAS Collaboration], Phys. Lett. B **716**, 1 (2012) [arXiv:1207.7214 [hep-ex]].
3. S. Chatrchyan *et al.* [CMS Collaboration], Phys. Lett. B **716**, 30 (2012) [arXiv:1207.7235 [hep-ex]].
4. [ATLAS Collaboration], ATLAS-CONF-2013-034.
5. G. Degrassi, S. Di Vita, J. Elias-Miro, J. R. Espinosa, G. F. Giudice, G. Isidori and A. Strumia, JHEP **1208**, 098 (2012) [arXiv:1205.6497 [hep-ph]].
6. O. Antipin, M. Gillioz, J. Krog, E. Mølgaard and F. Sannino, arXiv:1306.3234 [hep-ph].
7. O. Antipin, M. Gillioz, E. Mølgaard and F. Sannino, arXiv:1303.1525 [hep-th]. Published in Phys. Rev. D.
8. J. Beringer *et al.* [Particle Data Group Collaboration], Phys. Rev. D **86**, 010001 (2012).
9. [CMS Collaboration], CMS-PAS-HIG-13-005.
10. S. Dimopoulos and H. Georgi, Nucl. Phys. B **193**, 150 (1981).
11. S. Dimopoulos, S. Raby and F. Wilczek, Phys. Rev. D **24**, 1681 (1981).
12. O. Antipin, M. Mojaza, C. Pica and F. Sannino, JHEP **1306**, 037 (2013) [arXiv:1105.1510 [hep-th]].

13. S. Weinberg, Phys. Rev. D **19**, 1277 (1979).
14. L. Susskind, Phys. Rev. D **20**, 2619 (1979).
15. O. Buchmueller, R. Cavanaugh, A. De Roeck, M. J. Dolan, J. R. Ellis, H. Flacher, S. Heinemeyer and G. Isidori *et al.*, Eur. Phys. J. C **72**, 2020 (2012) [arXiv:1112.3564 [hep-ph]].
16. G. Altarelli, arXiv:1206.1476 [hep-ph].
17. R. Barbieri, arXiv:1212.3440 [hep-ph].
18. S. Weinberg, Phys. Rev. D **13**, 974 (1976).
19. S. Dimopoulos and L. Susskind, Nucl. Phys. B **155**, 237 (1979).
20. E. Eichten and K. D. Lane, Phys. Lett. B **90**, 125 (1980).
21. F. Sannino, arXiv:0804.0182 [hep-ph].
22. F. Sannino, Acta Phys. Polon. B **40**, 3533 (2009) [arXiv:0911.0931 [hep-ph]].
23. M. Harada, F. Sannino and J. Schechter, Phys. Rev. D **54**, 1991 (1996) [hep-ph/9511335].
24. S. Weinberg, arXiv:1303.0342 [hep-ph].
25. R. Foadi and F. Sannino, Phys. Rev. D **87**, 015008 (2013) [arXiv:1207.1541 [hep-ph]].
26. F. Sannino and K. Tuominen, Phys. Rev. D **71**, 051901 (2005) [hep-ph/0405209].
27. D. K. Hong, S. D. H. Hsu and F. Sannino, Phys. Lett. B **597**, 89 (2004) [hep-ph/0406200].
28. N. D. Christensen and R. Shrock, Phys. Lett. B **632**, 92 (2006) [hep-ph/0509109].
29. D. D. Dietrich and F. Sannino, Phys. Rev. D **75**, 085018 (2007) [hep-ph/0611341].
30. F. Sannino, Phys. Rev. D **79**, 096007 (2009) [arXiv:0902.3494 [hep-ph]].
31. B. Holdom, Phys. Rev. D **24**, 1441 (1981).
32. T. Appelquist and F. Sannino, Phys. Rev. D **59**, 067702 (1999) [hep-ph/9806409].
33. M. Kurachi and R. Shrock, JHEP **0612**, 034 (2006) [hep-ph/0605290].
34. F. Sannino, Phys. Rev. D **82**, 081701 (2010) [arXiv:1006.0207 [hep-lat]].
35. F. Sannino, Phys. Rev. Lett. **105**, 232002 (2010) [arXiv:1007.0254 [hep-ph]].
36. S. Di Chiara, C. Pica and F. Sannino, Phys. Lett. B **700**, 229 (2011) [arXiv:1008.1267 [hep-ph]].
37. K. Yamawaki, M. Bando and K. -i. Matumoto, Phys. Rev. Lett. **56**, 1335 (1986).
38. M. Bando, K. -i. Matumoto and K. Yamawaki, Phys. Lett. B **178**, 308 (1986).
39. D. D. Dietrich, F. Sannino and K. Tuominen, Phys. Rev. D **72**, 055001 (2005) [hep-ph/0505059].
40. T. Appelquist and Y. Bai, Phys. Rev. D **82**, 071701 (2010) [arXiv:1006.4375 [hep-ph]].
41. S. Catterall and F. Sannino, Phys. Rev. D **76**, 034504 (2007) [arXiv:0705.1664 [hep-lat]].
42. S. Catterall, J. Giedt, F. Sannino and J. Schneible, JHEP **0811**, 009 (2008) [arXiv:0807.0792 [hep-lat]].
43. A. J. Hietanen, J. Rantaharju, K. Rummukainen and K. Tuominen, JHEP **0905**, 025 (2009) [arXiv:0812.1467 [hep-lat]].
44. S. Catterall, J. Giedt, F. Sannino and J. Schneible, arXiv:0910.4387 [hep-lat].
45. L. Del Debbio, B. Lucini, A. Patella, C. Pica and A. Rago, Phys. Rev. D **82**, 014510 (2010) [arXiv:1004.3206 [hep-lat]].
46. J. B. Kogut and D. K. Sinclair, Phys. Rev. D **81**, 114507 (2010) [arXiv:1002.2988 [hep-lat]].
47. J. B. Kogut and D. K. Sinclair, Phys. Rev. D **85**, 054505 (2012) [arXiv:1111.3353 [hep-lat]].
48. F. Bursa, L. Del Debbio, D. Henty, E. Kerrane, B. Lucini, A. Patella, C. Pica and T. Pickup *et al.*, Phys. Rev. D **84**, 034506 (2011) [arXiv:1104.4301 [hep-lat]].
49. J. Giedt and E. Weinberg, Phys. Rev. D **85**, 097503 (2012) [arXiv:1201.6262 [hep-lat]].

50. Z. Fodor, K. Holland, J. Kuti, D. Nogradi, C. Schroeder and C. H. Wong, Phys. Lett. B **718**, 657 (2012) [arXiv:1209.0391 [hep-lat]].
51. R. Lewis, C. Pica and F. Sannino, Phys. Rev. D **85**, 014504 (2012) [arXiv:1109.3513 [hep-ph]].
52. A. Hietanen, C. Pica, F. Sannino and U. I. Sondergaard, Phys. Rev. D **87**, 034508 (2013) [arXiv:1211.5021 [hep-lat]].
53. S. B. Gudnason, C. Kouvaris and F. Sannino, Phys. Rev. D **74**, 095008 (2006) [hep-ph/0608055].
54. T. A. Ryttov and F. Sannino, Phys. Rev. D **78**, 115010 (2008) [arXiv:0809.0713 [hep-ph]].
55. R. Foadi, M. T. Frandsen and F. Sannino, Phys. Rev. D **80**, 037702 (2009) [arXiv:0812.3406 [hep-ph]].
56. M. T. Frandsen and F. Sannino, Phys. Rev. D **81**, 097704 (2010) [arXiv:0911.1570 [hep-ph]].
57. E. Del Nobile, C. Kouvaris and F. Sannino, Phys. Rev. D **84**, 027301 (2011) [arXiv:1105.5431 [hep-ph]].
58. A. Belyaev, M. T. Frandsen, S. Sarkar and F. Sannino, Phys. Rev. D **83**, 015007 (2011) [arXiv:1007.4839 [hep-ph]].
59. T. Appelquist and F. Sannino, Phys. Rev. D **59**, 067702 (1999) [hep-ph/9806409].
60. R. Foadi, M. T. Frandsen, T. A. Ryttov and F. Sannino, Phys. Rev. D **76**, 055005 (2007) [arXiv:0706.1696 [hep-ph]].
61. A. Belyaev, R. Foadi, M. T. Frandsen, M. Jarvinen, F. Sannino and A. Pukhov, Phys. Rev. D **79**, 035006 (2009) [arXiv:0809.0793 [hep-ph]].
62. J. R. Andersen, O. Antipin, G. Azuelos, L. Del Debbio, E. Del Nobile, S. Di Chiara, T. Hapola and M. Jarvinen *et al.*, Eur. Phys. J. Plus **126**, 81 (2011) [arXiv:1104.1255 [hep-ph]].
63. D. B. Franzosi and R. Foadi, arXiv:1209.5913 [hep-ph].
64. T. A. Ryttov and R. Shrock, Phys. Rev. D **84**, 056009 (2011) [arXiv:1107.3572 [hep-ph]].
65. T. A. Ryttov and R. Shrock, Phys. Rev. D **82**, 055012 (2010) [arXiv:1006.5477 [hep-ph]].
66. T. A. Ryttov and R. Shrock, Eur. Phys. J. C **71**, 1523 (2011) [arXiv:1005.3844 [hep-ph]].
67. T. A. Ryttov and R. Shrock, Phys. Rev. D **81**, 115013 (2010) [arXiv:1004.2075 [hep-ph]].
68. T. Appelquist, M. Piai and R. Shrock, Phys. Rev. D **69**, 015002 (2004) [hep-ph/0308061].
69. H. S. Fukano and F. Sannino, Phys. Rev. D **82**, 035021 (2010) [arXiv:1005.3340 [hep-ph]].
70. F. Sannino, Mod. Phys. Lett. A **26**, 1763 (2011) [arXiv:1102.5100 [hep-ph]].
71. F. Sannino, Phys. Rev. D **80**, 065011 (2009) [arXiv:0907.1364 [hep-th]].
72. F. Sannino, Nucl. Phys. B **830**, 179 (2010) [arXiv:0909.4584 [hep-th]].
73. M. Mojaza, M. Nardecchia, C. Pica and F. Sannino, Phys. Rev. D **83**, 065022 (2011) [arXiv:1101.1522 [hep-th]].
74. N. Seiberg, Phys. Rev. D **49**, 6857 (1994) [hep-th/9402044].
75. N. Seiberg, Nucl. Phys. B **435**, 129 (1995) [hep-th/9411149].
76. J. C. Pati and A. Salam, Phys. Rev. Lett. **31**, 661 (1973).
77. J. C. Pati and A. Salam, Phys. Rev. D **8**, 1240 (1973).
78. J. C. Pati and A. Salam, Phys. Rev. D **10**, 275 (1974) [Erratum-ibid. D **11**, 703 (1975)].

79. R. N. Mohapatra and R. E. Marshak, Phys. Rev. Lett. **44**, 1316 (1980) [Erratum-ibid. **44**, 1643 (1980)].
80. R. de Adelhart Toorop, F. Bazzocchi and L. Merlo, JHEP **1008**, 001 (2010) [arXiv:1003.4502 [hep-ph]].
81. C. Csaki, Y. Shirman and J. Terning, Phys. Rev. D **84**, 095011 (2011) [arXiv:1106.3074 [hep-ph]].

125 GeV Technidilaton at the LHC

Shinya Matsuzaki *

Maskawa Institute for Science and Culture, Kyoto Sangyo University, Motoyama, Kamigamo, Kita-Ku, Kyoto 603-8555, Japan.

The technidilaton (TD) is a composite scalar predicted in walking technicolor (WTC), arising as a pseudo Nambu-Goldstone boson associated with the spontaneous breaking of the approximate scale invariance. Due to the Nambu-Goldstone boson's nature, the TD can be as light as the LHC boson that has been discovered at around 125 GeV. We discuss the size of the TD mass and the coupling properties relevant to the LHC study. It turns out that the TD couplings to the standard model (SM) particles take the same form as those of the SM Higgs boson, except the essentially distinguishable two ingredients: i) the overall coupling strengths set by the decay constant related to the spontaneous breaking of the scale invariance, which is in general not equal to the electroweak scale; ii) the couplings to photons and gluons which can include extra contributions from technifermion loops and hence can be enhanced compared to the SM Higgs case. To be concrete, we take the one-family technicolor model to explore the TD LHC phenomenology at 125 GeV. It is shown that the TD gives the signal consistent with the currently reported LHC data, notably can explain the excess in the diphoton channel, due to the extra contributions to digluon and diphoton couplings coming from the one-family technifermion loops.

1. Introduction

On July 4, 2012, a new particle at around 125 GeV was discovered at the LHC.[1] Through measurements of the coupling property, spin and parity, the new particle has so far been almost consistent with the Higgs boson predicted as a key boson responsible for the origin of mass in the standard model (SM). One discrepancy from the SM Higgs has been, however, reported in the diphoton channel[2] where the observed signal event is about two times larger than the SM Higgs prediction. This would imply that the observed scalar boson is a SM-Higgs impostor concerning the underlying theory beyond the SM.

One such a possible impostor is the technidilaton (TD), a composite scalar boson, predicted in the walking technicolor (WTC)[3,4] with an approximately scale-invariant (conformal) gauge dynamics and a large anomalous dimension $\gamma_m = 1$. The TD is a pseudo Nambu-Goldstone boson for the spontaneous breaking of the approximate scale symmetry triggered by technifermion condensation and hence its lightness, say 125 GeV, would be protected by the approximate scale symmetry

*S.M. is currently at Institute for Advanced Research, Nagoya University, Nagoya 464-8602, Japan.

inherent to the WTC. Thus the discovery of TD would imply the discovery of the WTC.

The LHC signatures of the TD were studied in a couple of recent papers.[5–7] It was shown[6] that the 125 GeV TD is consistent with the currently reported diphoton signal as well as other signals such as WW^* and ZZ^*, etc.. It was emphasized that the TD is favored by the current data thanks to the presence of extra technifermion loop corrections to digluon and diphoton couplings.

This talk summarizes the 125 GeV TD phenomenology at the LHC and shows a theoretically interesting possibility that the TD can indeed be as light as the 125 GeV boson by certain nontrivial feature intrinsic to the dynamics of WTC. The stability of the TD mass is also discussed, where it turns out that the large TD decay constant plays an important role against corrections from the SM particles in a low-energy region to be relevant after the walking behavior.

We start with a review of the characteristic features of WTC including the spontaneous breaking of both chiral and scale symmetries and the mass generation of TD (Sec. 2). We then discuss the size of the TD mass following two approaches: straightforward nonperturbative computations in the ladder approximation[8–10] (Sec. 2.1) and a model estimate in a view of holography applied to WTC[7,11] (Sec.2.2). In contrast to the ladder estimate, the holographic model incorporates effects from the technigluon condensation, which turns out to play the crucial role to realize the TD mass on the order of 125 GeV.[7]

To make direct contact with the LHC study, we derive a low-energy effective Lagrangian[6] obtained by integrating out technifermions, which is governed by the lightest technihadron spectra involving the TD (Sec. 3). The Lagrangian is constructed based on nonlinear realization of the scale symmetry as well as the chiral symmetry coupled to the SM particles via the SM gauge interactions.

The TD couplings to the SM particles are read off from the effective Lagrangian. We then find that the couplings take essentially the same form as those of the SM particles, except two distinguishable ingredients: the overall scale of the couplings are set by the TD decay constant (F_ϕ) associated with the spontaneous breaking of the scale symmetry, not by the electroweak (EW) scale ($v_{\rm EW}$). Actually, F_ϕ turns out to be larger than $v_{\rm EW}$ from both ladder and holographic estimates. The other crucial discrepancy comes in the couplings to diphoton and digluon, which get effects from technifermion loop contributions to be either enhanced or suppressed, depending on models of WTC. To be concrete, we will take one-family model for the WTC to explicitly calculate those couplings. After understanding the TD coupling property from the effective Lagrangian, we pay attention to the stability of the TD mass against radiative corrections arising from the SM particle loops. The large TD decay constant F_ϕ then turns out to play the crucial role to keep the TD mass around 125 GeV without invoking heavy fine tunning, in contrast to the SM Higgs case.

Finally, we shall explore the LHC production and decay processes of the 125 GeV TD in the one-family model and estimate the signal strengths for the relevant channels to be compared with the experimental data on the Higgs searches (Sec. 4). It is shown that the TD signal is consistent with the currently reported data and can give a better fit than the SM Higgs, notably due to the presence of extra technifermion loop corrections.

2. Walking Technicolor and Technidilaton

Technicolor (TC)[12,13] accommodates the EW symmetry breaking by technifermion condensation, without invoking the fundamental Higgs boson, just like the quark condensation in QCD, and hence gives the dynamical explanation for the origin of mass. The original version of TC,[12] a naive scale-up version of QCD, was ruled out due to the excessive flavor-changing neutral currents.

A way out of this problem was suggested under a simple assumption of the existence of a large anomalous dimension for technifermion bilinear operator γ_m without any concrete dynamics and concrete value of the anomalous dimension.[14] It was the WTC[3,4] that exhibited a concrete dynamics based on a nonperturbative analysis of ladder Schwinger-Dyson (SD) equation with nonrunning (scale invariant/conformal) gauge coupling, $\alpha(p) \equiv \alpha$, yielding a concrete value of the anomalous dimension, $\gamma_m = 1$ in the chiral broken phase. Modern version of WTC[15–17] is based on the two-loop running coupling with the Caswell-Banks-Zaks infrared fixed point,[18] instead of the nonrunning one, in the improved ladder SD equation.

Another problem of the TC as a QCD scale-up is the EW constraints, so-called S and T parameters. This may also be improved in the WTC.[9,19] Even if WTC in isolation cannot overcome this problem, there still exist a possibility that the problem may be resolved in the combined dynamical system including the SM fermion mass generation such as the extended TC (ETC) dynamics,[20] in much the same way as the solution ("ideal fermion delocalization")[21] in the Higgsless models which simultaneously adjust S and T parameters by incorporating the SM fermion mass profile.

It would be better to understand the WTC through an illustration by the notion of gauge dynamics: Figure 1 shows a schematic view of the WTC in terms of the gauge coupling α (left panel) and its beta function $\beta(\alpha)$ (right panel). In a field theoretical sense, the "walking" can be realized by an accidental balance between gluonic and fermionic contributions to the running of gauge coupling of $SU(N_{\rm TC})$ gauge theory including $N_{\rm TF}$ fermions. In fact, the presence of walking region implies a pseudo infrared fixed point (α_*), a la Caswell-Banks-Zack infrared fixed point based on the two-loop beta function in the large N_f QCD.[18] During the walking region (region II in Fig. 1), the gauge coupling slowly reaches the critical coupling, which is slightly off from an infrared fixed point α_*, $\alpha_{\rm cr}(< \alpha_*)$, where the chiral symmetry is dynamically broken by technifermion condensation $\langle \bar{F}F \rangle \neq 0$ and hence technifermions get the dynamical mass $m_F = \mathcal{O}(4\pi F_\pi) \sim 1$ TeV, where F_π denotes the technipion decay constant associated with the chiral symmetry breaking. The walking regime (region II) expands in a wide range ($m_F < \mu < \Lambda_{\rm TC}$): below

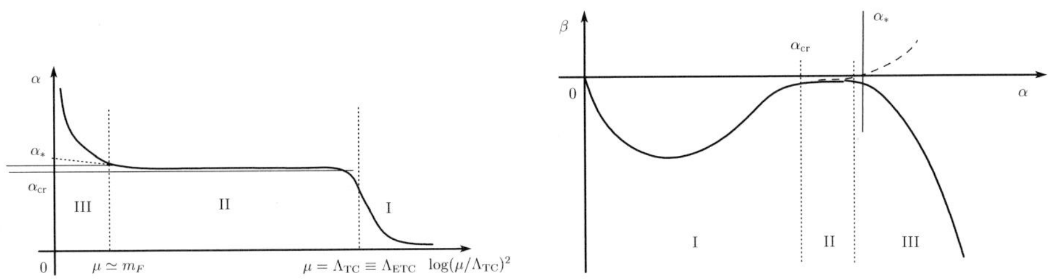

Fig. 1. A schematic view of WTC in terms of the running of the gauge coupling α (left panel) and the beta function β (right panel).

the infrared scale $m_F(\sim 1\,\mathrm{TeV})$, technifermions decouple and hence the balance with technigluon contributions gets lost, leading to the one edge of walking (region III), while above the ultraviolet scale $\Lambda_{\rm TC}(\sim 10^3 - 10^4$ TeV) the theory will be embedded into an ETC (region I).

The dynamical mass generation of technifermion is characterized by so-called Miransky scaling,[22] tied with the conformal phase transition,[17]

$$m_F \sim \Lambda_{\rm TC} e^{-\frac{\pi}{\sqrt{\alpha/\alpha_{\rm cr}-1}}}\,, \qquad \text{for} \qquad \alpha > \alpha_{\rm cr}\,. \tag{1}$$

Note that this scaling property implies the wide-range walking regime above $\alpha_{\rm cr}$ in region II of Fig. 1. The chiral condensate $\langle \bar{F}F \rangle$ is generated to scale with the large anomalous dimension $\gamma_m \simeq 1$ which solves the flavor-changing neutral current problem:[3]

$$\langle \bar{F}F \rangle_{\Lambda_{\rm TC}} \sim \frac{N_{\rm TC}}{4\pi^2} m_F \Lambda_{\rm TC}^2\,, \qquad \langle \bar{F}F \rangle_{\Lambda_{\rm TC}} \simeq \left(\frac{\Lambda_{\rm TC}}{m_F}\right)^{\gamma_m \simeq 1} \langle \bar{F}F \rangle_{m_F}\,. \tag{2}$$

Of particular interest is the fact that the WTC possesses the (approximate) scale invariance ($\beta(\alpha) \simeq 0$ for $m_F < \mu < \Lambda_{\rm TC}$), which is spontaneously broken by the technifermion condensation/mass generation. This implies presence of a (pseudo) Nambu-Goldstone boson ("dilaton") for the scale symmetry. This is a flavor-singlet composite scalar and may be a mixture of $\bar{F}F$ bound state and technigleuball. That is the TD, technidilaton.

The TD gets massive essentially due to the "nonperturbative" scale anomaly: According to the dynamical mass generation in Eq.(1), the gauge coupling α is renormalized, starts running and hence the "nonperturbative" beta function $\beta_{\rm NP}(\alpha)$ is generated:[23]

$$\beta_{\rm NP}(\alpha) = \Lambda_{\rm TC} \frac{\partial \alpha}{\partial \Lambda_{\rm TC}} = -\frac{2\alpha_{\rm cr}}{\pi} \left(\frac{\alpha}{\alpha_{\rm cr}} - 1\right)^{3/2}\,. \tag{3}$$

This induces the "nonperturbative" scale anomaly:

$$\partial_\mu D^\mu = \frac{\beta_{\rm NP}(\alpha)}{4\alpha^2} \left(\alpha G_{\mu\nu}^2\right) \neq 0\,, \tag{4}$$

where D_μ denotes the dilatation current and $G_{\mu\nu}$ the field strength of technigluon field. Thus the TD becomes massive due to the nonperturbative scale anomaly.[3,4]

2.1. *Technidilaton Mass: Ladder Estimate*

A straightforward calculation [8] based on the ladder SD equation and the ladder (homogeneous) Bethe-Salpeter (BS) equation in the walking regime indicates a relatively light scalar bound state (identified with TD):

$$\text{Ladder :} \qquad M_\phi \sim 4F_\pi \simeq 500\,\text{GeV}\,, \tag{5}$$

for the one-family model with $F_\pi \simeq 123$ GeV. This M_ϕ is much smaller than masses of the techni-vector/axial-vector mesons on TeV range, but still larger than the LHC boson at 125 GeV. This result is consistent with another calculation[9] based on the ladder SD equation and the ladder (inhomogeneous) BS equation, and also consistent with other indirect computation[10] based on the ladder gauged Nambu-Jona-Lasinio model.

In fact, the PCDC (partially conserved dilatation current) relation evaluated in the ladder approximation does not allow a very light TD unless the TD gets decoupled with divergent decay constant F_ϕ:[11,24] The PCDC relation reads

$$F_\phi^2 M_\phi^2 = -4\langle\theta_\mu^\mu\rangle = \frac{\beta(\alpha)}{\alpha}\langle G_{\mu\nu}^2\rangle\,, \tag{6}$$

where θ_μ^μ is the trace of energy-momentum tensor. To this relation, the ladder estimate gives $F_\phi^2 M_\phi^2 \simeq 3\eta\, m_F^4$ (near the conformal window) with $\eta \simeq \frac{N_{\rm TC}N_{\rm TF}}{2\pi^2} = \mathcal{O}(1)$.[24,25] This simply implies that the limit $M_\phi/m_F \to 0$, where the TD gets light compared with the weak scale $m_F (= \mathcal{O}(4\pi F_\pi))$, can only be realized when $F_\phi/m_F \to \infty$, i.e., a decoupled limit:

$$\text{Ladder :} \qquad \frac{M_\phi}{F_\pi} \to 0 \qquad \text{only when} \qquad \frac{F_\phi}{F_\pi} \to \infty\,. \tag{7}$$

2.2. *Technidilaton Mass: Holographic Estimate*

As suggested earlier,[26] however, there is a potential problem in the ladder approximation about the mass of the TD. Actually, the ladder approximation totally ignores non-ladder dynamics, say, most notably the full gluonic dynamics. Thus a possible way out would be to include fully *nonperturbative gluonic dynamics*. (Also a direct estimate of F_ϕ free from the ladder approximation and without invoking the PCDC (without referring to M_ϕ) is necessary to give more implications of the TD at the LHC.) One such a possibility besides lattice simulations would be a holographic computation based on the gauge-gravity duality.[27]

We shall work on a recently proposed holographic model,[11] which is based on deformation of a bottom-up approach for successful holographic-dual of QCD.[28,29] The model describes a five-dimensional gauge theory having $SU(N_{\rm TF})_L \times SU(N_{\rm TF})_R$ gauge symmetry, defined on the five-dimensional anti-de-Sitter space with the fifth direction being compactified on a finite interval. In addition to a bulk scalar dual to the technifermion bilinear operator $\bar{F}F$ with the anomalous dimension $\gamma_m = 1$, an extra bulk scalar field dual to technigluon condensate $\langle \alpha G_{\mu\nu}^2 \rangle$ is incorporated [a].

Following the well-known holographic recipe, one can easily calculate the scalar (or dilatation) current correlator and examine the lowest pole identified as the TD mass to get a formula in a light TD limit:

$$\text{Holography}: \qquad \frac{M_\phi}{4\pi F_\pi} \simeq \sqrt{\frac{3}{N_{\rm TC}} \frac{\sqrt{3}/2}{1+G}}\,, \qquad \text{where} \qquad G \sim \frac{\langle \alpha G_{\mu\nu}^2 \rangle}{F_\pi^4}\,. \tag{8}$$

(The details can be found in the published paper.[7]) This implies that, in contrast to the ladder approximation Eq.(5), the TD can be as light as the 125 GeV boson by a large gluonic effect G:

$$\text{Holography}: \qquad M_\phi \simeq 125\,\text{GeV} \qquad \text{at} \qquad G \simeq 10\,, \tag{9}$$

for the one-family model with $N_{\rm TC} = 3$. It is also interesting to note that the gluonic effect in QCD is much smaller, $G|_{\rm QCD} \simeq 0.25$.[11]

Similarly, one can calculate the TD decay constant F_ϕ by examining the pole residue of the dilatation current correlator to find a formula:[7]

$$\text{Holography}: \qquad \frac{F_\phi}{F_\pi} \simeq \sqrt{2N_{\rm TF}}\,. \tag{10}$$

This and Eq.(8) indicate that, in contrast to the ladder approximation Eq.(7), a very light/massless TD with the finite decay constant F_ϕ can be realized when G goes to infinity:

$$\text{Holography}: \qquad \frac{M_\phi}{F_\pi} \to 0 \qquad \text{and} \qquad \frac{F_\phi}{F_\pi} = \text{constant}\,, \qquad \text{as} \quad G \to \infty\,. \tag{11}$$

The large G limit actually corresponds to a "conformal" limit where $\beta(\alpha) \to 0$: Combining Eqs.(8) and (10) with the PCDC relation in Eq.(6) and calculating the technigluon condensate $\langle \alpha G_{\mu\nu}^2 \rangle$, one can evaluate the beta function as a function of G to find that $\beta(\alpha) \sim \frac{1}{G(1+G)^2} \to 0$ as $G \to \infty$.[7] Thus the holographic estimate suggests that a very light TD with the finite decay constant can be allowed due to the conformality tied with the nonperturbative gluonic effect G.

Though this holographic result is strong enough to indicate the presence of the 125 GeV TD, we would anticipate that lattice simulations will give a more conclusive answer.

3. Low-Energy Effective Lagrangian

Below the technifermion mass scale $m_F = \mathcal{O}(4\pi F_\pi) \sim 1$ TeV (See region III in Fig .1), the WTC can be viewed as an effective technihadron model as in the QCD case. Such an effective model will be described by a low-energy effective Lagrangian based on nonlinear realization of both chiral and scale symmetries, involving the

[a] The extra bulk scalar dual to $\langle \alpha G_{\mu\nu}^2 \rangle$ is actually necessary to properly reproduce the high energy behavior of vector/axial-vector current correlator including the gluon condensate term as predicted in the operator product expansion.[11]

125 GeV TD and the SM particles, which is most relevant for the LHC study [b]. The chiral/EW and scale invariant Lagrangian thus takes the form[6]

$$\mathcal{L}_{\rm inv} = \frac{v_{\rm EW}^2}{4}\chi^2 {\rm tr}[D_\mu U^\dagger D^\mu U] + \mathcal{L}_{\rm kin}(\chi) \,, \tag{12}$$

where $\chi = e^{\phi/F_\phi}$ is a nonlinear base of the scale symmetry, which parametrizes the TD field ϕ with the decay constant F_ϕ and has the scale dimension 1; $D_\mu U = \partial_\mu U - iW_\mu U + iUB_\mu$ with the $SU(2)_W$ and $U(1)_Y$ gauge fields W and B; $\mathcal{L}_{\rm kin}(\chi)$ denotes the scale invariant kinetic term of TD and U the usual chiral field parameterizing the (eaten) Nambu-Goldstone boson fields π as $U = e^{2i\pi/v_{\rm EW}}$; The $|D_\mu U|^2$ term gives the TD couplings to massive weak bosons:

$$g_{\phi WW/ZZ} = \frac{2m_{W/Z}^2}{F_\phi} \,. \tag{13}$$

As seen in Eqs.(3) and (4), the scale symmetry is actually broken explicitly as well as spontaneously by dynamical mass generation of technifermions, which has to be respected also in the nonlinear realization.[6] Such explicit breaking effects arise in the TD Yukawa couplings to the SM fermions, which reflect underlying ETC-induced four-fermion terms, and couplings to QCD gluons and photons related to the scale anomaly in the SM gauge sector. In order to incorporate these effects into the scale-invariant Lagrangian, we introduce a spurion field S having the scale dimension 1 coupled to the SM fermions, digluon gg and diphoton $\gamma\gamma$ in such a way that[6]

$$\begin{aligned}\mathcal{L}_S = & -m_f\left(\left(\frac{\chi}{S}\right)^{2-\gamma_m}\cdot\chi\right)\bar{f}f \\ & + \log\left(\frac{\chi}{S}\right)\left\{\frac{\beta_F(g_s)}{2g_s}G_{\mu\nu}^2 + \frac{\beta_F(e)}{2e}F_{\mu\nu}^2\right\} \,,\end{aligned} \tag{14}$$

where $G_{\mu\nu}$ and $F_{\mu\nu}$ respectively denote the field strengths for QCD gluon and photon fields; g_s and e are the QCD gauge and electromagnetic couplings, respectively; β_Fs are the beta functions only including the technifermion loop contributions.

The TD Yukawa coupling to the SM f-fermion arises from the first line of Eq.(14) as[4]

$$g_{\phi ff} = \frac{(3-\gamma_m)m_f}{F_\phi} \,, \tag{15}$$

along with scale dimension of technifermion bilinear operator $(3-\gamma_m)$. The anomalous dimension $\gamma_m \simeq 1$ in WTC, which is crucial to obtain the realistic mass of the SM fermions of the first and the second generations without suffering from the flavor-changing neutral current problems. However it was known for long time that it is not enough for the mass of the third-generation SM f-fermions like t, b, τ: A

[b]We focus only on the TD phenomenology with the SM particles and disregard technipions, some of which might participate in the low-energy model as discussed in a recent paper.[30]

simplest resolution would be the strong ETC model[31] having much larger anomalous dimension $1 < \gamma_m < 2$ due to the strong effective four-fermion coupling from the ETC dynamics in addition to the walking gauge coupling. Here we take $\gamma_m \simeq 2$, i.e., $(3-\gamma_m) \simeq 1$, as in the strong ETC model for the third-generation SM f-fermions like t, b, τ which are relevant to the current LHC data.

3.1. *Technidilaton Couplings to the SM particles*

We see from Eqs.(13) and (15) that the TD couplings to W and Z bosons and fermions are related to those of the SM Higgs by a simple scaling:

$$\begin{aligned} \frac{g_{\phi WW/ZZ}}{g_{h_{\rm SM}WW/ZZ}} &= \frac{v_{\rm EW}}{F_\phi}\,, \\ \frac{g_{\phi ff}}{g_{h_{\rm SM}ff}} &= \frac{v_{\rm EW}}{F_\phi}\,, \qquad \text{for} \quad f = t, b, \tau\,. \end{aligned} \tag{16}$$

In addition to the above scaling, the couplings to gluon and photon ($G_{\mu\nu}^2$ and $F_{\mu\nu}^2$ terms in Eq.(14)) involve the beta functions, $\beta_F(g_s)$ and $\beta_F(e)$, induced from F-technifermion loops.[5] We shall employ the one-family model as a concrete WTC setting and evaluate the betas $\beta_F(g_s)$ and $\beta_F(e)$ at the one-loop level in the perturbation [c]:

$$\text{one-family model}: \qquad \beta_F(g_s) = \frac{g_s^3}{(4\pi)^2}\frac{4}{3}N_{\rm TC}\,, \qquad \beta_F(e) = \frac{e^3}{(4\pi)^2}\frac{16}{9}N_{\rm TC}\,. \tag{17}$$

We thus find the scaling from the SM Higgs for the couplings to gg and $\gamma\gamma$, which can approximately be expressed at around 125 GeV as

$$\text{one-family model}: \qquad \begin{array}{lcl} \frac{g_{\phi gg}}{g_{h_{\rm SM}gg}} &\simeq& \frac{v_{\rm EW}}{F_\phi}\cdot(1+2N_{\rm TC}) \\ \frac{g_{\phi\gamma\gamma}}{g_{h_{\rm SM}\gamma\gamma}} &\simeq& \frac{v_{\rm EW}}{F_\phi}\cdot\left(1-\frac{32}{47}N_{\rm TC}\right) \end{array}, \tag{18}$$

where in estimating the SM contributions we have incorporated only the top and W boson loop contributions.

One could actually derive these TD couplings directly from the Ward-Takahashi identities for the dilatation symmetry (low-energy theorem).[6]

As seen from Eqs.(16) and (18), once the ratio $v_{\rm EW}/F_\phi$ is fixed, the TD LHC phenomenological study can be made just by quoting the SM Higgs coupling properties: We may estimate F_ϕ based on the ladder approximation:[6] The TD decay constant F_ϕ can actually be related to the TD mass M_ϕ through the PCDC – which is analogous to the PCAC (partially conserved axialvector current) relation for the QCD pion – involving the vacuum energy density $\mathcal{E}_{\rm vac}$:

$$F_\phi^2 M_\phi^2 = -16\mathcal{E}_{\rm vac}\,. \tag{19}$$

[c]This one-loop approximation can be justified in ladder estimate of the scale anomaly-related vertices ϕ-$\gamma(g)$-$\gamma(g)$.[6]

The vacuum energy density $\mathcal{E}_{\rm vac}$ is saturated by the technigluon condensation induced from the technifermion condensation, so is generically expressed as

$$\mathcal{E}_{\rm vac} = -\kappa_V \left(\frac{N_{\rm TC} N_{\rm TF}}{8\pi^2} \right) m_F^4 \,, \tag{20}$$

where m_F denotes the dynamical technifermion mass and $N_{\rm TF} = 2N_{\rm D} + N_{\rm EW-singlet}$ with N_D being the number of EW doublets formed by technifermions and the number of dummy technifermions, $N_{\rm EW-singlet}$, which are singlet under the SM gauges. The overall coefficient κ_V is determined once a straightforward nonperturbative calculation is made. The dynamical technifermion mass m_F can, on the other hand, be related to the technipion decay constant F_π:

$$F_\pi^2 = \kappa_F^2 \frac{N_{\rm TC}}{4\pi^2} m_F^2 \,, \tag{21}$$

with the overall coefficient κ_F and the property of $N_{\rm TC}$ scaling taken into account. The values of κ_V and κ_F may be quoted from the latest result[24] on a ladder Schwinger-Dyson analysis for a modern version of WTC:[15–17]

$$\kappa_V \simeq 0.7 \,, \qquad \kappa_F \simeq 1.4 \,, \tag{22}$$

where κ_F has been estimated based on the Pagels-Stokar formula.[32] In that case $N_{\rm TF}$ is fixed by the criticality condition for the walking regime as[16]

$$\frac{N_{\rm TF}}{4N_{\rm TC}} \simeq 1 \,. \tag{23}$$

The estimated values in Eqs.(22) and (23) are based on ladder approximation which are subject to certain uncertainties up to 30% observed for the critical coupling and hadron spectrum in QCD.[33] We may include this 30% uncertainty in estimation of each independent factor κ_V, κ_F^2 and the criticality condition $N_{\rm TF}/(4N_{\rm TC})$. Putting these all together, we thus estimate $v_{\rm EW}/F_\phi$ as

$$\text{one-family model}: \qquad \left.\frac{v_{\rm EW}}{F_\phi}\right|_{\rm Ladder} \simeq (0.1 - 0.3) \times \left(\frac{N_D}{4} \right) \left(\frac{M_\phi}{125\,{\rm GeV}} \right) \,, \tag{24}$$

Alternatively, using the holographic estimate in Eq.(10) and incorporating a typical ~30% correction into the holography coming from the next-to-leading order terms in $1/N_{\rm TC}$ expansion, we may estimate the TD decay constant F_ϕ to get[7]

$$\text{one-family model}: \qquad \left.\frac{v_{\rm EW}}{F_\phi}\right|_{\rm holo}^{+1/N_{\rm TC}} \simeq 0.2 - 0.4 \,. \tag{25}$$

which is coincidentally consistent with the ladder estimate in Eq.(24). Note that the two calculations are quite different qualitatively in a sense that the ladder calculation has no massless TD limit, while the holographic model including the nonperturbative gluonic dynamics does. Nevertheless, such a numerical coincidence may suggest that both models are reflecting some reality through similar dynamical effects for the particular mass region of the 125 GeV TD.

3.2. *Technidilaton Mass Stability*

Before proceeding to the LHC phenomenology, we shall briefly remark on stability of the light TD mass against radiative corrections. As a pseudo Nambu-Goldstone boson of scale invariance, the quadratic divergence is suppressed by the scale invariance for the walking regime $m_F < \mu < \Lambda_{\rm TC} (\sim \Lambda_{\rm ETC})$ (region II in Fig. 1). The scale symmetry breaking in the ultraviolet region $\mu > \Lambda_{\rm TC}$ (region I in Fig. 1) has no problem thanks to the naturalness as usual like in the QCD and the QCD-scale-up TC where the theory has only logarithmic divergences. Only possible source of the scale symmetry violation is from an effective theory for $\mu < m_F$ (region III in Fig. 1).

Below $\mu = m_F$ the dominant corrections to the TD mass M_ϕ come from the SM top quark loop [d], which can be estimated from the effective Lagrangian in Eqs.(12) and (14) as

$$\delta M_\phi^2 \simeq -\frac{3}{4\pi^2} \frac{m_t^2}{F_\phi^2} m_F^2 \,, \tag{26}$$

where the cutoff has been set by m_F. Due to the large TD decay constant $F_\phi \sim 1{\rm TeV}(\simeq m_F = \mathcal{O}(4\pi F_\pi))$ (see Eq.(25)), this correction only gives a tiny shift to the 125 GeV TD mass (Note numerically $m_t^2 \simeq 2M_\phi^2$):

$$\left| \frac{\delta M_\phi}{M_\phi(125{\rm GeV})} \right| \simeq \frac{3}{4\pi^2} \frac{m_F^2}{F_\phi^2} \simeq \mathcal{O}(10^{-2} - 10^{-1}) \,. \tag{27}$$

Thus the 125 GeV TD mass is fully stable against the radiative corrections, in contrast to the unnatural SM Higgs case.

4. 125 GeV Technidilaton Signal at the LHC

We now discuss the 125 GeV TD phenomenology at the LHC based on the coupling properties derived in the previous section. As is clear from Eqs.(16) and (25), the TD couplings to the SM particles are simply suppressed by the large decay constant F_ϕ compared to the SM Higgs case. As seen from Eq.(18), however, the gluon-gluon fusion (ggF) and diphoton couplings can be enhanced due to extra contributions from technifermions in the one-family model which compensate the overall suppression by F_ϕ. Thus the ggF process becomes mostly dominant in the TD LHC production, while other production processes such as vector boson fusion (VBF) and vector boson associate production (VH) are suppressed simply by the large TD decay constant F_ϕ. Similarly, the branching fraction is almost governed by the decay to digluon which becomes comparable with $b\bar{b}$ channel for the SM Higgs, while other channels such as the decay to WW^*, ZZ^* and fermion pairs are suppressed. (The

[d] Actually, a TD potential term involving the usual $\lambda\phi^4$ coupling could yield a similar correction to M_ϕ at one-loop level. This contribution, however, turns out to be negligible in magnitude compared to the top-loop correction in Eq.(26), due to the high suppression factor $(M_\phi/(4\pi F_\phi))^2$.[6]

TD total width at 125 GeV is comparable with the SM Higgs one.) More details on the production and decay properties can be found in the published papers.[6]

In order to make a direct comparison with the current LHC data,[1,2,34] we estimate the TD signal strengths normalized to the SM Higgs cross section,

$$\mu_X = \frac{\sigma_\phi(pp \to \phi) \times \mathrm{BR}(\phi \to X)}{\sigma_{h_{\mathrm{SM}}}(pp \to h_{\mathrm{SM}}) \times \mathrm{BR}(h_{\mathrm{SM}} \to X)} . \tag{28}$$

Taking the one-family model with $N_{\mathrm{TC}} = 4$ and $v_{\mathrm{EW}}/F_\phi = 0.2$ as a benchmark, in Table 1 we summarize the predicted signal strengths at 125 GeV for each category relevant to the current Higgs search at 8TeV LHC. As for VBF-tag categories, we have taken into account about 30% contamination from the ggF process as in the case of the SM Higgs.

Table 1. The predicted 8TeV LHC signal strengths of 125 GeV TD in the one-family model with $N_{\mathrm{TC}} = 4$ and $v_{\mathrm{EW}}/F_\phi = 0.2$ fixed.

channel	ggF-tag	VBF-tag	VH-tag
$\gamma\gamma$	1.4	0.4	0.02
ZZ^*	1.0	0.3	0.01
WW^*	1.0	0.3	0.01
$\tau^+\tau^-$	1.0	0.3	0.01
$b\bar{b}$	—	—	0.01

From Table 1 we see that the diphoton signal strength for the ggF-tag category is slightly enhanced and can explain the currently reported excess at the LHC,[1,2,34] while other channels in the ggF-tag categories are almost SM Higgs-like, not distinguishable from those for the SM Higgs. The characteristic feature of the 125 GeV TD in the one-family model is seen at the VBF- and VH-tag categories, which are significantly suppressed compared to the SM Higgs case, simply due to the overall suppression of the couplings by the large decay constant F_ϕ. This will be tested in the future, though the presently observed significance on such categories are still lower than those for the ggF-tag category.

Finally, we shall test the goodness-of-fit of the TD with use of the current LHC data,[1,2,34] based on the χ^2 function:

$$\chi^2 = \sum_{i \in \mathrm{events}} \left(\frac{\mu_i - \mu_i^{\mathrm{exp}}}{\sigma_i} \right)^2 , \tag{29}$$

where μ_i^{exp} denote the best-fit strengths for each channel reported and σ_i the corresponding one sigma errors. Taking (v_{EW}/F_ϕ) as a free parameter so as to satisfy the theoretically expected range in Eq.(24) (or Eq.(25)), in Fig. 2 we plot the χ^2 function for the 125 GeV TD in the one-family model with $N_{\mathrm{TC}} = 3, 4, 5$. In particular, the best-fit values for $N_{\mathrm{TC}} = 4, 5$ are as follows:

N_{TC}	$(v_{\mathrm{EW}}/F_\phi)_{\mathrm{best}}$	χ^2_{min}/d.o.f
4	0.22	$18/19 \simeq 0.95$
5	0.17	$18/19 \simeq 0.95$

, (30)

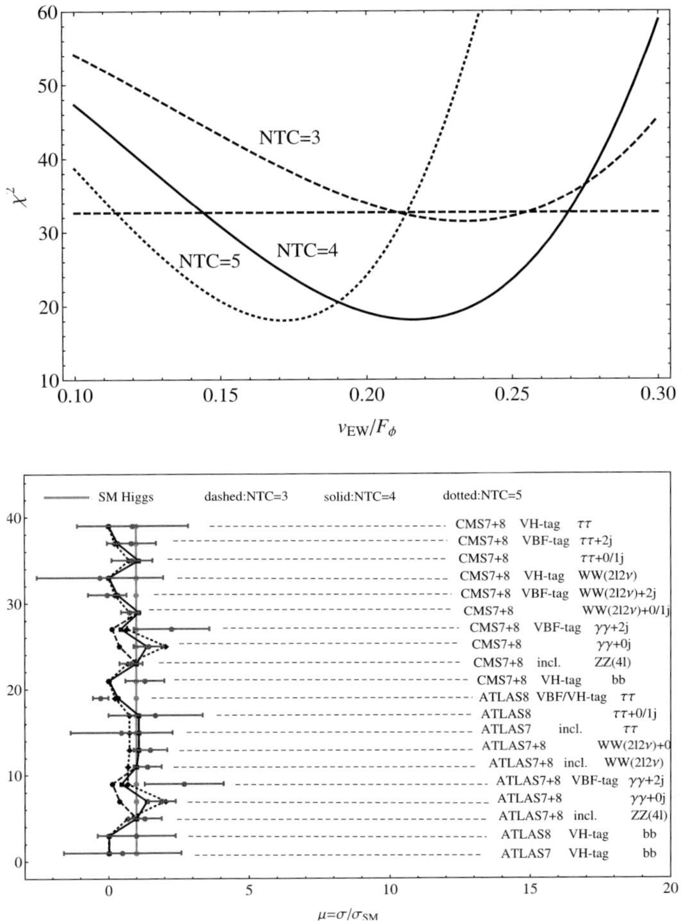

Fig. 2. Top panel: χ^2 as a function of $v_{\rm EW}/F_\phi$ for the 125 GeV TD in the one-family model with $N_{\rm TC} = 3, 4, 5$. The dotted straight line corresponds to the SM Higgs case. Bottom panel: the predicted signal strengths in comparison with the current LHC data.[1,2,34]

which are compared with the SM Higgs case, $\chi^2|_{\rm SM} \simeq 33/20 \simeq 1.6$, implying that the TD is more favorable than the SM Higgs ($\mu_i = 1$). This nice goodness of fit is due to the significant enhancement in the diphoton channel coming from the sector beyond the SM (technifermions): The technifermion loop contributions as in Eq.(18) become large enough to compensate the smallness of the overall $(v_{\rm EW}/F_\phi)$ in Eq.(24).

5. Conclusion

The TD is the characteristic light composite scalar in WTC. The mass can be 125 GeV, protected by the approximate scale invariance. The TD mass is also stable against the SM radiative corrections, where the large TD decay constant plays an important role. The couplings to the SM particles take essentially the same form as those for the SM Higgs, except couplings to diphoton and digluon which involve a high model-dependence. The 125 GeV TD in the one-family model gives the LHC signal consistent with the LHC data, notably can explain the currently reported diphoton excess. More precise measurements in VBF and VH categories will draw a definite conclusion that the TD is favored, or not.

Note added:

After the SCGT 12 workshop, the ATLAS and CMS have reported updated data in some channels for the Higgs search involving $WW^*, ZZ^*, \tau^+\tau^-$ and diphoton channels.[35,36] However, the signals in VBF- and VH-tag categories have not yet clearly been measured, which would actually be the most characteristic for the 125 GeV TD in the one-family model: The most significant channels with the significance over $\sim 3-4$ sigma are still only in the ggF-tag categories, as far as the amount of observed excess over the background is concerned. See Table 2. Thus the 125 GeV TD in the one-family model is still consistent with the latest data. Further update on measurements in the VBF- and VH-tag categories is needed to more definitely say that is the TD, or not.

Table 2. The predicted 8TeV LHC signal strengths of 125 GeV TD in the one-family model with $N_{\rm TC} = 4$ and $v_{\rm EW}/F_\phi = 0.2$ fixed, in comparison with the latest LHC data as of *Moriond 2013 EW and QCD sessions* (or *Aspen 2013-Higgs Quo Vadis*) in 2013.[35,36] The sizes of the observed significance for signal over background have also been displayed.

category	TD signal strength	ATLAS	significance	CMS	significance
ggF-tag					
$\mu_{\gamma\gamma}^{\rm ggF}$	1.4	1.6 ± 0.4	$\sim 6\sigma$	0.5 ± 0.5	$\sim 3\sigma$
$\mu_{WW^*}^{\rm ggF}$	1.0	0.8 ± 0.4	$\sim 4\sigma$	0.8 ± 0.2	$\sim 4\sigma$
$\mu_{ZZ^*}^{\rm ggF}$	1.0	$1.8^{+0.8}_{-0.5}$	$\sim 6\sigma$	$0.9^{+0.5}_{-0.4}$	$\sim 6\sigma$
$\mu_{\tau^+\tau^-}^{\rm ggF}$	1.0	$2.1^{+4.0}_{-3.0}$	$< 2\sigma$	0.7 ± 0.5	$\sim 3\sigma$
VBF-tag					
$\mu_{\gamma\gamma}^{\rm VBF}$	0.4	1.7 ± 0.9	$< 2\sigma$ with VH	$1.5^{+1.5}_{-1.1}$	$< 2\sigma$ with VH
$\mu_{WW^*}^{\rm VBF}$	0.3	1.7 ± 0.8	$< 2\sigma$	$0.04^{+0.77}_{-0.57}$	$< 2\sigma$
$\mu_{ZZ^*}^{\rm VBF}$	0.3	$1.2^{+3.8}_{-1.4}$	$< 2\sigma$ with VH	1.2 ± 5.6	$< 2\sigma$
$\mu_{\tau^+\tau^-}^{\rm VBF}$	0.3	$-0.4^{+3.0}_{-2.0}$	$< 2\sigma$ with VH	1.4 ± 0.6	$< 2\sigma$
VH-tag					
$\mu_{\gamma\gamma}^{\rm VH}$	0.02	$1.8^{+1.5}_{-1.3}$	$< 2\sigma$ with VBF	$1.5^{+1.5}_{-1.1}$	$< 2\sigma$ with VBF
$\mu_{WW^*}^{\rm VH}$	0.01	—	—	$-0.3^{+2.3}_{-2.0}$	$< 2\sigma$
$\mu_{ZZ^*}^{\rm VH}$	0.01	$1.2^{+3.8}_{-1.4}$	$< 2\sigma$ with VBF	—	—
$\mu_{\tau^+\tau^-}^{\rm VH}$	0.01	$-0.4^{+3.0}_{-2.0}$	$< 2\sigma$ with VBF	$0.8^{+1.5}_{-1.4}$	$< 2\sigma$
$\mu_{b\bar{b}}^{\rm VH}$	0.01	-0.4 ± 1.0	$< 2\sigma$	$1.3^{+0.7}_{-0.6}$	$< 2\sigma$

Acknowledgments

I would like to thank K. Yamawaki for collaborations for a couple of recent papers on the technidilaton phenomenology. This work was supported by the JSPS Grant-in-Aid for Scientific Research (S) #22224003.

References

1. G. Aad *et al.* [ATLAS Collaboration], Phys. Lett. B **716**, 1 (2012); S. Chatrchyan *et al.* [CMS Collaboration], Phys. Lett. B **716**, 30 (2012).
2. The ATLAS Collaboration, ATLAS-CONF-2012-168.
3. K. Yamawaki, M. Bando and K. Matumoto, Phys. Rev. Lett. **56**, 1335 (1986); M. Bando, T. Morozumi, H. So and K. Yamawaki, Phys. Rev. Lett. **59**, 389 (1987).
4. M. Bando, K. Matumoto and K. Yamawaki, Phys. Lett. B **178**, 308 (1986).
5. S. Matsuzaki and K. Yamawaki, Prog. Theor. Phys. **127**, 209 (2012); Phys. Rev. D **85**, 095020 (2012).
6. S. Matsuzaki and K. Yamawaki, Phys. Rev. D **86**, 035025 (2012); Phys. Lett. B **719**, 378 (2013).
7. S. Matsuzaki and K. Yamawaki, Phys. Rev. D **86**, 115004 (2012).
8. M. Harada, M. Kurachi and K. Yamawaki, Phys. Rev. D **68**, 076001 (2003); M. Kurachi and R. Shrock, JHEP **0612**, 034 (2006).
9. M. Harada, M. Kurachi and K. Yamawaki, Prog. Theor. Phys. **115**, 765 (2006).
10. S. Shuto, M. Tanabashi and K. Yamawaki, in *Proc. 1989 Workshop on Dynamical Symmetry Breaking*, Dec. 21-23, 1989, Nagoya, eds. T. Muta and K. Yamawaki (Nagoya Univ., Nagoya, 1990) 115-123; W. A. Bardeen, S. T. Love, Phys. Rev. **D45**, 4672-4680 (1992); M. S. Carena and C. E. M. Wagner, Phys. Lett. B **285**, 277 (1992); M. Hashimoto, Phys. Lett. B **441**, 389 (1998);
11. K. Haba, S. Matsuzaki and K. Yamawaki, Phys. Rev. D **82**, 055007 (2010).
12. S. Weinberg, Phys. Rev. D **13**, 974 (1976); L. Susskind, Phys. Rev. D **20**, 2619 (1979).
13. For reviews, see, e.g., E. Farhi and L. Susskind, Phys. Rept. **74**, 277 (1981); K. Yamawaki, Lecture at 14th Symposium on Theoretical Physics, Cheju, Korea, July 1995, arXiv:hep-ph/9603293; C. T. Hill and E. H. Simmons, Phys. Rept. **381**, 235 (2003) [Erratum-ibid. **390**, 553 (2004)]; F. Sannino, Acta Phys. Polon. **B40**, 3533-3743 (2009).
14. B. Holdom, Phys. Rev. D **24**, 1441 (1981).
15. K. D. Lane and M. V. Ramana, Phys. Rev. D **44**, 2678 (1991).
16. T. Appelquist, J. Terning and L. C. Wijewardhana, Phys. Rev. Lett. **77**, 1214 (1996); T. Appelquist, A. Ratnaweera, J. Terning and L. C. Wijewardhana, Phys. Rev. D **58**, 105017 (1998).
17. V. A. Miransky and K. Yamawaki, Phys. Rev. D **55**, 5051 (1997); Errata, **56**, 3768 (1997).
18. W. E. Caswell, Phys. Rev. Lett. **33**, 244 (1974); T. Banks and A. Zaks Nucl. Phys. B **196**, 189 (1982).
19. T. Appelquist and G. Triantaphyllou, Phys. Lett. B **278**, 345 (1992); R. Sundrum and S. D. H. Hsu, Nucl. Phys. B **391**, 127 (1993); T. Appelquist and F. Sannino, Phys. Rev. D **59**, 067702 (1999).
20. S. Dimopoulos and L. Susskind, Nucl. Phys. B **155**, 237 (1979); E. Eichten and K. D. Lane, Phys. Lett. B **90**, 125 (1980).
21. G. Cacciapaglia, C. Csaki, C. Grojean and J. Terning, Phys. Rev. D **71**, 035015 (2005); R. Foadi, S. Gopalakrishna and C. Schmidt, Phys. Lett. B **606**, 157 (2005);

R. S. Chivukula, E. H. Simmons, H. J. He, M. Kurachi and M. Tanabashi, Phys. Rev. D **72**, 016008 (2005).
22. V. A. Miransky, Nuovo Cim. A **90**, 149 (1985).
23. W. A. Bardeen, C. N. Leung and S. T. Love, Phys. Rev. Lett. **56**, 1230 (1986); Nucl. Phys. B **273**, 649 (1986).
24. M. Hashimoto and K. Yamawaki, Phys. Rev. D **83**, 016008 (2011).
25. V. A. Miransky and V. P. Gusynin, Prog. Theor. Phys. **81**, 426 (1989).
26. K. Yamawaki, Prog. Theor. Phys. Suppl. **167**, 127 (2007); Prog. Theor. Phys. Suppl. **180**, 1 (2010); Int. J. Mod. Phys. A **25**, 5128 (2010).
27. J. M. Maldacena, Adv. Theor. Math. Phys. **2**, 231 (1998) [Int. J. Theor. Phys. **38**, 1113 (1999)].
28. L. Da Rold and A. Pomarol, Nucl. Phys. B **721**, 79 (2005).
29. J. Erlich, E. Katz, D. T. Son and M. A. Stephanov, Phys. Rev. Lett. **95**, 261602 (2005).
30. J. Jia, S. Matsuzaki and K. Yamawaki, Phys. Rev. D **87**, 016006 (2013).
31. V. A. Miransky, K. Yamawaki, Mod. Phys. Lett. **A4**, 129-135 (1989); K. Matumoto, Prog. Theor. Phys. **81**, 277-279 (1989); T. Appelquist, M. Einhorn, T. Takeuchi, L. C. R. Wijewardhana, Phys. Lett. **B220**, 223 (1989).
32. H. Pagels and S. Stokar, Phys. Rev. D **20**, 2947 (1979).
33. T. Appelquist, K. D. Lane and U. Mahanta, Phys. Rev. Lett. **61**, 1553 (1988); K. - I. Aoki, T. Kugo and M. G. Mitchard, Phys. Lett. B **266**, 467 (1991); M. Harada, M. Kurachi and K. Yamawaki, Phys. Rev. D **70**, 033009 (2004).
34. The ATLAS Collaboration, ATLAS-CONF-2012-160; ATLAS-CONF-2012-169; The CMS Collaboration, HIG-PAS-12-041; HIG-PAS-12-042; HIG-PAS-12-043.
35. The ATLAS Collaboration, ATLAS-CONF-2013-012; ATLAS-CONF-2013-013; ATLAS-CONF-2013-030; The CMS Collaboration, HIG-PAS-13-001; HIG-PAS-13-002; HIG-PAS-13-003; HIG-PAS-13-004.
36. Recontres de Moriond 2013 Electroweak Session,
https://indico.in2p3.fr/conferenceDisplay.py?confId=7411;
QCD session,
http://moriond.in2p3.fr/QCD/2013/qcd.html;
Aspen 2013 - Higgs Quo Vadis,
http://indico.cern.ch/conferenceDisplay.py?confId=202554.

Holographic Technidilaton and LHC Searches

Maurizio Piai

Physics Department, Swansea University,
Swansea, SA2 8PP, United Kingdom.
E-mail: m.piai@swansea.ac.uk

The holographic techni-dilaton is a parametrically light composite scalar appearing in the spectrum of certain strongly-coupled models of electroweak symmetry breaking built within the context of gauge-gravity dualities. I focus on particular subclasses of such models, both in the rigorous top-down approach to holography, as well as in the phenomenological bottom-up approach. I review the calculation of the mass of such a light scalar state. I briefly discuss some phenomenological features emerging in bottom-up models, and the implications for these models of the current LHC measurements of the couplings of the recently discovered scalar with mass of approximately 125–126 GeV. I summarize the present status of development of top-down holographic models of holographic walking technicolor related to the truncation of Type-IIB supergravity on $T^{1,1}$.

Keywords: Dilaton, technicolor, holography.

1. Introduction and motivations

The possibility that electroweak symmetry breaking (EWSB) emerges from a new strongly-coupled theory is very appealing and natural.[1,2] While traditional QCD-like technicolor models are plagued by serious phenomenological drawbacks, in walking technicolor models[3] it is possible to accommodate the phenomenological constraints from flavor physics and electroweak precision measurements thanks to the non-trivial multi-scale underlying dynamics. It has been the subject of a large number of studies in the past to understand under what conditions walking theories yield a parametrically light composite scalar state in the spectrum, the techni-dilaton, associated with the spontaneous breaking of approximate scale invariance.[4] In the light of the recent discovery by the LHC collaborations of a new scalar state with mass of 125–126 GeV,[5] and couplings similar to those of the Higgs particle of the Standard Model, it is of utmost importance to understand what models contain a light dilaton, whether such particle could mimic the signatures of the Higgs boson, and what experimental (as well as theoretical) strategies would allow to distinguish unambiguously between the two.

Gauge-gravity dualities provide ideal tools for addressing this problem, since a well defined procedure exists for calculating physical quantities such as the mass spectrum of bound states, hence overcoming the intrinsic difficulty of working with strongly-coupled field theories. Some classes of models are built adopting the bottom-up approach to holography, a very flexible and comparatively easy to use tool. It is not difficult to find models of this class yielding a light techni-dilaton in the spectrum, and I will summarize the main results of the phenomenological studies in some of these models.[6–10]

Radically different classes of special models[11–15] emerge within the top-down approach to holography. The first example of rigorously constructed Type-IIB backgrounds admitting a light scalar bound state belongs to one of these classes, and it provides a proof of principle that there exist strongly-coupled dynamical models with this feature. The phenomenology of these models is under investigation, and a direct comparison to the experimental data requires further studies.

Before proceeding to describe the status of present studies of specific models, let us summarize what are the goals of this whole research program. The main goal is to construct the top-down gravity dual of a fully realistic walking technicolor model. The ideal such model should, in principle, satisfy the following requirements.

1. The background can be obtained by solving the coupled equations derived from a 5-dimensional sigma-model containing n scalars (and possibly vectors) coupled to gravity.
2. The action of the sigma-model is derived from consistent truncation of a known superstring theory in 10 dimensions, in the sense that an algebraic procedure exists for uplifting the 5-dimensional background to a solution of the equations of the relevant 10-dimensional supergravity.
3. There exists a clear identification of the dual four-dimensional field theory in terms of fundamental fields and interactions.
4. The background is asymptotically AdS, or close to it, in such a way that the dictionary of the AdS/CFT correspondence can be applied rigorously to the calculation of correlation functions and observable quantities of the dual theory. In particular, physical results do not depend on UV regulators.
5. The background has a natural end-of-space in the IR.
6. The uplift to 10-dimensions is non-singular.
7. The end of space can be interpreted in terms of the confinement of the dual gauge theory, at least in the sense that the calculation of the Wilson loop yields a linear behavior for the quark-antiquark static potential.
8. The background shows unambiguously the presence of a second dynamical scale parametrically larger than the confinement scale. In the (walking) energy range between these two, the dual gauge coupling runs slowly.
9. The calculation of the spectrum of glueballs of the dual field theory yields one parametrically light scalar state.
10. The couplings of this light state allow to identify it as a pseudo-dilaton.
11. A mechanism for modeling electroweak gauge symmetry and its breaking can be implemented rigorously, starting from 10-dimensional string theory.
12. The spectrum of mesons and the coefficients of the electroweak chiral Lagrangian can be computed rigorously, within the extension in 11.

13. The presence of the walking region produces a parametric separation between the confinement scale and the physical scale of electroweak symmetry breaking, suppressing precision electroweak parameters such as S.
14. There are regions of parameter space in which all the direct and indirect experimental measurements related to EWSB are successfully reproduced.
15. Predictions for the masses and couplings of new bound states (such as the techni-ρ mesons) can be made and are testable.

Each of these requirements is fulfilled by several different existing models, proving that achieving any of them is possible. A systematic construction of a model that might yield all of these features at once is under development, with several successful stages of this program being complete.

At the technical level, the most challenging and time-consuming part of this program is point 2, the consistent truncation of the KK reduction of supergravity down to five dimensions. There exists only a handful of such truncations in the literature, and the process of obtaining them is long and complicated. Also, the resulting theory usually involves highly non-trivial five-dimensional actions, a large number n of five-dimensional scalar fields, and the resulting coupled non-linear differential equations are hard to solve, even numerically. Hence, it is practically useful for phenomenological purposes to dispense with this requirement, and construct models by starting from an arbitrary five-dimensional action, with a limited number of scalars. What results cannot be thought of as the rigorous dual to a four-dimensional theory, and hence by doing so one is also explicitly renouncing to points 3, 6 and 7. Confinement is often crudely modeled with the introduction of a mass gap obtained by simply adding a hard cutoff in the space in the IR, hence renouncing also requirement 5. Indirectly, this also means that the calculation of the mass of the dilaton is incomplete, since it cannot include correctly the physics of confinement, and hence one has to exert caution in interpreting results related to points 9 and 10.

This type of simplified toy models, usually referred to as bottom-up constructions, are useful for phenomenological purposes. The reason being that in this context it is possible to explore effectively a vast space of possible models, discussing the physics of whole classes of theories at once, and focusing on the more phenomenological requirements 9–15. In particular, it is interesting to use these models to look for possible universal behaviors, which appear in many different models, and hence might reappear also in more rigorous constructions. Since a light scalar has been discovered by the LHC collaborations, this is also a pragmatic way of discussing the phenomenological features of models where the dilaton is present, temporarily dispensing with the theoretical question of whether this scenario is realized in any interacting four-dimensional gauge theory.

2. Bottom-up holographic models

The studies in[6–10] contain a large set of bottom-up models in which the sigma-model Lagrangian varies greatly. The generic five-dimensional action reads

$$\mathcal{S} = \int \mathrm{d}^5 y \sqrt{-g} \left[\frac{1}{4} R - \frac{1}{2} G_{ab} g^{MN} \partial_M \Phi^a \partial_N \Phi^b - V(\Phi) \right] , \tag{1}$$

where G_{ab} is the sigma-model metric for the n scalars Φ^a, and V the potential. Some of the models in the literature have simple polynomial potentials, others have potentials that are directly related to the simplest consistent truncations of Type-IIB on S^5. It is convenient to write the ansatz for the five-dimensional metric as

$$\mathrm{d}y^2 = e^{2A(r)} \mathrm{d}x_{1,3}^2 + \mathrm{d}r^2 , \tag{2}$$

with r the radial direction and x^μ the four-dimensional Minkowski coordinates. In the calculations, one assumes that there exist two boundaries $r_1 \ll r_2$ in the radial direction. The former should be interpreted as an IR regulator, but in most bottom-up models it is used as a rough implementation of a mass gap (confinement). The latter is a UV-cutoff, and must be removed by taking the limit $r_2 \to +\infty$. The function $A(r)$ is determined dynamically by solving the coupled differential equations derived from the action, subject to boundary conditions that are derived from appropriate localized boundary actions.

Details on these constructions and calculations can be found in the references, and in the literature. The three following general facts coherently emerge from bottom-up studies of the four-dimensional spectrum of scalar state.

- A parametrically light dilaton emerges in the spectrum for all the models in which the departure from AdS of the five-dimensional geometry is mainly due to a non-trivial profile for one of the background scalars Φ such that (in the UV) $\Phi \sim e^{-\Delta r}$ with $\Delta > 2$. The field-theory interpretation of such result is that when $\Delta > 2$, the dual field theory contains a VEV for an operator of dimension Δ. This VEV breaks spontaneously scale invariance and is the dynamical origin of the dilaton.
- Spectra computed with seemingly arbitrary polynomial potentials do not differ qualitatively from those obtained from more complicated, string-inspired potentials, hence indirectly justifying the use of the simplified treatment (compare for instance Fig. 8 in[6] with Fig. 6 in[7]).
- It is possible to remove the UV-cutoff, provided the geometry is asymptotically AdS. Removing the IR cutoff is more subtle. If the background equations yield a singular end-of-space for the five-dimensional geometry, a mass gap will be generated, and r_1 is just a regulator. But whether the IR regulator is removable depends on a case-by-case analysis.

What is most interesting about these models is the comparison of the mass of the light dilaton with other measurable physical quantities. It is possible to use these backgrounds to model electroweak symmetry breaking, and compute the couplings of the light scalar to photons, gluons, W and Z bosons. Also, in these models it is possible to compute the precision parameter S, and the mass of the techni-ρ mesons. By comparing the results to the experimental measurements of the precision parameters, and using the mass of the Z and the electroweak couplings to set the relevant parameters, one finds a surprising result: the decay constant F of the techni-dilaton and the mass of the techni-ρ mesons are related to each other, in such a way that in order to satisfy indicative bounds on the S-parameter one has to require that $M_\rho \gtrsim 2.4$ TeV and $F \gtrsim 1.1$ TeV. The result on the decay constant can be rewritten in terms of $v_W/F \lesssim 0.22$, with $v_W = 246$ GeV the electroweak scale.

The gauge bosons of the standard-model group are introduced by weakly gauging a global $SU(3) \times SU(2) \times U(1)$ symmetry of the strongly-coupled dual field theory. The details of the procedure for doing so can be found in the literature, but the main result is that the beta functions of the electromagnetic $U(1)$ and of the color $SU(3)$ are controlled by arbitrary coefficients. In field-theory terms, this follows from the fact that the new strongly coupled sector may contain large numbers of new charged and/or colored fundamental degrees of freedom. Consequently, the couplings of the light techni-dilaton resemble those of the Higgs particle of the minimal version of the Standard Model, with three fundamental differences, at the leading order. Besides the aforementioned fact that $a \equiv v_W/F \lesssim 0.22$, the couplings of the dilaton to photons and gluons are controlled by two free parameters, which we call c_γ and c_g, and parameterize the contributions of the new strongly-coupled sector to the beta functions of QED and QCD, respectively.

Phenomenological analysis appeared in the literature, questioning whether the experimental measurements carried on by TeVatron, ATLAS and CMS of the properties of the newly discovered scalar favor the standard-model values $(a, c_\gamma, c_g) = (1, 1, 1)$, or whether significant deviations are present (see for instance[16]). In,[10] one such study is performed, by a 3-parameter fit of all the experimental information. The results are, at present, inconclusive, substantial deviations from the standard-model expectations being still allowed at the 99.73% c. l. The hints of an enhanced number of decays into two photons favors dilaton models. Small values of a imply a strong suppression of all production mechanisms aside from gluon-gluon fusion. The data on such production mechanisms cannot discriminate this case with present accuracy. Hopefully, future updates from the experimental collaborations will clarify these points.

3. Top-down holographic models

The more fundamental question of whether there exist strongly-coupled field theories which yield a parametrically light dilaton in the spectrum of composite states must be answered within the rigorous top-down approach to holography, within 10-dimensional supergravity. Because one wants the model to describe a confining gauge theory, and because this is highly non-trivial to realize, it is sensible to start by generalizing known supergravity backgrounds which correctly capture the physics of confinement. At present, there exist three classes of models that exhibit the features of walking behavior, all of which are related to the consistent truncation of Type-IIB on $T^{1,1}$, have similar geometry in the IR, but different UV asymptotics.

- Class I:[11] the UV-asymptotics can be interpreted in terms of a special gauge theory deformed by the presence of dimension-8 operators.
- Class II:[12,13] the UV-asymptotics resemble the dual of a higher-dimensional compactified gauge theory, as in the Maldacena–Nunez model.[17]
- Class III:[14] the UV-asymptotics coincides with the one of the baryonic branch of Klebanov–Strassler.[18,19]

Solutions in class I satisfy the requirements 1, 2, 5 and 8 of the list in the Introduction. In particular, they are the first examples of models where the dual gauge coupling shows the feature of walking. A first encouraging attempt to build a realization of chiral symmetry breaking based on a particular embedding of $D7$-branes in backgrounds of class I exists,[20] indicating that the criteria 11–15 might be met, barring the major conceptual and technical limitations that come from the presence of higher-order operators.

Many such limitations are overcome by solutions in class II, which can explicitly be shown to satisfy also requirements 7[13] and 9:[12,15] the calculation of the Wilson loop yields the expected linear potential of a confining theory, and the spectrum of glueballs contains a parametrically light scalar.

Solutions of class III satisfy also the important formal requirement 4, hence it should be possible to verify also requirement 10, and to put on firm grounds the phenomenological implications 11-15. But at present no calculation of the spectrum and couplings of the glueballs exists within class III. Solutions of this class also admit a precise interpretation in terms of the dual field theory (requirement 3), although this is incomplete, because there is no clear field-theory dynamical origin for the VEV for a dimension-6 operator visible in the gravity background.

In conclusion, top-down models exist that exhibit walking dynamics, and in which the spectrum of glueballs contains a parametrically light scalar, which is most likely a psudo-dilaton. This is at present the only fully rigorous such calculation available in the literature, and provides a proof of principle that at least some walking theories admit such a parametrically light scalar in the spectrum.

This is still an ongoing research program. A framework exists within which all the calculations needed to test the 15 requirements listed in the introduction are doable, thanks to the construction in[14] of models in class III. Models of this class an inherently rather complicated and difficult to analyze, because the sigma-model countains 8 scalar fields, with a highly non-trivial action, but these studies are in progress. Once these calculations have been completed, this framework promises to yield the first realistic, controllable and calculable example of strongly coupled dynamical origin of electroweak symmetry breaking with walking dynamics, based on the rigorous top-down approach to holography.

Acknoledgments

I would like to thank D. Elander, P. Kumar, J. Gaillard, R Lawrance, D. Mateos, C. Nunez, I. Papadimitriou, A. Paredes and A. Rago for the collaborations this contribution is based upon, and the organizers of this Workshop for financial support. MP is supported in part by WIMCS and by the STFC grant ST/J000043/1.

References

1. S. Weinberg, Phys. Rev. D **19**, 1277 (1979); L. Susskind, Phys. Rev. D **20**, 2619 (1979); S. Weinberg, Phys. Rev. D **13**, 974 (1976).
2. R. S. Chivukula, arXiv:hep-ph/0011264; K. Lane, arXiv:hep-ph/0202255; C. T. Hill and E. H. Simmons, Phys. Rept. **381**, 235 (2003) [Erratum-ibid. **390**, 553 (2004)]; A. Martin, arXiv:0812.1841 [hep-ph]; F. Sannino, Acta Phys. Polon. B **40**, 3533 (2009). M. Piai, Adv. High Energy Phys. **2010**, 464302 (2010).
3. B. Holdom, Phys. Lett. B **150**, 301 (1985); K. Yamawaki *et al.* Phys. Rev. Lett. **56**, 1335 (1986); T. W. Appelquist *et al.* Phys. Rev. Lett. **57**, 957 (1986).
4. M. Bando *et al.* Phys. Lett. B **178**, 308 (1986); Phys. Rev. Lett. **56**, 1335 (1986); B. Holdom and J. Terning, Phys. Lett. B **187**, 357 (1987); Phys. Lett. B **200**, 338 (1988); C. Csaki *et al.*, Phys. Rev. D **76**, 125015 (2007); W. D. Goldberger *et al.*, Phys. Rev. Lett. **100**, 111802 (2008); L. Vecchi, arXiv:1002.1721 [hep-ph]; D. D. Dietrich *et al.*, Phys. Rev. D **72**, 055001 (2005); M. Kurachi and R. Shrock, JHEP **0612**, 034 (2006); T. Appelquist and Y. Bai, Phys. Rev. D **82**, 071701 (2010); L. Vecchi, arXiv:1007.4573 [hep-ph]; K. Haba *et al.*, Phys. Rev. D **82**, 055007 (2010); M. Hashimoto and K. Yamawaki, arXiv:1009.5482 [hep-ph];
5. S. Chatrchyan *et al.* [CMS Collaboration], Phys. Lett. B **716**, 30 (2012); G. Aad *et al.* [ATLAS Collaboration], Phys. Lett. B **716**, 1 (2012).
6. D. Elander and M. Piai, JHEP **1101**, 026 (2011).
7. S. P. Kumar *et al.*, JHEP **1105**, 008 (2011).
8. D. Elander and M. Piai, Nucl. Phys. B **864**, 241 (2012).
9. R. Lawrance and M. Piai, arXiv:1207.0427 [hep-ph].
10. D. Elander and M. Piai, arXiv:1208.0546 [hep-ph].
11. C. Nunez *et al.*,Int. J. Mod. Phys. A **25**, 2837 (2010);
12. D. Elander *et al.*, Phys. Lett. B **686**, 64 (2010);
13. C. Nunez *et al.*, arXiv:0909.0748 [hep-th].
14. D. Elander *et al.*, JHEP **1107**, 056 (2011).
15. D. Elander and M. Piai, arXiv:1212.2600 [hep-th].
16. K. Cheung and T. -C. Yuan, Phys. Rev. Lett. **108**, 141602 (2012); I. Low *et al.*, arXiv:1207.1093 [hep-ph]; T. Corbett *et al.*,arXiv:1207.1344 [hep-ph]; P. P. Giardino *et al.*, arXiv:1207.1347 [hep-ph]; J. Ellis and T. You, arXiv:1207.1693 [hep-ph]; M. Montull and F. Riva, arXiv:1207.1716 [hep-ph]; J. R. Espinosa *et al.*, arXiv:1207.1717 [hep-ph]; D. Carmi *et al.*,arXiv:1207.1718 [hep-ph]; D. Bertolini and M. McCullough, arXiv:1207.4209 [hep-ph]; S. Matsuzaki and K. Yamawaki, arXiv:1207.5911 [hep-ph]; Z. Chacko *et al.*,arXiv:1209.3259 [hep-ph]; G. Belanger *et al.*, arXiv:1212.5244 [hep-ph].
17. J. M. Maldacena and C. Nunez, Phys. Rev. Lett. **86**, 588 (2001); A. H. Chamseddine and M. S. Volkov, Phys. Rev. Lett. **79**, 3343 (1997).
18. I. R. Klebanov and M. J. Strassler, JHEP **0008**, 052 (2000).
19. A. Butti *et al.*, JHEP **0503**, 069 (2005).
20. L. Anguelova, Nucl. Phys. B **843**, 429 (2011); L. Anguelova *et al.*, Nucl. Phys. B **852**, 39 (2011); Nucl. Phys. B **862**, 671 (2012); T. E. Clark *et al.*, arXiv:1208.0817 [hep-th].

Composite Higgs and Techni-Dilaton at LHC

Deog Ki Hong*

Department of Physics, Pusan National University,
Busan, 609-735, Korea
**E-mail: dkhong@pusan.ac.kr*

In this talk I argue that a light techni-dilaton exists in models of walking technicolor near the conformal window. The light techni-dilaton then mixes with the composite Higgs, which is predicted to be narrow and light in walking technicolor models. It is an interesting possibility that the newly discovered boson at LHC might be a mixed state of techni-dilaton and the composite Higgs.

Keywords: Techni-dilaton; light composite Higgs; walking technicolor.

1. Introduction and Review

A new boson of mass 125 GeV has been discovered at the Large Hadron Collider (LHC).[1,2] Since the discovery was made, more data on the new boson has been accumulated to show that it is very much like the standard model Higgs.[3] The data is now consistent more or less with the standard model (SM) of elementary particles with deviation less than 2σ (See Fig. 1).

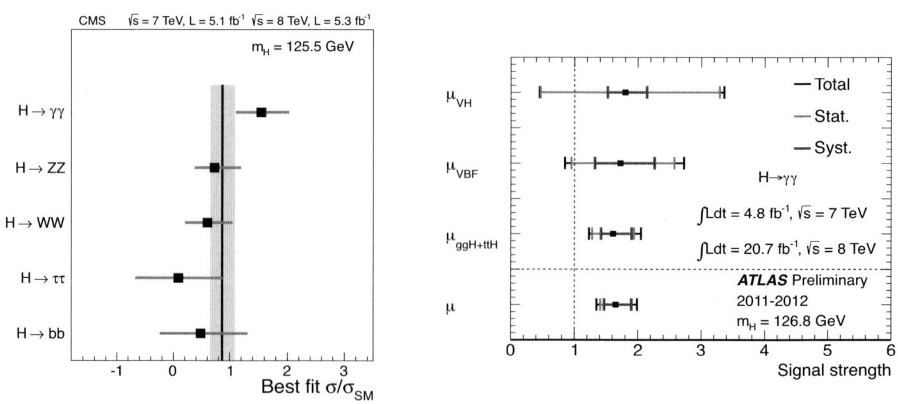

Fig. 1. The CMS results on the ratio of the cross sections of Higgs decay channels over the SM values (left panel) and the μ values of Higgs to two-photon channels, measured at ATLAS (right panel).

But, there are several convincing arguments that the SM is not complete and there should be new physics beyond the standard model (BSM). For instance, it does not explain the origin of dark matter, which constitutes 27% of total energy of our universe, according to the recent analysis of Planck 2013 data on the cosmic microwave background.[4] It is also known that the SM vacuum becomes meta-stable at energy 10^{10}-10^{11} GeV if we assume the SM is valid up to the Planck scale.[5] So, the fact that we have not seen any significant deviation of the SM and no new BSM particles at the LHC may just mean that the scale of new physics is much higher than the scale, probed at the LHC with the center of mass energy $\sqrt{s} = 8$ TeV.

In this talk I try to argue that the small deviation, though less than 2σ, seen at the LHC8 might be due to walking technicolor (WTC), a model of strong dynamics BSM, which breaks the electroweak symmetry dynamically by a new strong force. Especially, the newly discovered boson might be a mixture of two light scalars, namely light techni-dilaton and light composite Higgs, that WTC predicts.

2. Light Dilaton and PCDC

Walking technicolor is a model for dynamical electroweak symmetry breaking, where the electroweak symmetry is broken by the condensation of techni-quarks, with a nontrivial quasi infra-red (IR) fixed point.[6–8] The beta function of WTC and the scale dependence of the technicolor coupling are shown in Fig. 2.

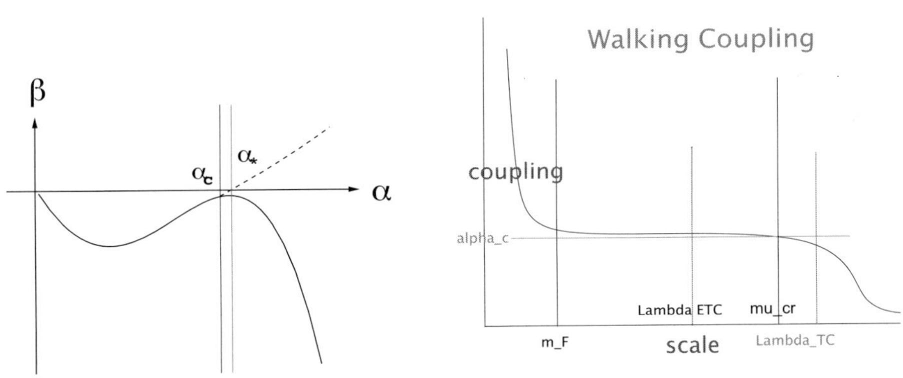

Fig. 2. The β function of WTC is shown in the left panel and the scale dependence of the coupling is plotted in the right panel.

If the condensation of techni-quarks were not to occur, the coupling would have an IR fixed point, $\beta(\alpha_*) = 0$, and theory will flow into a conformal window (CW). However, since in WTC we need to generate a scale at IR to break the electroweak symmetry, the theory should be slightly away from the conformal window so that the condensation of techni-quarks does occur. Namely the critical coupling for chiral symmetry breaking should be smaller than α_* and should be reached at a scale $\mu_{\rm cr}$ before the theory flows into the IR fixed point. If the critical coupling is very close to the would-be IR fixed point, $\alpha_c \equiv \alpha(\mu_{\rm cr}) \approx \alpha_*$, the β function almost vanishes and thus the coupling remains nearly constant for scales between $\mu_{\rm cr}$ and m_F, the dynamical mass of techni-quarks, which is of order of 1 TeV. Below the scale m_F the techni-quarks decouple and the coupling gets very strong to confine the techni-gluons.

The Schwinger-Dyson analysis for the techni-quark bilinear operator shows that its anomalous dimension is very close to 1 for $m_F < \mu \leq \mu_{\rm cr}$,

$$\gamma_m(\mu) = 1 + \sqrt{\alpha(\mu)/\alpha_c - 1} \approx 1\,. \tag{1}$$

Therefore, in WTC the techni-quark condensation is quite enhanced in the ultra-violet (UV) region or at the extended technicolor (ETC) scale, $\Lambda_{\rm ETC} < \mu_{\rm cr}$ so that ETC interactions give large masses to SM fermions, while suppressing the contributions of ETC interactions to the flavor-changing neutral currents by raising the ETC scale high enough, $\Lambda_{\rm ETC} \gtrsim 10^3\, m_F$:

$$\langle \bar{Q}Q \rangle\big|_{\Lambda_{\rm ETC}} = e^{\int_{m_F}^{\Lambda_{\rm ETC}} \frac{{\rm d}\mu}{\mu}\gamma_m(\mu)} \langle \bar{Q}Q \rangle\big|_{m_F} = \frac{\Lambda_{\rm ETC}}{m_F} \langle \bar{Q}Q \rangle\big|_{m_F} \approx m_F^2\, \Lambda_{\rm ETC}\,. \tag{2}$$

In a theory space, one can approach the conformal window by tuning the number of flavors. For SU($N_{\rm TC}$) gauge theories with N_f fermions in the fundamental representation, the two-loop calculation of β function shows that theories approach a conformal window for $x = N_f/N_{\rm TC} \approx 4$ in the large $N_{\rm TC}$ limit.[9] Another, but equivalent way of probing the theory space in the chirally broken phase is to change the coupling at the scale of dynamical mass, $\alpha(m_F)$. Miransky and Yamawaki has shown that the quantum phase transition that occurs, when $\alpha(m_F) \to \alpha_c$, exhibits an essential singularity, so-called Berezinsky-Kosterlitz-Thouless (BKT) or Miransky scaling (see Fig. 3).[10] Namely, the order parameter for the phase transition behaves near the phase transition as

$$m_F = \Lambda_{\rm TC} \exp\left(-\frac{\pi}{\sqrt{\frac{\alpha(m_F)}{\alpha_c} - 1}}\right), \tag{3}$$

where $\Lambda_{\rm TC}$ is the intrinsic scale of WTC, shown in Fig. 2.

Since the coupling is almost constant and $\beta(\alpha) \approx 0$, WTC has an approximate scale invariance, broken spontaneously, for scales $m_F < \mu < \mu_{\rm cr}$. Therefore there exists a dilatation current, $D_\mu = x^\nu \theta_{\mu\nu}$, where $\theta_{\mu\nu}$ is the so-called improved energy-momentum tensor. The dilatation current is, however, anomalous due to quantum effects. By an explicit calculation in the ladder approximation the scale anomaly is found to be proportional to m_F^4:

$$\langle \theta_\mu^\mu \rangle = \frac{\pi\beta(\alpha)}{\alpha^2} \left\langle \left(F_{\mu\nu}^a\right)^2 \right\rangle + \frac{g^2}{\mu_{\rm cr}^2} \left\langle \left(\bar{Q}Q\right)^2 \right\rangle \sim m_F^4. \tag{4}$$

Since the scale symmetry is spontaneously broken by the condensation of techni-quarks, a dilaton arises as a pseudo Nambu-Goldstone boson by Goldstone theorem:

$$\langle 0 | D^\mu | \sigma \rangle = i F_\sigma p^\mu e^{-ip\cdot x}\,, \tag{5}$$

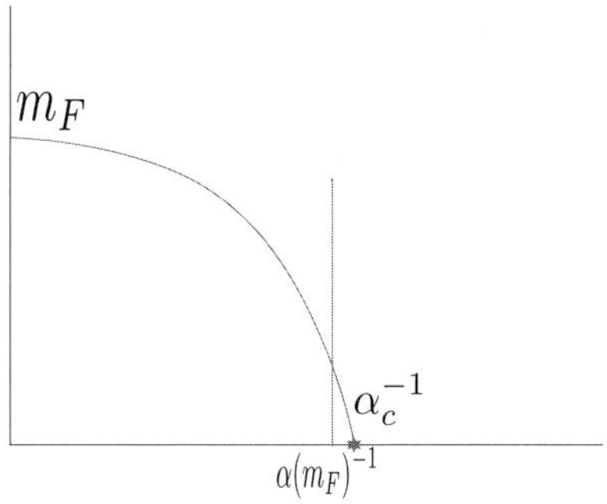

Fig. 3. The quantum (conformal) phase transition in a theory space as one varies $\alpha(m_F)$.

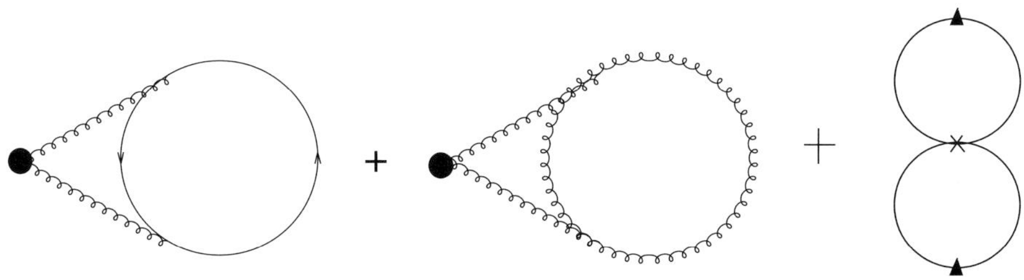

Fig. 4. The scale anomaly gets contributions from the $d = 4$ operators: the blob denotes the insertion of $(F^a_{\mu\nu})^2$ and the cross denotes the marginal four-Fermi operators, generated at scales below $\mu_{\rm cr}$. The solid lines are full propagators of techni-quarks and the curly lines are the full propagators of techni-gluons.

where F_σ is the dilaton decay constant. If the dilaton pole dominates at low energy, we may take $\partial_\mu D^\mu = F_\sigma m_\sigma^2\,\sigma$ to get

$$\langle \partial_\mu D^\mu \rangle = F_\sigma^2 m_\sigma^2 \simeq \kappa^2\, m_F^4\,, \tag{6}$$

where κ is a constant of order one. Therefore the dilaton can be quite light, $m_\sigma \simeq \kappa\, m_F^2/F_\sigma \ll m_F$, if $F_\sigma \gg m_F$. Whether the dilaton decay constant can be much bigger than m_F or not in WTC is not settled yet. But, I give a compelling argument that it is so. Let's assume that the chiral symmetry is not spontaneously broken but explicitly by a small mass, m_0. Then by the scale invariance the dilaton has to couple to the techni-quarks as

$$m_0\, e^{\sigma/F_\sigma} \bar{Q}Q\,. \tag{7}$$

If we integrate out the massive techni-quarks, we get an effective potential

$$V(\sigma) = 4\eta\, m_0^4 \left(\frac{\sigma}{F_\sigma}\right)^4 \left[\ln\left(\frac{\sigma}{F_\sigma}\right) - \frac{1}{4}\right], \tag{8}$$

which gives mass to the dilation

$$m_\sigma^2 = \eta\,\frac{m_0^4}{F_\sigma^2} \to 0, \quad \text{if} \;\; m_0 \to 0\,. \tag{9}$$

Since $\beta(\alpha) \approx 0$, the theory is scale-invariant even at quantum level, when the chiral condensation vanishes. The potential for the dilaton is therefore flat and the vacuum has a flat direction along the scale transformation. The dilaton decay constant, which is nothing but the vacuum expectation value of the dilaton fields, can take an arbitrary value. By taking m_0 to be parametrically much smaller than F_σ the dilaton can be made arbitrarily light. Therefore, even in the case the chiral symmetry is spontaneously broken, generating the dynamical mass for the techni-quarks, the dilaton can be made light, since the dilaton develops its vacuum expectation value at scale higher than μ_c as long as the theory is scale-invariant or $\beta(\alpha) \approx 0$.

3. Composite Higgs and Light Techni-Dilaton

Consider a following composite operator

$$\lim_{y \to x} Q(x)\bar{Q}(y) = (\mu\,|x-y|)^{\gamma_m}\, Q\bar{Q}(x)\,, \tag{10}$$

where Q is the techni-quark field and γ_m is the anomalous dimension of the techni-quark bilinear operator. Since the electroweak symmetry is broken by the condensation of techni-quarks, we may write the composite operator as

$$Q\bar{Q}(x) \sim H(x) = e^{i\pi_{\rm TC}/F_{\rm TC}} \begin{pmatrix} 0 \\ v + h(x) \end{pmatrix}, \tag{11}$$

where $v \equiv \langle H(x) \rangle = 247\ \mathrm{GeV}/\sqrt{N_f}$ is the vacuum expectation value of the composite field $H(x)$, which transforms like a bi-fundamental, (N_f, N_f) under the chiral symmetry of WTC. The techni-pions $\pi_{\rm TC}$ are the excitations along the vacuum degeneracy and three of them constitute the longitudinal components of the electroweak gauge bosons, $W^\pm$ and Z^0. The radial excitation $h(x)$, orthogonal to the techni-pion excitations, is the composite Higgs, which is similar to the σ or $f_0(500)$ isospin-singlet scalar meson.

In QCD the $f_0(500)$ meson is not so light and is quite broad. However, unlike the technicolor of scaled-up QCD the composite Higgs could be quite light in walking technicolor. For WTC with techni-quarks in a two-index higher-dimensional representation, the composite Higgs is shown to be quite light, compared to the technicolor scale ($\sim$ 1 TeV) in the large $N_{\rm TC}$ limit.[11] Similarly, Kutasov *et al.*[12] found in a holographic model of WTC the composite Higgs is quite lighter than techni-vector mesons in the conformal limit

$$\frac{m_h}{m_V} \approx 0.2\,. \tag{12}$$

Though composite Higgs and techni-dilaton have same quantum numbers at low-energy, they are created by two different composite operators, namely by dilatation currents and the techni-quark bilinear, respectively and transform differently under the scale transformation, $x^\mu \longrightarrow e^{-\lambda} x^\mu$, as following:

$$\sigma(x) \longrightarrow \sigma(x) + \lambda, \quad \text{while} \quad h(x) \longrightarrow e^{\lambda}\, h(x)\,. \tag{13}$$

Since they have same quantum numbers (except under the scale transformation), they do mix. The mixing can be read off from the effective Lagrangian for the composite Higgs:

$$\mathcal{L}_H = \frac{1}{2}\left|D_\mu H\right|^2 - \frac{1}{2} m_H^2\, e^{2\sigma/F_\sigma} H^\dagger H + \frac{1}{4}\lambda \left(H^\dagger H\right)^2 + \cdots . \tag{14}$$

The mixing angle is therefore given as for a large F_σ

$$\theta \sim \frac{m_h}{F_\sigma} . \tag{15}$$

By examining the recent data at LHC8, Matsuzaki and Yamawaki showed that $v/F_\sigma \simeq 0.2$ gives better χ^2 fit than the SM, if one disregards the channels with lower significances.[13] In this case the mixing is sizable and we have two light scalars at low energy. The newly discovered boson might be then a linear combination of both composite Higgs and techni-dilaton, which naturally explain both the enhancement in two-photon decay and the Tevatron data on $\bar{b}\,b$.

4. Conclusion

With discovery of new boson at LHC particle physics has now entered a new era. Though the current data is consistent with the SM, we believe there should be new physics beyond the SM. But, the current data indicates that the scale of new physics BSM might be much higher than the scale, proved at LHC8.

One of the good candidates for BSM is walking technicolor, which breaks the electroweak symmetry dynamically. WTC predicts a light techni-dilaton if it is close to the conformal window. In addition to the techni-dilaton WTC predicts a light and narrow composite Higgs, which mix with the techni-dilaton. Therefore the newly discovered boson might be a mixture of these two light scalars. The phomenology of WTC with two light scalars is quite interesting and is being investigated.

Acknowledgments

We would like to thank the organizers, especially Koichi Yamawaki, for the invitation to SCGT12 and the hospitality during the conference. This work was supported by the Korea Research Foundation Grant funded by the Korean Government KRF-2008-341-C00008 .

References

1. S. Chatrchyan *et al.* [CMS Collaboration], Phys. Lett. B **716**, 30 (2012) [arXiv:1207.7235 [hep-ex]].
2. G. Aad *et al.* [ATLAS Collaboration], Phys. Lett. B **716**, 1 (2012) [arXiv:1207.7214 [hep-ex]].
3. ATLAS and CMS collaborations, talks presented at Moriond, March 2013 and at Aspen, March 2013.
4. P. A. R. Ade *et al.* [Planck Collaboration], [arXiv:1303.5089 [astro-ph.CO]].

5. N. Arkani-Hamed, K. Blum, R. T. D'Agnolo and J. Fan, arXiv:1207.4482 [hep-ph].
6. B. Holdom, Phys. Rev. D **24** (1981) 1441.
7. K. Yamawaki, M. Bando and K. Matumoto, Phys. Rev. Lett. **56**, 1335 (1986).
8. T. Akiba and T. Yanagida, Phys. Lett. B **169**, 432 (1986); T. W. Appelquist, D. Karabali and L. C. R. Wijewardhana, Phys. Rev. Lett. **57**, 957 (1986); T. Appelquist and L. C. R. Wijewardhana, Phys. Rev. D **36**, 568 (1987).
9. T. Appelquist, J. Terning and L. C. R. Wijewardhana, Phys. Rev. Lett. **77**, 1214 (1996) [hep-ph/9602385].
10. V. A. Miransky and K. Yamawaki, Phys. Rev. D **55**, 5051 (1997) [Erratum-ibid. D **56**, 3768 (1997)] [hep-th/9611142].
11. D. K. Hong, S. D. H. Hsu and F. Sannino, Phys. Lett. B **597**, 89 (2004);
12. D. Kutasov, J. Lin and A. Parnachev, Nucl. Phys. B **858**, 155 (2012) [arXiv:1107.2324 [hep-th]]; *ibid* B **863** (2012) 361 [arXiv:1201.4123 [hep-th]].
13. S. Matsuzaki and K. Yamawaki, Phys. Lett. B **719**, 378 (2013) [arXiv:1207.5911 [hep-ph]].

Hadron Interactions from Lattice QCD

S. Aoki**

Graduate School of Pure and Applied Sciences, University of Tsukuba, Tsukuba, Ibaraki 305-8571, Japan
**E-mail: saoki@het.ph.tsukuba.ac.jp*

I explain the potential method in lattice QCD, recently proposed by the HAL QCD(Hadron to Atomic-nuclei from Lattice QCD) collaboration to extract hadron interaction in lattice QCD simulations.

Keywords: Hadron interaction; Lattice QCD; Nuclear potential; Nambu-Bethe-Salpeter wave function.

1. Introduction

A steady growth of computational powered as well as improvements of numerical algorithm make it possible to calculate static properties of light hadrons such as masses and decay constant at the physical quark masses in the continuum limit of lattice QCD.[1]

The nuclear force, a force between nucleons, is the most fundamental quantity in nuclear physics, which has been used to investigate hadronic many-body systems. Phenomenological nucleon-nucleon (NN) potentials have been constructed in 90's, so as to reproduce a large number of proton-proton and neutron-proton scattering data as well as deuteron properties. Fig. 1 (Left) shows some examples, all of which reflect characteristic features of the NN interaction: The repulsive core at short distance surrounded by the attractive well at medium and long distances. The nuclear saturation, the nuclear shell structure, the nuclear superfluidity and the structure of neutron stars are all related to the properties of the nuclear force.[2–4]

It has been a long-standing challenge in theoretical particle and nuclear physics to extract the hadron-hadron interactions from first principle. A framework suitable for such a purpose in lattice QCD is the Lüshcer finite volume method,[8] which relates the energy spectrum for two hadrons in a finite box with the elastic scattering phase shift of two hadrons in the infinite volume. The method has been applied to various two-hadrons systems.[9]

An alternative but closely related approach to hadronic interactions in lattice

**Adress after April 1st, 2013:* Yukawa Institute for Theoretical Physics, Kyoto University, Kitashirakawa Oiwakecho, Sakyo-ku, Kyoto 606-8502, Japan

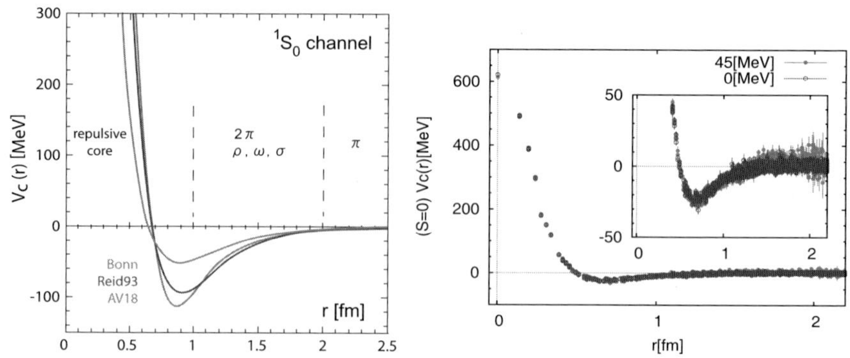

Fig. 1. (Left) Three examples of the modern NN potential in $^1S_0$ (spin-singlet and S-wave) channel: Bonn,[5] Reid93[6] and Argonne v_{18}.[7] Taken from Ref. 10. (Right) The isospin-triplet and spin-singlet central potential $V_C(r)$ obtained at $E_k \simeq 45$ MeV (red) and at $E \simeq 0$ (blue) in quenched QCD with $m_\pi \simeq 529$ MeV. Taken from Ref. 19.

QCD has recently been proposed and applied to the nucleon-nucleon (NN) system.[10–12] In the method, one first calculate the NN potential, and then extract physical observables such as the scattering phase shift by solving the Schrödinger equation with the potential obtained. See Refs. 13,14 for reviews on recent activities. In this talk, we discuss this potential method, called the HAL QCD method from the name of the collaboration.

2. Strategy

In quantum field theories such as QCD, it is non-trivial to define "potential", which is a concept in quantum mechanics. Furthermore, a potential itself can not be directly measured like a running coupling constant in QCD, so that it depends on how it is defined and therefore not unique as shown in Fig. 1. The potential, however, is still an useful tool to extract physical observables such as scattering phase shift. Using its non-uniqueness, one may adopt a convenient definition of the potential as long as it can reproduce physical observables in QCD correctly.

One such proposal is the HAL QCD method,[10–14] in which we first define the (equal-time) Nambu-Bethe-Salpeter (NBS) wave function as

$$\varphi_{\boldsymbol{k}}(\boldsymbol{x}) = \langle 0|N(\boldsymbol{r},0)N(\boldsymbol{r}+\boldsymbol{x},0)|NN, W_{\boldsymbol{k}}\rangle, \tag{1}$$

where $\langle 0|$ is the QCD vacuum state, $|NN, W_{\boldsymbol{k}}\rangle$ is the two-nucleon QCD eigenstate which has the relative momentum $\boldsymbol{k}$ and the center of mass energy $W_{\boldsymbol{k}} = 2\sqrt{\boldsymbol{k}^2 + m_N^2}$ with the nucleon mass m_N. The local interpolating operator of for nucleon $N(x)$ with $x = (\boldsymbol{x}, t)$ is given by $N(x) = \epsilon_{abc}(u^a(x)^T C\gamma_5 d^b(x))q^c(x)$ with color indices a, b, c, the charge conjugation matrix $C = \gamma_2\gamma_4$, $q^c(x) = (u^c(x), d^c(x))$ and u, d represent up and down quark fields, respectively. Spinor and flavor indices of $N(x)$ and $q(x)$ are implicit here.

It is important to note that, if the total energy is below the pion production threshold such that $W_{\boldsymbol{k}} < W_{\rm th} = 2m_N + m_\pi$ with the pion mass m_π, the radial part of the NBS wave function at large $r = |\boldsymbol{x}|$ for a given angular momentum L, the total spin S and the total isospin I behaves as[12,15]

$$\varphi_{\boldsymbol{k}}(r, LSI) \propto \frac{\sin(kr - L\pi/2 + \delta_{LSI}(k))}{kr} \tag{2}$$

where $\delta_{LSI}(k)$ with $k = |\boldsymbol{k}|$ is the NN scattering phase shift below the inelastic threshold, which appears in the S-matrix of QCD due to the unitary constraint.

We next define a non-local but energy-independent potential from the NBS wave function as[10–12]

$$[E_{\boldsymbol{k}} - H_0]\,\varphi_{\boldsymbol{k}}(\boldsymbol{x}) = \int d^3y\, U(\boldsymbol{x}, \boldsymbol{y})\,\varphi_{\boldsymbol{k}}(\boldsymbol{y}) \tag{3}$$

where $E_{\boldsymbol{k}} = \dfrac{k^2}{2\mu}$ with the reduced mass $\mu = \dfrac{m_N}{2}$, and $H_0 = \dfrac{-\nabla^2}{2\mu}$, and $U(\boldsymbol{x}, \boldsymbol{y})$ is a non-local potential, which is finite and renormalization scheme independent. Here non-relativistic "approximation" has never been introduced, though Lorentz covariance is lost by taking the equal time to define the NBS wave function in Eq. (1) and the potential is defined through the non-relativistic Schrödinger equation (3).

One of the most important properties here is that there exist a non-local but energy-independent potential . Indeed it is easy to construct such a potential as

$$U(\boldsymbol{x}, \boldsymbol{y}) = \sum_{\boldsymbol{k}_a, \boldsymbol{k}_b}^{|\boldsymbol{k}_{a,b}|<k_{\rm th}} [E_{\boldsymbol{k}_a} - H_0]\,\varphi_{\boldsymbol{k}_a}(\boldsymbol{x}) \mathcal{N}^{-1}_{\boldsymbol{k}_a,\boldsymbol{k}_b} \varphi^\dagger_{\boldsymbol{k}_b}(\boldsymbol{y}), \tag{4}$$

where $k_{\rm th}$ is the threshold momentum defined by $W_{\rm th} = 2\sqrt{k_{\rm th}^2 + m_N^2}$, and $\mathcal{N}^{-1}$ is the inverse of $\mathcal{N}_{\boldsymbol{k}_a,\boldsymbol{k}_b} = (\varphi_{\boldsymbol{k}_a}, \varphi_{\boldsymbol{k}_b})$ with the inner product $(f, g) \equiv \int d^3x\, f^\dagger(x) g(x)$. Note that $|\boldsymbol{k}_{a,b}| < k_{\rm th}$ for both $\mathcal{N}$ and $\mathcal{N}^{-1}$. It is then easy to see that the above $U(\boldsymbol{x}, \boldsymbol{y})$ satisfies Eq. (3) as

$$\begin{aligned}\int d^3y\, U(\boldsymbol{x}, \boldsymbol{y}) \varphi_{\boldsymbol{k}}(\boldsymbol{y}) &= \sum_{\boldsymbol{k}_a, \boldsymbol{k}_b}^{|\boldsymbol{k}_{a,b}|<k_{\rm th}} [E_{\boldsymbol{k}_a} - H_0]\,\varphi_{\boldsymbol{k}_a}(\boldsymbol{x}) \mathcal{N}^{-1}_{\boldsymbol{k}_a,\boldsymbol{k}_b} \mathcal{N}_{\boldsymbol{k}_b,\boldsymbol{k}} \\ &= [E_{\boldsymbol{k}} - H_0]\,\varphi_{\boldsymbol{k}}(\boldsymbol{x})\end{aligned} \tag{5}$$

for $k < k_{\rm th}$. Because of the property in Eq. (2), we can extract the phase shift $\delta_{LSI}(k)$ in QCD at $k < k_{\rm th}$, by solving the Schrödinger equation with this $U(\boldsymbol{x}, \boldsymbol{y})$. Of course, the non-local potential which satisfies Eq. (3) is not unique since one may add terms which affect Eq. (3) only at $k \geq k_{\rm th}$.

In the lattice QCD simulations, the NBS wave functions can be obtained only for a ground state and possibly a few low-lying excited states, so that a summation over $\boldsymbol{k}_{a,b}$ becomes inaccurate. For practical uses, we therefore expand non-local potential in terms of the velocity (or the derivative) as $U(\boldsymbol{x}, \boldsymbol{y}) = V(\boldsymbol{x}, \nabla)\delta^{(3)}(\boldsymbol{x} - \boldsymbol{y})$ with

$$V(\boldsymbol{x}, \nabla) = \underbrace{V_0(r) + V_\sigma(r)\boldsymbol{\sigma}_1 \cdot \boldsymbol{\sigma}_2 + V_T(r) S_{12}}_{=V^{\rm LO}(\boldsymbol{x})} + \underbrace{V_{\rm LS}(r)\boldsymbol{L}\cdot\boldsymbol{S}}_{V^{\rm NLO}(\boldsymbol{x})} + O(\nabla^2), \tag{6}$$

where $\boldsymbol{\sigma}_i$ is the Pauli-matrix acting on the spin index of the i-th nucleon, $S_{12} = 3(\boldsymbol{x}\cdot\boldsymbol{\sigma}_1)(\boldsymbol{x}\cdot\boldsymbol{\sigma}_2)/r^2 - \boldsymbol{\sigma}_1\cdot\boldsymbol{\sigma}_2$ is the tensor operator, $\boldsymbol{L} = \boldsymbol{x}\times\boldsymbol{p}$ is an orbital angular momentum, and $\boldsymbol{S} = (\boldsymbol{\sigma}_1 + \boldsymbol{\sigma}_2)/2$ is the total spin. Each coefficient function is further decomposed as $V_X(r) = V_X^0(r) + V_X^\tau(r)\boldsymbol{\tau}_1\cdot\boldsymbol{\tau}_2$ for $X = 0, \sigma, T, \mathrm{LS}, \cdots$, where $\boldsymbol{\tau}_i$ is the flavor matrix acting on the flavor index of the i-th nucleon.

The leading order(LO) potential $V^{\mathrm{LO}}(\boldsymbol{x})$ is then extracted from one NBS wave function as

$$V^{\mathrm{LO}}(\boldsymbol{x}) = \frac{[E_{\boldsymbol{k}} - H_0]\,\varphi_{\boldsymbol{k}}(\boldsymbol{x})}{\varphi_{\boldsymbol{k}}(\boldsymbol{x})}, \tag{7}$$

and we then solve the Schrödinger equation with this potential, in order to obtain phase shifts below the inelastic threshold. In this case, $\delta_{LSI}(k)$ is exact by construction, while $\delta_{LSI}(p \neq k)$ is approximated one due to the derivative expansion, whose expansion parameter is $(W_{\boldsymbol{p}} - W_{\boldsymbol{k}})/m_\pi$. It is noted that we can check a size of systematic errors at the LO in the expansion and can improve results by extracting higher order terms from two or more NBS wave functions.

3. Nuclear potentials

In lattice QCD the NBS wave function can be extracted from 4-pt correlation functions as

$$\begin{aligned} F(\boldsymbol{x}, t-t_0) = \langle 0|N(\boldsymbol{r},t)N(\boldsymbol{r}+\boldsymbol{x},t)\overline{\mathcal{J}}(t_0)|0\rangle &= \sum_n Z_n\varphi_{\boldsymbol{k}_n}(\boldsymbol{x})e^{-W_{\boldsymbol{k}_n}(t-t_0)} + \cdots \\ &\simeq Z_0\varphi_{\boldsymbol{k}_0}(\boldsymbol{x})e^{-W_{\boldsymbol{k}_0}(t-t_0)}, \qquad t-t_0\to\infty, \end{aligned} \tag{8}$$

where $\overline{\mathcal{J}}(t_0)$ is a source operator which creates two-nucleon states, $Z_n = \langle NN, W_{\boldsymbol{k}_n}|\overline{\mathcal{J}}(0)|0\rangle$, ellipses represent inelastic contributions, and $W_{\boldsymbol{k}_0}$ is the lowest energy of NN states.

The extraction of the NBS wave function in Eq. (8) relies on the ground state saturation for the correlation function, which can in principle be achieved by taking a large $t-t_0$. In practice, however, it is difficult to realize the ground state saturation for the NN system within reasonable errors, since $F(\boldsymbol{x}, t-t_0)$ becomes very noisy at large $t-t_0$.

An improved extraction of the NBS wave function has recently been proposed to overcome this difficulty.[16] Defining the normalized 4-point correlation function $R(\boldsymbol{x},t) \equiv F(\boldsymbol{x},t)/(e^{-m_N t})^2 \simeq \sum_n Z_n\,\varphi_{\boldsymbol{k}_n}(\boldsymbol{x})e^{-\Delta W_{\boldsymbol{k}_n} t}$, where $\Delta W_{\boldsymbol{k}_n} = W_{\boldsymbol{k}_n} - 2m_N = \boldsymbol{k}^2/m_N - (\Delta W_{\boldsymbol{k}})^2/(4m_N)$, and neglecting inelastic contributions, we obtain the time-dependent Schrödinger-like equation,

$$\left\{-H_0 - \frac{\partial}{\partial t} + \frac{1}{4m_N}\frac{\partial^2}{\partial t^2}\right\} R(\boldsymbol{x},t) = \int d^3y\, U(\boldsymbol{x},\boldsymbol{y})R(\boldsymbol{y},t) \simeq V(\boldsymbol{x})R(\boldsymbol{x},t), \tag{9}$$

which shows that the same $U(\boldsymbol{x},\boldsymbol{y})$ in Eq. (3) can be obtained from $R(\boldsymbol{x},t)$. An advantage of this method is that the ground state saturation is no more required for $R(\boldsymbol{x},t)$ to satisfy Eq. (9). For this method to work, however, t has to be large enough that elastic contributions dominate $R(\boldsymbol{x},t)$.

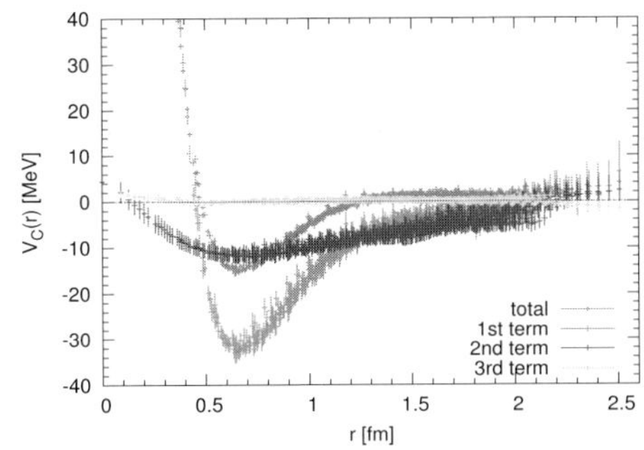

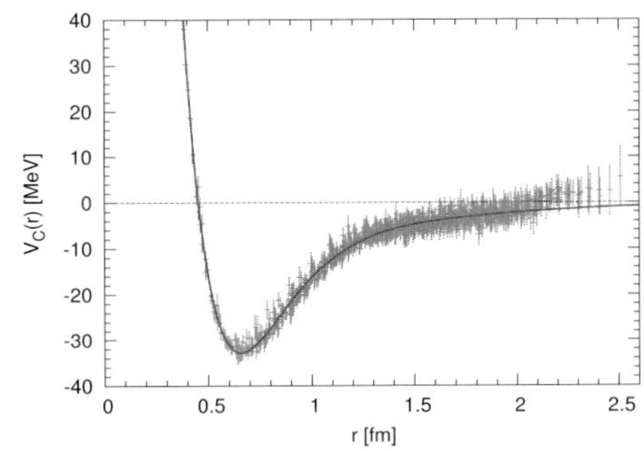

Fig. 2. (Left) Three contributions to $V_C(r)$ in Eq. (10) at $t = 9$. (Right) The multi-Gaussian fit of the central potential $V_C(r)$ with $N_{\rm Gauss} = 5$ at $t = 9$. Both figures are taken from Ref. 16.

We employ this improved method to calculate NN potential, using (2+1)-flavor QCD gauge configurations generated by PACS-CS collaboration[17] on $32^3 \times 64$ lattice with the RG improved Iwasaki gauge action at $\beta = 1.9$ and the non-perturbatively $O(a)$ improved Wilson quark action with $C_{\rm SW} = 1.715$, which corresponds to the lattice spacing $a \simeq 0.091$ fm ($a^{-1} = 2.176(31)$ GeV), the spatial extent $L = 32a \simeq 2.90$ fm. We calculate $R(\boldsymbol{x}, t)$ at a fixed value of light and strange quark mass combination, which corresponds to $m_\pi \simeq 701$ MeV, $m_K \simeq 789$ MeV and $m_N \simeq 1583$ MeV. In this calculation, we replace $e^{-m_N t}$ in the denominator of $R(\boldsymbol{x}, t)$ by the single-nucleon propagator $C_N(t) \equiv \sum_{\vec{x}} \langle 0|N(x)\bar{N}(0)|0\rangle$.

In the improved extraction, the central potential for the isospin-triplet and spin-singlet sector ($^1S_0$ in the standard notation) at the LO is given by

$$V_C(r) = -\frac{H_0 R(\boldsymbol{x}, t)}{R(\boldsymbol{x}, t;)} - \frac{(\partial/\partial t) R(\boldsymbol{x}, t)}{R(\boldsymbol{x}, t)} + \frac{1}{4m_N}\frac{(\partial/\partial t)^2 R(\boldsymbol{x}, t)}{R(\boldsymbol{x}, t)}, \tag{10}$$

each contribution of which is shown in Fig. 2 (Left) at $t = 9$. The first term in Eq. (10) (the red points) determines the overall shape of the potential, while the second term (the blue points) gives a major correction. The third term (the green points) corresponding to the relativistic correction, on the other hand, is negligible, showing that the non-relativistic approximation $\Delta W_{\boldsymbol{k}} \simeq \boldsymbol{k}^2/m_N$ works well. Note that the second term in Eq. (10) would be constant if the ground state saturation were achieved. A clear r-dependence of the second term tells us that contaminations of excited states indeed exist and are non-negligible.

The whole potential in Fig. 2 (Right) has a similar structure to the know phenomenological NN potentials, namely the repulsive core at short distance surrounded by the attractive well at medium and long distances, as already shown in Fig. 1 (Left).

The central potential $V_C(r)$ is fitted with multi-Gaussian functions as $g(r) \equiv \sum_{n=1}^{N_{\rm gauss}} V_n \cdot \exp(-\nu_n r^2)$, where V_n and $\nu_n (> 0)$ are used as fit parameters, $N_{\rm gauss}$ denotes the number of Gaussian functions. The fit is shown also in Fig. 2 (Right).

We then solve the Schrödinger equation in $^1S_0$ channel with this potential, in

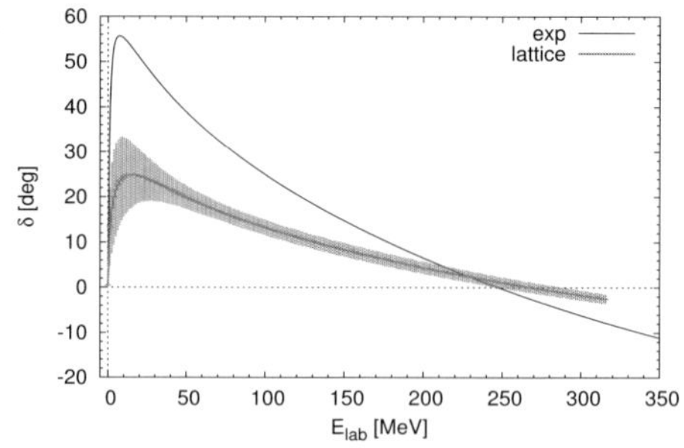

Fig. 3. The scattering phase in $^1S_0$ channel in the laboratory frame obtained from the lattice NN potential, together with experimental data.[18] Taken from Ref. 16.

order to extract the scattering phase $\delta(k)$, which is shown in Fig. 3, together with the experimental data for a comparison. A qualitative feature of the phase shift as a function of k is well reproduced, though the strength is weaker, most likely due to the heavy pion mass ($m_\pi \simeq 701$ MeV) in this calculation. The scattering length at $m_\pi \simeq 701$ MeV in the present method, calculated from the derivative of the scattering phase shift at $E_{\rm lab} = 0$, leads to $a(^1S_0) = \lim_{k\to 0} \tan\delta(k)/k = 1.6 \pm 1.1$ fm, which is still smaller than the experimental value at the physical point, $a^{(\rm exp)}(^1S_0) \sim 20$ fm, as seen from a comparison in Fig. 3.

Finally we discuss the convergence of the velocity expansion, Eq. (6). If the non-locality of the potential were absent, the LO potential would be exact at all energies below the inelastic threshold. Therefore, we can estimate a size of higher order terms in the velocity expansion, by comparing the LO potentials obtained at two energies.

Potentials shown so far are extracted with the periodic boundary condition in spatial directions for quark fields, which leads to the almost zero kinetic energy $E_{\boldsymbol{k}}$ for the ground state. With the anti-periodic boundary condition in spatial directions, on the other hand, we can significantly increase the ground state energy. To study the energy dependence of the LO potentials, we thus calculate the LO potentials at two different energies, $E_k \simeq 0$ MeV (periodic b.c.) and 45 MeV (anti-periodic b.c.), in quenched QCD at $m_\pi \simeq 529$ MeV and $L \simeq 4.4$ fm, by using the standard extraction of the potentials with the ground state saturation.[19] In Fig. 1 (Right), we compares the isospin-triplet and spin-singlet central potential $V_C(r)$ obtained at $E_k \simeq 45$ MeV (red) with that at $E_k \simeq 0$ MeV (blue). As seen from the figure, a good agreement between two determinations indicates that higher order contributions in the velocity expansion are small in the energy region between 0 MeV and 45 MeV. This also means that the local potential obtained at $E \simeq 0$ can be used to describe NN scattering phase shifts for the energy up to 45 MeV, at this pion mass in quenched QCD.

4. Conclusions

In this talk, I explain the HAL QCD method, recently proposed to extract hadronic interactions via lattice QCD simulations.

The equal-time NBS wave function for two nucleons plays a major role in the HAL QCD method, since it encodes the NN scattering phase shift in its asymptotic behavior at large separation. We therefore define the non-local but energy-independent potential through it, which, by construction, gives the correct phase shift for the NN scattering in QCD at all energies below the inelastic threshold. We then expand the non-local potential in terms of velocities and truncate it at the lowest few orders. Once the approximated potential is obtained in the velocity expansion, we can calculate the NN phase shift approximately at all energies below the inelastic threshold, by solving the Schrödinger equation with this potential.

The HAL QCD method is general and can be applied to other hadronic interactions for baryon-baryon, meson-baryon and three nucleon systems. See Refs. 13,14 for reviews on such investigations.

Acknowledgement

I would like to thank all members of the HAL QCD collaboration for useful discussions. This work is supported in part by the Grants-in-Aid for Scientific Research on Innovative Areas (No.2004: 20105001, 20105003) and SPIRE (Strategic Program for Innovative Research).

References

1. Z. Fodor and C. Hoelbling, *Rev. Mod. Phys.* **84** (2012) 449.
2. R. Tamagaki et al., *Prog. Theor. Phys. Suppl.* **112** (1993) 1.
3. H. Heiselberg and V. Pandharipande, *Ann. Rev. Nucl. Part. Sci.* **50** (2000) 481.
4. J. M. Lattimer and M. Prakash, *Phys. Rep.* **333** (2000) 121.
5. R. Machleidt, *Phys. Rev. C* **63** (2001) 024001 [nucl-th/0006014].
6. V. G. J. Stoks, R. A. M. Klomp, C. P. F. Terheggen and J. J. de Swart, *Phy. Rev. C* **49** (1994) 2950.
7. R. B. Wiringa, V. G. J. Stoks and R. Schiavilla, *Phy. Rev. C* **51** (1995) 38.
8. M. Lüscher, *Nucl. Phys. B* **354**, (1991) 531.
9. K. Orginos, *PoS* **LATTICE2011** (2011) 016.
10. N. Ishii, S. Aoki and T. Hatsuda, *Phys. Rev. Lett.* **99** (200) 022001.
11. S. Aoki, T. Hatsuda and N. Ishii, *Comput. Sci. Dis.* **1** (2008) 015009.
12. S. Aoki, T. Hatsuda and N. Ishii, *Prog. Theor. Phys.* **123** (2010) 89.
13. S. Aoki for HAL QCD Collaboration, *Prog. Part. Nucl. Phys.* **66** (2011) 687.
14. S. Aoki *et al.* [HAL QCD Collaboration], *Prog. Theor. Exp. Phys.* **2012**, (2012) 01A106.
15. N. Ishizuka, *PoS* **LAT2009** (2009) 119.
16. N. Ishii *et al.* [HAL QCD Collaboration], *Phys. Lett. B* **712** (2012) 437.
17. S. Aoki *et al.* [PACS-CS Collaboration], *Phys. Rev. D* **79** (2009) 034503.
18. http://www.nn-online.org/
19. K. Murano, N. Ishii, S. Aoki and T. Hatsuda, *Prog. Theor. Phys.* **125** (2011) 1225.

Dynamics with Isospin Symmetric Higgs Boson: The Quark Mass Hierarchy and the LHC Data

V. A. Miransky*

Department of Applied Mathematics, Western University,
London, Ontario N6A 5B7, Canada
**E-mail: vmiransk@uwo.ca*

The ATLAS and CMS Collaborations have recently discovered a new 125 GeV boson. We show that the properties of this particle, including the enhancement of its diphoton decay rate, can be explained in a model with an isospin symmetric Higgs boson. The predictions of the model relevant for future experiments are also discussed.

Keywords: Isospin symmetric Higgs boson, top-Higgs composite.

1. Introduction

The ATLAS and CMS Collaborations discovered a new boson h in the mass range 125-126 GeV, compatible to the Standard Model (SM) Higgs boson H.[1] In particular, the decay channels of $h \to ZZ^*$ and $h \to WW^*$ are fairly consistent with the SM. On the other hand, there are still many open questions concerning the values of the Yukawa couplings of the h boson to quarks and leptons. Because of that, it is still unclear whether the h boson is the only source of the fermion masses. Moreover, the data might already suggest existence of a new physics beyond the SM: the diphoton branching ratio $\mathrm{Br}(h \to \gamma\gamma)$ is about 1.6 times larger than the SM value.[a]

In this paper, I review the approach to electroweak symmetry breaking and the fermion mass hierarchy, based on the class of models with isospin symmetric (IS) electroweak Higgs boson suggested by Michio Hashimoto and the author in Refs. 4–6. As will be discussed below, the LHC data, including those for the enhanced diphoton branching ratio, can be explained in this approach. It is noticeable that these models also make several predictions, which can be checked at the LHC in the near future (see Sec. 6).

[a]To the contrary, the authors of Ref. 2 argue that none of the measured couplings deviates from its SM values significantly. Also, the QCD uncertainties are discussed in Ref. 3. Last but not least, with more data, the CMS Collaboration does not confirm the excess seen currently by the ATLAS. Thus, the observed deviations are not yet definitive.

2. IS Higgs Models

There is a large hierarchy between quark masses from different families.[7] Besides, the isospin violation in different families is also hierarchical. It is very strong in the third family, strong (although essentially weaker) in the second family, and mild in the first one: $\frac{m_t}{m_b} \simeq 41.5$, $\frac{m_c}{m_s} \simeq 13.4$, and $\frac{m_u}{m_d} \simeq 0.38$–$0.58$.[7] This is a big mystery: In the framework of the SM, it is required to introduce hierarchical Yukawa couplings by hand, e.g., $\frac{y_t^{\rm SM}}{y_b^{\rm SM}} \simeq 41.5$, and $\frac{y_c^{\rm SM}}{y_s^{\rm SM}} \simeq 13.4$.

One of our main motivations for introducing the IS Higgs models,[4,5] describing the hierarchies in the quark mass spectrum, was to find a dynamical mechanism that could shed light on the experimental fact that the isospin violation in the quark mass spectrum is essentially stronger in heavier families.

The main characteristics of these models are the following: a) It is assumed that the dynamics primarily responsible for electroweak symmetry breaking (EWSB) leads to the mass spectrum of quarks with no (or weak) isospin violation. *Moreover, it is assumed that the values of these masses are of the order of the observed masses of the down-type quarks.* b) The second (central) assumption is introducing the horizontal interactions for the quarks in the three families. As a first step, a *subcritical* (although nearcritical, i.e., strong) diagonal horizontal interactions for the top quark is utilized which lead to the observed ratio $\frac{m_t}{m_b} \simeq 41.5$. The second step is introducing *equal* strength (i.e., isospin symmetric) horizontal flavor-changing-neutral (FCN) interactions between the t and c quarks and the b and s ones.

All together, these interactions naturally provide the observed ratio $m_c/m_s \simeq 13.4$ in the second family.[4] As emphasized in Ref. 4, the choice of the IS masses being close to the values of the observed masses of the down-type quarks is crucial in this scenario. As to the mild isospin violation in the first family, it was studied together with the effects of the family mixing, reflected in the Cabibbo-Kobayashi-Maskawa (CKM) matrix[4] (see also Sec. 5 below).

In this scenario, besides the EWSB interactions, the dominant dynamics responsible for the form of the mass spectrum of quarks is connected with the diagonal horizontal interactions for the third family and the horizontal, isospin symmetric, FCN interactions between the second and third ones. One of the signatures of this scenario is the appearance of a composite top-Higgs doublet Φ_{h_t} (resonance) composed of the quarks and antiquarks of the third family, see Refs. 4,5.[b]

Thus, the main source of the isospin violation in this approach is only the strong top quark interactions. On the other hand, because these interactions are subcritical, the top quark plays a minor role in EWSB. The latter distinguishes this scenario from the top quark condensate model.[9–14] Note that unlike the topcolor assisted technicolor model (TC2),[15] this class of models utilizes subcritical dynamics for the top quark, so that without strong fine tuning, the bosons from the top-Higgs

[b]Such composites in the nearcritical regime in a symmetric phase of models with dynamical chiral symmetry breaking were studied by several authors.[8]

doublet Φ_{h_t} are heavy, say, of order 1 TeV, in general (compare with Ref. 5).

Although the concrete model in Refs. 4,5 utilized the fourth family of fermions[16,17] for generating EWSB, this choice is not crucial, as the authors emphasized in Ref. 4. In particular, the fourth family can be replaced by just a IS Higgs boson doublet Φ_h, without specifying its composite origin (if any).[6] In Ref. 6, it was considered just such a version in which the neutral scalar from the Φ_h doublet was identified with the 125 GeV h boson. Here we emphasize that while the neutral top-Higgs boson h_t has a large top-Yukawa coupling, the IS Higgs boson h does not, $y_t \simeq y_b \sim 10^{-2}$. On the other hand, the hWW^* and hZZ^* coupling constants are close to those in the SM (see below). Also, the mixing between Φ_h and much heavier Φ_{h_t} should be small (compare with Ref. 5). Let us now describe the decay processes of the IS Higgs h.

3. Decay Modes $h \to \gamma\gamma$, $h \to Z\gamma$, $h \to WW^*$, and $h \to ZZ^*$

Let us consider the diphoton branching ratio in the IS Higgs model. It is well known that the W-loop contribution to $H \to \gamma\gamma$ is dominant in the SM, while the top-loop effect is destructive against the W-loop. More concretely, the diphoton partial width in the SM reads:[18]

$$\Gamma(H \to \gamma\gamma) = \frac{\sqrt{2}G_F\alpha^2 m_H^3}{256\pi^3}\left|A_1(\tau_W) + N_c Q_t^2 A_{\frac{1}{2}}(\tau_t)\right|^2, \qquad \tau_W \equiv \frac{m_H^2}{4m_W^2}, \tau_t \equiv \frac{m_H^2}{4m_t^2}, \tag{1}$$

where G_F denotes the Fermi constant, $N_c = 3$ represents the number of color, and $Q_t = +2/3$ is the electric charge of the top quark. The loop functions A_1 and $A_{\frac{1}{2}}$ for W and t, respectively, are given by

$$A_1(\tau) \equiv -\frac{1}{\tau^2}\Big[2\tau^2 + 3\tau + 3(2\tau - 1)f(\tau)\Big], \tag{2}$$

and

$$A_{\frac{1}{2}}(\tau) \equiv \frac{2}{\tau^2}\Big[\tau + (\tau - 1)f(\tau)\Big], \tag{3}$$

with $f(\tau) \equiv \arcsin^2\sqrt{\tau}$ for $\tau \leq 1$. Then, the numerical values of the W- and t-loop functions read

$$A_1(\tau_W) = -8.32, \qquad A_{\frac{1}{2}}(\tau_t) = 1.38, \tag{4}$$

for $m_W = 80.385$ GeV,[7] $m_t = 173.5$ GeV,[7] and $m_H = 125$ GeV.

On the other hand, in the IS Higgs model, the Yukawa coupling between the top and the IS Higgs h is as small as the bottom Yukawa coupling, so that the top-loop contribution is strongly suppressed. The partial decay width of $h \to \gamma\gamma$ is thus enhanced without changing essentially $h \to ZZ^*$ and $h \to WW^*$. A rough estimate taking the isospin symmetric top and bottom Yukawa couplings $y_t \simeq y_b \approx 10^{-2}$ is as follows:

$$\frac{\Gamma^{\rm IS}(h \to \gamma\gamma)}{\Gamma^{\rm SM}(H \to \gamma\gamma)} \simeq 1.56, \qquad \frac{\Gamma^{\rm IS}(h \to WW^*)}{\Gamma^{\rm SM}(H \to WW^*)} = \frac{\Gamma^{\rm IS}(h \to ZZ^*)}{\Gamma^{\rm SM}(H \to ZZ^*)} = \left(\frac{v_h}{v}\right)^2 \simeq 0.96. \tag{5}$$

Here using the Pagels-Stokar formula,[19] we estimated the vacuum expectation value (VEV) of the top-Higgs h_t as $v_t = 50$ GeV, and the VEV v_h of the IS Higgs h is given by the relation $v^2 = v_h^2 + v_t^2$ with $v = 246$ GeV. Note that the values of the ratios in Eq. (5) are not very sensitive to the value of v_t, e.g., for $v_t = 40$–100 GeV, the suppression factor in the pair decay modes to WW^* and ZZ^* is 0.97–0.84 and the enhancement factor in the diphoton channel is 1.58–1.37. For the decay mode of $h \to Z\gamma$, this model yields

$$\frac{\Gamma^{\rm IS}(h \to Z\gamma)}{\Gamma^{\rm SM}(H \to Z\gamma)} \simeq 1.07 \tag{6}$$

(the data concerning this decay channel has not yet been reported).[c] Note that the total decay width is almost unchanged, so that Eqs. (5) and (6) indicate the suppression/enhancement factors of the corresponding branching ratios.

The values in Eq. (5) agree well with the data in the ATLAS and CMS experiments. However, obviously, the main production mechanism of the Higgs boson, the gluon fusion process $gg \to h$, is now in trouble. The presence of new chargeless colored particles, which considered by several authors[20] can help to resolve this problem. We pursue this possibility below.

4. Model with Colored Scalar

We utilize an effective theory near the EWSB scale. The model contains: (1) the IS Higgs doublet Φ_h, which is mainly responsible for the EWSB and couples to the top and bottom in the isospin symmetric way, (2) the top-Higgs doublet Φ_{h_t}, which is required to obtain the correct top mass, and (3) the colored scalar and/or fermions which are required to enhance $gg \to h$.

The items (1) and (2) above are essentially described in Refs. 4,5. The only difference is that the two composite Higgs doublets composed of the fourth family quarks should now be replaced by the IS Higgs. We will discuss this point later. Note that in this case the Lagrangian density $\mathcal{L}$ in the effective theory contains the IS Higgs quartic coupling λ_h, $\mathcal{L} \supset -\lambda_h|\Phi_h^\dagger\Phi_h|^2$, and the mass $m_h = 125$ GeV corresponds to a small λ_h via the relation $m_h^2 \simeq 2\lambda_h v_h^2$ like in the SM, because the mixing between Φ_h and much heavier Φ_{h_t} is tiny in the present model (compare with Refs. 4,5). However, unlike the case of the SM,[21] this does not imply that the theory keeps the perturbative nature up to some extremely high energy scale, as we will see below.

---

[c]These so different results for the $h \to \gamma\gamma$ and $h \to Z\gamma$ decays reflect the fact that in the SM the contribution of the top-loop diagram is suppressed for $H \to Z\gamma$ much stronger than for $H \to \gamma\gamma$.

As to a concrete realization of item (3), we may introduce a real scalar field S in the adjoint representation of the color $SU(3)_c$ and utilize the Higgs-portal model,[20] just as a benchmark case:

$$\mathcal{L} \supset \mathcal{L}_S = \frac{1}{2}(D_\mu S)^2 - \frac{1}{2}m_{0,S}^2 S^2 - \frac{\lambda_S}{4}S^4 - \frac{\lambda_{hS}}{2}S^2\Phi_h^\dagger\Phi_h,$$
$$\Phi_h = \begin{pmatrix} \omega^+ \\ \frac{1}{\sqrt{2}}(v_h + h + iz_0) \end{pmatrix}, \tag{7}$$

where $\omega^\pm$ and z_0 are the components eaten by $W^\pm$ and Z. The scalar field S is chosen to be assigned to the $(\mathbf{8}, \mathbf{1})_0$ representation of the $SU(3)_c \times SU(2)_W \times U(1)_Y$. Other representations, for example, a color triplet, are also possible. Note that we do not incorporate a Higgs-portal term between S and Φ_{h_t} and other possible cubic and quartic terms into Eq. (7), because they do not play any important role in the following analysis.

The mass-squared term for the scalar S is given by

$$M_S^2 = m_{0,S}^2 + \frac{\lambda_{hS}}{2}v_h^2, \tag{8}$$

and should be positive in order to avoid the color symmetry breaking. Typically, $M_S \sim 200$ GeV is allowed in the current data.[20] We will take a positive value for λ_{hS} and a classically (quasi-)scale invariant model with $m_{0,S}^2 \approx 0$, which is favorable to reproduce the SM like gluon fusion production.

Let us consider the contribution of the color octet S to the gluon fusion process $gg \to h$ in the leading order:

$$\frac{\sigma(gg \to h)}{\sigma^{\rm SM}(gg \to H)} \sim \frac{\Gamma(h \to gg)}{\Gamma^{\rm SM}(H \to gg)} = \left| \frac{C_A \lambda_{hS} \frac{v v_h}{2M_S^2} A_0(\tau_S)}{A_{\frac{1}{2}}(\tau_t)} \right|^2, \tag{9}$$

with $C_A = 3$, $\tau_S \equiv m_h^2/(4M_S^2)$, and

$$A_0(\tau) \equiv -\frac{1}{\tau^2}\left[\tau - f(\tau)\right]. \tag{10}$$

We find $A_0 \simeq 0.37$–0.34 for $M_S = 150$–400 GeV, so that an appropriate value of the Higgs-portal coupling is

$$\lambda_{hS} \simeq 2.5\text{–}2.7 \times \frac{M_S^2}{v v_h}. \tag{11}$$

As a typical value, we may take $\lambda_{hS} = 1.8$ for $M_S = 200$ GeV and $v_t = 50$ GeV. When $m_{0,S}^2 \approx 0$, i.e., $M_S^2 \approx \lambda_{hS}v_h^2/2$, we obtain $\Gamma(h \to gg) \approx 0.6 \times \Gamma^{\rm SM}(H \to gg)$, independently of the values of λ_{hS}. In order to stabilize the Higgs potential for S at the tree level, the relation $|\lambda_{hS}| < 2\sqrt{\lambda_S\lambda_h}$ is also required.

A comment concerning the IS Higgs quartic coupling λ_h is in order. In the SM, the Higgs mass 125 GeV suggests that the theory is perturbative up to an extremely high energy scale.[21] On the contrary, in the present model, when we take a large Higgs-portal coupling λ_{hS} that reproduces $gg \to h$ correctly, the quartic coupling

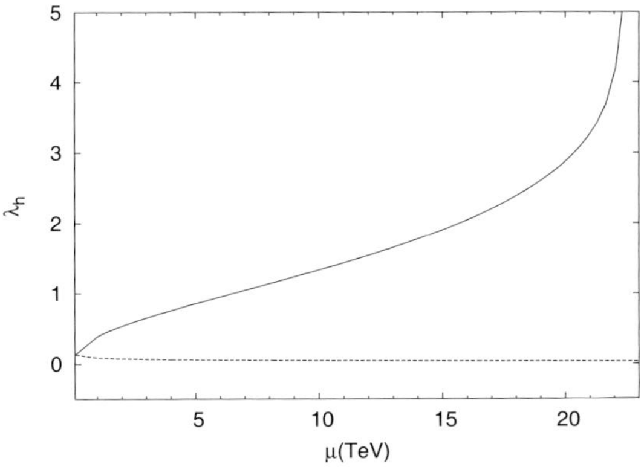

Fig. 1. The running behavior of the IS Higgs quartic coupling λ_h. The solid and dashed lines correspond to λ_h and the SM Higgs quartic coupling, respectively. We fixed the IS Higgs mass $m_h = \sqrt{2\lambda_h}v_h = 125$ GeV and took $\lambda_{hS} = 1.8$ and $\lambda_S = 1.5$. Unlike the SM, the IS Higgs quartic coupling grows up due to a large Higgs-portal coupling λ_{hS} and a small top-Yukawa coupling y_t.

λ_h will grow because the β-function for λ_h contains the λ_{hS}^2 term. Also, there is no large negative contribution to the β-function for λ_h from the top-Yukawa coupling $y_t \sim 10^{-2}$.

One can demonstrate such a behavior more explicitly by using the renormalization group equations. In Fig. 1, the running of the coupling λ_h is shown. The IS Higgs mass is $m_h = \sqrt{2\lambda_h}v_h$, and we take it to be equal 125 GeV. Taking a large Higgs-portal coupling $\lambda_{hS} = 1.8$ and the S^4-coupling $\lambda_S = 1.5$, it turns out that the coupling λ_h rapidly grows. Due to the running effects, the naive instability of the scalar potential at the tree level is resolved around the TeV scale in this case. The blowup scale strongly depends on the initial values of λ_{hS} and λ_S.

5. Quark mass matrices

Let us discuss the structure of the quark mass matrices in the present model. The Yukawa interactions are written by[4,5]

$$-\mathcal{L}_Y = \sum_{i,j} \bar{\psi}_L^{(i)} Y_D^{ij} d_R^{(j)} \Phi_h + \sum_{i,j} \bar{\psi}_L^{(i)} Y_U^{ij} u_R^{(j)} \tilde{\Phi}_h + y_{h_t} \bar{\psi}_L^{(3)} t_R \tilde{\Phi}_{h_t}, \tag{12}$$

with

$$\tilde{\Phi}_h \equiv i\tau_2 \Phi_h^*, \quad \tilde{\Phi}_{h_t} \equiv i\tau_2 \Phi_{h_t}^*, \quad \Phi_{h_t} = \begin{pmatrix} \omega_t^+ \\ \frac{1}{\sqrt{2}}(v_t + h_t + iz_t) \end{pmatrix},$$

$$\langle \Phi_h \rangle = \begin{pmatrix} 0 \\ \frac{v_h}{\sqrt{2}} \end{pmatrix}, \quad \langle \Phi_{h_t} \rangle = \begin{pmatrix} 0 \\ \frac{v_t}{\sqrt{2}} \end{pmatrix}, \tag{13}$$

$$Y_D \equiv \frac{\sqrt{2}}{v_h} M_D, \quad Y_U \equiv \frac{\sqrt{2}}{v_h} M_U, \tag{14}$$

and

$$M_D = \begin{pmatrix} m_0^{(1)} & \xi_{12} m_0^{(1)} & \xi_{13} m_0^{(1)} \\ \xi_{21} m_0^{(1)} & m_0^{(2)} + \delta \cdot m_b & \xi_{23} m_0^{(2)} \\ \xi_{31} m_0^{(1)} & \xi_{32} m_0^{(2)} & m_0^{(3)} \end{pmatrix}, \quad M_U = \begin{pmatrix} \eta_{11} m_0^{(1)} & \eta_{12} m_0^{(1)} & \eta_{13} m_0^{(1)} \\ \eta_{21} m_0^{(1)} & m_0^{(2)} + \delta \cdot m_t & \eta_{23} m_0^{(2)} \\ \eta_{31} m_0^{(1)} & \eta_{32} m_0^{(2)} & m_0^{(3)} \end{pmatrix}, \tag{15}$$

where $\psi_L^{(i)}$ denotes the weak doublet quarks from the i-th family, and $u_R^{(i)}$ and $d_R^{(i)}$ represent the right-handed up- and down-type quarks.

The top-Higgs part is responsible for the top mass, $m_t \simeq y_{h_t} \frac{v_t}{\sqrt{2}}$. The IS masses $m_0^{(i)}$ are the same mass scales as the down-type quarks, say, $m_0^{(3)} \sim 1$ GeV, $m_0^{(2)} \sim$ 100 MeV, and $m_0^{(1)} \sim 1$ MeV. The common one-loop factor $\delta \sim 1/100$ yields the correct mass hierarchy between m_s and m_c via the hierarchy between m_b and m_t. Also, the off-diagonal coefficients are assumed to be $\xi_{ij}, \eta_{ij} \sim \mathcal{O}(1)$, with some dynamical mechanism. (We kept η_{11} in the up sector for generality.) The CKM matrix is approximately determined by the down-type quark mass matrix,[5]

$$V_{\rm CKM} \approx \begin{pmatrix} 1 - \frac{|\xi_{12}|^2}{2}\left(\frac{m_0^{(1)}}{m_0^{(2)}}\right)^2 & \xi_{12}\frac{m_0^{(1)}}{m_0^{(2)}} & \xi_{13}\frac{m_0^{(1)}}{m_0^{(3)}} \\ -\xi_{12}^*\frac{m_0^{(1)}}{m_0^{(2)}} & 1 - \frac{|\xi_{12}|^2}{2}\left(\frac{m_0^{(1)}}{m_0^{(2)}}\right)^2 & \xi_{23}\frac{m_0^{(2)}}{m_0^{(3)}} \\ -(\xi_{13}^* - \xi_{12}^*\xi_{23}^*)\frac{m_0^{(1)}}{m_0^{(3)}} & -\xi_{23}^*\frac{m_0^{(2)}}{m_0^{(3)}} & 1 \end{pmatrix}. \tag{16}$$

We can then reproduce the CKM matrix, basically. For example, with the inputs, $m_0^{(1)} = 10$ MeV, $m_0^{(2)} = 68$ MeV, $m_0^{(3)} = 4.2$ GeV, $m_t = 173.5$ GeV, $\delta = 7 \times 10^{-3}$, $\xi_{12} = \xi_{21} = \eta_{12} = \eta_{21} = 2.0$, $\xi_{13} = \xi_{31} = \eta_{13} = \eta_{31} = 1.6$, $\xi_{23} = \xi_{32} = \eta_{23} = \eta_{32} = -2.5$, $\eta_{11} = \frac{1}{4}$, we obtain $m_d = 4.9$ MeV, $m_s = 95$ MeV, $m_b = 4.2$ GeV, $m_u = 2.2$ MeV, $m_c = 1.3$ GeV, $|V_{ud}| \simeq |V_{cs}| = 0.975$, $|V_{tb}| \simeq 1$, $|V_{us}| \simeq |V_{cd}| = 0.22$, $|V_{cb}| = 0.041$, $|V_{ts}| = 0.039$, $|V_{ub}| = 0.0042$, $|V_{td}| = 0.013$. These values fairly agree with the PDG ones.[7]

As was emphasized above in Sec. 2, because of the subcriticality dynamics in this class of models, the extra bosons from the top Higgs doublet Φ_{h_t} are heavy, say, $\mathcal{O}(1\text{TeV})$. Thus, their one-loop contributions to the B^0-$\bar{B}^0$ mixing, $b \to s\gamma$ and $Z \to b\bar{b}$ are suppressed. A tree FCNC term also appears in the up sector, so that the D^0–$\bar{D}^0$ mixing is potentially dangerous. However, because the FCNC coupling Y_{t-c-h_t} is found to be tiny, $Y_{t-c-h_t} \sim \frac{m_t}{v_t}\frac{m_u}{m_t}\frac{m_c}{m_t} = 10^{-6}$–$10^{-7}$, this does not cause any troubles.

6. Conclusion

The model with an IS Higgs boson yields not only an explanation of the ATLAS and CMS data, including the enhanced diphoton Higgs decay rate, but also makes several predictions. The most important of them is that the value of the top-Yukawa coupling h-t-$\bar{t}$ should be close to the bottom-Yukawa one. Another prediction relates to the decay mode $h \to Z\gamma$, which unlike $h \to \gamma\gamma$ is enhanced only slightly, $\Gamma^{\rm IS}(h \to Z\gamma) = 1.07 \times \Gamma^{\rm SM}(H \to Z\gamma)$. Last but not least, the LHC might potentially discover the top-Higgs resonance h_t, if lucky.

Acknowledgments

I am grateful to the organizers of the SGCT 12 Workshop, in particular, Prof. Koichi Yamawaki, for their warm hospitality. This work is supported by the Natural Sciences and Engineering Research Council of Canada.

References

1. ATLAS Collaboration, Phys. Lett. B **716**, 1 (2012); CMS Collaboration, Phys. Lett. B **716**, 30 (2012).
2. T. Plehn and M. Rauch, Europhys. Lett. **100**, 11002 (2012).
3. J. Baglio, A. Djouadi and R. M. Godbole, Phys. Lett. **716**, 203 (2012).
4. M. Hashimoto and V. A. Miransky, Phys. Rev. D **80**, 013004 (2009) [arXiv:0901.4354 [hep-ph]].
5. M. Hashimoto and V. A. Miransky, Phys. Rev. D **81**, 055014 (2010) [arXiv:0912.4453 [hep-ph]].
6. M. Hashimoto, V. A. Miransky, Phys. Rev. D **86**, 095018 (2012) [arXiv:1208.1305 [hep-ph]].
7. J. Beringer *et al.* [Particle Data Group], Phys. Rev. D **86**, 010001 (2012).
8. R. S. Chivukula, A. G. Cohen, and K. D. Lane, Nucl. Phys. B **343**, 554 (1990); T. Appelquist, J. Terning, and L. C. R. Wijewardhana, Phys. Rev. D **44**, 871 (1991); R. R. Mendel and V. A. Miransky, Phys. Lett. B **268**, 384 (1991); V. A. Miransky, Phys. Rev. Lett. **69**, 1022 (1992).
9. V. A. Miransky, M. Tanabashi and K. Yamawaki, Phys. Lett. B **221**, 177 (1989); Mod. Phys. Lett. A **4**, 1043 (1989).
10. Y. Nambu, Enrico Fermi Institute Report No. 89-08, 1989; in *Proceedings of the 1988 Kazimierz Workshop*, eds. Z. Ajduk et al. (World Scientific Publishing Co., Singapore, 1989).
11. W. J. Marciano, Phys. Rev. Lett. **62**, 2793 (1989); Phys. Rev. D **41**, 219 (1990).
12. W. A. Bardeen, C. T. Hill and M. Lindner, Phys. Rev. D **41**, 1647 (1990).
13. For a dynamical model with extra dimensions, see, e.g., M. Hashimoto, M. Tanabashi and K. Yamawaki, Phys. Rev. D **64**, 056003 (2001); *ibid* D **69**, 076004 (2004); V. Gusynin, M. Hashimoto, M. Tanabashi and K. Yamawaki, Phys. Rev. D **65**, 116008 (2002).
14. For a comprehensive review, see C. T. Hill and E. H. Simmons, Phys. Rept. **381**, 235 (2003); Erratum-ibid. **390**, 553 (2004).
15. C. T. Hill, Phys. Lett. B **345**, 483 (1995).
16. B. Holdom, Phys. Rev. Lett. **57**, 2496 (1986) [Erratum-*ibid.* **58**, 177 (1987)]; C. T. Hill, M. A. Luty and E. A. Paschos, Phys. Rev. D **43**, 3011 (1991); H. J. He, N. Polonsky and S. f. Su, Phys. Rev. D **64**, 053004 (2001); G. D. Kribs, T. Plehn, M. Spannowsky and T. M. P. Tait, Phys. Rev. D **76**, 075016 (2007); G. Burdman and L. Da Rold, JHEP **0712**, 086 (2007); M. Hashimoto, Phys. Rev. D **81**, 075023 (2010).
17. For a review, see P. H. Frampton, P. Q. Hung and M. Sher, Phys. Rept. **330**, 263 (2000).
18. J. R. Ellis, M. K. Gaillard and D. V. Nanopoulos, Nucl. Phys. B **106**, 292 (1976);

M. A. Shifman, A. I. Vainshtein, M. B. Voloshin and V. I. Zakharov, Sov. J. Nucl. Phys. **30**, 711 (1979) [Yad. Fiz. **30**, 1368 (1979)].

19. H. Pagels and S. Stokar, Phys. Rev. D **20**, 2947 (1979).
20. R. Boughezal and F. Petriello, Phys. Rev. D **81**, 114033 (2010); B. Batell, S. Gori and L. -T. Wang, JHEP **1206**, 172 (2012); B. A. Dobrescu, G. D. Kribs and A. Martin, Phys. Rev. D **85**, 074031 (2012); G. D. Kribs and A. Martin, arXiv:1207.4496 [hep-ph].
21. J. A. Casas, J. R. Espinosa and M. Quiros, Phys. Lett. B **342**, 171 (1995); *ibid* B **382**, 374 (1996); G. Isidori, G. Ridolfi and A. Strumia, Nucl. Phys. B **609**, 387 (2001); J. Elias-Miro, J. R. Espinosa, G. F. Giudice, G. Isidori, A. Riotto and A. Strumia, *ibid* B **709**, 222 (2012); S. Alekhin, A. Djouadi and S. Moch, arXiv:1207.0980 [hep-ph].

"Super"–Dilatation Symmetry of the Top-Higgs System

Christopher T. Hill

Fermi National Accelerator Laboratory P.O. Box 500
Batavia, Illinois 60510, USA
E-mail: hill@fnal.gov

The top-Higgs system, consisting of top quark (LH doublet, RH singlet) and Higgs boson kinetic terms, with gauge fields set to zero, has an *exact* symmetry where both fermion and Higgs fields are shifted and mixed in a supersymmetric fashion. The full Higgs-Yukawa interaction and Higgs-potential, including additional $\sim 1/\Lambda^2$ NJL-like interactions, also has this symmetry to $\mathcal{O}(1/\Lambda^4)$, up to null-operators. Thus the interaction Lagrangian can be viewed as a power series in $1/\Lambda^2$. The symmetry involves interplay of the Higgs quartic interaction with the Higgs-Yukawa interaction and implies the relationship, $\lambda = \frac{1}{2}g^2$ between the top–Yukawa coupling, g, and Higgs quartic coupling, λ, at a high energy scale $\Lambda \sim$ few TeV. We interpret this to be a new physics scale. The top quark is massless in the symmetric phase, satisfying the Nambu-Goldstone theorem. The fermionic shift part of the current is $\propto (1 - H^\dagger H/v^2)$, owing to the interplay of λ and g, and vanishes in the broken phase. Hence the Nambu-Goldstone theorem is trivially evaded in the broken phase and the top quark becomes heavy (it is not a Goldstino). We have $m_t = m_h$, subject to radiative corrections that can in principle pull the Higgs into concordance with experiment.

1. Introduction

The Higgs boson can be viewed as a "pseudo-dilaton" in a particular limit.[1] There is, of course, a fundamental distinction between a "scale invariant Higgs field" and a "dilatonic Higgs field." A scale invariant Higgs field has a vanishing mass term, but can have a nonvanishing gauge, quartic and Yukawa couplings. To qualify as a (pseudo) dilatonic Higgs boson, the Higgs potential must be (approximately) flat.

Consider the pure Higgs Lagrangian (no gauge fields):

$$\mathcal{L}_0 = \partial_\mu H^\dagger \partial^\mu H - \frac{\lambda}{2}(H^\dagger H - v^2)^2 \tag{1}$$

As usual, the groundstate has a minimum for:

$$\langle H^i \rangle = \theta^i, \qquad \text{where we can choose:} \qquad \theta^i = (v, 0), \tag{2}$$

where θ^i is an arbitrary orientation in gauge space and can be rotated under $SU(2)_L \times U(1)$.

In the limit of small λ, the Higgs potential plays the analogue role of an "applied external magnetic field" to a spin system, pulling $\langle H^i \rangle$ to the minimum VEV, v. If we then take $\lambda \to 0$ the Lagrangian acquires a global "shift symmetry,"

$$\delta H^i = \theta^i \epsilon \qquad \longrightarrow \qquad \delta\, \partial_\mu H^\dagger \partial^\mu H = 0 \tag{3}$$

The alignment, θ^i, is held fixed and the shift is parameterized by the variable ϵ. The Noether current is then:

$$J_\mu = \frac{\delta \mathcal{L}_0}{\delta \partial^\mu \epsilon} = \theta^\dagger \partial_\mu H + H^\dagger \partial_\mu \theta \tag{4}$$

We see that θ is a defining part of the current. If we view θ as co-rotating with H under the global $SU(2) \times U(1)$ transformations, the charge $\int d^3x\, J_0$ then commutes with the gauge group.

In the broken phase of the theory the current looks more "dilatonic":

$$J_\mu \to \sqrt{2} v \partial_\mu h \tag{5}$$

The dilatonic nature of the Higgs implies that fields that acquire masses proportional to v are "scale invariant" in the sense of spontaneous scale breaking. That is, we can perform an ordinary scale transformation which would normally shift mass terms, but we can then undo this by a compensating shift in h. To see this, consider the top quark mass term:

$$g\overline{\psi}_L t_R H + h.c. \quad \longrightarrow \quad m_t \bar{t}t \left(1 + \frac{h}{\sqrt{2}v}\right) \tag{6}$$

Under an ordinary infinitesimal scale transformation we have $t(x) \to (1-\epsilon)^{3/2} t(x')$ and $h(x) \to (1-\epsilon)h(x')$ where $x_\mu = (1+\epsilon){x'}_\mu$, $d^4x = (1+\epsilon)^4 d^4x'$. Hence the action transforms as:

$$\int d^4x\; m_t\bar{t}t(x)\left(1 + \frac{h(x)}{\sqrt{2}v}\right) \to \to \int d^4x' \left((1+\epsilon)m_t\bar{t}t(x') + m_t\bar{t}t(x')\frac{h(x')}{\sqrt{2}v}\right) \tag{7}$$

The latter expression exhibits the fact that under ordinary scale transformations the $d = 4$ Higgs-Yukawa interaction is scale invariant, while the $d = 3$ mass term is not invariant.

However, with the dilatonic shift symmetry we can compensate the rescaled mass term by a shift in the Higgs field:

$$h(x') \to h(x') - \sqrt{2}v\epsilon \tag{8}$$

and we see that:

$$\int d^4x' \left((1+\epsilon)m_t\bar{t}t(x') + m_t\bar{t}t(x')\frac{h(x')}{\sqrt{2}v}\right) \to \int d^4x' \left(m_t\bar{t}t(x') + m_t\bar{t}t(x')\frac{h(x')}{\sqrt{2}v}\right) \tag{9}$$

Hence, the simultaneous application of the scale transformation and Higgs shift symmetry allows us to maintain the symmetry of the top quark mass term. The scale symmetry can be viewed as spontaneously broken with the Higgs boson playing the role of the Nambu-Goldstone mode. The same invariance applies to the gauge fields, W and Z. Higgs self-interactions that involve nonzero λ would not be invariant under scale transformations with dilatonic shifts in h. The symmetry is also broken by scale anomalies (running couplings).

2. Generalize to a "Super-Dilatation"

The Higgs boson is thus a (pseudo) dilaton if the shift transformation is a (approximate) symmetry of the action. Fundamentally it stems from the exact shift or modular symmetry of the gaugeless Higgs kinetic term, as in Eq. (3):

$$\delta H^i = \theta^i \epsilon \tag{10}$$

A key point we wish to emphasize is that ϵ is the infinitesimal parameter of the transformation, while the orientation, θ^i, is held fixed. θ^i defines a "ray" and the shift moves the field along this direction in field space. We take θ^i to have the same mass dimension as the Higgs, *i.e.*, dimensions of mass and it is a normalized isospinor, $\theta^\dagger\theta = v^2$, where we conventionally choose the alignment $\theta^i = (v, 0)$. Eq. (10) is a symmetry of the gaugeless Higgs boson kinetic terms, $\partial H^\dagger \partial H$. In such a theory the shift symmetry is exact.

We now propose a generalization of dilatation symmetry for the Higgs boson that involves a "super"-symmetric relationship between the top and Higgs fields.[2] The shift in the Higgs boson field is now promoted to an operator. This symmetry is exact in the top, with bottom-left, and Higgs, kinetic terms in the gaugeless limit (up to total divergences). Consider the top and Higgs kinetic terms of the standard model with *gauge fields set to zero*:

$$\mathcal{L}_K = \overline{\psi}_L i\partial\!\!\!/ \psi_L + \overline{t}_R i\partial\!\!\!/ t_R + \partial H^\dagger \partial H \tag{11}$$

We define the infinitesimal transformation:

$$\begin{aligned}
\delta\psi_L^{ia} &= \theta_L^{ia}\eta\epsilon - i\frac{\partial\!\!\!/ H^i \theta_R^a}{\Lambda^2}\epsilon; &\qquad \delta\overline{\psi}_{L\,ia} &= \overline{\theta}_{L\,ia}\eta\epsilon + i\frac{\overline{\theta}_{Ra}\partial\!\!\!/ H_i^\dagger}{\Lambda^2}\epsilon;\\
\delta t_R^a &= \theta_R^a\eta\epsilon - i\frac{\partial\!\!\!/ H_i^\dagger \theta_L^{ia}}{\Lambda^2}\epsilon; &\qquad \delta\overline{t}_{Ra} &= \overline{\theta}_{Ra}\eta\epsilon + i\frac{\overline{\theta}_{Lia}\partial\!\!\!/ H^i}{\Lambda^2}\epsilon;\\
\delta H^i &= \frac{\overline{\theta}_{Ra}\psi_L^{ia} + \overline{t}_{Ra}\theta_L^{ia}}{\Lambda^2}\epsilon; &\qquad \delta H_i^\dagger &= \frac{\overline{\psi}_{Lai}\theta_R^a + \overline{\theta}_{Lai}t_R^a}{\Lambda^2}\epsilon.
\end{aligned} \tag{12}$$

where i (a) is an isospin (color) index. η is a relative normalization factor that we determine subsequently.

It is readily seen that Eq. (12) is an invariance of Eq. (11) up to total derivatives:

$$\begin{aligned}
\delta(\overline{\psi}_L i\partial\!\!\!/ \psi_L) &= \frac{(\overline{\psi}_L\theta_R)\cdot\partial^2 H}{\Lambda^2}\epsilon + h.c. + t.d.\\
\delta(\overline{t}_R i\partial\!\!\!/ t_R) &= \frac{(\overline{\theta}_L t_R)\cdot\partial^2 H}{\Lambda^2}\epsilon + h.c. + t.d.\\
\delta(\partial H^\dagger\partial H) &= -\frac{(\overline{\psi}_L\theta_R + \overline{\theta}_L t_R)\cdot\partial^2 H}{\Lambda^2}\epsilon + h.c. + t.d.\\
\text{hence,}\quad \delta\mathcal{L}_K &= 0 + t.d.
\end{aligned} \tag{13}$$

The symmetry of the gauge free kinetic terms makes no use of equations of motion or on-shell conditions. At this stage, the shifts in ψ_L and t_R by $\theta_{L,R}$ proportional to η play no role, but will be essential with the Higgs-Yukawa interaction and Higgs mass term. Indeed, shifting $\overline{\psi}_L i\partial\!\!\!/ \psi_L \to \overline{\psi}_L i\partial\!\!\!/ \theta_L$ yields a total divergence provided we have switched off the local gauge fields. The fermionic shift symmetry, however, raises an issue of consistency of a massive top quark with the Nambu-Goldstone theorem which has a remarkable resolution, as shown in Section 3.

This transformation exploits the interplay of the quantum numbers of ψ_L, t_R and H. It resembles a scalar supermultiplet transformation of component fields, where the Higgs field is treated as a superpartner of ψ_L.[3] We emphasize that this is not a representation of the supersymmetry algebra, as there is no "F" auxillary field.[3] This is essentially a scalar supermultiplet transformation with fixed $F = 0$ and the superparameters replaced by $\theta\epsilon$). With the assignment of scales of the $\theta_{L,R}$ and the presence of Λ the commutators of subsequent transformations for different $\theta_{L,R}$ cannot close. Also, the $\theta_{L,R}$ carry flavor and color quantum numbers, and the failure of the algebra to close into a superalgebra is presumably a supersymmetric extension of the Coleman-Mandula no-go theorem. In fact, this is a $U(1)$ symmetry with the transformation parameter, ϵ, for fixed background values of $\theta_{L,R}$. As such, the commutator trivially vanishes on the fields: $[\delta_{\epsilon'}, \delta_\epsilon](\psi, H, t_R) = 0$

We presently turn to the full Lagrangian of the top-Higgs system in the standard model with gauge fields turned off:

$$\begin{aligned}\mathcal{L}_H &= i\overline{\psi}_L \partial\!\!\!/ \psi_L + i\overline{t}_R \partial\!\!\!/ t_R + \partial H^\dagger \partial H \\ &\quad + g(\overline{\psi}_L t_R H + h.c.) - M_H^2 H^\dagger H - \frac{\lambda}{2}(H^\dagger H)^2 \end{aligned} \tag{14}$$

From Eq. (12) we compute the transformations:

$$\delta(-M_H^2 H^\dagger H) = -\frac{\epsilon}{\Lambda^2}\, M_H^2(\overline{\psi}_L \theta_R + \overline{\theta}_L t_R)\cdot H + h.c \tag{15}$$

$$\delta(-\frac{\lambda}{2}(H^\dagger H)^2) = -\frac{\epsilon}{\Lambda^2}\, \lambda(\overline{\psi}_L \theta_R + \overline{\theta}_L t_R)\cdot H H^\dagger H + h.c. \tag{16}$$

$$\begin{aligned}\delta(g\overline{\psi}_L t_R H + h.c.) &= g\eta\epsilon(\overline{\psi}_L\theta_R + \overline{\theta}_L t_R)H + g^2\frac{\epsilon}{2\Lambda^2}(\overline{\theta}_R\psi_L + \overline{t}_R\theta_L)\cdot(H^\dagger H^\dagger H) \\ &\quad + g\frac{\epsilon}{\Lambda^2}\overline{\psi}_L t_R(\overline{\theta}_R\psi_L + \overline{t}_R\theta_L) \\ &\quad + ig\frac{2\epsilon}{\Lambda^2}\overline{\psi}_L\gamma_\mu\frac{\tau^A}{2}\theta_L\left(H^\dagger \overleftrightarrow{\partial^\mu}\frac{\tau^A}{2}H\right) + ig\frac{\epsilon}{2\Lambda^2}\overline{\psi}_L\gamma_\mu\theta_L\left(H^\dagger\overleftrightarrow{\partial^\mu}H\right) \\ &\quad - ig\frac{\epsilon}{\Lambda^2}\overline{\theta}_R\gamma_\mu t_R\left(H^\dagger\overleftrightarrow{\partial^\mu}H\right) + h.c. + t.d. \end{aligned} \tag{17}$$

where we use the isospin Fierz identity, $[\tau^A]_{ij}[\tau^A]_{kl} = 2\delta_{il}\delta_{kj} - \delta_{ij}\delta_{kl}$, and: $\overleftrightarrow{\partial^\mu} = \frac{1}{2}(\overrightarrow{\partial^\mu} - \overleftarrow{\partial^\mu})$. Here we have applied *the fermionic equations of motion*:

$$i\partial\!\!\!/ t_R + g\psi_L\cdot H^\dagger = 0 \qquad\qquad i\partial\!\!\!/ \psi_L + g t_R H = 0 \tag{18}$$

and Eq. (17) follows.[2]

Notice in Eq. (17) we have generated a set of higher dimension operator terms of the form:

$$\frac{g\epsilon}{\Lambda^2}\overline{\psi}_L t_R(\overline{\theta}_R\psi_L + \overline{t}_R\theta_L) + i\frac{2g\epsilon}{\Lambda^2}\overline{\psi}_L\gamma_\mu\frac{\tau^A}{2}\theta_L(H^\dagger \overset{\leftrightarrow}{\partial^\mu} \frac{\tau^A}{2}H)$$
$$+i\frac{g\epsilon}{2\Lambda^2}\overline{\psi}_L\gamma_\mu\theta_L(H^\dagger \overset{\leftrightarrow}{\partial^\mu} H) - i\frac{g\epsilon}{\Lambda^2}\overline{\theta}_R\gamma_\mu t_R(H^\dagger \overset{\leftrightarrow}{\partial^\mu} H) + h.c. + t.d. \tag{19}$$

In analogy to a "bottoms up" derivation of a nonlinear chiral Lagrangian (see section (3) for a review), these terms can be cancelled by adding higher dimension operators to the original Lagrangian of the form:

$$\mathcal{L}_{d=6} = \frac{\kappa}{\Lambda^2}(\overline{\psi}_L t_R \overline{t}_R \psi_L) + \frac{2\kappa}{\Lambda^2}(\overline{\psi}_L\gamma_\mu\frac{\tau^A}{2}\psi_L)(H^\dagger i \overset{\leftrightarrow}{\partial^\mu} \frac{\tau^A}{2}H)$$
$$+\frac{\kappa}{2\Lambda^2}(\overline{\psi}_L\gamma_\mu\psi_L)(H^\dagger i \overset{\leftrightarrow}{\partial^\mu} H) - \frac{\kappa}{\Lambda^2}(\overline{t}_R\gamma_\mu t_R)(H^\dagger i \overset{\leftrightarrow}{\partial^\mu} H) \tag{20}$$

where we will presently relate the coupling constant, κ, to M_H, Λ and g below.

We thus obtain the effective Lagrangian,

$$\mathcal{L}_H = \overline{\psi}_L i\partial\!\!\!/ \psi_L + i\overline{t}_R\partial\!\!\!/ t_R + \partial H^\dagger \partial H$$
$$+g(\overline{\psi}_L t_R H + h.c.) - M_H^2 H^\dagger H - \frac{\lambda}{2}(H^\dagger H)^2$$
$$+\frac{\kappa}{\Lambda^2}(\overline{\psi}_L t_R \overline{t}_R \psi_L) + \frac{2\kappa}{\Lambda^2}(\overline{\psi}_L\gamma_\mu\frac{\tau^A}{2}\psi_L)(H^\dagger i \overset{\leftrightarrow}{\partial^\mu} \frac{\tau^A}{2}H)$$
$$+\frac{\kappa}{2\Lambda^2}(\overline{\psi}_L\gamma_\mu\psi_L)(H^\dagger i \overset{\leftrightarrow}{\partial^\mu} H) - \frac{\kappa}{\Lambda^2}(\overline{t}_R\gamma_\mu t_R)(H^\dagger i \overset{\leftrightarrow}{\partial^\mu} H) \tag{21}$$

Performing the super-dilatation transformation of Eq. (12) we now demand that:

$$\delta\mathcal{L}_H = 0 + \mathcal{O}\left(\frac{1}{\Lambda^4}\right) \tag{22}$$

The generated $\mathcal{O}(1/\Lambda^4)$ terms can be compensated by adding additional $1/\Lambda^4$ terms to the Lagrangian. By continued iteration of Eq. (12) we would generate a power series of contact interactions that are scaled by $\sim 1/\Lambda^{2n}$.

First we see that the transformation of the Higgs mass term of Eq. (21), from Eqs. (15–17), cancels against the first term of the transformed Higgs-Yukawa interaction, provided:

$$g\eta = \frac{M_H^2}{\Lambda^2} \tag{23}$$

(beware: M_H^2 is the negative Lagrangian (mass)^2, while $m_h^2 = -2M_H^2$ is the physical positive Higgs boson (mass)^2 in the broken phase, and we take the normalization $v^2 = \langle H^\dagger H\rangle = (175\ GeV)^2$) This establishes the normalization factor, η. It also establishes the relative sign (we assume Λ^2 positive). If we're in the symmetric (broken) phase, M_H^2 positive (negative), then we have $g\eta > 0$ ($g\eta < 0$). We have the freedom of choosing arbitrary η since the defining kinetic term invariance involves only ϵ.

One might think we can now take Λ^2 to be arbitrarily large compared to M_H^2 by adjusting $|\eta| << 1$. However, the second term of Eq. (17) must also cancel against the transformation of the first κ term appearing in Eq. (21). This requires that:

$$\eta\kappa = -g, \qquad \text{or, using Eq. (23):} \qquad \frac{\kappa}{\Lambda^2} = -\frac{g^2}{M_H^2} \tag{24}$$

This is a striking result: a seesaw relation between the weak scale and Λ-scale terms. In the $d = 6$ operators we have a Nambu-Jona-Lasionio component. The Nambu-Jona-Lasinio attractive interaction corresponds to $\kappa > 0$, and we see in Eq. (24) that the super-dilatation is then consistent only if $M_H^2 < 0$. Moreover, to make η small requires taking κ large.

Finally, the most interesting relationship, which is the analogue of the Goldberger-Treiman relationship in a chiral Lagrangian (see section 3), arises from the cancellation of the $\sim \epsilon(\overline{\psi}_L\theta_R + \overline{\theta}_L t_R)\cdot HH^\dagger H$ terms of Eqs. (16) and (17) under the super-dilatation symmetry:

$$0 = (\lambda - \frac{1}{2}g^2)\frac{\epsilon}{\Lambda^2}(\overline{\psi}_L\theta_R + \overline{\theta}_L t_R)\cdot HH^\dagger H + h.c \tag{25}$$

or,

$$\lambda = \frac{1}{2}g^2 \tag{26}$$

Note that his transformation does not involve η.

The $\lambda = g^2/2$ relationship refers to the coefficient of the $D = 6$ operator, $(\overline{\theta}_L t_R)\cdot HH^\dagger H + h.c.$ We therefore assume that it applies at the scale Λ. The low energy relationship between the couplings g^2 and λ then depends upon the renormalization group running from Λ to $v_{weak} \approx 175$ GeV. If we ignore the RG running then Eq. (26) would hold at the weak scale, and implies the supersymmetric relationship $m_h^2 = 2\lambda v_{weak}^2 = m_t^2$ in the broken phase. This is an improvement over the usual NJL result, $m_h^2 = 4m_t^2$ (though we were seeking $m_h^2 = m_t^2/2$).

Note that we can Fierz rearrange the first term of Eq. (20):

$$(\overline{\psi}^a_L t_{Ra})_i(\overline{t}_{Rb}\psi^b)^i \to -(\overline{\psi}_{iL}\gamma_\mu\frac{\lambda^A}{2}\psi^i_L)(\overline{t}_R\gamma^\mu\frac{\lambda^A}{2}t_R) + \mathcal{O}(\infty/\mathcal{N}) \tag{27}$$

where $N = 3$ is the number of colors. This term is a pure Nambu-Jona-Lasinio interaction as arises in topcolor[4–6] in the form of a (color current)$\times$(color current). Indeed, massive Yang–Mills boson exchange for a boson of mass M^2 and momentum exchange $q^2 < M^2$ produces the negative sign for (current)$\times$(current) interactions. A positive sign for the first term of Eq. (20) is the attractive sign for the Nambu-Jona-Lasinio model, and we thus see that the attractive sign corresponds to the correct (negative) sign for topgluon exchange. However, we see that the isospin (current)$\times$(current) interaction (second term of Eq. (20)) then has the wrong sign (positive) for a gauge boson exchange.

Since all of the higher dimension $d = 6$ operators are of the form (current)$\times$(current), they preserve the chirality of the fermions. That is, the terms of Eq. (20) contain no cross terms of the form $\overline{\psi}_L H t_R (H^\dagger H)^p$. They thus admit the discrete symmetry, $\psi_L \to (-1)^N \psi_L$ and $t_R \to (-1)^{N+1} t_R$. Operators of mixed chirality can therefore be excluded, or suppressed, on symmetry grounds. If they are introduced they can be small effects, controlled by the symmetry limit.

3. Analogy to a Chiral Lagrangian

For comparison, we quickly review a familiar derivation of a pion chiral Lagrangian from the "bottoms-up." Consider the kinetic terms:

$$\mathcal{L}_K = \overline{\psi}_L i\partial\!\!\!/ \psi_L + \overline{\psi}_R i\partial\!\!\!/ \psi_R + \frac{1}{2}\partial_\mu \pi \partial^\mu \pi \tag{28}$$

We'll consider the RH-chiral symmetry:

$$\begin{aligned} \delta\psi_L &= 0 \\ \delta\psi_R &= i\theta\psi_R \qquad \delta\overline{\psi}_R = -i\theta\overline{\psi}_R \\ \delta\pi &= f_\pi \theta \end{aligned} \tag{29}$$

and we demand the Lagrangian is invariant under this global transformation:

$$\delta\mathcal{L}_K = 0 \tag{30}$$

The RH-chiral current is:

$$-\frac{\delta\mathcal{L}_K}{\delta\, \partial^\mu\theta} = \overline{\psi}_R \gamma_\mu \psi_R - f_\pi \partial_\mu \pi \tag{31}$$

and we assume f_π is "determined from experiment," *e.g.*, $\pi \to \mu\nu$ (of course, in the real world this is the left-handed current).

Consider the interactions consisting of a massive "nucleon" coupled to pion:

$$\mathcal{L}_V = M\overline{\psi}\psi - ig\pi\overline{\psi}\gamma^5\psi = M\overline{\psi}_L\psi_R - ig\pi\overline{\psi}_L\psi_R + h.c. \tag{32}$$

We perform the RH-chiral transformation transformation:

$$\delta\mathcal{L}_V = (i\theta M - igf_\pi\theta + g\pi f_\pi\theta)\,\overline{\psi}_L\psi_R + h.c. \tag{33}$$

so, invariance requires:

$$\delta\mathcal{L}_V = 0 \qquad \longrightarrow \qquad g = \frac{M}{f_\pi} \tag{34}$$

which is the Goldberger-Treiman relation.

However, we must also cancel the "higher order term" $\propto \pi\theta\overline{\psi}_L\psi_R$. We thus include an $\mathcal{O}(\pi^2)$ term:

$$\mathcal{L}_V \to M\left(1 - \frac{i\pi}{f_\pi} + c\frac{\pi^2}{f_\pi^2}\right)\overline{\psi}_L\psi_R + h.c. \tag{35}$$

Now:

$$\delta\mathcal{L}_V \to M\left(i\theta - i\frac{f_\pi\theta}{f_\pi} + \frac{\pi\theta}{f_\pi} + 2c\frac{\pi}{f_\pi^2}f_\pi\theta + ic\frac{\pi^2}{f_\pi^2}f_\pi\theta\right)\overline{\psi}_L\psi_R h.c. \tag{36}$$

so:

$$\delta\mathcal{L}_V = 0 \qquad \longrightarrow \qquad g = \frac{M}{f_\pi}, \qquad c = -\frac{1}{2} \tag{37}$$

But, now we must cancel higher order term $\propto \pi^2\theta\overline{\psi}_L\psi_R$ which implies an $\mathcal{O}(\pi^3)$ interaction, and so-forth.

We can sum the resulting power series and we find, iteratively, the solution:

$$\mathcal{L}_V = M\overline{\psi}_L U\psi_R + h.c. \qquad\qquad U = \exp(i\pi/f_\pi) \tag{38}$$

whence, the symmetric Lagrangian is:

$$\mathcal{L} = \overline{\psi}_L i\partial\!\!\!/ \psi_L + \overline{\psi}_R i\partial\!\!\!/ \psi_R + \frac{f_\pi^2}{2}\partial_\mu U^\dagger \partial^\mu U + M\overline{\psi}_L U\psi_R + h.c. \tag{39}$$

and we have obtained the"nonlinear σ-model Lagrangian."

Our present strategy is similar. We begin with the super-dilatational invariance of the top-Higgs kinetic terms. We then analyze the transformation of the Higgs-Yukawa, Higgs mass and quartic interactions. We demand overall invariance of the Lagrangian. We thus find the "Goldberger-Treiman" relationship, $\lambda = g^2/2$, which implies $m_t = m_h$ in the broken phase. This induces higher dimension operators. Ultimately, we expect to sum the tower of operators, though in the present case we expect that these arise via new dynamics, such as heavy recurrences of composite Higgs bosons and vector–like top quarks.

4. Current Structure and the Nambu-Goldstone Theorem

The critical aspect of our construction is that the operator shift of δH in the quartic Higgs interaction is cancelling against the super-rotation (*i.e.*, the "twist") of $\delta\psi$ in the Higgs-Yukawa interaction. Moreover, the pure fermionic shift in $\delta\psi \sim \eta\epsilon\theta$, in the Higgs-Yukawa interaction, *i.e.*, proportional to η, cancels against the δH shift in the Higgs mass term. This ties the transformations together into a single structure.

The Nambu-Goldstone theorem for a fermionic shift $\delta\psi \sim \eta\epsilon\theta$, which would naively imply a massless top quark (a "Goldstino"), is evaded in the broken phase. How does our theory evade the existence of a zero-mode associated with the fermionic shift? Naively, this would seem to prohibit a massless top quark. In fact, this happens in a subtle way. One must carefully construct the currents given our use of equations of motion in $\delta(g\overline{\psi}_L t_R H + h.c.)$. We therefore wish to clarify the the relationship to the Nambu-Goldstone theorem in the present set up.

We consider, for technical simplicity, a simpler "minimal" transformation defined by $\theta_L = 0$:

$$\delta\psi_L^{ia} = -i\frac{\partial\!\!\!/ H^i\theta_R^a}{\Lambda^2}\epsilon; \qquad \delta\overline{\psi}_{L\,ia} = i\frac{\overline{\theta}_{Ra}\partial\!\!\!/ H_i^\dagger}{\Lambda^2}\epsilon; \tag{40}$$

$$\delta t_R^a = \theta_R^a\eta\epsilon; \qquad \delta\overline{t}_{Ra} = \overline{\theta}_{Ra}\eta\epsilon; \tag{41}$$

$$\delta H^i = \frac{\overline{\theta}_{Ra}\psi_L^{ia}}{\Lambda^2}\epsilon; \qquad \delta H_i^\dagger = \frac{\overline{\psi}_{Lai}\theta_R^a}{\Lambda^2}\epsilon. \tag{42}$$

The parameter η is still fixed by the symmetry interplay of the Higgs mass term and Yukawa interaction as in Eq. (23),

$$g\eta = \frac{M_H^2}{\Lambda^2} = -\frac{\lambda v^2}{\Lambda^2} \tag{43}$$

Consider the top and Higgs system of the standard model with gauge fields set to zero:

$$\mathcal{L}_H = \mathcal{L}_K + g(\overline{\psi}_L t_R H + h.c.) - M_H^2 H^\dagger H - \frac{\lambda}{2}(H^\dagger H) \tag{44}$$

$$\mathcal{L}_K = \overline{\psi}_L i\partial\!\!\!/\, \psi_L + \overline{t}_R i\partial\!\!\!/\, t_R + \partial H^\dagger \partial H \tag{45}$$

It is readily seen that Eqs. (40–42) is a global invariance of Eqs. (45) up to total derivatives. We presently allow ϵ to be a local function of spacetime $\epsilon(x)$ (note that the derivatives in Eq. (40) act only upon H and not upon $\epsilon(x)$). We have:

$$\begin{aligned}
\delta(\overline{\psi}_L i\partial\!\!\!/\, \psi_L) &= \frac{(\overline{\psi}_L\theta_R)\cdot\partial^2 H}{\Lambda^2}\epsilon + \frac{(\overline{\psi}_L\gamma_\mu\partial\!\!\!/\, H\theta_R)}{\Lambda^2}\partial^\mu\epsilon + h.c. + t.d. \\
\delta(\overline{t}_R i\partial\!\!\!/\, t_R) &= i(\overline{t}_R\partial\!\!\!/\,\theta_R)\eta\epsilon + i(\overline{t}_R\gamma_\mu\theta_R)\eta\partial^\mu\epsilon + h.c. + t.d. \\
\delta(\partial H^\dagger\partial H) &= -\frac{(\overline{\psi}_L\theta_R)\cdot\partial^2 H}{\Lambda^2}\epsilon + \frac{(\overline{\psi}_L\theta_R)\cdot\partial_\mu H}{\Lambda^2}\partial^\mu\epsilon + h.c. + t.d.
\end{aligned} \tag{46}$$

The kinetic terms thus lead to a Noether current:

$$J_\mu^K = \frac{\delta\mathcal{L}_K}{\delta\partial_\mu\epsilon} = i(\overline{t}_R\gamma_\mu\theta_R)\eta + \frac{(\overline{\psi}_L\gamma_\mu\partial\!\!\!/\, H\theta_R)}{\Lambda^2} + \frac{(\overline{\psi}_L\theta_R)}{\Lambda^2}\partial_\mu H + h.c. \tag{47}$$

The symmetry of the full action, as we have emphasized, involves a cancellation of the shift of Eqs. (42) in the Higgs quartic interaction against the "twist" of Eq. (40) in the Higgs-Yukawa interaction. In calculating the transformation of the Higgs-Yukawa interaction, however, we make use of an "integration by parts" and discard total divergences (and subsequently use the fermion equations of motion). This integration by parts in the "twist" of Eq. (40) causes the derivative to act upon the parameter $\epsilon(x)$:

$$\delta(-M_H^2 H^\dagger H) = -\frac{\epsilon}{\Lambda^2}\, M_H^2(\overline{\psi}_L\theta_R)\cdot H + h.c \tag{48}$$

$$\delta(-\frac{\lambda}{2}(H^\dagger H)^2) = -\frac{\epsilon}{\Lambda^2}\,\lambda(\overline{\psi}_L\theta_R)\cdot HH^\dagger H + h.c. \tag{49}$$

$$\begin{aligned}
\delta(g\overline{\psi}_L t_R H + h.c.) &= g\eta\epsilon(\overline{\psi}_L\theta_R)H + \frac{g^2\epsilon}{2\Lambda^2}(\overline{\theta}_R\psi_L)\cdot(H^\dagger H^\dagger H) \\
&\quad + \frac{g\epsilon}{\Lambda^2}(\overline{\psi}_L t_R)(\overline{\theta}_R\psi_L) - i\frac{g\epsilon}{\Lambda^2}\overline{\theta}_R\gamma_\mu t_R\left(H^\dagger \overleftrightarrow{\partial^\mu} H\right) \\
&\quad - i\frac{g}{2\Lambda^2}\overline{\theta}_R\gamma_\mu t_R(H^\dagger H)\partial^\mu\epsilon + h.c. + t.d.
\end{aligned} \tag{50}$$

The last term in Eq. (50) shows explicitly that the result of the integration by parts leads to an additional term $\propto \partial^\mu\epsilon$. This, in turn, modifies the current, which now becomes:

$$J_\mu = \frac{\delta \mathcal{L}_H}{\delta \partial_\mu \epsilon} = i(\bar{t}_R \gamma_\mu \theta_R)\left(\eta + \frac{gH^\dagger H}{2\Lambda^2}\right) + \frac{(\overline{\psi}_L \gamma_\mu \not{\partial} H \theta_R)}{\Lambda^2} + \frac{(\overline{\psi}_L \theta_R)}{\Lambda^2} \partial_\mu H + h.c. \quad (51)$$

Using the relationship of Eq. (43), $g\eta = -\lambda v^2/\Lambda^2$, and the "Goldberger-Treiman" analogue, $\lambda = g^2/2$, the current can be written:

$$J_\mu = \frac{\delta \mathcal{L}_H}{\delta \partial_\mu \epsilon} = i\eta(\bar{t}_R \gamma_\mu \theta_R)\left(1 - \frac{H^\dagger H}{v^2}\right) + \frac{(\overline{\psi}_L \gamma_\mu \not{\partial} H \theta_R)}{\Lambda^2} + \frac{(\overline{\psi}_L \theta_R)}{\Lambda^2} \partial_\mu H + h.c. \quad (52)$$

The modification of the current occurs in the first term which is associated with the fermionic "shift"-symmetry of t_R. Again, this arises from the crucial linkage of the δH shift in the quartic Higgs interaction to the $\delta\psi_L$ shift in the Higgs-Yukawa interaction.

Note that, in the broken phase where $\langle H \rangle = v$ the modification of the current has the effect of "turning off" the fermionic shift. Indeed, we will now see that this has a remarkable effect in evading the Nambu-Goldstone theorem in the broken phase, and permitting the top quark to be massive.

Consider the two-point function of our current of Eq. (52) with t_R:

$$S(y) = \int d^4x e^{iq\cdot x}\, \partial^\mu \langle 0 | T^* J_\mu(x)\, t_R(y) |0\rangle \quad (53)$$

(T^* implies anti-commutation in the ordering of fermion fields). Formally, with $\partial^\mu J_\mu = 0$, we have, from the ∂_0 acting upon the T^* ordering a $\delta(x^0 - y^0)$, and:

$$\begin{aligned} S(y) &= \int d^3x\, e^{iq\cdot x} \langle 0| \{J_0(x),\, t_R(y)\} |0\rangle \\ &= \int d^3x\, e^{-i\vec{q}\cdot\vec{x}} \langle 0| \{J_0(\vec{x}),\, t_R(\vec{y})\}_{e.t.} |0\rangle \\ &= \langle 0| \{Q,\, t_R(\vec{y})\} |0\rangle \end{aligned} \quad (54)$$

where the charge operator Q is:

$$Q = \int d^3x\, J_0(\vec{x}). \quad (55)$$

In the symmetric phase of the standard model we have the Higgs VEV, $\langle H \rangle = 0$, and we can neglect all terms in the current that involve H. The charge operator then involves only the first term in $J^K_\mu = i\eta \bar{t}_R \gamma_\mu \theta_R + h.c.$, whence it generates a shift in the fermion field:

$$\langle 0| \{Q,\, t_R(\vec{y})\} |0\rangle = \eta\theta_R \quad (56)$$

On the other hand we have:

$$\begin{aligned} S(y) &= \int d^4x\, e^{iq\cdot x}\, \partial^\mu \langle 0| T^* i\bar{t}_R(x)\gamma_\mu \theta_R \eta \quad t_R(y) |0\rangle \\ &= -\int d^4x\, e^{iq\cdot x}\, i\partial^\mu \gamma_\mu S_F(x-y)\theta_R \eta \\ &= \left.\frac{q^2 + \not{q} m}{q^2 - m^2}\theta_R \eta\right|_{q\to 0} \end{aligned} \quad (57)$$

In the $q^2 \to 0$ limit the consistency of Eq. (57) with Eqs. (54, 56) requires that the fermion mass satisfy $m = 0$. This is the fermionic Nambu-Goldstone theorem and it informs us that any fermionic action which has a pure fermionic shift symmetry, must correspond to a massless fermion. This is, indeed, the case in the symmetric phase in which the top quark is massless and $\langle H \rangle = 0$.

Naively we might conclude that the top quark is forced by our symmetry to be a Goldstino and remain massless, even in the broken phase. However, we have seen that the current is modified in a significant way in the present case:

$$J_\mu^{shift} = i\eta(\bar{t}_R \gamma_\mu \theta_R) \left(1 - \frac{H^\dagger H}{v^2}\right) + h.c. \tag{58}$$

In the broken phase, when $\langle H \rangle = v \neq 0$, this implies that the pure fermionic shift operator in the current "turns off:"

$$J_\mu^{shift} = 0 \tag{59}$$

This is a consequence of the interplay between the quartic interaction and the Higgs-Yukawa interaction in our construction. It implies that there can exist dynamical situations in which a Goldstino is massless in a symmetric phase, but acquires mass in a broken phase of a theory. The underlying fermionic shift, $\delta\psi = \eta\theta$ is intact, but the current is modified dynamically to evade the naive Nambu-Goldstone theorem. One might wonder what happens for $\langle H^\dagger H \rangle \neq v^2$ and $\neq 0$? This is, of course, and unstable vacuum and the S-matrix derivation fails. Of course, the above current algebra analysis serves only as a consistency check on our original Lagrangian analysis, which showed that a vacuum with massive top and Higgs, with $m_t = m_h$, exists.

Indeed, the symmetry yields $\lambda = g^2/2$ in both phases of the Standard Model. However, the fermionic shift part of the current is nontrivially modified by the quartic-Yukawa interplay and is $\propto (1 - H^\dagger H/v^2)$. It thus vanishes in the broken phase with $\langle H \rangle = v$. Therefore, the top quark becomes massive in the broken phase in the usual way, with the relationship $m_h = m_t$. This is consistent with the Nambu-Goldstone theorem that would otherwise naively force the top quark to be a massless Goldstino. This relationship $\lambda = g^2/2$ is the analogue of a Goldberger-Treiman" relationship. It holds at a high scale, Λ, and is subject to renormalization group and higher dimension operator effects that can bring the physical masses into concordance with $m_h^2 \approx m_t^2/2$.

5. UV-Completion and Conclusions

The effective Lagrangian we obtain from the minimal transformation of Eq. (40) is fairly simple:

$$\begin{aligned}\mathcal{L}_H &= \overline{\psi}_L i\partial\!\!\!/ \psi_L + i\overline{t}_R \partial\!\!\!/ t_R + \partial H^\dagger \partial H \\ &+ g(\overline{\psi}_L t_R H + h.c.) - M_H^2 H^\dagger H - \frac{\lambda}{2}(H^\dagger H)^2 \\ &+ \frac{\kappa}{\Lambda^2}(\overline{\psi}_L t_R \overline{t}_R \psi_L) - \frac{\kappa}{\Lambda^2}(\overline{t}_R \gamma_\mu t_R)(H^\dagger i \overleftrightarrow{\partial^\mu} H) + \mathcal{O}\left(\frac{1}{\Lambda^2}\right) \end{aligned} \quad (60)$$

We retain the relationships of Eqs. (23,24,26).

Note the structure of the higher dimension operators. One of these is a Nambu-Jona-Lasinio four-fermion interaction, as expected in topcolor.[4] The other is a current-current interaction of the Higgs with the top quark. With Λ sufficiently large, κ must become large in accord with the seesaw relationship of Eq. (24).

This is suggestive of a dynamics in which a boundstate recurrence of the Higgs boson, composed of $\overline{t}t$ is generated via the NJL interaction.[4–6] The $d = 6$ Higgs-top interactions can likewise generate a composite Dirac (vectorlike) fermion, composed of $\sim H^\dagger t_R$ with gauge quantum numbers of the LH top quark.[7] The Dirac fermion has a RH component that is an s-wave, $\sim H^\dagger t_R$, and a LH component that is a p-wave, $\sim H^\dagger i\partial\!\!\!/ t_R/\Lambda$.

We believe the tower of operators $\sim 1/\Lambda^{2n}$ may be determined, and the dynamical model admitting our super-dilatation may be understood as a full solution to this dynamics. This may be facilitated by introducing the composite fields explicitly as auxilliary fields. These and related issues are under further investigation.[7]

While the usual superalgebra of SUSY does not permit the Higgs to be the superpartner of the (t, b) quarks, the symmetry we present here does accomplish this. We emphasize that the "super"-dilatation symmetry is not a conventional super–algebra, *i.e.*, it is not a grading of the Lorentz Group, and is not associated with a nontrivial nonabelian closed superalgebra (at least not in our present exploratory formulation). The symmetry is a generalization of a dilatonic shift symmetry for the Higgs, in which the shift is promoted to an operator, and the transformation contains a single bosonic parameter ϵ, and we have a $\sim U(1)$, invariance, which closes trivially.

Our symmetry is remniscent of a "reparameterization invariance," *e.g.*, as occurs in heavy quark effective field theory (HQET).[8–10] In the latter case one considers an $M \to \infty$ limit for a heavy quark and constructs a field theoretic Lagrangian for a given four-velocity "supersector," v_μ. The Lagrangian takes the form of a series expansion in higher dimension operators weighted by powers of $1/M$. The leading terms in the theory display heavy-spin symmetry (*e.g.*, degenerate 0^- and 1^- mesons). The reparameterization invariance is a residual symmetry that constrains the full operator structure and relates the coefficients of the terms in the Lagrangian to higher orders of $1/M$. The reparameterization invariance is essentially the vestige of the underlying hidden Lorentz invariance.[10]

References

1. See: S. Iso and Y. Orikasa, "TeV Scale B-L model with a flat Higgs potential at the Planck scale - in view of the hierarchy problem -," arXiv:1210.2848 [hep-ph].
   T. Abe, R. Kitano, Y. Konishi, K. -y. Oda, J. Sato and S. Sugiyama, "Minimal Dilaton Model," arXiv:1209.4544 [hep-ph].
   K. A. Meissner and H. Nicolai, Acta Phys. Polon. B **40**, 2737 (2009) [arXiv:0809.1338 [hep-th]]. and references therein.
2. C. T. Hill, "'Super'-Dilatation Symmetry of the Top-Higgs System," arXiv:1211.2773 [hep-ph], to appear in Phys. Rev. D.
3. J. Wess and J. Bagger, Princeton, USA: Univ. Pr. (1992) 259 p; see eqs(3.3-3.9).
4. C. T. Hill, Phys. Lett. B **266**, 419 (1991); C. T. Hill, Phys. Lett. B **345**, 483 (1995).
5. V. A. Miransky, M. Tanabashi, and K. Yamawaki, Phys. Lett. **B221**, 177 (1989); Mod. Phys. Lett. **A4**, 1043 (1989);
   W. J. Marciano, Phys. Rev. Lett. **62**, 2793 (1989);
   Y. Nambu, Enrico Fermi Institute Report No. 89-08, 1989 (unpublished); in *Proceedings of the 1989 Workshop on Dynamical Symmetry Breaking*, edited by T. Muta and K. Yamawaki (Nagoya University, Nagoya, Japan, 1990).
6. W. A. Bardeen, C. T. Hill, and M. Lindner, Phys. Rev. **D41**, 1647 (1990).
7. B. Dobrescu, and C. T. Hill (unpublished); B. Dobrescu, and C. T. Hill and E. Neil, work in progress.
8. H. Georgi, Nucl. Phys. **B348**, 293 (1991); Phys. Lett. B **240**, 447 (1990);
   M. Neubert, Phys. Rept. **245**, 259 (1994)
   A. V. Manohar and M. B. Wise, Camb. Monogr. Part. Phys. Nucl. Phys. Cosmol. **10**, 1 (2000).
9. M. E. Luke and A. V. Manohar, Phys. Lett. B 286, 348 (1992).
10. J. Heinonen, R. J. Hill and M. P. Solon, arXiv:1208.0601 [hep-ph].

Higgs Boson Mass and the Scale of New Physics

M. Shaposhnikov*

Institut de Théorie des Phénomènes Physiques,
École Polytechnique Fédérale de Lausanne,
CH-1015 Lausanne, Switzerland
**E-mail: mikhail.shaposhnikov@epfl.ch*
http://archiveweb.epfl.ch/itp.epfl.ch/page58722.html

I will discuss the prediction of the Higgs boson mass coming from asymptotic safety of the Standard Model (SM), and related lower bounds on it from Higgs inflation and absolute stability of SM vacuum.

Keywords: Higgs boson, inflation, asymptotic safety, vacuum stability

1. Introduction

On July 4, 2012 the ATLAS and CMS collaborations at CERN announced a discovery of a new particle,[1,2] which looks like a Higgs boson of the Standard Model. The mass of this new resonance is

$$M_H = 125.5 \pm 0.6\,\text{GeV},$$

according to ATLAS, with almost the same number according to CMS:

$$M_H = 125.8 \pm 0.6\,\text{GeV}.$$

Suppose that this is indeed the SM Higgs boson. What does the value of its mass mean for high energy physics?

In this talk I will argue that possible answers to this question are:

- Electroweak scale is determined by Planck physics.
- There is no new energy scale between the Fermi and Planck scales.
- New physics responsible for dark matter, baryon asymmetry of the universe and neutrino masses is hidden below the Fermi scale.

The paper is organised as follows. In Section 2, we will discuss different theoretical constraints on the Higgs boson mass. In Section 3 we will discuss the prediction of the Higgs mass from the idea of asymptotic safety of the Standard Model. In Section 4 we will review the proposal of the Higgs inflation and constraints on the Higgs boson mass coming from it. In Section 5 we provide some arguments in favour of absence of new energy scales between the Fermi and Planck scales. In Section 6 we will describe shortly a proposal for new physics below the electroweak scale, which allows to explain simultaneously neutrino masses and oscillations, baryon asymmetry of the universe and dark matter. Section 7 is conclusions.

2. Constraints on the Higgs mass

Within the SM the mass of the Higgs boson is an arbitrary parameter which can have any value in the interval

$$M_{\rm meta} < M_H < M_{\rm Landau}, \tag{1}$$

where $M_{\rm meta} \simeq 111$ GeV is the mass of the Higgs boson for which our vacuum has a life-time equal to that of the Universe[3–5] (metastability bound). The value $M_{\rm Landau} \simeq 1$ TeV is the mass of the Higgs boson for which the Landau pole in the scalar self-coupling λ coincides with M_H (triviality bound).[6–8] If the Higgs mass happened to be smaller than $M_{\rm meta}$, we would be forced to conclude that there must be some new physics beyond the SM, which stabilizes the SM vacuum. If the Higgs mass happened to be larger than $M_{\rm Landau}$, we would be forced to conclude that there must be some new physics beyond the SM, which removes the Landau pole. For experimental value of the Higgs mass the lifetime of our vacuum exceeds by many orders of magnitude the life-time of the Universe, whereas the Landau pole in λ occurs at energies by many orders of magnitude larger than the quantum gravity scale associated with the Planck mass M_P. In other words, the discovery of the Higgs boson with this particular value of mass tells that the SM is a consistent weakly coupled effective theory all the way up to the Planck scale, and that no new physics above the Fermi scale is needed from this point of view.

In discussion of the bounds on the mass of the Higgs boson, yet another number (discussed in more detail below), $M_{\rm min}$ will be important in what follows. It corresponds to the boundary of absolute stability of the electroweak vacuum: for $M_H \geq M_{\rm min}$ it is stable, while if $M_H < M_{\rm min}$ there exists yet another minimum of the effective potential with smaller energy.[9–14] It is important to note that if the mass of the Higgs boson is in the interval $M_{\rm meta} < M_H < M_{\rm min}$, this does not invalidate the SM but simply means that we live in the metastable state with a very long lifetime.

3. Higgs mass from asymptotic safety

Though the mass of the Higgs boson cannot be predicted within the Standard Model alone, embedding it in the larger framework may fix it. Of course, all predictions of the Higgs mass are highly speculative as they rely on untested physics and use strong assumptions.

In this section we will describe a scenario which is based on the idea of asymptotic safety of the SM[15] due to quantum gravitational effects.

The most minimalistic approach to quantum gravity is associated with asymptotic safety.[16] Though General Relativity is non-renormalizable by perturbative methods, it may exist as a field theory non-perturbatively, exhibiting a non-trivial ultraviolet fixed point (for a review see[17]). If true, the asymptotically safe gravity plus the SM may happen to be the ultimate theory of all interactions (with some reservations discussed below in Section 6). However, to be the case, the ultraviolet behaviour of the the U(1) gauge coupling $g' \equiv g_1$, the Higgs self-interaction λ, and Yukawa couplings (most notably, that of the top-quark, y_t), which, are not asymptotically free in the SM, must be converted to the regular behaviour due to gravitational effects.

The prediction of the Higgs boson mass from the requirement of asymptotic safety of the SM is found as follows.[15] Consider the SM running of the coupling constants and add to the β-functions extra terms coming from gravity, deriving their structure from dimensional analysis:

$$\beta_h^{\text{grav}} = \frac{a_h}{8\pi} \frac{\mu^2}{M_P^2(\mu)} h, \tag{2}$$

where a_1, a_2, a_3, a_y, and a_λ are some constants (anomalous dimensions) corresponding to the gauge couplings of the SM g, g', g_s, the top Yukawa coupling y_t, and the Higgs self-coupling λ. In addition,

$$M_P^2(\mu) \simeq M_P^2 + 2\xi_0\mu^2 \tag{3}$$

is the running Planck mass with $\xi_0 \approx 0.024$ following from numerical solutions of functional RG equations[18–20] (in fact, the numerical value of ξ_0 is not essential for what follows). Now, require that the solution for all coupling constants is finite for all μ and that λ is always positive. The SM can only be asymptotically safe if a_1, a_2, a_3, a_y are all negative, leading to asymptotically safe behavior of the gauge and Yukawa couplings. For $a_\lambda < 0$ we are getting the interval of admissible Higgs boson masses, $M_{\text{min}}^{\text{safety}} < M_H < M_{\text{max}}^{\text{safety}}$. The value of $M_{\text{max}}^{\text{safety}}$ is around 175 GeV and corresponds to the Higgs mass at which the Landau pole in the Higgs self-coupling coincides with the Planck scale.[3,5,21–23] The value of $M_{\text{min}}^{\text{safety}}$ will be given somewhat later.

However, if $a_\lambda > 0$, as follows from computations of[19,20] only one value of the Higgs boson mass $M_H = M_{\text{min}}^{\text{safety}}$ leads to asymptotically safe behavior of λ. As is explained in,[15] this behavior is only possible provided $\lambda(M_P) \approx 0$ and $\beta_\lambda(\lambda(M_P)) \approx 0$[a]. The evolution of the Higgs self-coupling for the case of $a_h < 0$ is shown in Fig. 1, and for the case $a_h > 0$ in Fig. 2. Remarkably, the numerical value of $M_{\text{min}}^{\text{safety}}$ depends very weakly on unknown parameters a_1, a_2, a_3, a_y, ξ_0, practically does not depend on the specific form of the power correction in (2) (for instance, the factor (μ^2/M_P^2) can be raised to any power), and is very stable even if one vary M_P within 3 orders of magnitude.

To make a precision computation of $M_{\text{min}}^{\text{safety}}$ and a number of other critical bounds, it is sufficient to find a solution of two equations in the Standard Model,

[a]Interestingly, almost the same conditions determine the transition from the absolutely stable electroweak vacuum to the metastable one. Ref.[24] proposed that the existence of two degenerate vacua - multiple point principle - is to be used for fixing the mass of the Higgs boson.

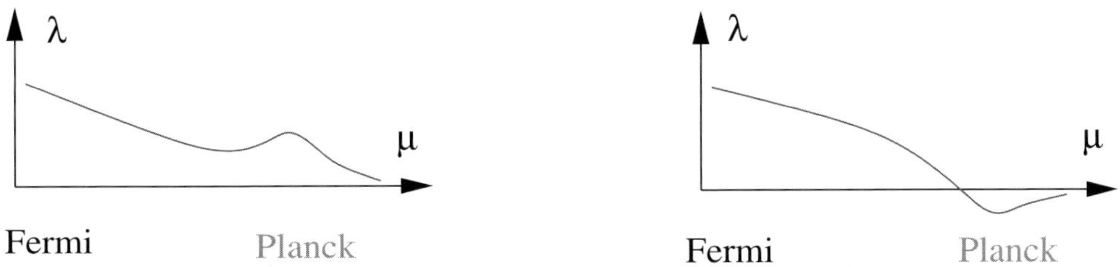

Fig. 1. Schematic depiction of the behavior of the scalar self-coupling if $a_h < 0$ for $M_{\rm min}^{\rm safety} < M_H < M_{\rm max}^{\rm safety}$ (left) and $M_H < M_{\rm min}^{\rm safety}$ (right). In both cases gravity leads to asymptotically free behavior of the scalar self-coupling. Negative λ lead to instability and thus excluded.

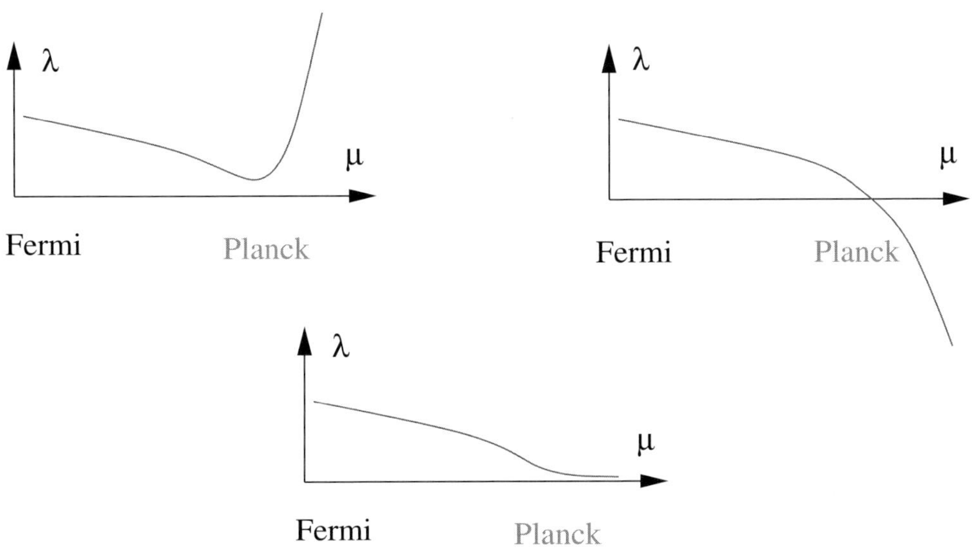

Fig. 2. Schematic depiction of the behavior of the scalar self-coupling if $a_h > 0$ for $M_H > M_{\rm min}^{\rm safety}$, leading to Landau-pole behavior (left), $M_H > M_{\rm min}^{\rm safety}$, leading to instability (right) and $M_H = M_{\rm min}^{\rm safety}$, asymptotically safe behavior (middle). Only this choice is admissible.

without inclusion of gravity[25] :

$$\lambda(\mu_0) = 0, \quad \beta_\lambda(\lambda(\mu_0)) = 0, \tag{4}$$

where β_λ is the β-function governing the renormalisation group (RG) running of λ. The mass of the Higgs boson found from eq. (4) will be called the benchmark mass $M_{\rm min}$. Here we define all the couplings of the SM in the $\overline{\rm MS}$ renormalisation scheme which is used de-facto in the most of the higher-loop computations. Clearly, if any other renormalization scheme is used, the equations $\lambda = \beta_\lambda = 0$ will give another benchmark mass, since the definition of all the couplings are scheme dependent. The benchmark mass $M_{\rm min}$ coinsides, within $100 - 200$ MeV with $M_{\rm min}^{\rm safety}$, with the absolute vacuum stability bound on the Higgs mass, and with the bound on the Higgs mass from the Higgs inflation (see Section 4).

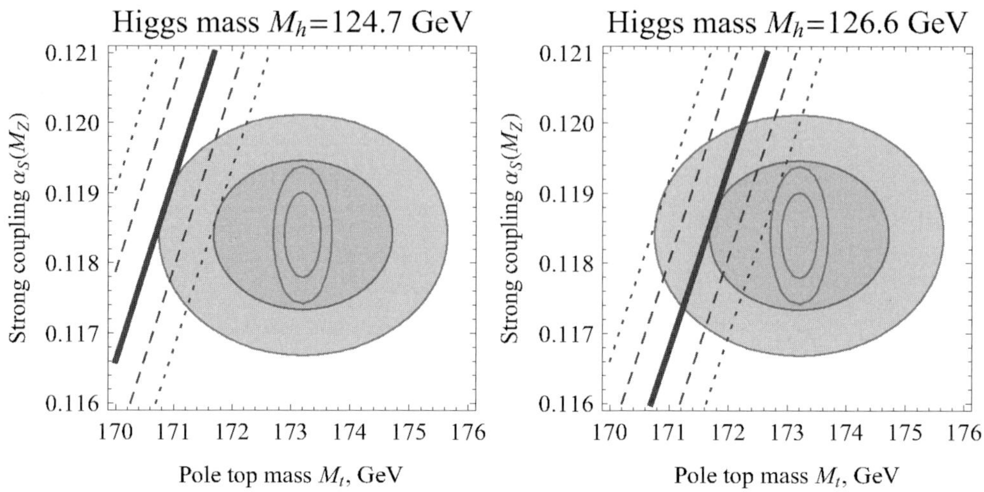

Fig. 3. The values of the strong coupling constant α_s and top mass M_t corresponding to several minimal Higgs boson mass $M_{\rm min}$ within the LHC accuracy. The 68% and 95% experimentally allowed regions for α_s and M_t are given by shaded areas. The dashed (dotted) lines correspond to 1 GeV (2 GeV) uncertainty in the $M_{\rm min}$ theoretical determination. The red lines in the center correspond to the expected precision from a e^+e^- collider.

The procedure of computing $M_{\rm min}$ is very clean and transparent. Take the standard $\overline{\rm MS}$ definition of all coupling constants of the SM, fix all of them at the Fermi scale given the experimentally known parameters such as the mass of the top quark, QCD coupling, etc., and consider the running Higgs self-coupling $\lambda(\mu)$ depending on the standard t'Hooft-Veltman parameter μ. Then, adjust $M_{\rm min}$ in such a way that equations (4) are satisfied at some μ_0.

Most recent computations of $M_{\rm min}$ were presented in.[25,26] In[25] the matching between physical and $\overline{\rm MS}$ parameters was computed at two-loop level with accounting of $\mathcal{O}(\alpha\alpha_s)$ corrections, and the three-loop running of coupling constants, computed recently in[27,28] were used. A somewhat later Ref.[26] accounted for extra $\mathcal{O}(\lambda^2, \lambda y_t^2, y_t^4)$ corrections in the matching, which resulted in a $\mathcal{O}(100-200)$ MeV shift in the value of the Higgs mass, and reduced the theoretical uncertainties to ~ 0.8 GeV (if summed up quadratically), or to ~ 1.5 GeV (if summed up linearly). The central value depends on the pole mass of the top quark and on the value of the strong coupling constant as follows:[25,26]

$$m_{crit} = [129.6 + \frac{m_t - 173.2}{0.9} \times 1.8 - \frac{\alpha_s - 0.1184}{0.0007} \times 0.56] \text{ GeV} , \qquad (5)$$

The comparison of this formula with experimental data is shown in Fig. 3, taken from.[25]

The prediction coming from asymptotic safety of the SM is thus $1 - 2\ \sigma$ away from the experimental value!

The uncertainties in prediction can be decreased by a number of theoretical and experimental means :

- By computation of remaining two-loop $\mathcal{O}(\alpha^2)$ corrections to pole - $\overline{\text{MS}}$ matching for the Higgs mass and top masses. Theoretical uncertainty can reduced at most to ~ 0.5 GeV, due to irremovable non-perturbative contribution $\sim \Lambda_{QCD}$ to top quark mass.
- By a better measurement of t-quark mass (present error in M_H due to this uncertainty is $\simeq 4$ GeV at 2σ level). For this end a construction of t-quark factory – e^+e^- or $\mu^+\mu^-$ collider with energy $\simeq 175 \times 175$ GeV would be necessary (see also[29]).
- By a better measurement of α_s (present error in M_H due to this uncertainty is $\simeq 1$ GeV at 2σ level).

4. Higgs mass from inflation

An important constraint on the Higgs mass can be derived from the requirement that the Higgs boson of the Standard Model plays the role of the inflaton.[30]

The inclusion of a non-minimal interaction of the Higgs field with gravity, given by the Lagrangian $\xi|\phi|^2 R$, where R is the Ricci scalar, changes drastically the behavior of the Higgs potential in the region of large Higgs fields $\phi > M_{\text{inflation}} \simeq M_P/\sqrt{\xi}$.[30] Basically, the potential becomes flat at $\phi > M_{\text{inflation}}$, keeping the value it acquired at $\phi \simeq M_P/\sqrt{\xi}$. This feature leads to a possibility of Higgs-inflation: if the parameter ξ is sufficiently large, $700 < \xi < 10^5$,[23] the Higgs boson of the SM can make the Universe flat, homogeneous and isotropic, and can produce the necessary spectrum of primordial fluctuations. The possibility of the Higgs inflation is also strongly related to the value of the Higgs boson mass: the successful inflation can only occur if $M_{\text{min}}^{\text{inflation}} < M_H < M_{\text{max}}^{\text{inflation}}$.[22,23,31,32] The upper limit $M_{\text{max}}^{\text{inflation}}$ comes from the requirement of the validity of the SM up to the inflation scale $M_{\text{inflation}}$. Near $M_{\text{min}}^{\text{inflation}}$ the behavior of the effective potential in the Einstein frame changes as shown in Fig. 4: if $M_H < M_{\text{min}}^{\text{inflation}}$ the "bump" in the Higgs potential prevents the system to go to the SM vacuum state. As in the previous case, these bounds can be formulated with the use of the Higgs self-coupling λ. Basically, it must be perturbative and positive for all energy scales below $M_{\text{inflation}}$. Though any Higgs boson mass in the interval $M_{\text{min}}^{\text{inflation}} < M_H < M_{\text{max}}^{\text{inflation}}$ can lead to successful inflation, the value $M_{\text{min}}^{\text{inflation}}$ is somewhat special. For the lower part of the admitted interval the value of the non-minimal coupling ξ reaches its minimal value $\xi \simeq 700$, extending the region of applicability of perturbation theory.[23,32,33]

The computation of the lower bound on the Higgs boson mass from inflation is more complicated. It is described in detail in.[22,23] Basically, one has to compute the Higgs potential in the chiral electroweak theory associated with large values of the Higgs field and find when the slow-roll inflation in this potential can give the large-scale perturbations observed by the COBE satellite. The outcome of these computations, however, can be formulated in quite simple terms: for inflationary bound find $M_{\text{min}}^{\text{inflation}}$ from the condition $\lambda(\mu) > 0$ for all $\mu < M_{\text{inflation}}$.[23] A priori, the inflationary bound could have been very different from M_{min} and thus from

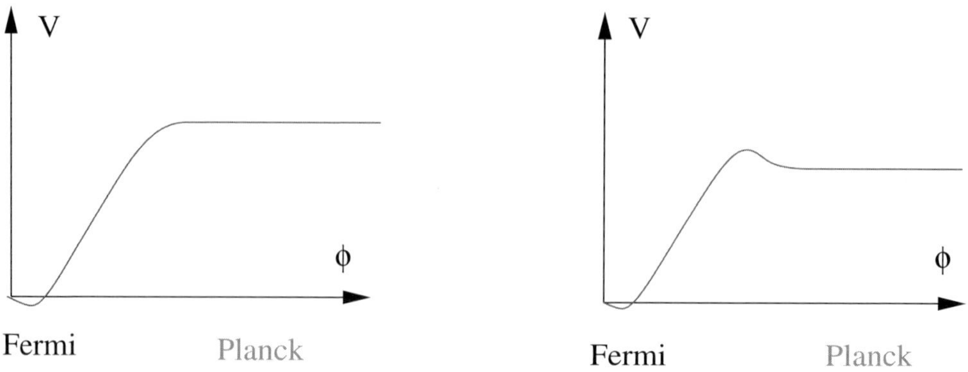

Fig. 4. Schematic depiction of the effective potential V for the Higgs field in the Higgs-inflationary theory in the Einstein frame for $M_H > m_{\rm min}^{\rm inflation}$ (left) and $M_H < m_{\rm min}^{\rm inflation}$ (right).

$M_{\rm min}^{\rm stability}$. Indeed, both $M_{\rm min}$ and $M_{\rm min}^{\rm stability}$ know *nothing* about the Planck scale and are defined entirely within the SM, whereas the inflationary bound does use M_P. However, the remarkable numerical coincidence, between μ_0 and M_P, discussed in Section 5, makes $M_{\rm min}$ and $M_{\rm min}^{\rm inflation}$ practically the same. The coupling constant λ evolves very slowly near the Planck scale, so that the regions for the Higgs boson mass following from the conditions $\lambda(\mu) > 0$ for $\mu < M_P$ and $\mu < M_{\rm inflation}$ are almost identical. This leads to the result that $\Delta m^{\rm inflation} \equiv M_{\rm min}^{\rm inflation} - M_{\rm min} \simeq -0.1 - 0.2\,{\rm GeV}$.

One must note, that a scale analysis made in the SM vacuum state leads to violation of perturbation theory at energies above M_P/ξ.[34–36] This means, that calculations in the Higgs inflationary models should be done with some additional assumptions about the high energy physics, formulated in,[33] specifically the approximate scale invariance at high field backgrounds. As is explained in,[33] adding to the SM higher-dimensional operators with a Higgs-field dependent cutoff modifies the lower bound on the Higgs boson mass in Higgs inflation. If these operators are coming with "natural" power counting coefficients (for exact definition see[33]) the sensitivity of the Higgs boson mass bound to unknown details of ultraviolet physics is rather small $\Delta M_{\rm min}^{\rm inflation} \simeq 0.6\,{\rm GeV}$.[33]

5. New physics between the Fermi and Planck scales?

Our definition of the "benchmark" Higgs boson mass consists of the solution of the two equations (4) and gives, in addition to $M_{\rm min}$, the value of the scale μ_0 at which the scalar self-coupling and its β-function vanish simultaneously. The central value for μ_0 is 2.9×10^{18} GeV and is quite stable if m_t and α_s are varied in their confidence intervals (see Fig. 5). One can see that there is a remarkable coincidence between μ_0 and the (reduced) Planck scale $M_P = 2.44\times10^{18}\,{\rm GeV}$. The physics input in the computation of μ_0 includes the parameters of the SM only, while the result gives the gravity scale. A possible explanation is related to the asymptotic safety

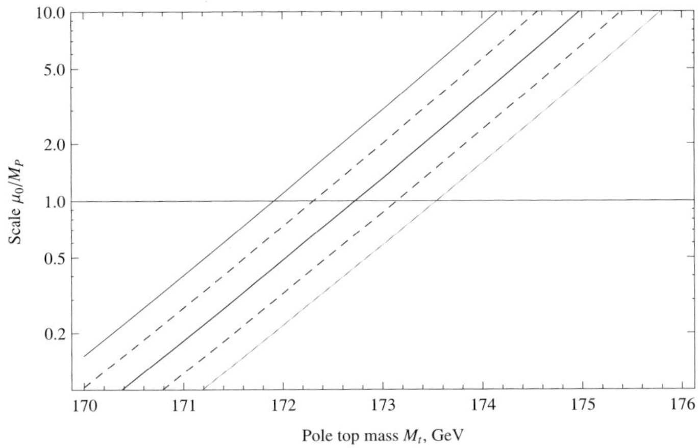

Fig. 5. The scale μ_0 (solution of (4)) depending on the top mass M_t. The dashed lines correspond to 1σ uncertainty in the α_s. The yellow shaded region corresponds to adding the α_s experimental error and the theoretical uncertainty in the matching of the top Yukawa y_t and top pole mass.

of the SM, discussed above. It remains to be seen if this is just the random play of the numbers or a profound indication that the electroweak symmetry breaking is related to Planck physics. If real, this coincidence indicates that there should be no new energy scales between the Planck and Fermi scales, as they would remove this coincidence unless some conspiracy is taking place.

6. New physics below the Fermi scale

Even if the SM and gravity are asymptotically safe and the Higgs field drives inflation, the Standard Model plus gravity cannot be an ultimate theory of Nature. This is ruled out by:

- Observations of neutrino oscillations (in the SM neutrinos are massless and do not oscillate).
- Evidence for Dark Matter (SM does not have particle physics candidate for DM).
- Absence of antimatter in the Universe (baryon asymmetry of the Universe is too small in the SM).

However, a somewhat less conservative hypothesis of the absence of intermediate energy scales between the Fermi scale 100 GeV and the Planck scale 10^{18} GeV works very well. In fact, all confirmed observational signals in favour of physics beyond the SM, mentioned above, can be associated with new physics *below* the electroweak scale, for reviews see[37–40] and references therein. The minimal model – νMSM, contains, in addition to the SM particles, 3 relatively light singlet Majorana fermions. These fermions could be responsible for neutrino masses, dark matter and baryon asymmetry of the Universe. The νMSM predicts that the LHC will continue to confirm the Standard Model and see no deviations from it. At the same time, new experiments at the high-intensity frontier, discussed in,[41] may be needed to uncover the new physics below the Fermi scale. In addition, new observations in astrophysics, discussed in,[38] may shed light to the nature of Dark Matter. As the running of couplings in the νMSM coincides with that in the SM, all results of the present discussion are equally applicable to the νMSM.

7. Conclusions

If the Higgs-like particle found at the LHC is indeed the Higgs boson of the Standard Model, there is no necessity for a new energy scale between the Fermi and Planck scales. The EW theory remains in a weakly coupled region all the way up to M_P, whereas the SM vacuum state lives longer than the age of the Universe. If the SM Higgs boson mass will be found to *coincide* with $M_{\min}$ given by (5), this would put a strong argument in favor of the *absence* of such a scale and indicate that the electroweak symmetry breaking may be associated with the physics at the Planck scale.

The experimental precision in the Higgs boson mass measurements at the LHC can eventually reach 200 MeV and thus be much smaller than the present theoretical ($\sim$ 1–2 GeV) and experimental ($\sim$ 5 GeV, 2σ) uncertainties in determination of $M_{\min}$. The largest uncertainty comes from the measurement of the mass of the top quark. It does not look likely that the LHC will substantially reduce the error in the top quark mass determination. Therefore, to clarify the relation between the Fermi and Planck scales a construction of an electron-positron or muon collider with a center-of-mass energy of $\sim 175 + 175$ GeV (Higgs and t-quark factory) would be needed. This would be decisive for setting up the question about the necessity for a new energy scale besides the two ones already known—the Fermi and the Planck scales. In addition, this will allow to study in detail the properties of the two heaviest particles of the Standard Model, potentially most sensitive to any types on new physics.

References

1. G. Aad *et al.*, *Phys.Lett.* **B716**, p. 1 (2012).
2. S. Chatrchyan *et al.*, *Phys.Lett.* **B716**, p. 30 (2012).
3. J. R. Espinosa, G. F. Giudice and A. Riotto, *JCAP* **0805**, p. 002 (2008).
4. G. Isidori, V. S. Rychkov, A. Strumia and N. Tetradis, *Phys.Rev.* **D77**, p. 025034 (2008).
5. J. Elias-Miro, J. R. Espinosa, G. F. Giudice, G. Isidori, A. Riotto *et al.*, *Phys.Lett.* **B709**, p. 222 (2012).
6. L. Maiani, G. Parisi and R. Petronzio, *Nucl. Phys.* **B136**, p. 115 (1978).
7. N. Cabibbo, L. Maiani, G. Parisi and R. Petronzio, *Nucl. Phys.* **B158**, p. 295 (1979).
8. M. Lindner, *Zeit. Phys.* **C31**, p. 295 (1986).
9. N. V. Krasnikov, *Yad. Fiz.* **28**, p. 549 (1978).
10. P. Q. Hung, *Phys. Rev. Lett.* **42**, p. 873 (1979).
11. H. D. Politzer and S. Wolfram, *Phys. Lett.* **B82**, p. 242 (1979).

12. G. Altarelli and G. Isidori, *Phys. Lett.* **B337**, p. 141 (1994).
13. J. A. Casas, J. R. Espinosa and M. Quiros, *Phys. Lett.* **B342**, p. 171 (1995).
14. J. A. Casas, J. R. Espinosa and M. Quiros, *Phys. Lett.* **B382**, p. 374 (1996).
15. M. Shaposhnikov and C. Wetterich, *Phys. Lett.* **B683**, p. 196 (2010).
16. S. Weinberg, *in: General Relativity: An Einstein Centenary Survey, Cambridge University Press* , p. 790 (1979).
17. M. Niedermaier and M. Reuter, *Living Rev.Rel.* **9**, p. 5 (2006).
18. M. Reuter, *Phys.Rev.* **D57**, p. 971 (1998).
19. R. Percacci and D. Perini, *Phys.Rev.* **D68**, p. 044018 (2003).
20. G. Narain and R. Percacci, *Class.Quant.Grav.* **27**, p. 075001 (2010).
21. T. Hambye and K. Riesselmann, *Phys. Rev.* **D55**, p. 7255 (1997).
22. F. L. Bezrukov, A. Magnin and M. Shaposhnikov, *Phys. Lett.* **B675**, p. 88 (2009).
23. F. Bezrukov and M. Shaposhnikov, *JHEP* **07**, p. 089 (2009).
24. C. D. Froggatt and H. B. Nielsen, *Phys.Lett.* **B368**, p. 96 (1996).
25. F. Bezrukov, M. Y. Kalmykov, B. A. Kniehl and M. Shaposhnikov, *JHEP* **1210**, p. 140 (2012).
26. G. Degrassi, S. Di Vita, J. Elias-Miro, J. R. Espinosa, G. F. Giudice *et al.*, *JHEP* **1208**, p. 098 (2012).
27. L. N. Mihaila, J. Salomon and M. Steinhauser, *Phys. Rev.* **86**, p. 096008 (2012).
28. K. G. Chetyrkin and M. F. Zoller, *JHEP* **1206**, p. 033 (2012).
29. S. Alekhin, A. Djouadi and S. Moch, *Phys. Lett.* **B716**, p. 214 (2012).
30. F. L. Bezrukov and M. Shaposhnikov, *Phys. Lett.* **B659**, p. 703 (2008).
31. A. De Simone, M. P. Hertzberg and F. Wilczek, *Phys. Lett.* **B678**, p. 1 (2009).
32. A. O. Barvinsky, A. Y. Kamenshchik, C. Kiefer, A. A. Starobinsky and C. Steinwachs, *JCAP* **0912**, p. 003 (2009).
33. F. Bezrukov, A. Magnin, M. Shaposhnikov and S. Sibiryakov, *JHEP* **1101**, p. 016 (2011).
34. C. Burgess, H. M. Lee and M. Trott, *JHEP* **0909**, p. 103 (2009).
35. J. Barbon and J. Espinosa, *Phys.Rev.* **D79**, p. 081302 (2009).
36. C. P. Burgess, H. M. Lee and M. Trott, *JHEP* **1007**, p. 007 (2010).
37. M. Shaposhnikov, *hep-th/0708.3550* (2007).
38. A. Boyarsky, O. Ruchayskiy and M. Shaposhnikov, *Ann. Rev. Nucl. Part. Sci.* **59**, p. 191 (2009).
39. L. Canetti, M. Drewes and M. Shaposhnikov, *Phys. Rev. Lett.* **110**, p. 061801 (2013).
40. L. Canetti, M. Drewes, T. Frossard and M. Shaposhnikov, *arXiv:1208.4607* (2012).
41. D. Gorbunov and M. Shaposhnikov, *JHEP* **10**, p. 015 (2007).

What Can We Learn from the 126 GeV Higgs Boson for the Planck Scale Physics ? - Hierarchy Problem and the Stability of the Vacuum -

Satoshi Iso

Theory Center, Institute of Particles and Nuclear Studies
High Energy Accelerator Research Organization (KEK)
1-1 Oho, Tsukuba, Ibaraki, Japan 305-0801
**E-mail: satoshi.iso@kek.jp*

The discovery of the Higgs particle at around 126 GeV has given us a big hint towards the origin of the Higgs potential. The running quartic self-coupling decreases and crosses zero somewhere in the very high energy scale. It is usually considered as a signal of the instability of the standard model (SM) vacuum, but it can also indicate a link between the physics in the electroweak scale and the Planck scale. Furthermore, the LHC experiments as well as the flavor physics experiments give strong constraints on the physics beyond the SM. It urges us to reconsider the widely taken approach to the physics beyond the SM (BSM), namely the approach based on the gauge unification below the Planck scale and the resulting hierarchy problem. Motivated by the recent experiments, we first revisit the hierarchy problem and consider an alternative appoach based on a classical conformality of the SM without the Higgs mass term.

In this talk, I review our recent proposal of a B-L extension of the SM with a flat Higgs potential at the Planck scale.[1,2] This model can be an alternative solution to the hierarchy problem as well as being phenomenologically viable to explain the neutrino oscillations and the baryon asymmetry of the universe. With an assumption that the Higgs has a flat potential at the Planck scale, we show that the B-L symmetry is radiatively broken at the TeV scale via the Coleman-Weinberg mechanism, and it triggers the electroweak symmetry breaking through a radiatively generated scalar mixing. The ratio of these two breaking scales is dynamically determined by the B-L gauge coupling.

Keywords: Hierarchy problem, Stability of the vacuum, B-L gauge symmery, Classical conformality.

1. Central dogma of particle physics

In the LHC era, we acquired various hints towards the physics beyond the SM. The first hint is of course the mass of the recently discovered Higgs-like particle. The value of 126 GeV is quite interesting because it is close to the boarder of the stability bound. Given the vev of the Higgs at 246 GeV, its mass gives an information of the curvature of the potential at the minimum. As the mass 126 GeV is smaller than 246 GeV, the Higgs potential is rather shallow and unstable against the radiative corrections. The (in)stability of the SM vacuum can be investigated explicitly by looking at the running bahaviour of the quartic coupling. The beta function of the quartic Higgs coupling λ_H is given by

$$\beta_H = \frac{1}{16\pi^2}\left(24\lambda_H^2 - 6Y_t^4 + \frac{9}{8}g^4 + \frac{3}{8}g_Y^4\right). \tag{1}$$

Y_t is the top Yukawa coupling and g, g_Y are $SU(2)_L, U(1)_Y$ gauge couplings. It is either positive or negative whether the negative contribution by the large top Yukawa coupling is compensated by the gauge couplings and the Higgs quartic coupling. The corresponding quartic coupling to the 126 GeV Higgs boson does not suffice to compensate it and the beta function is negative. So the running quartic coupling crosses zero somewhere at the UV energy scale. It is very suggestive that the observed value of the Higgs mass is close to the stability bound up to the Planck scale[3]

$$M_h[\text{GeV}] > 129.2 + 1.8\left(\frac{M_t[\text{GeV}] - 173.2}{0.9}\right) - 0.5\left(\frac{\alpha_s(M_Z) - 0.1184}{0.0007}\right) \pm 1.0_{th}. \tag{2}$$

When the Higgs mass is lighter than the above bound, new physcis must appear below the Planck scale. But if it lies just on the border of the stability bound, it gives a big hint to the origin of the Higgs potential at the Planck scale.[4]

Another important information is that the LHC results are almost consistent with the SM. Furhtermore the precision experiments of the flavor physics, Babar, Belle and LHCb, gave stringent constraints on the physics beyond the SM. Of course, in spite of the above rather unexpectedly good agreement with the SM, there exist phenomena which cannot be explained within the SM. Nuetrino oscillation requires the dimension 5 operator $l\phi l\phi$ and a new scale beyond the SM must be introduced. The baryon asymmetry of the universe also requires an additional source of the CP violation. The SM anyway needs to be extended to explain these phenomana.

The most common appoach to go beyond the SM is based on a unification of the gauge couplings below the Planck scale, i.e. the GUT scale. Then we need a natural explanation why the electroweak scale is much smaller than the GUT scale. In order to solve the hierarchical structure of the scales, the supersymmetry is introduced. Here I call the sequence of ideas from GUT to the hierarchy problem and the low energy supersymmetry *the central dogma of particle physics.* In addition to solving the hierarchy problem, it can improve the gauge coupling unification as well as providing candidates of the dark matter particles. But as the bonus we get or as the price we pay, it predicts many new particles at the TeV scale and the recent experiments have already given strong constraints on the models with low energy supersymmetry.

In such circumstances, it may be a good time to reconsider the central dogma of particle physics. In this note, we take an approach to the hierarchy problem suggested by Bardeen. In the next section, we interpret the Bardeen's argument in terms of the renormalization group. If we adopt the argument, the most natural mechanism to break the electreweak symmetry is the Coleman-Weinberg (CW) mechanism. But we know that the CW mechanism does not work within the SM because of the large top Yukawa coupling, so we need to extend the SM. In section 3, we introduce our model, a classically conformal $B-L$ extension of the SM and then discuss the dynamics of the model.

2. Bardeen's argument of the hierarchy problem

We pay a special attention to the almost scale invariance of the SM. At the classical level, the SM Lagrangian is conformal invariant except for the Higgs mass term. Bardeen has argued[5] that once the classical conformal invariance and its minimal violation by quantum anomalies are imposed on the SM, it may be free from the quadratic divergences.

Bardeen's argument on the hierarchy problem is interpreted as follows.[6] We classify divergences of the scalar mass term in the SM into the following 3 classes,

- quadratic divergences: Λ^2
- logarithmic divergences with a small coefficient: $m^2 \log(\Lambda/\mu)$
- logarithmic divergences with a large coefficient: $M^2 \log(\Lambda/\mu)$

The logarithmic divergences are operative both in the UV and the IR. In particular, it controls a running of coupling constants and is observable. On the other hand, the quadratic divergence can be always removed by a subtraction. Once subtracted, it no longer appears in observable quantities. In this sense, it gives a boundary condition of a quantity in the IR theory at the UV energy scale where the IR theory is connected with a UV completion theory. Indeed, the RGE of a Higgs mass term m^2 in the SM

$$V(H) = -m^2 H^\dagger H + \lambda_H (H^\dagger H)^2 \tag{3}$$

is approximately given by

$$\frac{dm^2}{dt} = \frac{m^2}{16\pi^2}\left(12\lambda_H + 6Y_t^2 - \frac{9}{2}g^2 - \frac{3}{2}g_Y^2\right). \tag{4}$$

The quadratic divergence is subtracted by a boundary condition either at the IR or UV scale. Once the initial condition of the RGE is given at the UV scale, it is no longer operative in the IR. The RGE shows that the mass term m^2 is multiplicatively renormalized. If it is zero at a UV scale M_{UV}, it continues to be zero at lower energy scales. In this sense, the quadratic divergence is not the issue in the IR effective theory, but the issue in the UV completion theory. Hence if the SM (and its extension at the TeV scale) is directly connected with a UV completion theory at the Planck scale physics, the hierarchy problem turns out to be a problem of the boundary condition at the UV scale. If the UV completion theory is an ordinary field theory, it will be difficult to protect the masslessness of the scalar particle against radiative corrections by massive particles of the UV scale unless we introduce, e.g. the low-energy supersymmetry. But in the string theory, symmetry is sometimes enhanced on a moduli space and massless scalars can survive even without supersymmetry.

Also discrete symmetry like T-duality, which is invisible in the low energy effective theory, may prohibit a generation of potential at the string scale.

The multiplicative renormalization of the Higgs mass term is violated by a mixing with a massive field in the loop. If the massive field aquires its mass in a different mechanism with the EWSB, the Higgs mass has a logarithmic divergence

$$\delta m^2 \sim \frac{\lambda^2 M^2}{16\pi^2} \log(\Lambda^2/m^2) \tag{5}$$

which modifies the RGE as

$$\frac{dm^2}{dt} = \frac{m^2}{16\pi^2}\left(12\lambda_H + 6Y_t^2 - \frac{9}{2}g^2 - \frac{3}{2}g_Y^2\right) + \frac{M^2}{8\pi^2}\lambda^2. \tag{6}$$

The last term corresponds to the logarithmic divergence with a large coefficient. The coefficient M^2 has nothing to do with the mass of the Higgs m^2, and it violates the multiplicativity of the Higgs mass. Thus the hierarchy problem, namely the stability of the EWSB scale, is caused by such a mixing of relevant operators (mass terms) with hierarchical energy scales $m \ll M$. In the Bardeen's argument, he also imposes an absence of intermediate scales above the EW scale. The logarithmic divergence with a large coefficient (5) is sometimes confused with the quadratic divergence, but if the UV completion theory is something like a string theory, they should be distinguished.

From the above considerations, the hierarchy problem can be solved by imposing the following two different conditions;

- Correct boundary condition at the UV (Planck) scale M_{pl}
- Absence of mixings in intermediate scales below M_{pl}

The first condition subtracts the quadratic divergence at the Planck scale. It must be solved in the UV completion theory such as the string theory. The most natural boundary condition is that scalar fields which appear in the low energy physics are massless at the Planck scale. On the other hand, the second condition assures the absence of logarithmic divergences with large coefficients. Even if the scalars are massless at the Planck scale, they receive large radiative corrections from the mixing with other relevant operators. Without a cancellation mechanism like the supersymmetry, we need to impose an absence of intermediate scales between EW (or TeV) and Planck scales. Hence all symmetries are broken either at the Planck scale or near the EW scale. Especially, the breakings of the supersymmetry or the grand unification of gauge coupling should occur at the Planck scale. This second condition is also emphasized in the Bardeen's argument.[5] In such a scenario, Planck scale physics is directly connected with the electroweak physics.[8]

Hence a natural boundary condition of the mass term at the UV cut-off scale, e.g. M_{Pl}, is

$$m^2(M_{Pl}) = 0. \tag{7}$$

This is the condition of the classical conformality of the BSM. It (7) must be justified in the UV completion theory, and from the low energy effective theory point of view, it is just imposed as a boundary condition.

3. Flat potential at the Planck scale

The condition (7) restricts the form of the Higgs potential as

$$V(H) = \lambda_H (H^\dagger H)^2. \tag{8}$$

Here λ_H is the running coupling and the RG improved effecitve potential is given by making the coupling $\lambda_H(H)$ field dependent. The mass term is not generated even in the IR as discussed in the previous section once the boundary condtion (7) is imposed at the boundary with the UV completion theory. The mass of the Higgs at 126 GeV suggests that the running coupling becomes asymptotically vanishing near the Planck scale. The current bounds (2) is a bit heavier than the experimental data, but in this note, we assume that the Higgs quartic coupling vanishes at the Planck scale. Hence

$$V(H) = 0 \text{ at the Planck scale.} \tag{9}$$

The condition may connect the SM in the IR with the string theory in the UV. Now we have to solve two problems. The first is whether we can construct a phenomenologically viable model starting from the condtion of the flatness of the Higgs potential (9), and the second is to derive such a boundary condtion from the UV completion theory such as a string theory. Supersymmetry or grand unification, if exists, are broken at the Planck scale. In the following we focus on the first problem by proposing a B-L extension of the SM with a flat potential at the Planck scale. The second issue is left for future investigations.

Since the IR theory is assumed to have the boundary condition (9), the electroweak symmetry breaking should occur radiatively, namely the Coleman-Weinberg mechanism. However, it is now well-known that the CW mechanism cannot occur within the SM because of the large top-Yukawa coupling. Indeed, the CW mechanism is realized only when the beta-function of the quartic scalar coupling is positive and the running quartic coupling crosses zero somewhere in the IR. But as we saw, the beta function of the quartic Higgs coupling is positive in the SM and its behavior is opposite to the CW mechanism. Hence, in order to realize the EWSB, we need an additional sector in which the symmetry is broken radiatively by the CW mechanism and whose symmetry breaking triggers the EWSB. In the next section, we introduce our model, namely a B-L extension of the SM with a flat potential at the Planck scale.

4. B-L extension of the SM with flat potential at Planck

The idea to utilize the CW mechanism to solve the hierarchy problem was first modelled by Meissner and Nicolai[7] (see also[9]). In addition to the SM particles, they

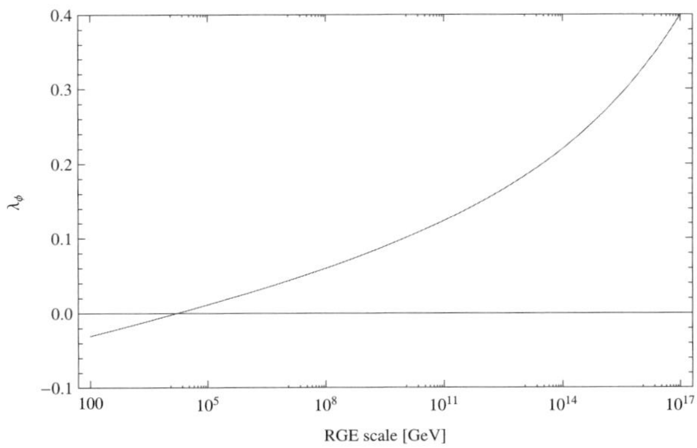

Fig. 1. RG evolution of the self-coupling λ_ϕ of a SM singlet scalar ϕ. Since the β function is positive, the running coupling crosses zero at a lower energy scale.

introduced right-handed neutrinos and a SM singlet scalar Φ. Inspired by the work, we proposed a minimal phenomenologically viable model.[2] It is the minimal B-L model,[10] but with a classical conformality. The model is similar to the one proposed by Meissner and Nicolai,[7] but the difference is whether the B-L symmetry is gauged or not. In a recent paper we further showed that by imposing the flatness (9) of the Higgs potential at the Planck scale the B-L breaking scale is related with the EWSB scale. The ratio of two scales is dynamically determined by the B-L gauge coupling and the B-L breaking scale is naturally constrained to be around TeV scale[11] for a not so small B-L gauge coupling.

Besides the SM particles the model consists of the B-L gauge field with the gauge coupling g_{B-L}, right-handed nuetrinos ν_R^i ($i = 1, 2, 3$ denotes the generation index) and a SM singlet complex scalar field Φ with two units of the B-L charge. The model is anomaly free. The Lagrangian contains Majorana Yukawa coupling $\sim Y_N^i \Phi \overline{\nu}_R^{ic} \nu_R^i$, and the see-saw mechanism gives masses to the left-handed neutrinos once the scalar Φ acquires vev.

5. Symmetry breakings of B-L and EW

Since the B-L gauge symmetry is broken by the CW mechanism, the breaking scale is correlated with the quartic coupling λ_Φ at the UV scale. Its running is described by

$$\frac{d\lambda_\Phi}{dt} = \frac{1}{16\pi^2}\left(20\lambda_\Phi^2 - \frac{1}{2}Tr\left[Y_N^4\right] + 96 g_{B-L}^4 + \cdots\right). \tag{10}$$

If the Mayorana Yukawa coupling is not so large, the beta function is positive. The typical behavior of the running λ_ϕ is drawn in Fig. 1. It crosses zero at a lower energy scale M_0, then the B-L symmetry is broken at $M_{B-L} \sim M_0 \exp(-1/4)$ through the CW mechanism.[1]

As shown in the paper,[2] the ratio of the scalar boson mass to the B-L gauge boson mass is given by

$$\left(\frac{m_\Phi}{m_{Z'}}\right)^2 \sim \frac{6}{\pi}\alpha_{B-L} \ll 1. \tag{11}$$

The condition that the B-L gauge coupling does not diverge up to the Planck scale requires $\alpha_{B-L} < 0.015$ at M_{B-L}. Hence the scalar boson becomes lighter than the B-L gauge boson, $m_\Phi^2 < 0.03\ m_{Z'}^2$. Such a very light scalar boson is a general prediction of the CW mechanism.

The EWSB is triggered by the B-L breaking. The flatness condition (9) of the Higgs potential predicts an absence of the scalar mixing at the Planck scale. Hence B-L and EW sectors are decoupled each other in the UV. But since the matter fields are coupled to both $U(1)_Y$ and $U(1)_{B-L}$, these two sectors become mixed through the $U(1)$-mixing. As a result, the scalar mixing term $\lambda_{mix}(H^\dagger H)(\Phi^\dagger \Phi)$ appears in the IR. It is interesting that a very small negative mixing λ_{mix} is always generated irrespective of the details of other parameters once we assume the flatness condition $\lambda_{mix}(M_{pl}) = 0$. By solving the RGE[1] we showed that the scalar mixing term is dynamically generated around $\lambda_{mix} \sim -4 \times 10^{-4}$. If the Φ field acquires a VEV $\langle\Phi\rangle = M_{B-L}$, the mixing term $\lambda_{mix}(H^\dagger H)(\Phi^\dagger \Phi)$ gives an effective mass term of the Higgs field. Since the coefficient λ_{mix} is negative, the EWSB is triggered and the Higgs VEV is given by

$$v = \langle H \rangle = \sqrt{\frac{|\lambda_{mix}|}{\lambda_H}} M_{B-L} \tag{12}$$

This gives a ratio between the EWSB scale to the B-L symmetry breaking scale. The scalar mixing is determined in terms of the gauge couplings, so the ratio of two breaking scales is also determined dynacamilly in terms of the gauge coupling g_{B-L}.

6. Model predictions

The dyanamics of the model is controlled by two parameters, g_{B-L} and λ_Φ, which determines the two breaking scales of B-L and EW. The experimental input $v = 246$ GeV gives a relation between these two and the dynamics of the model is essentially described by a single parameter. The figure 2 shows the prediction of our model. The vertical axis is the strength of α_{B-L} and the horizontal axis is the mass of the B-L gauge boson. The black line (from top left to down right) shows the prediction of our model. If an extra $U(1)$ gauge boson and a SM singlet scalar are found in the future, the prediction of our model is the mass relation (11), e.g., $m_\phi \sim 0.1\ m_{Z'}$ for $\alpha_{B-L} \sim 0.005$. The CW mechanism in the B-L sector predicts a much lighter SM singlet Higgs boson than the extra $U(1)$ gauge boson. It is different from the ordinary TeV scale B-L model where the symmetry is broken by a negative squared mass term.

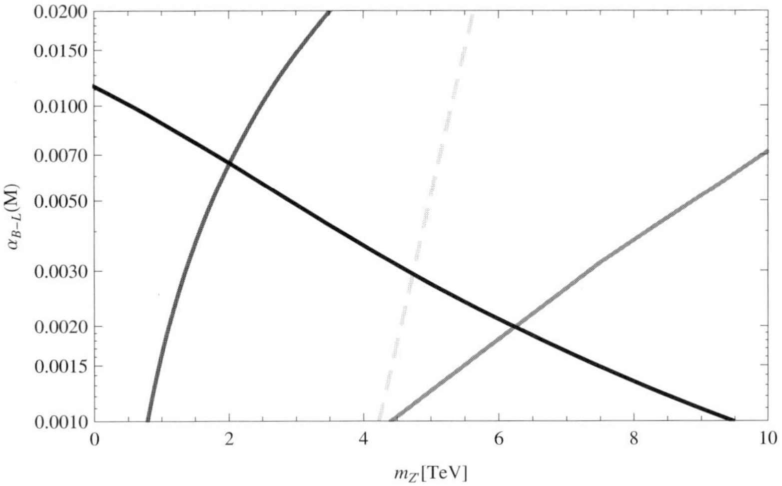

Fig. 2. Model prediction is drawn in the black line (from top left to down right). The $B-L$ gauge coupling α_{B-L} and the gauge boson mass $m_{Z'}$ are related because of the flat potential assumption at the Planck scale. The left side of the most left solid line in blue has been already excluded by the LEP experiment. The left of the dashed line can be explored in the 5-σ significance at the LHC with $\sqrt{s}$=14 TeV and an integrated luminosity 100 fb^{-1}. The left of the most right solid line (in red) can be explored at the ILC with $\sqrt{s}$=1 TeV, assuming 1% accuracy.

Nuetrino oscillation is realized by the type I see-saw mechanism with small neutrino Yukawa couplings. Baryon number asymmetry of the universe may be generated through the TeV scale leptogenesis with almost degenerate Majorana masses.[12] Furhter phenomenological issues such as $U(1)$ mixing or the lepton number violation at the TeV scale are discussed in a separate paper.

References

1. S. Iso and Y. Orikasa, PTEP **2013**, 023B08 (2013)
2. S. Iso, N. Okada and Y. Orikasa, Phys. Lett. B **676**, 81 (2009) Phys. Rev. D **80**, 115007 (2009)
3. J. Elias-Miro, J. R. Espinosa, G. F. Giudice, G. Isidori, A. Riotto and A. Strumia, Phys. Lett. B **709**, 222 (2012) S. Alekhin, A. Djouadi and S. Moch, Phys. Lett. B **716**, 214 (2012) I. Masina,
4. C. P. Burgess, V. Di Clemente and J. R. Espinosa, JHEP **0201**, 041 (2002) M. Shaposhnikov and C. Wetterich, Phys. Lett. B **683**, 196 (2010) M. Holthausen, K. S. Lim and M. Lindner, JHEP **1202** (2012) 037 F. Bezrukov, M. Y. .Kalmykov, B. A. Kniehl and M. Shaposhnikov,
5. W. A. Bardeen, FERMILAB-CONF-95-391-T
6. H. Aoki and S. Iso, Phys. Rev. D **86**, 013001 (2012)
7. K. A. Meissner and H. Nicolai, Phys. Lett. B **648**, 312 (2007) Phys. Lett. B **660**, 260 (2008) Eur. Phys. J. C **57**, 493 (2008)
8. M. Shaposhnikov, arXiv:0708.3550 [hep-th]. T. Asaka, S. Blanchet and M. Shaposhnikov, Phys. Lett. B **631**, 151 (2005) M. Shaposhnikov, Nucl. Phys. B **763**, 49 (2007)
9. A. G. Dias, Phys. Rev. D **73**, 096002 (2006), R. Hempfling, Phys. Lett. B **379**, 153

(1996), W. F. Chang, J. N. Ng and J. M. S. Wu, Phys. Rev. D **75**, 115016 (2007), R. Foot, A. Kobakhidze, K. L. McDonald and R. R. Volkas, Phys. Rev. D **76**, 075014 (2007) M. Holthausen, M. Lindner and M. A. Schmidt, Phys. Rev. D **82**, 055002 (2010), L. Alexander-Nunneley and A. Pilaftsis, JHEP **1009**, 021 (2010) K. Ishiwata, Phys. Lett. B **710**, 134 (2012) J. S. Lee and A. Pilaftsis, Phys. Rev. D **86**, 035004 (2012)

10. R. N. Mohapatra and R. E. Marshak, Phys. Rev. Lett. **44**, 1316 (1980); R. E. Marshak and R. N. Mohapatra, Phys. Lett. B **91**, 222 (1980); C. Wetterich, Nucl. Phys. B **187**, 343 (1981); A. Masiero, J. F. Nieves and T. Yanagida, Phys. Lett. B **116**, 11 (1982); R. N. Mohapatra and G. Senjanovic, Phys. Rev. D **27**, 254 (1983); W. Buchmuller, C. Greub and P. Minkowski, Phys. Lett. B **267**, 395 (1991).
11. S. Khalil, J. Phys. G **35**, 055001 (2008) M. Abbas and S. Khalil, JHEP **056**, 804 (2008) S. Khalil and A. Masiero, Phys. Lett. B **665**, 374 (2008) L. Basso, A. Belyaev, S. Moretti and C. H. Shepherd-Themistocleous, Phys. Rev. D **80**, 055030 (2009) L. Basso, A. Belyaev, S. Moretti, G. M. Pruna and C. H. Shepherd-Themistocleous, Eur. Phys. J. C **71**, 1613 (2011)
12. S. Iso, N. Okada and Y. Orikasa, Phys. Rev. D **83**, 093011 (2011) N. Okada, Y. Orikasa and T. Yamada, Phys. Rev. D **86**, 076003 (2012)

RG Cycles, Scale vs Conformal Invariance, and All That...

J-F. Fortin,

Theory Division, Department of Physics, CERN, CH-1211 Geneva 23, Switzerland
Stanford Institute for Theoretical Physics, Department of Physics, Stanford University, Stanford, CA 94305, USA

B. Grinstein and A. Stergiou

Department of Physics, University of California, San Diego, La Jolla, California 92130, USA

Two long-standing questions in Quantum Field Theory have been recently answered, at least in perturbation theory. The first one is, are there recursive flows as solutions of the renormalization group equation (answer: yes). The second one is, can a theory display invariance under dilatations but not be conformally invariant? (answer: no). We review these results. In so doing we give a hopefully pedagogic derivation of the trace anomaly equation and of the Weyl consistency conditions of Jack and Osborn.

Keywords: c-theorem; renormalization group; limit cycles

1. Introduction

In QFT an arbitrary reference dimensional parameter, the renormalization scale μ, must be introduced to define the theory. No-non-sense requires that physical results be independent of this parameter. This is encoded in the renormalization group equation (RGE):

$$\frac{dg^i}{dt} = \beta^i(g), \tag{1}$$

where $t = \ln \mu$, $g^1, g^2, \ldots$ are coupling constants and the β functions depend on all the g^i. Kenneth Wilson[1] first looked at the possible behavior of flows in QFT as $t \to \infty$. He realized there were two cases of particular interest, which cover almost all possibilities: (1) A flow can terminate at a "fixed point," and (2) A flow can tend towards a limiting cycle. He pointed out there is also a third case, "others," which however is too difficult to analyze in general. Case (2) means that there is in fact a flow on the limit cycle that is recurrent. One can contemplate yet another version of (2) in which the cycle is replaced by a trajectory that comes arbitrarily close to the starting point provided one waits a long enough RG "time" t. We call these flows "ergodic." We now appreciate that one is also interested in the limiting form of flows as $t \to -\infty$, the limit ascribed to the long distance behavior of QFTs, and these flows can be classified analogously.

In spite of this early classification, until recently examples of QFT with recurrent flows were nowhere to be found. The first result I report is the existence of relativistic, four-dimensional (one time, three space) unitary QFTs that exhibit limit cycles and ergodic flows.[2,3] The theories are not particularly special. They contain scalar, fermion and vector fields, are renormalizable and unitary. The flows can be established using perturbation theory in the small coupling constants (so that the beta functions can be exhibited to arbitrary accuracy in an expansion in powers of the coupling constants). An interesting observation is that supersymmetric theories do not exhibit recurrent flows.[4,5]

The interpretation of a fixed point is well known: at a fixed point the theory is invariant under scale and conformal transformations. A long standing open question is whether a reasonable QFT may be scale but not conformal invariant. By "reasonable" we mean a theory that is unitary, well defined and well behaved — for example, that the correlations of the stress-energy tensor are well defined. The opposite, that a conformal QFT is necessarily scale invariant, is easily established: the algebra of the generators of conformal transformations closes only if we include the generators of scale and of Poincare transformations. Using the methods of Zamolodchikov's proof of a c-theorem in two-dimensional QFTs, Joe Polchinski proved that in "reasonable" two-dimensional QFTs scale invariance does indeed imply conformal invariance.[6] But the question has remained open for theories in more than two dimensions. The second result I report here is a perturbative proof that in four-dimensional "reasonable" QFTs scale implies conformal invariance.[7,8]

Progress has been fast and turbulent. As we came across these remarkable results, we rushed to publish. This resulted in seven rapid-fire publication in the course of just over one year.[2–4,7,9–11] It may not be surprising then that, unfortunately, some of our early claims were wrong. Our starting point was Polchinski's conditions for scale without conformal invariance in four spacetime dimensions at lowest order in the loop expansion. What we did not understand at first was that the extension of these conditions to higher order in the loop expansion requires additional care, as I will explain below.

2. Limit Cycles

Let us review briefly the condition for fixed points in RG flows. A fixed point g_*^i is such that the LHS of (1) vanishes. You find a fixed point by solving the set of nonlinear simultaneous equations $\beta^i(g_*) = 0$. What about limit cycles? Consider a solution to the equation

$$\beta^i(g_*) = \mathsf{Q}^{ij} g_*^j, \tag{2}$$

with Q a constant matrix and a sum over the repeated index implied. Solving this equation is more complicated than finding fixed points, because one has to determine both the values of the couping constants g_* *and* the fixed matrix Q. But armed with a solution we can immediately find limit cycles. The key observation, justified below, is that $M(t) = \exp(t\mathsf{Q})$ is a one parameter element of an $SO(n)$ group, $n \geq 2$, of which Q is a generator, and that $\beta(M(t)g_*) = M(t)\beta(g_*)$. It follows that $g(t) = M(t)g_*$ is a solution to (1). Moreover, since $SO(n)$ is compact, $g(t)$ must describe a recurrent flow. One may think of the eigenvalues of Q as frequencies ω_k. If the frequencies are commensurate the trajectory will close on itself and we have a limit cycle, else one has an ergodic flow.

The first example we found of a limit cycle was for a model in $4-\epsilon$ dimensions. While unphysical, the ϵ expansion is instructive in the study of conformal theories.[12] Moreover, it often gives quite accurate results when evaluated at $\epsilon = 1$.[13] The model consists of $n_s = 2$ real scalars ϕ_a and $n_f = 2$ Weyl fermions ψ_i with canonical kinetic terms and interactions described by

$$\begin{aligned} V = {} & \tfrac{1}{24}\lambda_1\phi_1^4 + \tfrac{1}{24}\lambda_2\phi_2^4 + \tfrac{1}{4}\lambda_3\phi_1^2\phi_2^2 + \tfrac{1}{6}\lambda_4\phi_1^3\phi_2 + \tfrac{1}{6}\lambda_5\phi_1\phi_2^3 + (\tfrac{1}{2}y_1\phi_1\psi_1\psi_1 \\ & + \tfrac{1}{2}y_2\phi_2\psi_1\psi_1 + \tfrac{1}{2}y_3\phi_1\psi_2\psi_2 + \tfrac{1}{2}y_4\phi_2\psi_2\psi_2 + y_5\phi_1\psi_1\psi_2 + y_6\phi_2\psi_1\psi_2 + \text{h.c.})\,. \end{aligned} \tag{3}$$

The limit cycle that we found is[11]

$$\frac{d\vec{\lambda}}{dt} = \mathsf{Q}\vec{\lambda}, \qquad \frac{dy_{2n-1}}{dt} = q y_{2n}, \qquad \frac{dy_{2n}}{dt} = -q y_{2n-1} \tag{4}$$

where the matrix Q is given by

$$\mathsf{Q} = q \begin{pmatrix} 0 & 0 & 0 & 4 & 0 \\ 0 & 0 & 0 & 0 & -4 \\ 0 & 0 & 0 & -2 & 2 \\ -1 & 0 & 3 & 0 & 0 \\ 0 & 1 & -3 & 0 & 0 \end{pmatrix}. \tag{5}$$

This is a limit cycle: there is a single frequency, q, so the RG-trajectory is closed. The frequency starts at order ϵ^3 in an expansion in ϵ,

$$q = \frac{\sqrt{323(757 - 3\sqrt{52\,953})}}{2\,057\,700}\epsilon^3 + \ldots \approx 7 \times 10^{-5}\epsilon^3 + \ldots \tag{6}$$

And to complete the specification we need to give one point on the cycle:

$$\lambda_1 = \frac{8(7087+357\sqrt{52\,953})}{102\,885}\pi^2\epsilon + \frac{2(490\,537\,743\,519+468\,277\,825\sqrt{52\,953})}{408\,605\,205\,375}\pi^2\epsilon^2 + \cdots,$$

$$\lambda_2 = \frac{64(6346+9\sqrt{52\,953})}{102\,885}\pi^2\epsilon + \frac{17(11\,340\,943\,081+57\,223\,077\sqrt{52\,953})}{136\,201\,735\,125}\pi^2\epsilon^2 + \cdots,$$

$$\lambda_3 = -\frac{272(\sqrt{52\,953}-57)}{102\,885}\pi^2\epsilon + \frac{291\,302\,437\,755-3\,043\,364\,867\sqrt{52\,953}}{817\,210\,410\,750}\pi^2\epsilon^2 + \cdots,$$

$$\lambda_4 = \frac{32\sqrt{323(757-3\sqrt{52\,953})}}{102\,885}\pi^2\epsilon + \frac{13\sqrt{\begin{array}{l}190\,447\,787(13\,924\,269\,796\,644\,128\,925\,781 \\ \quad -49\,509\,459\,494\,439\,826\,531\sqrt{52\,953})\end{array}}}{55\,843\,528\,611\,660\,750}\pi^2\epsilon^2 + \cdots,$$

$$\lambda_5 = \frac{272\sqrt{323(757-3\sqrt{52\,953})}}{102\,885}\pi^2\epsilon + \frac{\sqrt{\begin{array}{l}571\,343\,361(652\,474\,762\,867\,234\,518\,381\,407 \\ \quad -663\,663\,219\,013\,252\,691\,017\sqrt{52\,953})\end{array}}}{19\,709\,480\,686\,468\,500}\pi^2\epsilon^2 + \cdots,$$

$$y_1 = -y_3 = \frac{2\sqrt{10}}{5}\pi\epsilon^{1/2} + \frac{\sqrt{10}(175\,503+442\sqrt{52\,953})}{3\,249\,000}\pi\epsilon^{3/2} + \cdots ,$$

and $y_{2,4,5,6} = 0$. The explicit form teaches us a few things. For example, that the points on the cycle are as far from the origin as they could be expected to be in a reliable perturbative calculation: the loop expansion counts powers of $\lambda \sim |y|^2 \sim \epsilon$. And also, that the scalar potential on the cycle is bounded from below.

There are other cycles in this model.[a] There is one with $y_{3,4,5,6} = 0$ so that ψ_2 is free on it. In fact, this is the first example we found, as we were studying the $n_s = 2$ $n_f = 1$ model.[2] But on this cycle the scalar potential is unbounded from below. This may or may not signal an instability: to settle this one should look at the effective potential rather than the tree level potential. It is not known how to do this with sufficient accuracy (we need a 3-loop RG-improved effective potential for two scalar fields). In any case, I do not see this instability as a problem. It is well known that models with unbounded potential have perturbative stable vacua. Consider $\lambda\phi^4$ theory in four dimensions with wrong sign potential (negative λ). The state $\phi = 0$ is perturbatively stable. It can decay via non-perturbative process and the lifetime is exponentially long for small coupling (estimated through the usual instanton calculus[14]). During the exponentially long life of the $\phi = 0$ pseudovaccuum, questions addressed perturbatively can be unequivocally answered.

The solution to (1) displayed above is obtained using the beta functions in dimensional regularization with minimal subtraction at third order in the loop expansion. Going to this order is necessary because the leading term in the perturbative expansion for the frequency q is of third order in the loop expansion. The beta functions are not known to third order in the loop expansion, but we managed to compute the terms in the beta function that can possibly affect the determination of q. Fortunately this requires only 12 three-loop Feynman diagrams. The result is robust; that is, using the beta functions to fourth order in the loop expansion will only modify the frequency by a term of order ϵ^4. Interestingly, the first evidence of the cycle, in Ref.,[2] came from a computation using the two-loop beta functions (those given in[15]). Since the equations are non-linear one obtains a result for q which is formally of third order in the loop expansion, even though the beta functions are only of second order. Including the three-loop correction to the beta functions merely shifts the value of q on the solution, but one must check that this shift does not turn a non-vanishing q from the two-loop computation into a vanishing one from the complete three loop computation.

The peculiar structure of the matrix Q in (5), that is, the location of the zeroes and the specific integral multiples of q, follows from simple group theoretic considerations. The same goes for the justification of the statement that $M\beta(g) = \beta(Mg)$. To explain this let us consider a description of the model that is better suited to

[a]There are many fixed points too.

generalization to arbitrary n_s and n_f. The action integral is

$$S_0 = \int d^d x \left(\frac{1}{2}\partial_\mu\phi_a\partial^\mu\phi_a + i\bar{\psi}_i\bar{\sigma}^\mu\partial_\mu\psi_i - \frac{\lambda_{abcd}}{4!}\phi_a\phi_b\phi_c\phi_d - \frac{y_{a|ij}}{2}\phi_a\psi_i\psi_j - \frac{y^*_{a|ij}}{2}\phi_a\bar{\psi}_i\bar{\psi}_j \right). \tag{7}$$

Note that λ_{abcd} is completely symmetric under permutations of the four indexes and $y_{a|ij} = y_{a|ji}$. The Lagrangian is formally invariant under an infinitesimal rotation in flavor space of both fields and couplings,

$$\delta\phi_a = \omega_{ab}\phi_b, \qquad \delta\psi_i = 0, \qquad \delta\lambda_{abcd} = (\omega\lambda)_{abcd}, \qquad \delta y_{a|ij} = \omega_{ab}y_{b|ij}, \tag{8}$$

where the infinitesimal parameter of the transformation is the matrix $\omega = -\omega^T$ and we have introduced the shorthand $(\omega\lambda)_{abcd} = \omega_{ae}\lambda_{ebcd} + \omega_{be}\lambda_{aecd} + \omega_{ce}\lambda_{abed} + \omega_{de}\lambda_{abce}$. The matrices ω are elements of the algebra of $SO(n_s)$ and ϕ_a and $y_{a|ij}$ transform in the fundamental representation while λ_{abcd} transforms as a four index symmetric tensor. There is of course an additional analogous $U(n_f)$ symmetry which rotates the spinors but not the scalars. Now, the beta functions $d\lambda_{abcd}/dt = \beta_{abcd}$ and $dy_{a|ij}/dt = \beta_{a|ij}$ are given by sums of monomials that contain several powers of the coupling constants in which repeated flavor indexes are summed. From a group-theoretic point of view then monomials in β_{abcd} transform in the 4-index symmetric tensor representation, and hence so does the beta function. Likewise for $\beta_{a|ij}$. Therefore under the $SO(n_s)$ transformation (8) we have $\delta\beta_{abcd} = (\omega\beta)_{abcd}$ and $\delta\beta_{a|ij} = \omega_{ab}\beta_{b|ij}$ which is the infinitesimal version of $M\beta(g) = \beta(Mg)$. It goes without saying (or maybe it must be said?) that the precise form of the cycle condition (2) is

$$\beta_{abcd} = (Q\lambda)_{abcd} \qquad \text{and} \qquad \beta_{a|ij} = Q_{ab}y_{b|ij}, \tag{9}$$

with Q an $n_s \times n_s$ real anti-symmetric matrix. There is only one frequency in the $n_s = 2$ case because there is only one independent entry in a 2×2 anti-symmetric matrix, $Q_{12} = -Q_{21} = q$. Moreover, the form of Q in (5) follows. For example, $d\lambda_1/dt = d\lambda_{1111}/dt = (Q\lambda)_{1111} = 4q\lambda_{1112} = 4\lambda_4$. For cycles we need scalars, $n_s \geq 2$, but one may ask, why are spinors necessary at all? The answer is that nontrivial solutions to (9) start at three-loop order for the models with $n_s \geq 2$ and $n_f \geq 1$, but for models with $n_f = 0$ one can show that cycles, if they exist, must have solutions that start at five-loop order.

There are also cycles in strictly four dimensional theories. We still need $n_s \geq 2$ and $n_f \geq 1$ so as to have enough structure in the beta functions for a three-loop solution, but now the driving force provided by the classical term of the beta function in the $4 - \epsilon$ case must be provided instead by an interaction with a vector field. We look for a solution to the cycle condition (9) supplemented by $\beta^g = 0$ for the beta function of the gauge coupling constant g. The example we found is the simplest we could think of.[3] It has an $SU(3)$ gauge group with two singlet real scalars, ϕ_1 and ϕ_2, two pairs of fundamental and antifundamental active Weyl fermions, $(\psi, \tilde{\psi})_{1,2}$, as well as $\frac{1}{2}(29 - 3\varepsilon)$ pairs of fundamental and antifundamental sterile Weyl fermions. The kinetic terms are canonical and the interactions are given by

$$V = \tfrac{1}{24}\lambda_1\phi_1^4 + \tfrac{1}{24}\lambda_2\phi_2^4 + \tfrac{1}{4}\lambda_3\phi_1^2\phi_2^2 + \tfrac{1}{6}\lambda_4\phi_1^3\phi_2 + \tfrac{1}{6}\lambda_5\phi_1\phi_2^3 + \left(\phi_1\,(\psi_1\ \psi_2)\begin{pmatrix} y_1 & y_2 \\ y_3 & y_4\end{pmatrix}\begin{pmatrix}\tilde{\psi}_1 \\ \tilde{\psi}_2\end{pmatrix} + \phi_2\,(\psi_1\ \psi_2)\begin{pmatrix} y_5 & y_6 \\ y_7 & y_8\end{pmatrix}\begin{pmatrix}\tilde{\psi}_1 \\ \tilde{\psi}_2\end{pmatrix} + \text{h.c.}\right). \quad (10)$$

The sterile Weyl spinors do not interact with the scalars, only with the vectors through minimal coupling. Sterile fermions are required for a perturbative solution to $\beta^g = 0$ *à la* Caswell–Banks–Zaks. The smallest value of ε for which our theory is physical is $\varepsilon = \frac{1}{3}$, but anticipating extending this work to arbitrary number of colors, we treat ε as an expansion parameter and only take $\varepsilon \to \frac{1}{3}$ at the end. The 12 calculated terms of the three-loop beta functions in the vector-less theory now must be supplemented with an additional 8 terms. I will not quote here the parameters of the cycle we discovered. Suffice it to say that the expansion parameters are bounded on the cycle: $|\lambda|/16\pi^2 < 5\%$, $|y|^2/16\pi^2 < 1\%$, and $g^2/16\pi^2 = 0.46\%$. The frequency is small, $q \approx 2 \times 10^{-7}$. The scalar potential on the cycle is not bounded from below. As I explained above I do not think this is a real concern. But in any case, experience with the $4 - \epsilon$ case suggests that more complicated models have cycles with bounded-from-below potentials. This is currently under investigation.

3. Three Surprises

The following three statements are wrong

(1) The trace anomaly is $\mathcal{L}_{\text{int}} = \beta^i \mathcal{O}_i$ (up to equations of motion).
(2) A theory is conformal if and only if $\beta^i = 0$.
(3) Scale invariance does not imply conformal invariance

In the rest of this paper I will explain these claims.

4. Revisiting the Trace Anomaly

The trace anomaly is by now QFT-textbook stuff. Everyone knows that a classically scale invariant theory has vanishing trace of the stress-energy tensor, $T^\mu{}_\mu = 0$. But quantum mechanically this is not generally true. Write the interaction terms of the Lagrangian density as $\mathcal{L}_{\text{int}} = g^i \mathcal{O}_i$, where g^i are coupling constants, dimensionless by assumption. Everyone knows that instead of a vanishing trace of the stress-energy tensor one has

$$T^\mu{}_\mu = \beta^i \mathcal{O}_i, \quad (11)$$

where $\beta^i = dg^i/dt$ is the beta function of g^i. What almost everyone does *not* know is that this is wrong! In all generality there may be additional terms on the RHS of (11). The omitted terms I am referring to are not the trivial ones, that is, terms that vanish by the equations of motion (EOM): if Δ is the matrix of dimensions of the fields ϕ,

$$T^{\mu}{}_{\mu} = \beta^i \mathcal{O}_i + (\Delta\phi) \cdot \frac{\delta S}{\delta\phi}. \qquad (12)$$

Neither am I referring to terms that involve the Riemann curvature tensor when in a curved background:

$$T^{\mu}{}_{\mu} = \beta^i \mathcal{O}_i + (\Delta\phi) \cdot \frac{\delta S_0}{\delta\phi} - \frac{a}{16\pi^2}G + \frac{c}{16\pi^2}F - \frac{b}{16\pi^2}R^2, \qquad (13)$$

where $F = R^{\mu\nu\rho\sigma}R_{\mu\nu\rho\sigma} - 4/(d-2)R^{\mu\nu}R_{\mu\nu} + 2/(d-2)(d-1)R^2$ is the square of the Weyl tensor in d dimensions and $G = 2/(d-3)(d-2)\left[R^{\mu\nu\rho\sigma}R_{\mu\nu\rho\sigma} - 4R^{\mu\nu}R_{\mu\nu} + R^2\right]$ is the Euler density. To discover the terms I speak of let's review the derivation of (12).

We consider a QFT on flat space with Lagrangian density $\mathcal{L}$ and action integral $S_0 = \int d^d x \mathcal{L}$ that is classically scale invariant (has only dimensionless coupling constants). We use dimensional regularization, $d = 4 - \epsilon$, with minimal subtraction. Computing the improved stress-energy tensor[16] from the regulated action, and taking the trace,

$$T^{\mu}{}_{\mu} = \epsilon\mathcal{L} - (1)\phi\frac{\delta S_0}{\delta\phi}. \qquad (14)$$

The factor "(1)" in the last term is appropriate for a scalar field ϕ, and would have to be replaced by (3/2) for a spinor. The right hand side fails to vanish even after using EOM because a scale invariant Lagrangian in $d = 4$ dimensions fails to be scale invariant in $d = 4 - \epsilon$. We can make the equation above useful by expressing the bare Lagrangian in terms of renormalized quantities. To this end use the RGE for $\mathcal{L}$,

$$\left[\hat{\beta}(g)\frac{\partial}{\partial g} - \hat{\gamma}\phi\frac{\partial}{\partial\phi} - \epsilon\right]\mathcal{L} = 0, \qquad (15)$$

to rewrite the trace in the more familiar form

$$T^{\mu}{}_{\mu} = \hat{\beta}(g)\frac{\partial\mathcal{L}}{\partial g} - (1 + \hat{\gamma}(g))\phi\frac{\delta S_0}{\delta\phi}. \qquad (16)$$

The hat over the beta function and anomalous dimensions mean that we have included the "classical" term that arises when the number of space-time dimensions is not strictly 4, $\hat{\beta}(g) = -\epsilon g + \beta(g)$ and $\hat{\gamma}(g) = -\epsilon + \gamma(g)$. We have written this as if there were only one coupling and one scalar field, but it is straightforward to generalize to several couplings and fields. If the action integral is that of (7), then

$$T^{\mu}{}_{\mu} = \hat{\beta}_I(g)\frac{\partial\mathcal{L}}{\partial g_I} - [(d_\phi + \hat{\gamma}(g))\phi] \cdot \frac{\delta S_0}{\delta\phi}. \qquad (17)$$

Here d_ϕ is the classical dimension of the field ϕ, $d_\phi = 3/2$ for spinors and $d_\phi = 1$ for scalars and vectors. The index I stands for either $(abcd)$ or $(a|ij)$ when referring to the scalar quartic couplings λ_{abcd} or the Yukawa couplings $y_{a|ij}$, respectively. The classical term in $\hat{\beta}_I$ is $-\epsilon k_I g_I$ (no sum on I) with $k_I = 1$ for quartic couplings and $k_I = 1/2$ for Yukawa and gauge couplings.

So what is the problem with this? Nothing. The equation is correct. Both the LHS and the RHS are finite as $\epsilon \to 0$. But the finite limit on the RHS comes about by a cancellation between (1) the classical term in the beta function times ϵ-poles in the insertion of the operators and (2) the product of the finite $d = 4$ beta functions with the corresponding finite renormalized operators. The surprise is that the operators $\mathcal{O}_I = \partial\mathcal{L}/\partial g_I$ on the RHS of (17) are not generally finite when inserted in Green functions. We can easily characterize by how much these operators fail to be renormalized.

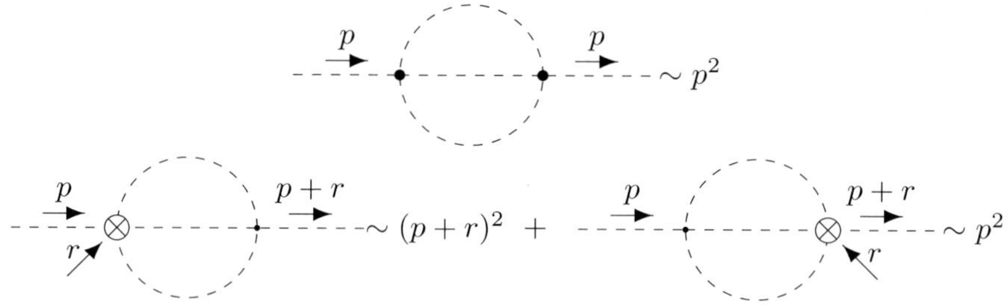

Fig. 1. Top: Feynman diagram for the lowest order contribution to the scalar field renormalization in a scalar field theory. The counterterm is the operator $\frac{1}{2}(\partial_\mu\phi)^2$. Bottom: Two point function with one insertion of the operator $\mathcal{O}_I = \frac{1}{4!}\phi_a\phi_b\phi_c\phi_d$. Momentum r flows into the diagram from the local operator.

Take a g_I-derivative of the renormalized partition function $Z = e^{iW} = \int [d\phi] e^{iS_0}$. Since W is a finite function of the renormalized coupling g_I, the derivative is finite:

$$\frac{\partial}{i\partial g_I}\int [d\phi] e^{iS_0} = \langle \int d^d x\, O_I \rangle = \text{finite} \tag{18}$$

So the space-time integral of the operator is finite, but what about the *local* operator? It can fail to be finite by something that integrates to zero, that is, a total divergence. Following Jack and Osborn[17] we introduce the notation $[\mathcal{O}_I]$ for the additionally subtracted finite operator. We have just argued that the divergence a "current" J_I renders the operator finite locally:

$$\langle [O_I] \rangle \equiv \langle O_I - \partial_\mu J_I^\mu \rangle = \text{finite}. \tag{19}$$

Which begs the question, what is this current J_I^μ? For example, for scalars fields we can have

$$J_I^\mu = -\frac{1}{2}(N_I)_{ab}\phi_a \overleftrightarrow{\partial} \phi_b. \tag{20}$$

Here the matrices N_I are functions of the couplings designed to remove the divergences in $\mathcal{O}_I$, so they have divergent expansions,

$$N_I = \frac{N_I^1}{\epsilon} + \frac{N_I^2}{\epsilon^2} + \cdots \tag{21}$$

We will only be concerned with the anti-symmetric part of this, so we will assume $N_I^T = -N_I$; we write $J_I^\mu = \partial^\mu \phi^T N_I \phi$ for short. The symmetric part does not play a role in our discussion: since it can be expressed as a total derivative, it's divergence is a two-derivative term that can be absorbed into a redefinition of the stress energy tensor.[16] Similarly, spinors can also contribute a term $(N_I^\psi)_{ij}\bar{\psi}_i i\bar{\sigma}^\mu \psi_j$. But vector fields cannot contribute: there are no gauge invariant dimension-3 currents constructed of gauge fields. It may well be that the reason we are unfamiliar with the corrections to the trace anomaly we are about to expose is that the investigation of conformal and scaling dynamics is almost exclusively focused on gauge theories for which this current correction is absent.

Before completing the correction to the trace anomaly, let's take a minute to see that this current is really needed to renormalize these operators. This correction was first introduced, as far as we know, in work of Jack and Osborn where the local operator is computed by functional differentiation with respect to space-time dependent coupling "constants" for a theory in a curved background.[17] We believe their work has been under-appreciated because it is fairly complicated. So we'd like to present their result in simple terms. Consider a scalar field theory. Since J_I^μ is quadratic in ϕ we search for Feynman diagrams with one insertion of $\mathcal{O}_I$ and with two external legs. The upper half of Fig. 1 shows the familiar Feynman diagram giving the lowest order contribution to wave-function renormalization of the scalar field. It has as counterterm the kinetic energy term, $\frac{1}{2}(\partial_\mu \phi)^2$. Without doing a calculation we know that the integral depends only on the external momentum p and that by dimensional analysis and renormalizability the infinite part is proportional to p^2. The bottom diagram in Fig. 1 is for the two point function with an insertion of the local operator $\mathcal{O}_I = \frac{1}{4!}\phi_a\phi_b\phi_c\phi_d$, denoted by an encircled-cross at the vertex. Momentum r flows into the diagram form this operator. Since it is given by the same integral as the wave-function renormalization, the left and right hand bottom diagrams are proportional to $(p+r)^2$ and p^2 respectively. An insertion of the kinetic energy counterterm $\frac{1}{2}(\partial_\mu \phi)^2$ is proportional to $p \cdot (p+r)$, which cannot fully subtract $(p+r)^2 + p^2$. Hence a counterterm of the form $\partial_\mu J_I^\mu$ is required. A simpler computation in a slightly more involved theory is the one loop diagram for two external scalars with one insertion of the Yukawa operator in the model of (7). I leave it to the reader to verify that it gives

$$(N_{c|ij})_{ab} = -\frac{1}{16\pi^2\epsilon}\frac{1}{2}(y^*_{a|ij}\delta_{bc} - y^*_{b|ij}\delta_{ac}). \tag{22}$$

Back to the equation for the trace of the stress-energy tensor: adding and subtracting $\partial_\mu J_I^\mu$ to (17) we finally have the correct version of the trace anomaly

$$T^\mu{}_\mu = \beta_I[\mathcal{O}_I] + \partial_\mu J^\mu - (1+\gamma)\phi\frac{\delta S_0}{\delta\phi} \tag{23}$$

where $J^\mu \equiv \hat{\beta}_I J_I^\mu$. Note that we have removed the hat from the beta function in the first term since the beta function is now multiplying a finite operator so the classical term in beta can be neglected as the regulator is removed. Similarly the

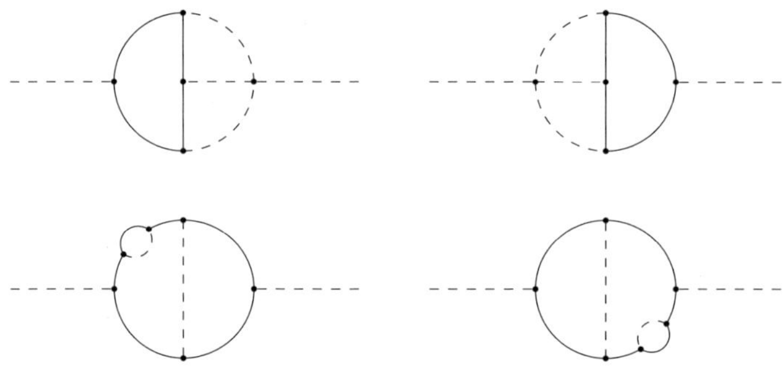

Fig. 2. Diagrams that contribute to N_I at three loops that are not symmetric under $a \leftrightarrow b$ and can therefore give a non-vanishing contribution to S.

operator $J_\mu = \partial_\mu \phi^T \hat{\beta}_I N_I \phi$ must be finite since everything else is in (23). Recalling the expansion (21), the product $\hat{\beta}_I N_I$ has an explicitly finite term and a counterterm that renders the current finite,

$$J_\mu = \partial_\mu \phi^T \left(S + N_I (Sg)_I\right) \phi, \qquad \text{where} \quad S = -k_I g_I N_I^1. \tag{24}$$

Note that S is a matrix, $S^T = -S$. It is an element of the Lie Algebra of $SO(n_s)$ and in (24) we used the symbol $(Sg)_I$, introduced earlier, that denotes the action of a generator of $SO(n_s)$ on a tensor. There is an alternative, convenient way of presenting the trace anomaly (23), which follows by using the EOM on $\partial_\mu J^\mu$,

$$T^\mu{}_\mu = B_I[\mathcal{O}_I] - (1 + \gamma + S)\phi \frac{\delta S_0}{\delta \phi} \tag{25}$$

where

$$B_I = \beta_I - (Sg)_I. \tag{26}$$

The correct version of the trace anomaly and the finite expression for the current J^μ in terms of the counterterm N_I was known to Jack and Osborn twenty years ago[17] who also computed S perturbatively and found it to vanish to second order in the loop expansion for the theory in (7), and to third order in the loop expansion for a theory of scalars only. Had they computed to three loops in the fermion-scalar model they would have found it to be non-vanishing.

4.1. *Compute S*

Let us compute S for the theory in (7). Recall $S = -k_I g_I N_I^1$ and to one loop N_I is given by (21). Taking care to add the complex conjugate, one easily checks that $S = 0$ at this order. It is easy to understand why this happens. Diagrammatically the one loop Feynman diagrams that contribute to the correlator $\langle \phi_a \phi_b \mathcal{O}_I \rangle$ are symmetric under $a \leftrightarrow b$. The diagrams contributing at two loops are still symmetric and one has to go to three loops to find non-symmetric diagrams; see Fig. 2.

A lengthy calculation gives[7]

$$
\begin{aligned}
(16\pi^2)^3(N_I^1)_{ab}\partial^\mu g_I \supset & -\tfrac{1}{2}\,\mathrm{tr}(y_a\partial^\mu y_c^* y_d y_e^*)\lambda_{bcde} - \tfrac{1}{3}\,\mathrm{tr}(y_a y_c^*\partial^\mu y_d y_e^*)\lambda_{bcde} \\
& - \tfrac{1}{2}\,\mathrm{tr}(y_a y_c^* y_d\partial^\mu y_e^*)\lambda_{bcde} - \tfrac{5}{24}\,\mathrm{tr}(y_a y_c^* y_d y_e^*)\partial^\mu\lambda_{bcde} - \tfrac{1}{24}\,\mathrm{tr}(y_b\partial^\mu y_c^* y_d y_e^*)\lambda_{acde} \\
& - \tfrac{5}{24}\,\mathrm{tr}(y_b y_c^*\partial^\mu y_d y_e^*)\lambda_{acde} - \tfrac{1}{24}\,\mathrm{tr}(y_b y_c^* y_d\partial^\mu y_e^*)\lambda_{acde} - \tfrac{5}{24}\,\mathrm{tr}(\partial^\mu y_b y_c^* y_d y_e^*)\lambda_{acde} \\
& - \tfrac{7}{32}\,\mathrm{tr}(y_a\partial^\mu y_c^* y_d y_d^* y_b y_c^*) - \tfrac{7}{96}\,\mathrm{tr}(y_a y_c^*\partial^\mu y_d y_d^* y_b y_c^*) - \tfrac{23}{96}\,\mathrm{tr}(y_a y_c^* y_d\partial^\mu y_d^* y_b y_c^*) \\
& - \tfrac{7}{96}\,\mathrm{tr}(y_a y_c^* y_d y_d^*\partial^\mu y_b y_c^*) - \tfrac{7}{32}\,\mathrm{tr}(y_a y_c^* y_d y_d^* y_b\partial^\mu y_c^*) + \tfrac{1}{16}\,\mathrm{tr}(y_a\partial^\mu y_c^* y_c y_d^* y_b y_d^*) \\
& - \tfrac{5}{48}\,\mathrm{tr}(y_a y_c^*\partial^\mu y_c y_d^* y_b y_d^*) - \tfrac{1}{48}\,\mathrm{tr}(y_a y_c^* y_c\partial^\mu y_d^* y_b y_d^*) - \tfrac{7}{96}\,\mathrm{tr}(y_a y_c^* y_c y_d^*\partial^\mu y_b y_d^*) \\
& + \tfrac{1}{16}\,\mathrm{tr}(y_a y_c^* y_c y_d^* y_b\partial^\mu y_d^*) + \text{h.c.} - \{a\leftrightarrow b\},
\end{aligned}
$$

where only the contributions of the diagrams in Fig. 2 are included. The derivative-on-couplings notation is simply a bookkeeping device that allows to write the result compactly. Also y_a stands for the symmetric matrix $(y_a)_{ij} = y_{a|ij}$. This is the first ever computation of non-vanishing S:

$$(16\pi^2)^3 S_{ab} = \tfrac{5}{8}\,\mathrm{tr}(y_a y_c^* y_d y_e^*)\lambda_{bcde} + \tfrac{3}{8}\,\mathrm{tr}(y_a y_c^* y_d y_d^* y_b y_c^*) + \text{h.c.} - \{a\leftrightarrow b\}. \tag{27}$$

It is quite remarkable that the lengthy computation yields such a simple result! An even simpler result is obtained for supersymmertic theories, for which it is easy to show that $S = 0$ to all orders in perturbation theory.[4,5]

5. Conditions for Scale without Conformal Invariance

For a theory to be scale invariant a dilatation current D^μ must be conserved, $\partial_\mu D^\mu = 0$.A dilatation current has a simple form in terms of the improved stress-energy tensor $D^\mu = x_\nu T^{\mu\nu}$. Therefore

$$\partial_\mu D^\mu = T^\mu{}_\mu \tag{28}$$

so a vanishing trace of the improved stress-energy tensor is sufficient for scale invariance.

The currents associated with the four generators of special conformal transformations can be assembled into a two index tensor, $K^{\mu\nu}$. These currents are conserved if the stress-energy tensor is traceless,

$$\partial_\mu K^{\mu\nu} = -x^\nu T^\mu{}_\mu \tag{29}$$

Tracelessness of this stress-energy tensor is not necessary for conformal invariance: if $T^\mu{}_\mu = \partial_\mu\partial_\nu L^{\mu\nu}$ for some two index symmetric tensor then one can improve again the stress energy tensor and the generators of special conformal transformations so that in terms of them $\partial_\mu K^{\mu\nu} = -x^\nu T^\mu{}_\mu = 0$.[6]

One would be tempted to conclude that requiring scale invariance gives conformal invariance since the former is guaranteed by $T^\mu{}_\mu = 0$. Polchinski pointed out, however, that it may be possible to construct a conserved dilatation current out of a non-conserved canonical dilatation current and a non-conserved internal transformation (a current without explicit space-time dependence). This strategy is not unfamiliar. In the Standard Model (SM) of electroweak interactions baryon number B and lepton number L fail to be conserved by anomalies, but the combination $B-L$ is conserved. So consider then a dilatation current of the form

$$D^\mu = x_\nu T^{\mu\nu} - V^\mu \tag{30}$$

where the "virial current" V^μ is non-conserved and does not depend explicitly on spacetime coordinates. Moreover, we require that it is not of the form $V^\mu = \partial_\nu L^{\mu\nu}$. Then if we can find a virial current such that

$$\partial_\mu D^\mu = T^\mu{}_\mu - \partial_\mu V^\mu = 0 \tag{31}$$

we will have found the theory is scale invariant while not conformal if

$$T^\mu{}_\mu = \partial_\mu V^\mu \neq 0. \tag{32}$$

Remarkably one can cast the conditions for scale without conformal invariance in purely algebraic terms. For this we need a candidate virial current, since we do not know a priori what it may be. We have in mind the theory of (7), possibly including gauge interactions. But the choices are limited: we need vector operators that are gauge invariant and classically of dimension 3:

$$V^\mu = R_{ab}\partial^\mu \phi_a \phi_b + iP_{ij}\bar{\psi}_i \bar{\sigma}^\mu \psi_j, \tag{33}$$

where $P^\dagger = -P$, so that V^μ is real, and $R^T = -R$ since the symmetric part gives a total derivative (hence a contribution to the virial current of the form $\partial_\mu L^{\mu\nu}$). Then with (25) and using the EOM in (33), the conditions for scale without conformal invariance (31) and (32) give

$$\begin{aligned} \beta_{abcd} - (S\lambda)_{abcd} &= (R\lambda)_{abcd} \neq 0, \\ \beta_{a|ij} - S_{ab}y_{b|ij} &= R_{ab}y_{b|ij} \neq 0. \end{aligned} \tag{34}$$

I have set $P=0$ in the above since in our investigations P seems to play no role (but we have no general argument for this). Alternatively we can write the conditions (34) as

$$\beta_{abcd} = (Q\lambda)_{abcd},\ \beta_{a|ij} = Q_{ab}y_{b|ij}, \quad \text{with} \quad R = Q - S \neq 0. \tag{35}$$

What is meant by solving Eqs. (35)? These are not functional equations. Rather, they are similar in spirit to the equations that give fixed points: one must specify values of coupling constants *and* of the fixed matrix Q that satisfy the equations. In fact, a fixed point corresponds to a solution with $Q=0$.

These equations, without the contribution from S, were first given by Polchinski for the scalar case and by Dorigoni and Rychkov for the scalar-fermion case.[18] These authors went on to show that the only solutions at one loop are fixed points. Our interest in the problem commenced from the realization that solutions exist for $Q \neq 0$ if one goes beyond one loop.

6. S is Q

The alert reader may have noticed that the equations (35) are precisely the conditions for recursive RG flows (9). Moreover, the same alert reader noticed that both Q and S start at third order in the loop expansion and are lie algebra valued and appeared in connection with a scalar current. It begs the question, are they related?

Of course, they are not generally the same. S is a function of the couplings defined everywhere in theory space, while Q is a constant matrix, defined as a solution to (9) at a particular point in theory space, and there may exist, and typically do, several distinct solutions. But we can prove a theorem, that in perturbation theory

- On cycles $S = Q$ (up to a symmetry generator)
- On fixed points $S = 0$ (up to a symmetry generator)

Outline of the proof:

(1) On cycles a quantity $\tilde{A}$ remains constant simply because it is $SO(n_s)$ invariant
(2) $\tilde{A}$ satisfies a strong c-theorem in perturbation theory, it grows like the square of $B_I = \beta_I - (S\lambda)_I$.
(3) On a cycle, $\beta_I = (Q\lambda)_I$
(4) Putting (1)-(3) together, $(Q\lambda) - (S\lambda)_I = 0$
(5) Hence $S = Q$ up to a symmetry, $[(S - Q)\lambda]_I = 0$

The "magic" of the proof resides in the Weyl consistency conditions of Jack and Osborn,[17,19] which can be brought to the form

$$8B_I\partial_I\tilde{A} = \chi^g_{IJ}B_IB_J. \tag{36}$$

Here $\tilde{A} \equiv a + \frac{1}{8}w_IB_I$. The quantities w_I, a and χ^g_{IJ} are functions of couplings defined by Jack and Osborn that arise in the trace anomaly equation on a curved background when the couplings are promoted to space-time dependent spurions. a is well known, it is the coefficient of the Euler density in the Weyl anomaly that Cardy proposed as a candidate for a c-theorem;[20] see Eq. (13). From Eq. (36) one can show that, at least perturbatively, it is in fact $\tilde{A}$ that gives a c-theorem. The reason for the caveat that the result follows only perturbatively is that the "metric in theory space" χ^g_{IJ} is known to be positive definite only because it is so at lowest order in the loop expansion in unitary theories. And this is why our theorem that $S = Q$ on recursive flows is established only perturbatively.

Let's flesh out the outline of the proof a little bit. Since $\tilde{A}$ has no free flavor indices it is a scalar under $SO(n_s)$ transformations (8). This implies $(\omega g)_I\partial_I\tilde{A} = 0$ for any ω in the Lie algebra of $SO(n_s)$. On a cycle or fixed point $B_I = \beta_I - (Sg)_I \to [(Q - S)g]_I$ is of the form $(\omega g)_I$. So on a cycle or fixed point the LHS of (36) vanishes. Hence $\chi^g_{IJ}B_IB_J = 0$ on recursive flows or fixed points. Now we invoke the perturbative positivity of χ^g_{IJ} to show that $B_I = 0$. It follows that $(S\lambda) - (Q\lambda)_I = 0$ and hence $S = Q + \Delta Q$ where $(\Delta Qg)_I = 0$, that is, ΔQ generates a symmetry.

7. Scale IS Conformal

It is now simple exercise to show that in $d = 4$ spacetime dimensions, a local, renormalizable, unitary, Poincare invariant interacting QFT scale invariance implies invariance under special conformal transformations. Recall the condition for scale without conformal invariance is (35):

$$\beta_{abcd} = (Q\lambda)_{abcd}, \; \beta_{a|ij} = Q_{ab}y_{b|ij}, \quad \text{with} \quad R = Q - S \neq 0. \tag{37}$$

But we have just showed that the solution to $\beta_I = (Q\lambda)_I$ requires $S = Q + \Delta Q$, and therefore that $R = Q - S = \Delta Q$. Had we been more careful we would have written the condition on R as $(Rg)_I \neq 0$, since we need a non-vanishing divergence of the virial current.

An alternative argument that scale implies conformal invariance was given independently and simultaneously by Luty, Polchinski and Rattazzi.[8] The vanishing of $\tilde{A}$ gives $B_I = 0$ which from (26) gives a vanishing trace of the stress-energy tensor.[b]

8. Conclusions and Outlook

We have discovered recursive flows in 4-dimensional QFTs. They correspond to conformal theories. The existence of cycles rules out the possibility of gradient flows in a generic subtraction scheme, but they are not inconsistent with a c-theorem since c (specifically $\tilde{A}$) remains constant as it flows on the cycle. We have shown that in unitary, renormalizable, 4-dimensional Poincare invariant theories, scale invariance requires also invariance under special conformal transformations. At least in perturbation theory.

One can show that there is a subtraction scheme for which the beta function is B_I, related to the beta function of dimensional regularization with minimal subtraction by $B_I = \beta_I - (Sg)_I$, and the anomalous dimensions of the scalar fields is Γ given in terms of the MS-dim-reg one by $\Gamma = \gamma + S$. In this scheme "S" vanishes. But its effect remains in that it gives an antisymmetric part to the anomalous dimension Γ (γ is symmetric). In this scheme cycles are characterized by having a vanishing beta function, so it is natural that they correspond to conformal theories. Incidentally, the scheme change does not correspond to a transformation $g_I \to f_I(g)$, which explains why a vanishing beta function in one scheme does not give one in another scheme. In spite of the non-symmetric anomalous dimension matrix of the fundamental scalar fields, the two point functions of the theory are completely normal and fail to display any oscillations (one can show that conformal invariance requires that the eigenvalues of the anomalous dimension matrix remain real).

---

[b]While the first version of their work makes the correct claim that scale implies conformal, the proof there is incorrect: it is based on the demonstrably false impossibility of RG-cycles. In the first version of our paper[7] we criticized the arguments in,[8] but we failed to appreciate that $B_I = 0$ means an RG-cycle is conformal. The two groups communicated and agreed to post simultaneous but independent second versions of their papers.

Whether scale implies conformal non-perturbatively remains an open question. Recently Komargodsky and Schwimmer outlined a proposal for a non-perturbative proof of the weak version of a c-theorem in four dimensions.[21] This suggests that there may be a non-perturbative proof, using methods similar to theirs, that the metric in theory space χ^g_{IJ} is positive definite, and therefore that scale implies conformal beyond perturbation theory.

Our explicit computation of S allows future discovery of cycles much more effectively. Since S starts at third order in the loop expansion, one merely needs to find zeroes of the beta function to one loop order (two loops for the gauge coupling) and plug the value of the "fixed point" couplings into S. If $S = 0$ the solution is likely a fixed point, while if $S \neq 0$ then it is certainly a cycle, with $Q = S$. This will be put to use in searching for cycles in 4-dimensions with bounded scalar potential.

Lots of additional open question remain. Are there flows connecting fixed points to cycles? What is the interpretation of the cycles in $d = 4-\epsilon$? (The proof that scale implies conformal fails in that case). Does the $\epsilon \to 1$ limit of $d = 4 - \epsilon$ cycles define non-perturbative cycles in 3 spacetime dimensions, and if so, are these theories conformally invariant?

Acknowledgments

The work of BG and AS supported in part by the U.S.DOE under contract DOE-FG03-97ER40546 and that of JFF by the ERC grant BSMOXFORD No. 228169.

References

1. K. G. Wilson, *Phys.Rev.* **D3**, p. 1818 (1971).
2. J.-F. Fortin, B. Grinstein and A. Stergiou, *Phys.Lett.* **B704**, 74 (2011).
3. J.-F. Fortin, B. Grinstein and A. Stergiou, *JHEP* **1212**, p. 112 (2012).
4. J.-F. Fortin, B. Grinstein, C. W. Murphy and A. Stergiou, *Phys.Lett.* **B719**, 170 (2013).
5. Y. Nakayama (2012).
6. J. Polchinski, *Nucl.Phys.* **B303**, p. 226 (1988).
7. J.-F. Fortin, B. Grinstein and A. Stergiou, *JHEP* **1301**, p. 184 (2013).
8. M. A. Luty, J. Polchinski and R. Rattazzi, *JHEP* **1301**, p. 152 (2013).
9. J.-F. Fortin, B. Grinstein and A. Stergiou, *JHEP* **1207**, p. 025 (2012).
10. J.-F. Fortin, B. Grinstein and A. Stergiou, *Phys.Lett.* **B709**, 408 (2012).
11. J.-F. Fortin, B. Grinstein and A. Stergiou, *JHEP* **1208**, p. 085 (2012).
12. K. G. Wilson, *Phys.Rev.* **D7**, 2911 (1973).
13. K. G. Wilson and M. E. Fisher, *Phys.Rev.Lett.* **28**, 240 (1972).
14. L. Lipatov, *Sov.Phys.JETP* **45**, 216 (1977).
15. I. Jack and H. Osborn, *Nucl.Phys.* **B249**, p. 472 (1985).
16. J. Callan, Curtis G., S. R. Coleman and R. Jackiw, *Annals Phys.* **59**, 42 (1970).
17. I. Jack and H. Osborn, *Nucl.Phys.* **B343**, 647 (1990).
18. D. Dorigoni and V. S. Rychkov (2009).
19. H. Osborn, *Nucl.Phys.* **B363**, 486 (1991).
20. J. L. Cardy, *Phys.Lett.* **B215**, 749 (1988).
21. Z. Komargodski and A. Schwimmer, *JHEP* **1112**, p. 099 (2011).

Top-Seesaw Assisted Technicolor Model with 126 GeV Higgs Boson

Hidenori S. Fukano* and Kimmo Tuominen

Department of Physics, University of Jyväskylä, P.O.Box 35, FIN-40014 Jyväskylä, Finland
Helsinki Institute of Physics, P.O.Box 64, FIN-00014 University of Helsinki, Finland

We discuss a model which involves the top quark condensation and the walking technicolor. We focus on the scalar boson in such a model from the viewpoint of the observed scalar boson at the LHC.

1. Introduction

The ATLAS and CMS collaborations at the LHC have announced the discovery of a new scalar boson with mass $m_h \simeq 126$ GeV.[2,3] The production cross section and decay rates of this new boson appear to be consistent with the prediction of the Higgs boson in the Standard Model (SM). Therefore the next logical step is to try to uncover its properties more precisely and to see how well it fits in with various extensions of the SM. A scenario based on the dynamical electroweak symmetry breaking still remains as a viable alternative, although the discovery of a light scalar boson is a severe obstruction for traditional technicolor models.[4]

There are generally at least two different alternatives based on the dynamical electroweak symmetry breaking. First possibility is walking technicolor which is controlled by the quasiconformal gauge dynamics.[5] The walking technicolor model has an approximate scale symmetry and a scalar boson emerges as the pseudo Nmabu-Goldstone boson after the approximate scale symmetry is broken. This scalar boson, so-called techni-dilaton, might be light and explain the observed new boson at the LHC.[6,7] Second possibility is the top quark condensation[8,9] where the electroweak symmetry breaking is triggered by the Nambu-Jona-Lasinio (NJL) dynamics.[10] The NJL dynamics leads to emergence of a scalar boson with mass $m_h = 2\Sigma$ where Σ is the dynamical fermion mass. This scalar boson is described as a SM like electroweak scalar doublet. This is different from the description of the techni-dilaton.[6]

In this talk we consider the new observed scalar boson from the viewpoint of the top quark condensation model. In the top quark condensation model, scalar boson is composed of the top quark and its mass is related to the top quark mass by the NJL dynamics, i.e. $m_h = 2m_t$. This is not suitable to explain the observed scalar boson with $m_h \simeq 126\,\text{GeV}$ at the LHC. However, it might be possible to explain

*speaker,E-mail : hidenori.f.sakuma@jyu.fi

the observed scalar boson mass in the top quark condensation model by sharing the top quark mass with another dynamical sector, e.g. extended technicolor.[11] This scenario is the top-seesaw assisted technicolor model[1,12] where we have used the top-seesaw model[13] which might be a promising model among several models based on the top quark condensation under the present experimental constraints.

2. Top-seesaw assisted technicolor (TSSTC) model

In this section, we consider the 126 GeV scalar boson in a model which involves walking technicolor and top-seesaw dynamics simultaneously. However, in this talk, we concentrate only on the top-seesaw sector to focus on a scalar boson originating from the electroweak doublet like the Higgs boson in the SM. Particle contents are summarized in Table 1. Here $Q^{1,2,3}$ are the usual SM chiral quarks, but $U^{(4)}, D^{(4)}$ are vector-like, i.e. the electroweak singlets. The usual SM leptons are sufficient to avoid the gauge anomalies, and we do not show them explicitly in the Table 1. We assume that $SU(3)_1$ topcolor gauge coupling is stronger than $SU(3)_2$ topcolor

Table 1. Particle content and charge assignments.

field	SU(3)$_1$	SU(3)$_2$	SU(2)$_L$	$U(1)_Y$
$Q_L^{(3)}$	3	1	2	1/6
$U_R^{(3)}, D_R^{(3)}$	1	3	1	(2/3 , -1/3)
$U_L^{(4)}, D_L^{(4)}$	1	3	1	(2/3, -1/3)
$U_R^{(4)}, D_R^{(4)}$	3	1	1	(2/3,-1/3)
$Q^{(1,2)}$	1	3	SM	SM

gauge coupling and the topcolor breaking, $SU(3)_1 \times SU(3)_2 \to SU(3)_c$, is triggered at a scale Λ. The unbroken $SU(3)_c$ is the usual color gauge group. The topcolor breaking provides following four fermion interactions at scale Λ:

$$\mathcal{L}^{4f} = G_b \left(\overline{D}_R^{(4)} Q_L^{(3)}\right)^2 + G_t \left(\overline{U}_R^{(4)} Q_L^{(3)}\right)^2 + G_{tb} \left(\overline{Q}_L^{(3)} U_R^{(4)}\right) \left(\overline{D}_R^{(4)c} i\tau_2 Q_L^{(3)c}\right) , \quad (1)$$

where the superscript ^{c} implies charge conjugation. The diagonal terms, G_b and G_t arise from the exchange of color octet massive gauge bosons with mass $\sim \Lambda$ originating from the topcolor breaking. The off diagonal term G_{tb} may arise from e.g. the topcolor instantons.[14] Using the fermion bubble sum approximation,[9] the low energy Lagrangian at $\mu < \Lambda$ is given by

$$\mathcal{L}_{\rm TSS}(\Phi_1, \Phi_2) = \sum_{i=1,2} |D_\mu \Phi_i|^2 + \mathcal{L}_M + \mathcal{L}_{\rm yukawa}^{\rm TSS} - V_{\rm TSS}(\Phi_1, \Phi_2) , \quad (2)$$

where $\mathcal{L}_{\rm yukawa}^{\rm TSS}$ consists of the Yukawa interaction terms for the third family quarks and their vector-like partner quarks and is given explicitly by

$$\mathcal{L}_{\rm yukawa}^{\rm TSS} = -y_1 \overline{Q}_L^{(3)} \Phi_1 D_R^{(4)} - y_2 \overline{Q}_L^{(3)} \tilde{\Phi}_2 U_R^{(4)} + \text{h.c.} . \quad (3)$$

The potential $V_{\rm TSS}(\Phi_1, \Phi_2)$ is given by

$$V_{\rm TSS}(\Phi_1,\Phi_2) = M_{11}^2|\Phi_1|^2 + M_{22}^2|\Phi_2|^2 - M_{12}^2\left[\Phi_1^\dagger\Phi_2 + {\rm h.c.}\right] \quad (4)$$
$$+\frac{1}{2}\lambda_1(\Phi_1^\dagger\Phi_1)^2 + \frac{1}{2}\lambda_2(\Phi_2^\dagger\Phi_2)^2 + \lambda_3(\Phi_1^\dagger\Phi_1)(\Phi_2^\dagger\Phi_2) + \lambda_4(\Phi_1^\dagger\Phi_2)(\Phi_2^\dagger\Phi_1)\,.$$

Finally, $\mathcal{L}_M$ in Eq.(2) is the electroweak singlet mass term and is given by

$$\mathcal{L}_M = -\overline{U}_R^{(3)} M_{34}^U U_L^{(4)} - \overline{U}_R^{(4)} M_{44}^U U_L^{(4)} + [U \to D] + {\rm h.c.}\,. \quad (5)$$

The doublets $\Phi_{1,2}$ are parametrized as $\Phi_i = \left(\pi_i^+\,,\ (v_i + h_i^0 - i\pi_i^0)/\sqrt{2}\right)^T$ $(i = 1,2)$ and the covariant derivatives for Φ_i are of the same form as for the SM Higgs doublet. The NJL dynamics based on Eq.(1) is rewritten by the renormalization group equations for y_i, λ_i in Eq.(3) together with suitable compositeness conditions.[9,14] In Fig. 1 the resultant dynamical (left) fermion mass and (right)

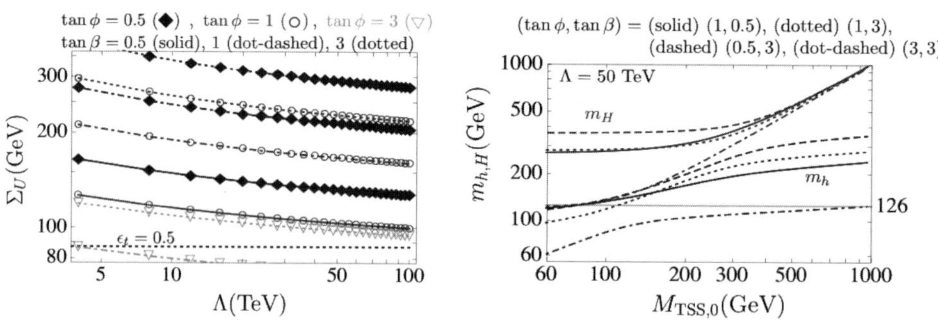

Fig. 1. The dynamical mass for (left) the top quark sector and (right) the CP-even higgs boson with $\tan\phi = 0.5, 1, 3$ and $\tan\beta = 0.5, 1, 3$. The horizontal dotted line in the left panel corresponds to $m_t({\rm TSS}) = 0.5 m_t$. The horizontal line in the right panel corresponds to $m_h = 126\,{\rm GeV}$.

CP-even Higgs bosons are shown, and $\tan\beta\,, \tan\phi$ are given by $\tan\beta \equiv v_2/v_1$, $\tan\phi \equiv v_{\rm TC}/\sqrt{v_1^2 + v_2^2}$ and $v_{\rm EW}^2 = (246\,{\rm GeV})^2 \equiv v_1^2 + v_2^2 + v_{\rm TC}^2$. The dynamical fermion mass for the top quark sector is $\Sigma_U \equiv y_2 v_2/\sqrt{2}$. The parameter $M_{\rm TSS,0}$ is defined as $M_{\rm TSS,0} \equiv M_{12}^2/(\sin\beta\cos\beta)$ and $\epsilon_t \equiv m_t({\rm ETC})/m_t$ where $m_t({\rm ETC})$ is contribution to the top quark mass from the extended technicolor sector. We fix $\epsilon_t = 0.5$ here. From the right panel in Fig. 1, we find several candidates of parameters to realize the $m_h = 126\,{\rm GeV}$, for example,

$$M_{\rm TSS,0} = 77\,{\rm GeV} \ \ {\rm for} \quad \tan\phi = 1 \quad , \quad \tan\beta = 0.5\,, \quad (6)$$

$$M_{\rm TSS,0} = 960\,{\rm GeV} \ {\rm for} \quad \tan\phi = 3 \quad , \quad \tan\beta = 3\,. \quad (7)$$

The case of Eq.(6) leads to a problematic mass difference among physical Higgs bosons: $m_h, m_A \simeq 126\,{\rm GeV} < m_H \simeq m_{H^\pm} \simeq 300\,{\rm GeV}$ where $A, H^\pm$ are the CP-odd and charged Higgs boson respectively. This mass difference gives a large contribution to the Peskin-Takeuchi T-parameter[15] and it is disfavored by the electroweak precision tests. On the other hand, the case Eq.(7) provides $m_h \simeq 126\,\text{GeV} < m_H \simeq m_A \simeq m_{H^\pm} \simeq 1\,\text{TeV}$ which is not constrained by the electroweak precision tests, since the degenerate heavy higgs bosons do not result in a large contribution to the T-parameter. Furthermore $R_b \equiv \Gamma(Z \to b\bar{b})/\Gamma(Z \to \text{hadrons})$ constraint also favors the case of Eq.(7) rather than Eq.(6).[1] Therefore, we conclude that the case of Eq.(7) is favorable parameter set in the present model from the viewpoint of both the electroweak precision tests and the observed new scalar boson mass.

3. 126 GeV Higgs boson in the TSSTC at the LHC

In this section, we compare the Higgs boson in the present model for Eq.(7) to the recent LHC SM Higgs boson search data. For this purpose, we vary $\epsilon_b \equiv m_b(\text{ETC})/m_b$ in a range $0.1 \le \epsilon_b \le 1$ which does not affect the previous discussions. We focus on the signal strength defined as

$$\mu_X^{\text{ggF}} \equiv \frac{\sigma_{\text{ggF}}(h_{\text{TSSTC}})}{\sigma_{\text{ggF}}(h_{\text{SM}})} \times \frac{\text{Br}(h_{\text{TSSTC}} \to X)}{\text{Br}(h_{\text{SM}} \to X)}\,, \tag{8}$$

where σ_{ggF} is the production cross section of the Higgs boson by the gluon fusion and $\text{Br}(h \to X)$ $(X = \gamma\gamma, WW^*, ZZ^*)$ is the branching ratio of the Higgs boson. Furthermore, h_{TSSTC} denotes the lightest CP-even Higgs boson in the present model and h_{SM} denotes the Higgs boson in the SM. In the present model, the yukawa coupling between the top quark and the Higgs boson becomes larger than the SM one since h_{TSS} is composed of the top quark and its vector-like partner. Thus it is possible to obtain the enhancement of the gluon fusion production process: $\sigma_{\text{ggF}}(h_{\text{TSSTC}})/\sigma_{\text{ggF}}(h_{\text{SM}}) \simeq 2$ in Eq.(8). However, we find that the hVV-coupling in the present model for Eq.(7) is much smaller than the SM Higgs boson case. This is so, since $v_2/v_{\text{EW}} = 0.1$ is satisfied for the case Eq.(7). This fact ensures that we need not take into account the production process via the vector boson fusion in Eq.(8) when we compare μ_X $(X = \gamma\gamma, WW^*, ZZ^*)$ to the LHC data. However, the signal strength of the fermionic decay mode is given by

$$\mu_{bb} \equiv \frac{\sigma_{\text{WH}}(h_{\text{TSSTC}}) + \sigma_{\text{ZH}}(h_{\text{TSSTC}})}{\sigma_{\text{WH}}(h_{\text{SM}}) + \sigma_{\text{ZH}}(h_{\text{SM}})} \times \frac{\text{Br}(h_{\text{TSSTC}} \to b\bar{b})}{\text{Br}(h^{\text{SM}} \to b\bar{b})}\,, \tag{9}$$

and we find that the suppression factor $v_2/v_{\text{EW}} = 0.1$ affects the production cross section part in the signal strength. Furthermore, we should bear in mind that the suppression factor $v_2/v_{\text{EW}} = 0.1$ also affects the decay process which includes the hVV-coupling. All signal strengths in the present model are shown in the left panel in Fig. 2 as a function of ϵ_b. For comparison, we also present the values of $\mu_{\gamma\gamma,ZZ^*,WW^*}$ from the ATLAS and CMS data.[2,3,16,17] One can see that all signal strengths are very small ($\mu_X \ll 1$) in the present model due to the above suppression factor if $\epsilon_b \simeq 0.1 - 0.6$. However, μ_{VV^*} changes drastically for $\epsilon_b \simeq 0.8$. This is because a large ϵ_b around $\epsilon_b = 1$implies that the bottom-yukawa coupling in the present model is smaller than the SM one and its smallness brings i) small $\mathrm{Br}(h_{\mathrm{TSSTC}} \to b\bar{b})$ and ii) large $\mathrm{Br}(h_{\mathrm{TSSTC}} \to VV^*)$ compared to the SM Higgs boson case.

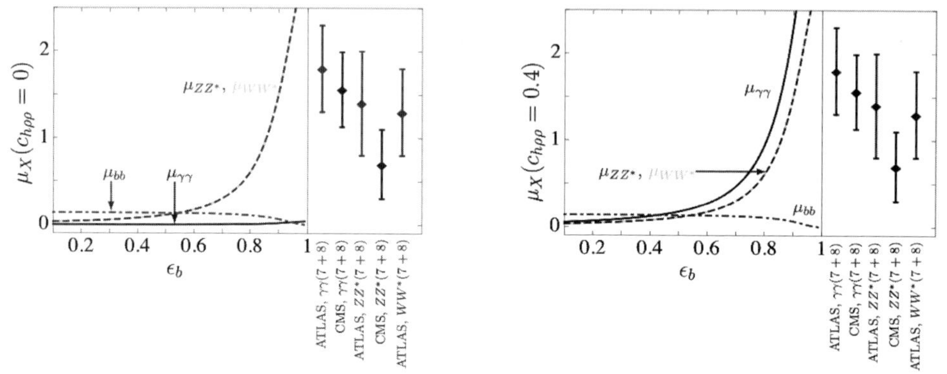

Fig. 2. The signal strength $\mu_X(X = \gamma\gamma, WW^*, ZZ^*, bb)$ as a function of ϵ_b in the present model. The blue solid, green dotted, red dashed, magenta dot-dashed curves correspond to $\mu_{\gamma\gamma}, \mu_{WW^*}, \mu_{ZZ^*}, \mu_{bb}$, respectively. In both panels, the LHC combined results for $\gamma\gamma$, ZZ^*, WW^* in[2,3,16,17] are shown together. The left panel shows the signal strength for a case with $C_{h\rho\rho} = 0$ and the right panel shows the signal strength for a case with $C_{h\rho\rho} = 0.4$ with $M_\rho = 1\,\mathrm{TeV}$

In the present model it is possible to explain the observed diphoton excess, without influencing the other decay channels, by adding a color-singlet isotriplet vector meson $\rho_\mu^{0,\pm}$. In fact, if this vector meson couples to h_{TSS} as $\mathcal{L}_{h\rho\rho} = g_{h\rho\rho} \cdot h\rho^{+\mu}\rho_\mu^-$, then the decay width $\Gamma(h_{\mathrm{TSSTC}} \to \gamma\gamma)$ gets a contribution

$$\Gamma(h_{\mathrm{TSSTC}} \to \gamma\gamma) \simeq \frac{\alpha^2 g^2}{1024\pi^3} \frac{m_h^3}{M_W^2} (-7)^2 \times \left(\cos\phi \sin(\beta - \alpha) + C_{h\rho\rho}\right)^2 , \qquad (10)$$

where $C_{h\rho\rho} \equiv [g_{h\rho\rho}/(gM_W)] \times M_W^2/M_\rho^2$. Here we impose that the isotriplet vector meson mass is much larger than m_h: $M_\rho \gg 2m_h$ and $\tan\alpha \simeq 1$, which is the mixing angle in the CP-even higgs boson sector in the present model. In Eq.(10) we present the approximate expression for the vector boson contributions to the diphoton decay channel and drop the fermion contributions since the fermion contributions are smaller than the vector boson contributions. Now, we do not specify the origin of the isotriplet vector meson and we treat $(C_{h\rho\rho}, M_\rho)$ as free parameter. In the right panel in Fig. 2, we show the signal strength μ_X for $C_{h\rho\rho} = 0.4$ and $M_\rho = 1\,\mathrm{TeV}$ as a function of ϵ_b. We find that this modification, i.e. adding $\mathcal{L}_{h\rho\rho}$, gives a large contribution to $\Gamma(h_{\mathrm{TSSTC}} \to \gamma\gamma)$ but $\mathrm{Br}(h_{\mathrm{TSSTC}} \to WW^*/ZZ^*/b\bar{b})$ are not affected by this addition as expected. Therefore it is possible to explain the large diphoton excess in the present model with keeping other decay channels be consistent with the LHC SM Higgs data.

4. Summary

In this talk we have considered a model which involves both the top-seesaw and the technicolor simultaneously. Especially, we have focused on the top-seesaw sector which provides the SM Higgs-like scalar boson by the NJL dynamics. We conclude that it may be possible to realize the 126 GeV Higgs boson in such model by sharing the top quark mass with another dynamical sector. Furthermore we have found that the Higgs boson in the top-seesaw assisted technicolor model may be consistent with the LHC data if the new isotriplet vector meson exists and couples to the Higgs boson in the top-seesaw assisted technicolor model.

References

1. H. S. Fukano and K. Tuominen, arXiv:1210.6756 [hep-ph].
2. G. Aad *et al.* [ATLAS Collaboration], Phys. Lett. B **716**, 1 (2012) [arXiv:1207.7214 [hep-ex]].
3. S. Chatrchyan *et al.* [CMS Collaboration], Phys. Lett. B **716**, 30 (2012) [arXiv:1207.7235 [hep-ex]].
4. L. Susskind, Phys. Rev. D **20**, 2619 (1979). ; S. Weinberg, Phys. Rev. D **19**, 1277 (1979).
5. B. Holdom, Phys. Lett. B **150**, 301 (1985) ; K. Yamawaki, M. Bando and K. Matumoto, Phys. Rev. Lett. **56**, 1335 (1986). ; T. Akiba and T. Yanagida, Phys. Lett. B **169**, 432 (1986). ; T. W. Appelquist, D. Karabali and L. C. R. Wijewardhana, Phys. Rev. Lett. **57**, 957 (1986).
6. S. Matsuzaki and K. Yamawaki, arXiv:1207.5911 [hep-ph].
7. D. Elander and M. Piai, arXiv:1208.0546 [hep-ph].
8. V. A. Miransky, M. Tanabashi and K. Yamawaki, Phys. Lett. B **221**, 177 (1989); ; V. A. Miransky, M. Tanabashi and K. Yamawaki, Mod. Phys. Lett. A **4**, 1043 (1989).
9. W. A. Bardeen, C. T. Hill and M. Lindner, Phys. Rev. D **41**, 1647 (1990).
10. Y. Nambu and G. Jona-Lasinio, Phys. Rev. **122**, 345 (1961).
11. S. Dimopoulos and L. Susskind, Nucl. Phys. B **155**, 237 (1979); E. Eichten and K. D. Lane, Phys. Lett. B **90**, 125 (1980).
12. H. S. Fukano and K. Tuominen, arXiv:1102.1254 [hep-ph] ; H. S. Fukano and K. Tuominen, Phys. Rev. D **85**, 095025 (2012) [arXiv:1202.6296 [hep-ph]].
13. B. A. Dobrescu, C. T. Hill, Phys. Rev. Lett. **81**, 2634-2637 (1998). [hep-ph/9712319]; R. S. Chivukula, B. A. Dobrescu, H. Georgi, C. T. Hill, Phys. Rev. **D59**, 075003 (1999). [hep-ph/9809470] .
14. H. J. He, C. T. Hill and T. M. P. Tait, Phys. Rev. D **65**, 055006 (2002) [arXiv:hep-ph/0108041].
15. M. E. Peskin, T. Takeuchi, Phys. Rev. Lett. **65**, 964-967 (1990); M. E. Peskin, T. Takeuchi, Phys. Rev. **D46**, 381-409 (1992).
16. CMS-PAS-HIG-12-015, http://cdsweb.cern.ch/record/1460419/files/HIG-12-015-pas.pdf
17. CMS-PAS-HIG-12-016, http://cdsweb.cern.ch/record/1460664/files/HIG-12-016-pas.pdf
18. J. F. Gunion, H. E. Haber, G. L. Kane and S. Dawson, Front. Phys. **80**, 1 (2000).

Minimal Dilaton Model

Tomohiro Abe*

Institute of Modern Physics and Center for High Energy Physics, Tsinghua University, Beijing 100084, China
E-mail: tomohiro_abe@tsinghua.edu.cn

Ryuichiro Kitano

Department of Physics, Tohoku University, Sendai 980-8578, Japan
E-mail: kitano@tuhep.phys.tohoku.ac.jp

Yasufumi Konishi[a] and Joe Sato[b]

Department of Physics, Saitama University, Saitama 355-8570, Japan
[a]*E-mail: konishi@krishna.th.phy.saitama-u.ac.jp*
[b]*E-mail: joe@phy.saitama-u.ac.jp*

Kin-ya Oda

Department of Physics, Kyoto University, Kyoto 606-8502, Japan
E-mail: odakin@gauge.scphys.kyoto-u.ac.jp

Shohei Sugiyama

Institute for Cosmic Ray Research (ICRR), University of Tokyo, Kashiwa, Chiba 277-8582, Japan
E-mail: shohei@icrr.u-tokyo.ac.jp

We construct a minimal calculable model of a light dilaton based on the scenario where only top and Higgs sectors are involved in a quasi scale invariant dynamics. The model consistently accommodate the electroweak precision tests even when the Higgs boson is very heavy, thereby allows one to consider a possibility that the particle at around 125 GeV, discovered at the LHC experiments, is identified as the light dilaton rather than the Higgs boson. We find that the current LHC data allow distinct parameter regions where the observed particle is either mostly the Higgs boson or the dilaton.

Keywords: Dilaton; LHC; di-photon excess.

*speaker

1. Introduction

It has been reported recently that both the ATLAS and the CMS experiments observed resonances in the $\gamma\gamma$, ZZ, and WW channels at around 125 GeV.[1,2] The resonance can naturally be interpreted as a signal of the Higgs boson in the Standard Model (SM). The mass range suggests that the Higgs sector of the SM is fairly weakly coupled, perfectly consistent with the precision measurements at the LEP experiments.

There is a logical possibility that the excess is not due to the Higgs boson but rather a Nambu-Goldstone boson associated with an approximate scale invariance in the dynamical sector. It has been proposed that the dilaton in the walking technicolor[3] may explain the signals.[4] There are also discussion for an effective theory description of dilaton systems.[5] The radion in the Randall-Sundrum model has been discussed as a possible candidate.[6,7]

It has been reported that the excess of events at around 126.5 (125) GeV at ATLAS (CMS) in the diphoton final state is enhanced from the expectation of the SM by a factor 1.8 ± 0.5 (1.6 ± 0.4).[1,2] Though this tendency is not to the level of significance, it is noteworthy that the above stated models can naturally account for the trend.

If the excesses at around 125 GeV are due to such a non-Higgs particle, one needs to consider a consistency with the constraints on the Peskin-Takeuchi S and T parameters.[8] This non-Higgs scenario requires the real Higgs boson (if exists) to be heavier than ~ 600 GeV in order to be consistent with the LHC data, and such a heavy Higgs boson does not give a good fit to the precision measurements. It is required to have some other contributions to the S and T parameters.

In this paper, we construct an effective (minimal) model of such a framework and see if there is a viable parameter region. The model consists of a massive vector-like top-partner fermion. The top-partner mass M represents a mass gap in the dynamical sector, to which the dilaton naturally couples in order to recover the scale invariance: $M \to Me^{-\phi/f}$. The coupling in turn provides interactions between the dilaton ϕ and the photons/gluons through loop diagrams of the top-partners, explaining the excesses at the ATLAS and CMS. The top-partner can also contribute to the Peskin-Takeuchi S, T parameters, which turns out to be a good direction to come back to the allowed region, canceling the heavy SM Higgs boson contribution to the T parameter.

We discuss whether such a dilaton explanation of the excesses is viable and how such a scenario can be discriminated by measuring various cross sections and decay branching ratios at the LHC experiments. The minimal model we study below catches essential features of dilaton/radion models, and the model parameters we discuss can easily be translated into those in other models.

2. The Minimal Dilaton Model

We write down the following Lagrangian for the dilaton system:

$$\begin{aligned}\mathcal{L} = \mathcal{L}_{\rm SM} - \frac{e^{-2\phi/f}}{2}\partial_\mu\phi\partial^\mu\phi - \overline{T}\left(D + Me^{-\phi/f}\right)T - \left[y'\overline{T}_R(q_{3L}\cdot H) + \text{h.c.}\right] \\ -V(\phi, H),\end{aligned} \tag{1}$$

where T is the heavy vector-like top partner representing the resonance, q_{3L} is the left-handed (top and bottom) quark doublet, and H is the SM Higgs doublet field. The Lagrangian $\mathcal{L}_{\rm SM}$ is the Standard Model, and the term with a coupling constant y' provides a mixing between the top quark and T. The Lagrangian has a non-linearly realized scale invariance except for the scalar potential term $V(\phi, H)$ which contains terms with small explicit breaking of the scale invariance. The potential terms provide mass terms for ϕ as well as a mixing between ϕ and the Higgs boson. We choose the origin of the field ϕ so that $\langle\phi\rangle = 0$. A mass term of $\overline{u}_{3R}T_L$, with u_{3R} being the right-handed top quark, can be eliminated by an appropriate field redefinition.

We can simplify Eq. (1) by a field redefinition;

$$S = fe^{-\phi/f}. \tag{2}$$

Then the Lagrangian is given as follows:

$$\mathcal{L} = \mathcal{L}_{\rm SM} - \frac{1}{2}\partial_\mu S\partial^\mu S - \overline{T}\left(D + \frac{M}{f}S\right)T - \left[y'\overline{T}(q_{3L}\cdot H) + \text{h.c.}\right] - \tilde{V}(S, H). \tag{3}$$

The scale invariance is now linearly realized. The potential $\tilde{V}(S, H)$ should be arranged so that $\langle S\rangle = f$ and $\langle H^0\rangle = v/\sqrt{2}$. We propose this Lagrangian as a minimal effective description of an approximately scale invariant theory of (dynamical) electroweak symmetry breaking involving the only top and Higgs sectors.

In general, two scalar fields have mixing terms. So the mass eigenstates of physical scalar fields, s and h, are given by

$$S = f + s\cos\theta_H - h\sin\theta_H, \tag{4}$$

$$H = \frac{1}{\sqrt{2}}\begin{pmatrix} 0 \\ v + s\sin\theta_H + h\cos\theta_H\end{pmatrix}, \tag{5}$$

where θ_H is the mixing angle between the dilaton and the Higgs. Hereafter we assume the mass of s is about 125 GeV and h is heavier than it.

If the mixing between the dilaton and the Higgs boson is small and if the dilaton is the one which explains the observed resonance at the LHC, the Higgs boson must be heavier than about 600 GeV in order to be consistent with the Higgs boson searches at the LHC. With such a heavy Higgs boson, constraints from the Peskin-Takeuchi S, T parameters require a new contribution to come back to the ellipse in the S-T plane. As we will see shortly, such a contribution is already there in this model since the loop diagrams of the t' field can push S and T parameters towards the right direction. This Lagrangian therefore provides a compact realistic model of the light dilaton, which can be used for LHC studies.

3. Constraints from precision measurements

We calculate S and T parameters at one-loop level.

$$S \simeq \sin^2\theta_L \frac{N_c}{6\pi}\left(\frac{1}{3} - \cos^2\theta_L\right)\ln\frac{m_t^2}{m_{t'}^2} - \frac{\cos^2\theta_H}{12\pi}\ln\frac{m_{\rm href}^2}{m_h^2} - \frac{\sin^2\theta_H}{12\pi}\ln\frac{m_{\rm href}^2}{m_s^2}, \quad (6)$$

$$T \simeq \sin^4\theta_L \frac{N_c}{16\pi}\frac{1}{s_W^2 c_W^2}\frac{m_{t'}^2}{m_Z^2} + \frac{3\cos^2\theta_H}{16\pi c_W^2}\ln\frac{m_{\rm href}^2}{m_h^2} + \frac{3\sin^2\theta_H}{16\pi c_W^2}\ln\frac{m_{\rm href}^2}{m_s^2}, \quad (7)$$

where $m_{\rm href}$ is the reference Higgs boson mass, and θ_L is a mixing angle in the top sector. We find t and t' loop give positive contributions to T parameter. This is a good feature because now we have two scalars at 125 GeV and 600 GeV, and have to compensate the contributions from the 600 GeV particle by something. This is achieved by t and t' loop contributions.

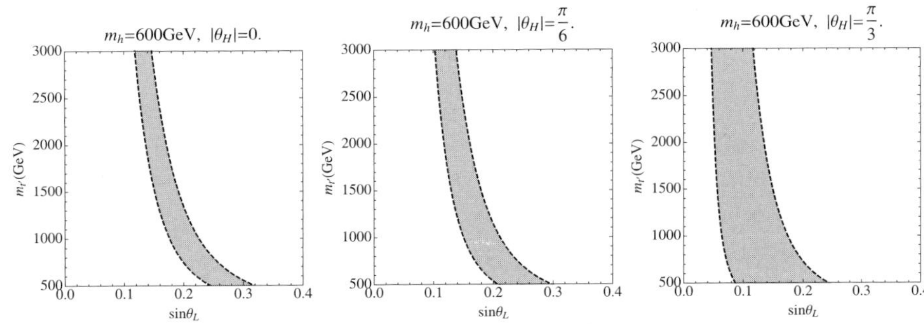

Fig. 1. Favored region plot from the Peskin-Takeuchi S, T. White regions are excluded at 95% CL.

4. Dilaton at the LHC

We introduce the following variable:

$$R_X = \frac{\sigma_X}{\sigma_X^{SM}}. \quad (8)$$

This is nothing but the ratio of the production cross section to the SM one. We calculate three production processes, gluon fusion process ($R_{\rm GF}$), vector boson fusion process ($R_{\rm GF}$), and vector boson associated production process ($R_{\rm VH}$). In this model, they are given as follows:

$$R_{\rm GF} \simeq (\eta\cos\theta_H + \sin\theta_H)^2, \quad (9)$$

$$R_{\rm VBF} = \sin^2\theta_H, \quad (10)$$

$$R_{\rm VH} = \sin^2\theta_H, \quad (11)$$

where

$$\eta = \frac{v}{f}. \quad (12)$$

We also introduce the following variable which is the ratio of partial decay width to the SM one.

$$R(s \to X) = \frac{\Gamma(s \to X)}{\Gamma(h \to X)^{\rm SM}} \tag{13}$$

Main decay processes are given as follows:

$$R(s \to ff) = \sin^2\theta_H, \tag{14}$$

$$R(s \to WW) = \sin^2\theta_H, \tag{15}$$

$$R(s \to ZZ) = \sin^2\theta_H, \tag{16}$$

$$R(s \to gg) \simeq (\eta\cos\theta_H + \sin\theta_H)^2, \tag{17}$$

$$R(s \to \gamma\gamma) \simeq \left(\eta\frac{A_{t'}}{A_{\rm SM}}\cos\theta_H + \sin\theta_H\right)^2, \tag{18}$$

where

$$A_{t'} \simeq \frac{16}{9}, \quad A_{\rm SM} \simeq -6.5. \tag{19}$$

In Fig. 2, we show $\mu(gg \to s \to \gamma\gamma)$ (the left panel) and $\mu(gg \to s \to WW/ZZ/ff)$ (the right panel). In $\theta_H \sim 0$ region, we find $\mu(gg \to s \to \gamma\gamma) > 1$. On the other hand, other processes are suppressed in this region. So there is a tension between di-photon excess and consistency of the other processes with current LHC data.

We combine signal strength and study the consistency with the LHC data. The results are shown in Fig. 3. We see that both experiments have allowed dilaton-like region $|\theta_H| < \pi/4$ within 90% confidence interval though the ATLAS disfavors the purely dilatonic case $\theta_H \simeq 0$ outside the 95% confidence interval.

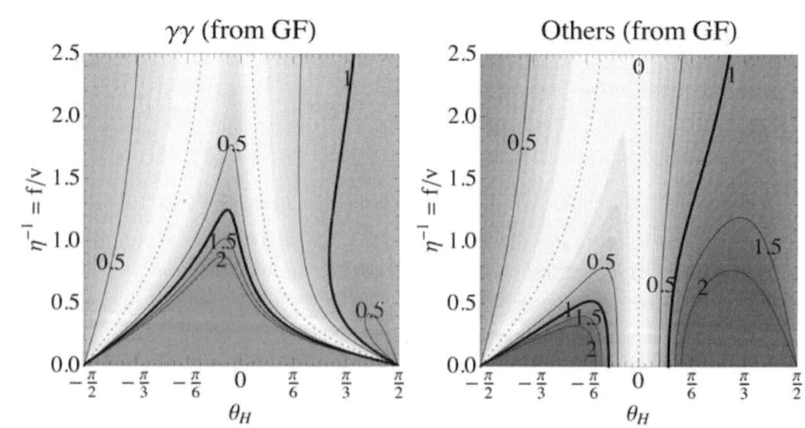

Fig. 2. Signal strength of s produced via gluon fusion. The left panel shows $\mu(gg \to s \to \gamma\gamma)$, and the right panel shows $\mu(gg \to s \to WW/ZZ/ff)$.

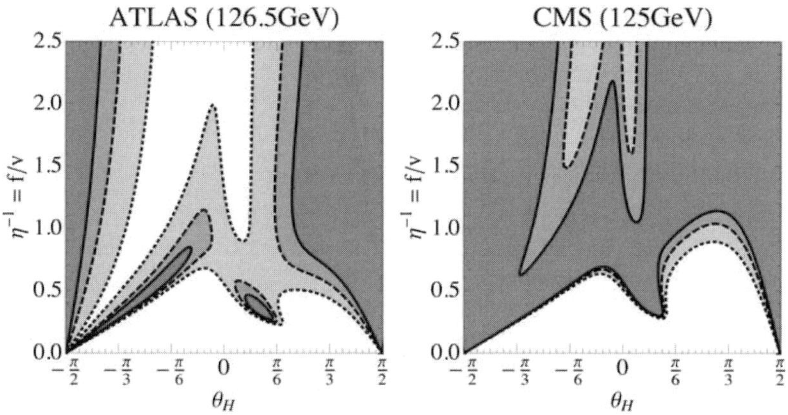

Fig. 3. Favored regions within 90, 95 and 99% confidence intervals, enclosed by solid, dashed, and dotted lines, respectively. Density (area) of favored region decreases (increases) in according order. Results are shown for ATLAS (left), CMS (center), and combined (right). See text for details.

5. Summary

We have considered a possibility that the Higgs-like excesses observed at the LHC experiments are actually the signals of the dilaton associated with spontaneously broken scale invariance. We have constructed a minimal model of the dilaton which can be produced through the gluon fusion process at the LHC, and can decay into two photons. The effective coupling is obtained through the loop diagrams of a new vector-like state T that has the same gauge charges as the right-handed top quark. The T field contributes to the Peskin-Takeuchi S, T parameters in the electroweak precision tests. This contribution allows one to push the Higgs boson mass above the experimental constraint, 600 GeV, providing a consistent framework for the light dilaton plus a heavy Higgs scenario.

We find that the current experimental data allows distinct parameter regions where the excesses are either Higgs-like or dilaton-like. Once the excesses are confirmed with more statistics, it is possible to distinguish two scenarios.

References

1. G. Aad *et al.* [ATLAS Collaboration], *"Observation of a new particle in the search for the Standard Model Higgs boson with the ATLAS detector at the LHC,"* Phys. Lett. B [arXiv:1207.7214 [hep-ex]].
2. S. Chatrchyan *et al.* [CMS Collaboration], *"Observation of a new boson at a mass of 125 GeV with the CMS experiment at the LHC,"* Phys. Lett. B [arXiv:1207.7235 [hep-ex]].
3. K. Yamawaki, M. Bando and K. -i. Matumoto, *"Scale Invariant Technicolor Model and a Technidilaton,"* Phys. Rev. Lett. **56**, 1335 (1986);
   M. Bando, K. -i. Matumoto and K. Yamawaki, *"Technidilaton,"* Phys. Lett. B **178**, 308 (1986).

4. S. Matsuzaki and K. Yamawaki, *"Techni-dilaton signatures at LHC,"* Prog. Theor. Phys. **127** (2012) 209 [arXiv:1109.5448 [hep-ph]];
S. Matsuzaki and K. Yamawaki, *"Techni-dilaton at 125 GeV,"* Phys. Rev. D **85** (2012) 095020 [arXiv:1201.4722 [hep-ph]];
S. Matsuzaki and K. Yamawaki, *"Discovering 125 GeV techni-dilaton at LHC,"* arXiv:1206.6703 [hep-ph];
S. Matsuzaki and K. Yamawaki, *"Is 125 GeV techni-dilaton found at LHC?,"* arXiv:1207.5911 [hep-ph];
D. Elander and M. Piai, *"The decay constant of the holographic techni-dilaton and the 125 GeV boson,"* arXiv:1208.0546 [hep-ph];
S. Matsuzaki and K. Yamawaki, *"Holographic techni-dilaton at 125 GeV,"* arXiv:1209.2017 [hep-ph].
5. W. D. Goldberger, B. Grinstein and W. Skiba, *"Distinguishing the Higgs boson from the dilaton at the Large Hadron Collider,"* Phys. Rev. Lett. **100** (2008) 111802 [arXiv:0708.1463 [hep-ph]];
J. Fan, W. D. Goldberger, A. Ross and W. Skiba, *"Standard Model couplings and collider signatures of a light scalar,"* Phys. Rev. D **79** (2009) 035017 [arXiv:0803.2040 [hep-ph]].
6. K. Cheung and T. -C. Yuan, *"Could the excess seen at 124-126 GeV be due to the Randall-Sundrum Radion?,"* Phys. Rev. Lett. **108**, 141602 (2012) [arXiv:1112.4146 [hep-ph]].
7. B. Grzadkowski, J. F. Gunion and M. Toharia, *"Higgs-Radion interpretation of the LHC data?,"* Phys. Lett. B **712** (2012) 70 [arXiv:1202.5017 [hep-ph]];
Y. Tang, *"Comment on 'Could the Excess Seen at 124-126 GeV Be due to the Randall-Sundrum Radion?',"* arXiv:1204.6145 [hep-ph];
H. Kubota and M. Nojiri, *"Radion-higgs mixing state at the LHC with the KK contributions to the production and decay,"* arXiv:1207.0621 [hep-ph].
8. M. E. Peskin and T. Takeuchi, *"Estimation of oblique electroweak corrections,"* Phys. Rev. D **46** (1992) 381;
M. E. Peskin and T. Takeuchi, *"A New constraint on a strongly interacting Higgs sector,"* Phys. Rev. Lett. **65** (1990) 964.

Partially Composite Higgs in Supersymmetry*

Ryuichiro Kitano

Department of Physics, Tohoku University, Sendai 980-8578, Japan
E-mail: kitano@tuhep.phys.tohoku.ac.jp

In light of the discovery of the new particle at 125 GeV and the strong lower limits on the masses of superparticles from LHC, we discuss a possible picture of weak scale supersymmetry.

1. Introduction

The searches for superparticles at the LHC have been putting severe constraints on weak scale supersymmetry. The lower bound on the gluino mass from searches for jets with missing energy is roughly,[1,2]

$$m_{\tilde{g}} \gtrsim 1 - 1.5 \text{ TeV}. \tag{1}$$

A similar constraint applies for squarks in the first generation. The constraints on the masses of the scalar top and bottom quarks (stop and sbottom) are slightly weaker such as

$$m_{\tilde{t},\tilde{b}} \gtrsim 500 - 600 \text{ GeV}, \tag{2}$$

for a non-degenerate neutralino.[3,4]

The new particle at 125 GeV[5,6] is also giving an important constraint on SUSY models. If the particle is interpreted as the lightest Higgs boson in the minimal supersymmetric standard model (MSSM), the stop mass is required to be

$$m_{\tilde{t}} \gtrsim (\text{a few}) \times \text{ TeV}. \tag{3}$$

In light of these constraints, we discuss what kind of theoretical framework is implied by the LHC data if SUSY is the solution to the naturalness problem.

2. The Log Problem

In softly broken SUSY models, the quadratic divergence is canceled, and that is supposed to be the solution to the hierarchy problem. As one can see below, that is not quite enough any more. We are facing the situation where one also needs to eliminate the log divergence.

*Based on collaborations with Markus Luty and Yuichiro Nakai.

There is a logarithmic divergence in the quantum corrections to mass parameters in supersymmetric standard model. Once we put a large gluino mass, $m_{\tilde{g}}$, at a high scale M, the stop mass squared and the quadratic term in the Higgs potential (which is the Higgs mass squared times $(-1/2)$), receives quantum corrections proportional to $m_{\tilde{g}}^2 \log(M/m_{\tilde{g}})$. If the log factor is large and if there is no significant cancellation among parameters, we naively expect

$$m_h^2 \sim m_{\tilde{t}}^2 \sim m_{\tilde{g}}^2. \tag{4}$$

This is clearly inconsistent with $m_h \sim 125$ GeV and $m_{\tilde{g}} \gtrsim 1$ TeV. This is the SUSY fine-tuning problem. Especially, in models which are friendly with the grand unification, such as gravity mediation, the log factor tends to be large. In the MSSM where the Higgs mass is obtained by the gauge coupling times the VEV of the Higgs field at tree level, the above relation either predicts an unacceptably large Higgs VEV or light gluino/stop that is no longer allowed by the LHC data.

In order to correct the wrong prediction in Eq. (4), the logarithmic quantum correction should be cut-off at a scale close to the TeV scale. In that case, the size of the quantum corrections are estimated to be

$$\delta m_h^2 \sim \frac{y_t^2 N_c}{8\pi^2} m_{\tilde{t}}^2 \sim (0.2)^2 m_{\tilde{t}}^2 \tag{5}$$

$$\delta m_{\tilde{t}}^2 \sim \frac{g_3^2}{4\pi^2} \frac{N_c^2 - 1}{N_c} m_{\tilde{g}}^2 \sim (0.3)^2 m_{\tilde{g}}^2. \tag{6}$$

Therefore, a mild hierarchy such as

$$m_h \sim 0.2 m_{\tilde{t}}, \quad m_{\tilde{t}} \sim 0.3 m_{\tilde{g}}, \tag{7}$$

is possible without fine-tuning. For example,

$$m_{\tilde{g}} \sim 2 \text{ TeV}, \quad m_{\tilde{t}} \sim 600 \text{ GeV}, \quad m_h \sim 120 \text{ GeV}, \tag{8}$$

is realized naturally. The spectrum is consistent with the Higgs mass and barely satisfies the LHC searches for superparticles. At this stage, one needs to give up the MSSM since the spectrum is inconsistent with Eq. (3).

Therefore, in light of various news from LHC, a theoretical framework to realize natural SUSY requires:

- the cut-off of the quantum correction at TeV, and
- a new contribution to the Higgs boson mass beyond the stop loop.

Both of the requirements actually say that we should have a new physics beyond the MSSM at the TeV energy scale.

3. Naive Dimensional Analysis and Partially Composite Higgs

One of the most drastic ideas for the TeV new physics beyond the MSSM is the scenario where the Higgs boson is (partially) a composite particle. A TeV scale strong dynamics is assumed to provide an MSSM like theory as the low energy effective theory. Because of the drastic change of the description at the TeV scale, one may expect a cut-off of the log divergence and a new contribution to the Higgs mass, simultaneously.

In fact, modern efforts of SUSY model building have started with this type of models.[7–9] SUSY is introduced to protect the quadratic divergence of the Higgs potential whereas the electroweak scale is generated by some dynamics at TeV as in the technicolor model. Such models are now well motivated to be considered again given that there is a light Higgs boson which is not quite as light as the MSSM predictions.

Once we assume that there is some dynamics at a scale Λ responsible for the electroweak symmetry breaking, the naive dimensional analysis[10,11] says

$$\Lambda \sim 4\pi v \sim 3 \text{ TeV}, \quad m_h \sim \Lambda. \tag{9}$$

where $v = 246$ GeV. This is the typical prediction of the technicolor models. Now, for having a light Higgs boson, one can assume that the Higgs fields are slightly weakly coupled to the dynamics at the scale Λ. By introducing a dimensionless parameter ϵ which measures the weakness of the coupling compared to the naive estimates, we have

$$\Lambda \sim 4\pi\epsilon v, \quad m_h \sim \epsilon\Lambda \sim m_H, \tag{10}$$

where m_H is the size of the soft SUSY breaking contribution to the Higgs potential. In the MSSM language, $\epsilon\Lambda$ is the μ-term and m_H is the soft SUSY breaking mass. From Eq. (10), we have

$$\frac{m_h}{v} \sim 4\pi\epsilon^2. \tag{11}$$

By putting $m_h = 125$ GeV, we obtain $\epsilon \sim 0.2$ and $\Lambda \sim 600$ GeV. If the top quark is also involved in the dynamics, the estimate of the dynamically generated top quark mass is

$$m_t \sim \epsilon_t^2 \Lambda, \tag{12}$$

where ϵ_t again measures the weakness of the coupling to the strong sector. For $m_t \sim 170$ GeV, we have

$$\epsilon_t \sim 0.5. \tag{13}$$

From these estimates, we have a rough picture:

- there is some dynamics at $\Lambda \sim 600$ GeV,
- the Higgs fields (and possibly the top quark) are weakly coupled to the dynamics with a suppression factor $\epsilon \sim 0.2$ (and $\epsilon_t \sim 0.5$), and
- above the scale Λ, the picture drastically changes so that the log divergences of the soft SUSY breaking parameters are cut-off.

The Higgs quartic coupling and the Higgsino mass are both dominated by the dynamically generated potential. This is very different from the MSSM case where the Higgs potential is mainly from the $SU(2)_L \times U(1)_Y$ gauge interactions (D-term) and the μ-term is added by hand.

The parameters ϵ and ϵ_t can be thought of as the degree of compositeness. The relations among v, m_h, and m_t suggest that the Higgs/top sector is not fully composite, but maybe partially composite.[12] The partially composite scenario is realized only when there is some reason for the absence of the tree-level Higgs potential of $O(\Lambda)$. SUSY is giving a good reason for it since one can naturally have an approximately flat potential even at the quantum level. The (approximately) massless composite particle is ubiquitous in SUSY gauge theories.

4. A Model for Dynamical Electroweak Symmetry Breaking

A pretty simple model of partially composite Higgs can be constructed.[13,14] The dynamical sector is an $\mathcal{N} = 1$ SUSY $U(2)$ gauge theory with five flavors of chiral superfields. The assignment of the quantum numbers of the standard model gauge group, $(SU(3), SU(2))_Y$, for the matter fields is

$$f_A : (1,2)_0, \quad \overline{f}_A : (1,1)_{\pm 1/2}, \quad (A = 1,2), \tag{14}$$

$$f'_\alpha : (3,1)_{1/6}, \quad \overline{f}'_\alpha : (\overline{3},1)_{-1/6}, \quad (\alpha = 3,4,5). \tag{15}$$

This theory is in the conformal window.[15] The Higgs field, $H = (H_u, H_d)$, and the top quark, q, t^c, can couple to this CFT through the superpotential terms:

$$W = \lambda_H \overline{f} H f + \lambda_q \overline{f}' q f + \lambda_t \overline{f} t^c f'. \tag{16}$$

The gauge coupling g and the coupling constants λ_H, λ_q, and λ_t flow to IR fixed points whose values are estimated by the a-maximization[16] such as:

$$\frac{g}{4\pi} \sim 0.4, \quad \frac{\lambda_{H_d}}{4\pi} \sim 0.1, \quad \frac{\lambda_{H_u}}{4\pi} \sim 0.3, \tag{17}$$

$$\frac{\lambda_q}{4\pi} \sim 0.3, \quad \frac{\lambda_t}{4\pi} \sim 0.3. \tag{18}$$

By adding a mass to f' and $\overline{f}'$ by superpotential:

$$W = \Lambda \overline{f}' f', \tag{19}$$

the $SU(2)$ gauge interaction becomes strong and confines at the scale Λ. The coupling constants λ's receive multiplicative renormalization, $\lambda \to \lambda k$ with $k > 1$. By turning on SUSY breaking terms by

$$\Lambda \to \Lambda(1 + m_{\rm SUSY}^2 \theta^2), \tag{20}$$

the Higgs potential are generated. For $\Lambda \sim m_{\rm SUSY}$, the lightest Higgs field (which is mainly H_d) obtain a mass:

$$m_h^2 \sim \left(\frac{\lambda_{H_d}}{4\pi}\right)^2 \Lambda^2 \sim m_h^2|_{\rm MSSM} + \frac{\lambda_{H_d}^4 v^2}{(4\pi)^2}, \tag{21}$$

and the Higgsino mass is also generated as $m_{\tilde{H}} \sim (\lambda_{H_u}\lambda_{H_d}/(4\pi)^2)\Lambda$. The dynamically generated top quark mass is

$$m_t \sim \frac{\lambda_q \lambda_t \lambda_{H_d}}{(4\pi)^2} v. \tag{22}$$

The correspondence to discussion in the previous section is

$$\epsilon \sim \frac{\lambda_{H_d}}{4\pi}, \quad \epsilon_t^2 \sim \frac{\lambda_q \lambda_t}{(4\pi)^2}. \tag{23}$$

We see that the required value $\epsilon \sim 0.2$ and $\epsilon_t^2 \sim (0.5)^2$ is roughly consistent with the fixed point values with the k factor ~ 2.

This model should be thought of as the effective theory around the scale Λ. In order to cut-off the log divergence proportional to the gluino mass squared, the semi-weakly coupled description above should be replaced by a more strongly coupled one. For example, one can assume that the gauge symmetry is extended to $SU(3)$ above the scale Λ, which makes the theory a strongly coupled CFT. In this case, the stop mass parameter is also strongly renormalized such as

$$\frac{d}{d\log\mu} m_{\tilde{t}}^2 = c m_{\tilde{t}}^2 - \frac{g_3^2}{(4\pi)^2} \frac{N_c^2 - 1}{N_c} m_{\tilde{g}}^2 + \cdots, \tag{24}$$

where c is of $O(1)$. The quasi IR fixed point of the stop mass is then roughly

$$m_{\tilde{t}}^2 \sim \frac{g_3^2}{(4\pi)^2 c} \frac{N_c^2 - 1}{N_c} m_{\tilde{g}}^2, \tag{25}$$

which is equivalent to estimate without the log divergence. The same is true for the effect of the stop mass to the Higgs mass parameters. One can consider a more drastic change of the theory such as the appearance of an extra dimension at a scale Λ. See Ref.[14] for a proposal to obtain composite Higgs fields couple to the $U(2)$ model above from a higher dimensional QCD.

5. Why SUSY?

We see that there is a simple solution to the SUSY fine-tuning problem once we cut-off the theory at the TeV scale. But once we give up the SUSY desert, why do we need SUSY?

Aside from the theoretical beauty or calculability, we should remember that the presence of the light Higgs boson already provides a good motivation. The naive dimensional analysis says that relation between the decay constant v and the mass of the 0^+ resonance, m_h, is estimated to be

$$m_h \sim 4\pi v \sim 3 \text{ TeV}, \tag{26}$$

which is quite different from the situation in the electroweak sector. On the other hand, the MSSM provides a relation

$$m_h \sim m_Z \ll 4\pi v. \tag{27}$$

Although this turns out to be too small, SUSY provides an excellent starting point to explain 125 GeV; $m_Z \sim 91$ GeV rather than 3 TeV.

6. Lessons from QCD

The QCD is a natural place to look for a hint for the theory of electroweak symmetry breaking. Other than the electroweak, that is the only known example of the spontaneous symmetry breaking, actually happening in the vacuum. Indeed, the patterns of the symmetry breaking in the chiral symmetry breaking and the electroweak symmetry breaking are identical; $SU(2)_L \times U(1)_Y \to U(1)_{\rm EM}$. The only big difference is the mass of 0^+ resonances, *i.e.*, the Higgs boson mass relative to the order parameter v or f_π.

When we look up the table in the PDG,[17] we actually find relatively light scalar mesons:

$$f_0(600), \quad f_0(980), \tag{28}$$

in which $f_0(600)$ is a pretty broad resonance. These are the candidates of the Higgs boson counterparts in QCD. There are also light and narrow vector resonances

$$\rho(770), \quad \omega(782). \tag{29}$$

Of course, there are pions, π^0 and $\pi^\pm$ associated with the chiral symmetry breaking, corresponding to the longitudinal modes of Z^0 and $W^\pm$ in the electroweak symmetry breaking.

One can construct an effective theory to describe the interactions among these light hadrons as the Higgs model,[18,19] which is the (linearly realized) Hidden Local Symmetry.[20] The model is based on a $U(2)$ gauge theory where we introduce Higgs fields:

$$H_L^{iA}, \quad H_R^{iA}, \quad (i, A = 1, 2). \tag{30}$$

Both are 2×2 matrices. One of the indices is for $U(2)$ gauge group and the other is for flavor symmetry, $SU(2)_L \times SU(2)_R$. The H_L and H_R fields transform under $SU(2)_L$ and $SU(2)_R$, respectively. By requiring the flavor symmetry in the Lagrangian and setting up the potential so that the minimum is at

$$H_L = H_R \propto \mathbf{1}, \tag{31}$$

the chiral symmetry is spontaneously broken down to a diagonal group, and provides massless pions. The radial direction corresponds to the Higgs boson, $f_0(980)$ (or $f_0(600)$), and the $U(2)$ gauge fields obtain masses which correspond to $\rho(770)$ and $\omega(782)$.

This simple linear sigma model contains a pretty interesting object, a string associated with the gauge symmetry breaking. By setting the Lagrangian parameters so that the masses and the couplings of hadrons are reproduced, one can estimate the tension of the string as:[19]

$$\sigma \sim (400 \text{ MeV})^2. \tag{32}$$

In QCD, there is a natural object to be identified with this string. It is the QCD string which stretches between color sources to cause confinement. The tension of the QCD string is estimated by the lattice simulations and also from the quarkonium spectrum, and both give approximately $\sigma \sim (430 \text{ MeV})^2$, which is very consistent with Eq. (32).

If the string in the linear sigma model corresponds to the QCD string, there appears an interesting interpretation of the Hidden Local Symmetry. The QCD string is supposed to carry the color flux whereas the string in the sigma model is carrying the magnetic flux of the ρ and ω mesons. The identification then says that the ρ and ω mesons are the magnetic gauge bosons in QCD. Since the Higgs fields H_L and H_R are charged under $U(2)$, they are (non-abelian) magnetic monopoles. The Higgs phenomena of the magnetic gauge theory, that is also describing the chiral symmetry breaking, is now identified as the color confinement in QCD.

There is a theoretical support for this interpretation. It has been demonstrated in Ref.,[18] that by making a deformation of a SUSY QCD model, one can see that the Hidden Local Symmetry appears as the magnetic theory via the Seiberg duality. See also earlier discussion[21–23] on the interpretation of the ρ meson as the magnetic gauge boson.

The Seiberg dual picture of QCD is a $U(2)$ gauge theory with five flavors, that is in fact identical to the model discussed in Section 4 for the model of dynamical electroweak symmetry breaking. The Higgs fields introduced above, H_L and H_R correspond to f and $\overline{f}$, respectively, in the model of Section 4 which arise as dual quarks transforming under both the magnetic gauge group and the flavor group. By adding soft SUSY breaking terms, the dual quarks obtain VEVs and the magnetic gauge group is Higgsed while the chiral symmetry breaking happens. This Higgs mechanism in the magnetic picture corresponds to the confinement.

Interestingly, a model of the low energy QCD can be regarded as the magnetic picture, and the magnetic picture has the same structure as the model of partially composite Higgs for the electroweak symmetry breaking. From this view, physics behind the electroweak symmetry breaking may actually be very similar to (or the same as) QCD, and the quantitative difference, the mass of the Higgs boson, may originate simply from whether SUSY is a good (approximate) symmetry at the dynamical scale.

7. Predictions

There are two kinds of predictions in the framework we discussed. One is the SUSY spectrum in Eq. (8). Since these are the estimates from (probably unavoidable) finite quantum corrections, the naturalness suggests that the estimates roughly give the upper bounds. The light stop should be observed at the LHC soon.

Another interesting prediction is the presence of the relatively light resonances from the strong sector. As we have discussed, the light Higgs fields actually suggest a low dynamical scale such as $\Lambda \sim 600$ GeV, from the naive dimensional analysis. Since the dynamics should have $SU(2)_L \times U(1)_Y$ as a part of the global symmetry, one expects vector resonances, W' and Z', associated with the current operators of the global symmetry. These are the ρ meson counterparts of QCD. In the explicit model of Section 4, they corresponds to $U(2)$ gauge bosons. Finding such resonances will be very important for the understanding of the physics of electroweak symmetry breaking.

8. Summary

The limits on SUSY particles from LHC and the 125 GeV new particle may be suggesting a very interesting possibility, a dynamical electroweak symmetry breaking much below $4\pi v \sim 3$ TeV. In such a case, we will see many resonances and also superparticles at the LHC, opening up a new era of particle physics to look for the theory behind the electroweak symmetry breaking, just as it has been done to look for QCD since the days of the discovery of the pions.

Acknowledgments

I would like to thank organizers of SCGT12. This work is supported in part by the Grant-in-Aid for Scientific Research 23740165 of JSPS.

References

1. The ATLAS Collaboration, ATLAS-CONF-2012-109.
2. S. Chatrchyan *et al.* [CMS Collaboration], Phys. Rev. Lett. **109**, 171803 (2012) [arXiv:1207.1898 [hep-ex]].
3. The ATLAS Collaboration, ATLAS-CONF-2013-001, ATLAS-CONF-2012-165.
4. The CMS Collaboration, CMS PAS SUS-12-023, CMS PAS SUS-12-028.
5. G. Aad *et al.* [ATLAS Collaboration], Phys. Lett. B **716**, 1 (2012) [arXiv:1207.7214 [hep-ex]].
6. S. Chatrchyan *et al.* [CMS Collaboration], Phys. Lett. B **716**, 30 (2012) [arXiv:1207.7235 [hep-ex]].
7. E. Witten, Nucl. Phys. B **188**, 513 (1981).
8. M. Dine, W. Fischler and M. Srednicki, Nucl. Phys. B **189**, 575 (1981).
9. S. Dimopoulos and S. Raby, Nucl. Phys. B **192**, 353 (1981).
10. A. Manohar and H. Georgi, Nucl. Phys. B **234**, 189 (1984).
11. M. A. Luty, Phys. Rev. D **57**, 1531 (1998) [hep-ph/9706235].
12. R. Kitano, M. A. Luty and Y. Nakai, JHEP **1208**, 111 (2012) [arXiv:1206.4053 [hep-ph]].
13. H. Fukushima, R. Kitano and M. Yamaguchi, JHEP **1101**, 111 (2011) [arXiv:1012.5394 [hep-ph]].
14. R. Kitano and Y. Nakai, arXiv:1212.2726 [hep-ph].

15. N. Seiberg, Nucl. Phys. B **435**, 129 (1995) [hep-th/9411149].
16. K. A. Intriligator and B. Wecht, Nucl. Phys. B **667**, 183 (2003) [hep-th/0304128].
17. J. Beringer *et al.* [Particle Data Group Collaboration], Phys. Rev. D **86**, 010001 (2012).
18. R. Kitano, JHEP **1111**, 124 (2011) [arXiv:1109.6158 [hep-th]].
19. R. Kitano, M. Nakamura and N. Yokoi, Phys. Rev. D **86**, 014510 (2012) [arXiv:1202.3260 [hep-ph]].
20. M. Bando, T. Kugo, S. Uehara, K. Yamawaki and T. Yanagida, Phys. Rev. Lett. **54**, 1215 (1985).
21. N. Seiberg, Int. J. Mod. Phys. A **16**, 4365 (2001) [hep-th/9506077].
22. M. Harada and K. Yamawaki, Phys. Rev. Lett. **83**, 3374 (1999) [hep-ph/9906445].
23. Z. Komargodski, JHEP **1102**, 019 (2011) [arXiv:1010.4105 [hep-th]].

Metastable D-term Dynamical SUSY Breaking

Nobuhito Maru

Department of Physics and Research and Education Center for Natural Sciences, Keio University, Yokohama, 223-8521, Japan

We present the mechanism of the dynamical supersymmetry (SUSY) breaking at the metastable vacuum in the $\mathcal{N} = 1$ $U(N)$ SUSY gauge theory with adjoint superfields. The dynamical SUSY breaking is triggered by the non-vanishing D-term coupled to the observable sector, and is realized by the self-consistent Hartree-Fock approximation of the NJL type while it eventually brings us the non-vanishing F-term as well. We numerically check the local stability of our metastable vacuum.

Keywords: Dynamical SUSY Breaking, D-term, Self-consistent Hartree-Fock approximation

1. Introduction

Spontaneous breaking of SUSY occurs much less frequent compared with that of internal symmetry in quantum field theory and has attracted much interest[1] of theorists for over the three decades. Mass hierarchy in elementary particle physics indicates that it is most desirable to break $\mathcal{N} = 1$ SUSY dynamically. In fact, the non-renormalization theorem[2] protects the generation of holomorphic operator in perturbation theory and instanton generated nonperturbative superpotentials have been the major source of dynamical SUSY breaking (DSB).

We focus our attention on general $\mathcal{N} = 1$ theory in four dimensions consisting of vector superfields and chiral superfields in the adjoint representation with a non-canonical gauge kinetic function. It has recently been shown in refs.[3] that, in this general situation, SUSY is dynamically broken in the metastable vacuum. The mechanism that triggers the DSB is the condensate of the Dirac bilinear, forcing one of the order parameters D of SUSY to be non-vanishing. This is very much reminiscent of the Nambu-Jona Lasinio (NJL) theory[4] of broken chiral symmetry and hence the BCS superconductivity,[5] being formulated in terms of the effective action of the auxiliary field whose stationary value is the order parameter. The method of approximation employed is the self-consistent Hartree-Fock (HF) approximation where the tree and the one-loop contributions are regarded as comparable. Once this mechanism operates, non-vanishing F-term is shown to be induced and contributes, for instance, to the mass of the fermions. The mechanism requires massive adjoint scalars, in particular, the scalar gluons and, together with the feature that the D term triggers the breaking, is quite distinct from the previous proposals[6,7] of DSB both from theoretical and experimental perspectives. The overall $U(1)$ where the non-vanishing D and the Nambu-Goldstone fermion (NGF) reside serves as the hidden sector and no messenger field is necessary[3] as non-vanishing third prepotential derivatives connect the $U(1)$ sector with the observable $SU(N)$ sector.[8]

In the next section, we review the original reasoning that has led us to the D-term triggered DSB. We set up the background field formalism to be used in the subsequent sections, separating the three kinds of background from the fluctuations. In section three, we elaborate upon our treatment of the effective potential with the three kinds of background fields as well as the point of the HF approximation in refs. 3. In section four, we present a qualitative argument of our variational analyses of the effective potential. Treating F-term as an induced perturbation, we demonstrate that the stationary values $(D_*, \varphi_*, \overline{\varphi_*})$ are determined by the intersection of the two real curves. Numerical analysis is provided that demonstrates the existence of such solution as well as the self-consistency of our analysis. The second variation of the scalar potential is computed and the local stability of the vacuum is shown from the numerical data. Summary is given in the last section.

2. The action, assumptions and some properties

The lagrangian we consider is

$$\mathcal{L} = \int d^4\theta K(\Phi^a, \overline{\Phi}^a) + (gauging) + \int d^2\theta \mathrm{Im}\frac{1}{2}\tau_{ab}(\Phi^a)\mathcal{W}^{\alpha a}\mathcal{W}^b_\alpha + \left(\int d^2\theta W(\Phi^a) + c.c.\right). \tag{1}$$

At the lowest order in perturbation theory, there is no source which gives a vacuum expectation value (vev) to the auxiliary field D^0: $\langle D^0\rangle_{\mathrm{tree}} = 0$. The $U(N)$ gaugino is massless at the tree level while the fermionic partner of the scalar gluon receives the tree level mass $m_a = m_0 = \langle g^{00}\partial_0\partial_0 W\rangle_{\mathrm{tree}}$.

It is useful to summarize here a set of assumptions.

(1) a general $\mathcal{N} = 1$ SUSY action of chiral superfield Φ^a in the adjoint representation and the vector superfield V^a with the Kähler potential $K(\Phi^a, \overline{\Phi}^a)$ with its gauging, the gauge kinetic superfield $\tau_{ab}(\Phi^a)$ following from the second derivatives of a generic holomorphic function $\mathcal{F}(\Phi^a)$, and the superpotential $W(\Phi^a)$.
(2) third derivatives of $\mathcal{F}(\Phi^a)$ at the scalar vev's are non-vanishing.
(3) the superpotential preserves $\mathcal{N} = 1$ SUSY at tree level.
(4) the gauge group is $U(N)$ and the vacuum is taken to be in the unbroken phase of $U(N)$.

In refs. 3, it was shown that the vacuum develops a non-vanishing vev of an auxiliary field D^0 in the HF approximation. The theory, therefore, realizes the D-term DSB. The relatively simple estimate has shown that the vacuum can be made long lived. Let us recall a few more key aspects.

The part of the lagrangian which produces the fermion mass matrix of size $2N$ is

$$-\frac{1}{2}(\lambda^a,\psi^a)\begin{pmatrix} 0 & -\frac{\sqrt{2}}{4}\mathcal{F}_{abc}D^b \\ -\frac{\sqrt{2}}{4}\mathcal{F}_{abc}D^b & \partial_a\partial_c W \end{pmatrix}\begin{pmatrix}\lambda^c \\ \psi^c\end{pmatrix} + (c.c.). \tag{2}$$

It was observed that the auxiliary D^a field, which is an order parameter of $\mathcal{N}=1$ SUSY, couples to the fermionic (but not bosonic) bilinears through the third prepotential derivatives: the non-vanishing vev of D^0 immediately gives a Dirac mass to the fermions. Equation of motion for D-term implies

$$\langle D^0\rangle = -\frac{1}{2\sqrt{2}}\langle g^{00}\left(\mathcal{F}_{0cd}\psi^d\lambda^c + \overline{\mathcal{F}}_{0cd}\overline{\psi}^d\overline{\lambda}^c\right)\rangle, \tag{3}$$

telling us that the condensation of the Dirac bilinear is responsible for $\langle D^0\rangle \neq 0$.

We diagonalize the holomorphic part of the mass matrix:

$$M_{Fa} \equiv \begin{pmatrix} 0 & -\frac{\sqrt{2}}{4}\langle\mathcal{F}_{0aa}D^0\rangle \\ -\frac{\sqrt{2}}{4}\langle\mathcal{F}_{0aa}D^0\rangle & \langle\partial_a\partial_a W\rangle \end{pmatrix}. \tag{4}$$

Note that the non-vanishing third prepotential derivatives are $\mathcal{F}_{0aa}$ where a refers to the generators of the unbroken gauge group. The two eigenvalues of eq. (4) for each generator are

$$\Lambda^{(\pm)}_{a\mathbf{11}} = \langle\partial_a\partial_a W\rangle\lambda^{(\pm)}_{a\mathbf{11}}, \quad \lambda^{(\pm)}_{a\mathbf{11}} \equiv \frac{1}{2}\left(1\pm\sqrt{1+\Delta^2_{\mathbf{11}}}\right), \quad \Delta^2_{a\mathbf{11}} \equiv \frac{\langle\mathcal{F}_{0aa}D^0\rangle^2}{2\langle\partial_a\partial_a W\rangle^2}. \tag{5}$$

It was also shown in refs. 3 that the non-vanishing F^0 term is induced by the consistency of our procedure of computation. (See also ref. 9). This is because the stationary value of the scalar fields gets shifted upon the variation (the vacuum condition). The final mass formula for the $SU(N)$ fermions is to be read off from

$$\begin{aligned}\mathcal{L}^{(holo)}_{\text{mass}} &= -\frac{1}{2}\langle g_{0a,a}\rangle\langle\overline{F}^0\rangle\psi^a\psi^a + \frac{i}{4}\langle\mathcal{F}_{0aa}\rangle\langle F^0\rangle\lambda^a\lambda^a - \frac{1}{2}\langle\partial_a\partial_a W\rangle\psi^a\psi^a \\ &\quad + \frac{\sqrt{2}}{4}\langle\mathcal{F}_{0aa}\rangle\psi^a\lambda^a\langle D^0\rangle.\end{aligned} \tag{6}$$

The mass matrix is read off as

$$\mathcal{M}_a = \begin{pmatrix} -\frac{i}{2}g^{aa}\mathcal{F}_{0aa}F^0, & -\frac{\sqrt{2}}{4}\sqrt{g^{aa}(\mathrm{Im}\mathcal{F})^{aa}}\mathcal{F}_{0aa}D^0 \\ -\frac{\sqrt{2}}{4}\sqrt{g^{aa}(\mathrm{Im}\mathcal{F})^{aa}}\mathcal{F}_{0aa}D^0, & g^{aa}\partial_a\partial_a W + g^{aa}g_{0a,a}\overline{F}^0 \end{pmatrix} = \begin{pmatrix} m^a_{\lambda\lambda} & m^a_{\lambda\psi} \\ m^a_{\psi\lambda} & m^a_{\psi\psi}\end{pmatrix}. \tag{7}$$

We parametrize this matrix such that, in the case of $F^0 = \overline{F}^0 = 0$, its form reduces to that of refs. 3. The quantities with multiple indices such as $\mathcal{F}_{0aa}$ receive $U(N)$ invariant expectation values: $\langle\mathcal{F}_{0aa}\rangle = \langle\mathcal{F}_{000}\rangle$ e.t.c. We suppress the indices as we work with the unbroken $U(N)$ phase in this paper. The two eigenvalues of the holomorphic mass matrix are written as

$$\Lambda^{(\pm)} \equiv (\mathrm{tr}\mathcal{M})\lambda^{(\pm)}, \quad \lambda^{(\pm)} = \frac{1}{2}\left(1\pm\sqrt{(1+if)^2+\left(1+\frac{i}{2}f\right)^2\Delta^2}\right) \tag{8}$$

where $\Delta \equiv -2m_{\lambda\psi}/m_{\psi\psi}$, $f \equiv 2im_{\lambda\lambda}/\mathrm{tr}\mathcal{M}$.

3. The effective potential in the HF approximation

In the HF approximation, one begins with considering the situation where one-loop corrections in the original expansion in $\hbar$ become large and are comparable to the tree contribution. In this section, we start the analysis of this kind for our effective potential. There are three constant background fields as arguments of the effective potential: $\varphi \equiv \varphi^0$ (complex), $U(N)$ invariant background scalar, $D \equiv D^0$ (real) and $F \equiv F^0$ (complex). The latter two are the order parameters of $\mathcal{N}=1$ SUSY.

Let us denote our effective potential by V.

$$V = V_{\text{tree}} + V_{\text{c.t.}} + V_{1-\text{loop}}. \tag{9}$$

The first term is the tree contributions, the second one is the SUSY counterterm and the last one is the one-loop contributions, which are explicitly given by

$$V_{\text{tree}} = -gF\overline{F} - \frac{1}{2}(\text{Im}\mathcal{F}'')D^2 - FW' - \overline{FW'}, \tag{10}$$

$$V_{\text{c.t.}} = -\frac{1}{2}\text{Im}\int d^2\theta \Lambda \mathcal{W}^{0\alpha}\mathcal{W}_{0\alpha} = -\frac{1}{2}(\text{Im}\Lambda)D^2, \tag{11}$$

$$V_{1-\text{loop}} = \frac{N^2|\text{tr}\mathcal{M}|^4}{32\pi^2}\left[A(\varepsilon,\gamma)\left(|\lambda^{(+)}|^4 + |\lambda^{(-)}|^4 - \left|\frac{m_s}{\text{tr}\mathcal{M}}\right|^4\right)\right.$$
$$\left. -|\lambda^{(+)}|^4\log|\lambda^{(+)}|^2 - |\lambda^{(-)}|^4\log|\lambda^{(-)}|^2 + \left|\frac{m_s}{\text{tr}\mathcal{M}}\right|^4\log\left|\frac{m_s}{\text{tr}\mathcal{M}}\right|^4\right] \tag{12}$$

where the scalar gluon mass is $m_s(\varphi,\overline{\varphi}) \equiv g^{-1}(\varphi,\overline{\varphi})W''(\varphi)$. The one-loop potential is calculated by the dimensional reduction scheme and the divergence reside in

$$A(\varepsilon,\gamma) = \frac{1}{2} - \gamma + \frac{1}{\varepsilon}, \qquad \varepsilon = 2 - \frac{d}{2}. \tag{13}$$

The divergence is subtracted by a counterterm associated with $\text{Im}\mathcal{F}''$ by setting up a renormalization condition

$$\left.\frac{1}{N^2}\frac{\partial^2 V}{(\partial D)^2}\right|_{D=0,\varphi=\varphi_*,\overline{\varphi}=\overline{\varphi}_*} = 2c. \tag{14}$$

Note that this condition is set up at $D=0$ and the stationary point of the scalar.

4. Stationary conditions, gap equation and stability of our false vacuum

Now we turn to our variational problem. It is stated as

$$\frac{\partial V}{\partial D} = 0, \tag{15}$$

$$\frac{\partial V}{\partial F} = 0 \text{ and its complex conjugate}, \tag{16}$$

$$\frac{\partial V}{\partial \varphi} = 0 \text{ and its complex conjugate}. \tag{17}$$

We will regard the solution to be obtained by considering eqs. (15) and (17) first and solving D and φ for F and $\overline{F}$:

$$D = D_*(F, \overline{F}), \quad \varphi = \varphi_*(F, \overline{F}), \quad \overline{\varphi} = \overline{\varphi}_*(F, \overline{F}). \tag{18}$$

Eq. (16) is then

$$\left. \frac{\partial V(D = D_*(F, \overline{F}), \varphi = \varphi_*(F, \overline{F}), \overline{\varphi} = \overline{\varphi}_*(F, \overline{F}), F, \overline{F})}{\partial F} \right|_{D, \varphi, \overline{\varphi}, \overline{F} \text{ fixed}} = 0 \tag{19}$$

and its complex conjugate. These will determine $F = F_*, \overline{F} = \overline{F}_*$.

Here, we are going to work in the region where the strength $|F_*|$ is small and can be treated perturbatively. This means that, in the leading order, the problem posed by eq. (15) and eq. (17) becomes

$$\frac{\partial V(D, \varphi, \overline{\varphi}, F = 0, \overline{F} = 0)}{\partial D} = 0, \tag{20}$$

$$\frac{\partial V(D, \varphi, \overline{\varphi}, F = 0, \overline{F} = 0)}{\partial \varphi} = \frac{\partial V(D, \varphi, \overline{\varphi}, F = 0, \overline{F} = 0)}{\partial \overline{\varphi}} = 0 \tag{21}$$

and this problem does not involve the tree potential eq. (10) except the D^2 term, as F and $\overline{F}$ are set zero. Eq. (20) is nothing but the gap equation, while eq. (21) is the stationary conditions for the scalar. This is the variational problem which we analyze. A set of stationary values $(D_*, \varphi_*, \overline{\varphi}_*)$ is determined as the solution.

In this paper, we do not write down the explicit expressions for the gap equation and the stationary conditions for the adjoint scalar because of their complicated forms (See the last paper in refs. 3 for details) and the limitation of the space to describe. Instead, we give a qualitative argument how these conditions are solved and numerical results in the simplified case.

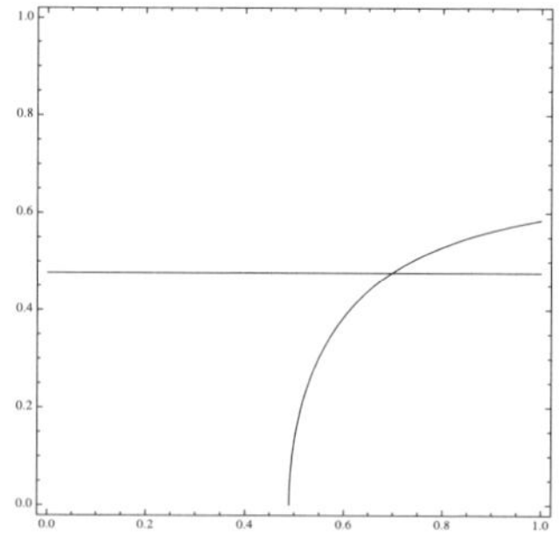

Fig. 1. The schematic picture of the intersection of the two curves which represent the solution to the gap equation (the red one) and the φ flat condition (the blue one). The horizontal axis is denoted by φ/M and the vertical one by Δ_0. The values at the stationary point $(\Delta_{0*}, \varphi_* = \overline{\varphi}_*)$ are read off from the intersection point.

The stationary values $(D_*, \varphi_*, \overline{\varphi}_*)$ are determined by the gap equation and the stationary condition. The solution to the stationary condition in the Δ_0 profile is determined as the point of intersection of the potential with the quadratic term having $\varphi = \overline{\varphi}$ dependent coefficients. Actually, it is a real curve in the full $(\Delta_0, \varphi = \overline{\varphi})$ plane. Likewise, the solution to the gap equation provides us with another real curve in the $(\Delta_0, \varphi = \overline{\varphi})$ plane. The values $(\Delta_{0*}, \varphi_* = \overline{\varphi}_*)$ are the intersection of these two. The schematic figure of the intersection is displayed in Figure 1. By tuning our original input functions, it is possible to arrange such intersection.

We study some numerical solutions to the gap equation and the stationary condition for φ in the real Δ_0 case. In order to find φ_* explicitly, the forms of the prepotential $\mathcal{F}$ and the superpotential W must be specified. Here, we take a simple prepotential and a superpotential as

$$\mathcal{F} = \frac{i}{2N}\mathrm{tr}\varphi^2 + \frac{1}{3!MN}\mathrm{tr}\varphi^3, \quad W = \frac{m^2}{N}\mathrm{tr}\varphi + \frac{1}{3!N}\mathrm{tr}\varphi^3, \tag{22}$$

where m, M carry dimensions. In particular, M is a cutoff scale of the theory.

Some numerical solutions to the gap equation and the stationary condition for φ in some parameter points are listed in the Table below. In these examples, we have taken some values of $-\frac{N^2}{\mathrm{Im}(i+\Lambda)}$ and m just for an illustration and the ratio $|F_*/D_*|$ is evaluated. We can find that the F-term is smaller than the D-term in these examples. Also, the scalar gluon mass squared are positive[a], which ensures a local stability of our vacuum. As a summary of our understanding, a schematic

Δ_{0*}	φ_*/M $(-\frac{N^2}{\mathrm{Im}(i+\Lambda)})$	$\lvert F_*/D_*\rvert$	m_φ^2
0.477	0.707 (10000)	0.524 ($m \ll M$)	0.4998
1.3623	0.8639 (2000)	0.224 ($m \ll M$)	0.7463
1.3623	0.5464 (5000)	0.142 ($m \ll M$)	0.2986
1.3623	0.3863 (10000)	0.100 ($m \ll M$)	0.1492

figure is drawn in Fig. 2, which illustrates the local stability of the scalar potential at the vacuum of dynamically broken SUSY in comparison with the well-known NJL potential.

5. Summary

We have proposed a new mechanism of DSB at the metastable vacuum in a $\mathcal{N} = 1$ SUSY $U(N)$ gauge theory with an adjoint chiral super field and non-canonical gauge kinetic term, in which non-vanishing D-term vev is developed in the self-consistent HF approximation of NJL type. Non-zero F-term is also induced by such a D-term as well. We analyzed a gap equation for D-term and the stationary conditions for φ and F-term in the case that F-term can be treated as a perturbation comparing

[a] In the Table, only the leading term of the scalar gluon mass is shown. See the last paper in refs. 3 for more details.

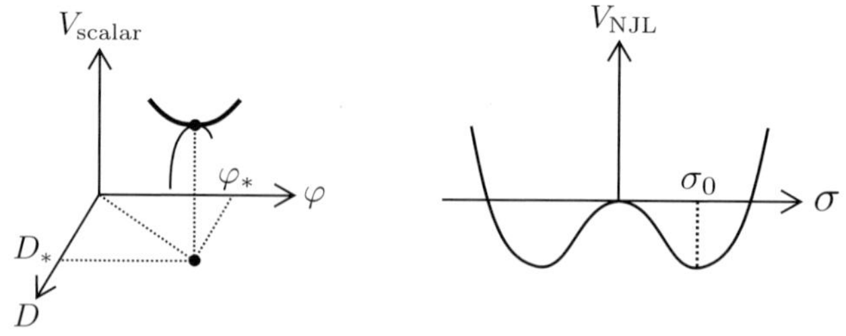

Fig. 2. Comparison of V_{scalar} around the stationary value (D_*, φ_*) with V_{NJL}.

to D-term. The solutions are obtained as the intersection point of two curves representing the gap equation and the stationary conditions. Numerical examples of the solutions are found in the simplified case and the local stability of vacua we found is also confirmed by the numerical data.

Acknowledgements

The work of the author is supported in part by the Grant-in-Aid for Scientific Research from the Ministry of Education, Science and Culture, Japan (24540283) and by Keio Gijuku Academic Development Funds.

References

1. E. Witten, Nucl. Phys. B **188**, 513 (1981); Nucl. Phys. B **202**, 253 (1982).
2. M. T. Grisaru, W. Siegel and M. Rocek, Nucl. Phys. B **159**, 429 (1979).
3. H. Itoyama and N. Maru, Int. J. Mod. Phys. A **27**, 1250159 (2012); arXiv:1207.7152 [hep-ph]; arXiv:1301.7548 [hep-ph].
4. Y. Nambu and G. Jona-Lasinio, Phys. Rev. **122** (1961) 345; Phys. Rev. **124** (1961) 246.
5. J. Bardeen, L. N. Cooper and J. R. Schrieffer, Phys. Rev. **108**, 1175 (1957); Y. Nambu, Phys. Rev. **117**, 648 (1960).
6. I. Affleck, M. Dine and N. Seiberg, Nucl. Phys. B **241**, 493 (1984); Phys. Lett. B **140**, 59 (1984); Nucl. Phys. B **256**, 557 (1985).
7. Y. Meurice and G. Veneziano, Phys. Lett. B **141**, 69 (1984); Y. Meurice and G. Veneziano, D. Amati, K. Konishi, Y. Meurice, G. C. Rossi and G. Veneziano, Phys. Rept. **162** (1988) 169.
8. K. Fujiwara, H. Itoyama and M. Sakaguchi, Prog. Theor. Phys. **113** (2005) 429; Nucl. Phys. B **723** (2005) 33.
9. M. Dine, A. E. Nelson, Y. Nir and Y. Shirman, Phys. Rev. D **53**, 2658 (1996); L. M. Carpenter, P. J. Fox and D. E. Kaplan, hep-ph/0503093.

Generalized Skyrmions and The Mass of the Lightest Electroweak Baryon*

John Ellis

Theoretical Particle Physics and Cosmology Group, Physics Department, King's College London, London WC2R 2LS, UK, and TH Division, Physics Department, CERN, CH-1211 Geneva 23, Switzerland
E-mail: John.Ellis@cern.ch

Marek Karliner

Raymond and Beverly Sackler School of Physics and Astronomy, Tel Aviv University, Israel
E-mail: marek@proton.tau.ac.il

Michal Praszalowicz

Smoluchowski Institute of Physics, Jagiellonian University, 4 Reymonta Str., 30-059 Kraków, Poland
E-mail: michal@if.uj.edu.pl

In general, effective low-energy Lagrangian models of composite electroweak symmetry breaking (EWSB) have soliton solutions that may be identified with technibaryons. The masses of such states may be related to the coefficients of fourth-order terms in the effective Lagrangian, as in QCD. We show how the current theoretical and phenomenological constraints on the corresponding fourth-order coefficients in the electroweak theory could be used to estimate qualitative lower and upper bounds on the lightest electroweak baryon mass. We point out that in strongly-interacting models of electroweak symmetry breaking, non-Skyrmion-like solitons are also possible. Future data from the LHC as well as dark matter experiments will constrain this possibility. With the currently available data the upper bound on the mass of the electroweak soliton is between 18 and 59 TeV. With improved information that LHC is likely to provide on quartic terms in the electroweak Lagrangian, these bounds might be reduced to 4.6 to 8.1 TeV.

Keywords: Higgs Physics, Beyond Standard Model, Technicolor and Composite Models, Skyrmions

1. Introduction

Strongly-coupled theories generically exhibit spontaneous breaking of their global symmetries. Physics below the natural scale of such theories is then described by an effective Lagrangian that describes the dynamics of the relevant (quasi)Goldstone

*Based on Refs. [1] and [2].

bosons. Such Lagrangians typically exhibit soliton solutions.[3,4] Chiral soliton physics in QCD is the best-known example of this phenomenon, which has also been studied in many other theories.[5] In QCD the prototype of this approach has been based the pioneering work of Skyrme,[6] yielding considerable phenomenological successes.[7,8] It has been known for a long time, however, that the Skyrme model is just the simplest representative of a large family of models providing effective descriptions of QCD dynamics at low energy. Moreover, we have good reasons to expect that effective dynamics of other strongly-coupled theories might be quite different from QCD.

The idea that electroweak symmetry breaking might be due to an underlying strong dynamics has a long history.[9] The discovery of a Higgs candidate at the LHC[10] has triggered renewed interest in this subject. Even though the Higgs candidate has many properties that are consistent with a standard model Higgs boson, many alternative scenarios described by low-energy effective chiral Lagrangians have not been ruled out.[11] Among these, there is considerable interest in the possibility that the Higgs candidate might be the pseudo-dilaton of some nearly-conformal strongly-interacting electroweak sector.[12] It is important to discriminate between the various scenarios of EWSB and study of their soliton solutions might provide a useful tool to this effect.[1]

The properties of the chiral solitons, and their masses in particular, depend on the coefficients of the the various terms in the effective Lagrangians, which are usually expanded in a series of derivatives of increasing order. The quadratic terms are fixed by the VEV and the symmetry, so the quartic terms are the leading non-universal terms that constrain the mass.[4] The simplest case is $SU(2) \times SU(2) \to SU(2)$ that can describe both QCD and EWSB with two possible quartic terms, as discussed in Sec. 2. One of them was originally discussed by Skyrme,[6] and has been extensively used in Skyrmion phenomenology.[7] It is clear, however, that the other term can also appear. It has been studied in context of QCD.[13–17]

On a quantitative level, it is useful to recall that in the context of QCD keeping only the Skyrme quartic term, defined in Sec. 2, reproduces experimental data to within 20–30%, which is adequate for our purpose. With the Skyrme term alone the nucleon mass usually comes out too large, and our analysis indicates that including the non-Skyrme term widens the range somewhat. On this basis we expect the effective Lagrangian with both quartic terms to provide a useful estimate of upper limits on the electroweak soliton mass.

After obtaining the classical solution one usually performs semiclassical quantization of the collective coordinates. When non-Skyrme terms are present, in principle one then needs to deal with higher-order time derivatives, but in QCD such terms are suppressed in the $1/N_c$ expansion[4] and can therefore be neglected in the leading order. Similar considerations apply to many strongly-interacting models of EWSB, but each case needs to be examined individually. Here were examine only the classical approximation to the soliton mass and also neglect the Casimir contribution to the mass.[19,20]

As discussed in detail in the following, stability of the solitons is determined by the ratio of the coefficients of the two quartic terms. There is a generic range, including the original Skyrmion, where the solitons are stable, at least within the class of spherically symmetric configurations. There is also a generic range where there are no stable solitons.[13–17] For QCD, large-N_c favours the stable range of quartic coefficients, near to the Skyrme limit. This range is also favoured by phenomenological analysis of chiral Langrangians for QCD.[21,22]

Our knowledge of the quartic coefficients in the the electroweak case is very limited at the moment and their ratio might turn out to be very different from QCD, perhaps even in the range where there are no stable solitons. On the other hand, if a strongly-coupled underlying theory is responsible for EWSB, with constituents binding to form 'electroweak baryons', we expect such baryons to be approximately described by stable solitons of the effective chiral Lagrangian, even if they are quite unlike the baryons and Skyrmions in QCD-like theories.[1]

In the following we explore the quartic couplings' parameter space.[13–17] Within the classical approximation we discuss the stability of the solitons and their masses, focusing on the electroweak case. This provides an important constraint for comparison with experiment of various strongly-interacting models of EWSB. We implement this constraint by confronting current limits and possible future measurements of electroweak baryon masses with current[23,24] and possible future limits[23] on the quartic terms in the electroweak Lagrangian. We recall that if stable electroweak solitons exist, they would survive from the Big Bang as cosmic relics and hence as a component of dark matter.[25] It would be interesting to check theoretically if their abundance is consistent with estimates of cold dark matter density and if the cross-section for their elastic scattering on nuclei[26] is consistent with upper limits from direct searches.[28]

The structure of this talk is as follows. We start by discussing in Section 2 the mass and stability of the classical $SU(2)$ soliton with both quartic terms present. These results, together with current and future expected data and limits on quartic couplings are then used in Sec. 3 to obtain bounds on the possible masses of QCD and electroweak solitons. Section 4 focuses on dark matter aspects of electroweak solitons and Section 5 is devoted to a summary and conclusions.

2. Classical Mass and Stability of an SU(2) Soliton

The effective chiral Lagrangian corresponding to some strongly-interacting sector with spontaneously-broken SU(2)×SU(2) chiral symmetry is usually organized in terms of (even) powers of derivatives of the chiral field:

$$U(x) = \exp\left(i\frac{\vec{\tau}\cdot\vec{\pi}(x)}{v}\right), \tag{1}$$

where the fields $\vec{\pi}(x)$ correspond to the Nambu-Goldstone bosons (pions in the case of QCD, longitudinal polarization states of massive gauge bosons in the case of a strongly-interacting electroweak sector) and v is a typical symmetry breaking scale

($v = F_\pi = 93$ MeV in QCD, $v \sim 246$ GeV in the case of electroweak theory). The derivative expansion is usually truncated at the fourth order, which could be reliable at energies below the characteristic strong-interaction scale, $\sim$ GeV in QCD and $\sim$ TeV or more in the electroweak theory:

$$\mathcal{L}_{\text{eff}} = \mathcal{L}_2 + \mathcal{L}_4 \,, \tag{2}$$

where

$$\mathcal{L}_2 = \frac{v^2}{4} \text{Tr}\left(\partial_\mu U \partial^\mu U^\dagger\right). \tag{3}$$

In the SU(2) case there are two independent invariants containing just four space-time derivatives:[13–17]

$$\mathcal{L}_4 = 2s\,\text{Tr}\left[(R_\mu R_\nu)(R^\mu R^\nu) - (R_\mu R^\mu)^2\right] + 2t\,\text{Tr}\left[(R_\mu R_\nu)(R^\mu R^\nu) + (R_\mu R^\mu)^2\right] \,, \tag{4}$$

where

$$R_\mu = \partial_\mu U\, U^\dagger \,. \tag{5}$$

The parameters s and t can be in principle calculated from the underlying strongly-interacting theory by integrating out the constituent degrees of freedom (quarks and gluons in the case of QCD, and yet to be determined in an electroweak theory) or – in a phenomenological approach – can be extracted from the data on the scattering of the Nambu-Goldstone bosons or massive gauge bosons[23] [a]. In QCD, as we discuss in more detail later, the large-N_c expansion and pion data indicate that $|t| \ll |s|$, so that the effective Lagrangian contains (in a first approximation) only the kinetic term (3) and the first term in (4), as in Skyrme's original work. For this reason, we refer to this as the Skyrme term, and refer to other term in (4), that with coefficient t, as the non-Skyrme term.

Over fifty years ago already, Skyrme[6] observed that (2) with $t = 0$ possesses solitonic solutions that, due to the fact that $\pi_3(\text{SU}(2)) = \mathbb{Z}$, carry an integer-valued topological quantum number

$$B = \frac{1}{24\pi^2} \int d^3x \epsilon^{ijk} \text{Tr}\left[(U^\dagger \partial_i U)(U^\dagger \partial_j U)(U^\dagger \partial_k U)\right]. \tag{6}$$

that can be interpreted as baryon number (see also [3]).

It is straightforward to calculate the classical contribution to the mass of the baryon in terms of the parameters v, s and t. In principle, all terms in the derivative expansion make significant contributions to the soliton mass. We have no information on the possible magnitudes of higher-order terms, but they are not expected [2] to change qualitatively the estimates presented here. Our primary interest here will be the classical soliton mass in the presence of a non-negligible non-Skyrme term, and how the possibility that $t \neq 0$ affects the (approximate) mass bounds provided by the Skyrme calculation with $t = 0$ discussed in [1].

---

[a] For rigorous lower bounds from Lorentz invariance, analyticity, unitarity and crossing, see [24].

Classical soliton solutions of the chiral field equations are usually found within the spherically-symmetric 'hedgehog' Ansatz for a static field configuration:

$$U(\vec{r}) = \exp\left(i\frac{\vec{\tau}\cdot\vec{r}}{r}P(r)\right), \tag{7}$$

where the profile function $P(r)$ is a solution to the Euler-Lagrange equation of motion, which for the Lagrangian (2)–(4) takes the following form:

$$P''(\rho) = -\frac{G(\rho)}{F(\rho)} \tag{8}$$

with boundary conditions

$$P(0) = \pi,\ P(\infty) = 0\,. \tag{9}$$

Here

$$\begin{aligned} F(\rho) &\equiv \rho^2\left(1 - 96\,tP'^2\right) + 64\,s\sin^2 P\,, \\ G(\rho) &\equiv -64\,t\rho P'^3 + 32\,s\sin 2P\,P'^2 + 2\rho P' \\ &\quad - \sin 2P\left(1 + 32\,(s-t)\,\frac{\sin^2 P}{\rho^2}\right), \end{aligned} \tag{10}$$

and the dimensionless radial variable ρ is defined as

$$\rho \equiv rv\,, \tag{11}$$

where $v = F_\pi,\ \simeq 246$ GeV in the QCD and electroweak cases, respectively.

Following the pioneering paper,[4] classical solutions of the field equations were studied numerically long ago[13] and stability regions in the (s,t) parameter space were analyzed in [14–17]. We have repeated their analyses for the purpose of the present work with the following results.

The total classical soliton mass

$$M_{\rm sol} = 4\pi v\,(M_2 + M_4) \tag{12}$$

where

$$\begin{aligned} M_2 &= \frac{1}{2}\int_0^\infty d\rho\left(\rho^2 P'^2 + 2\sin^2 P\right), \\ M_4 &= 16s\int_0^\infty d\rho\left(2P'^2\sin^2 P + \frac{\sin^4 P}{\rho^2}\right) - 8t\int_0^\infty dr\left(\rho^2 P'^4 + 2\frac{\sin^4 P}{\rho^2}\right), \end{aligned} \tag{13}$$

where the contribution M_2 corresponds to $\mathcal{L}_2$ and is clearly always positive, and the contribution M_4 corresponds to $\mathcal{L}_4$. By changing variable once more: $\rho \to \rho/\sqrt{|s|}$ or $\rho \to \rho/\sqrt{|t|}$, one can show that the solutions depend only on the ratio t/s. By rescaling: $\rho \to \lambda\rho$, one can show that solutions exist only when M_4 is positive. Inspecting directly M_4 in (13), we see that in the fourth quadrant of the (s,t) plane $(s > 0,\ t < 0)$ M_4 is always positive, so that solutions of (8) always exist, whereas in the second quadrant ($s < 0$, $t > 0$) M_4 is always negative, so there are no solutions of (8). In the first and third quadrants positivity bounds can be derived.[13,14] It turns out that in the first quadrant M_4 is negative for $t/s > 2$, so the soliton is unstable, and in the third quadrant M_4 stays positive for $t/s > 2$, so the soliton is stable. This is illustrated in Fig. 1, where the regions of positive and negative M_4 are displayed by green shading (diagonal squares) and red shading (vertical squares), respectively.

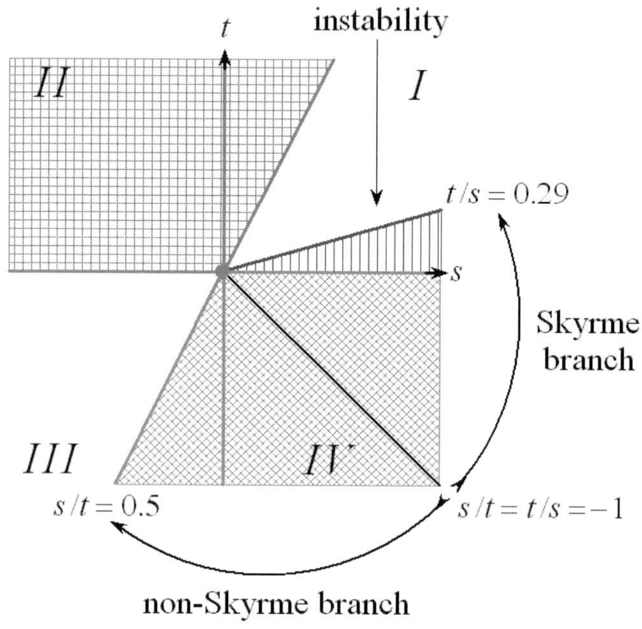

Fig. 1. The (s, t) parameter plane, indicating with green shading (diagonal squares) the range of parameters satisfying the positivity bound $M_4 > 0$ and hence admitting a solution of the differential equation (8), and with red shading (vertical squares) the range where solutions do not exist due to the negativity of the fourth order term: $M_4 < 0$. In the range $t/s < 0.29$ within the first quadrant, indicated in blue (vertical bars), solutions with positive M_4 exist, but no positivity bound can be derived.

In the remaining parts of quadrants I and III, no positivity bounds can be derived and the positivity of M_4 has to be checked numerically. It turns out that in the third quadrant $M_{\rm sol} > 0$ and $\to 0^+$ when $t/s \to 2^+$, whereas no solution exists for $t/s < 2$. The situation is different in the first quadrant, where solutions with positive M_4 exist for small t/s below the line $t/s \simeq 0.29$, where the function $F(\rho)$ vanishes and equation (8) cannot be solved. Interestingly, $F(\rho)$ is equal to the second variation of $M_{\rm sol}$, so the positivity of $F(\rho)$ is a necessary condition for $M_{\rm sol}$ to be the minimum with respect to the variations of $P(r)$. Below the line $t/s \sim 0.29$

the solution is classically stable, and above this line it is classically unstable.

However, it has been shown in [13] that there is no lower limit on the soliton mass anywhere in the first quadrant. This means that, when $t/s < 0.29$, the classical solution discussed above, though stable against local spherically-symmetric perturbations, can only be metastable at best, and might be unstable against non-spherically-symmetric perturbations. This region is shaded blue (with vertical bars) in Fig. 1.

As already remarked, soliton solutions depend only on the ratio t/s (or s/t). We analyze solutions starting from the Skyrme-like case with $t = 0$ and the non-Skyrme-like case with $s = 0$, and have divided the allowed parameter space into two branches on either side of the line $t/s = -1$, as also depicted in Fig. 1. In order to solve equation (8) within the Skyrme branch where $s \neq 0$, we have rescaled $\rho \to \rho/\sqrt{s}$ and evaluated the soliton mass in units of $4\pi v\sqrt{s}$, whereas within the non-Skyrme branch where $t \neq 0$ we have rescaled $\rho \to \rho/\sqrt{-t}$ and evaluated the soliton mass in units of $4\pi v\sqrt{-t}$. The results are presented in Table I as functions of $\varepsilon \equiv t/s$ for the Skyrme branch and $\varepsilon \equiv s/t$ for the non-Skyrme branch.

ϵ	$E_{\rm Sk}(\epsilon = t/s)$	$E_{\rm nSk}(\epsilon = s/t)$
−1.00	40.76	40.76
−0.90	40.04	39.40
−0.80	39.32	38.00
−0.70	38.58	36.55
−0.60	37.82	35.03
−0.50	37.04	33.44
−0.40	36.25	31.77
−0.30	35.43	30.00
−0.20	34.56	28.11
−0.10	33.71	26.08
0.00	32.82	23.88
0.10	31.88	21.44
0.20	30.89	18.65
0.30	–	15.33
0.40	–	10.99
0.45	–	7.96

Tab. I. Skyrmion mass for the Skyrme branch (second column) in units of $4\pi v\sqrt{s}$ and for the non-Skyrme branch (third column) in units of $4\pi v\sqrt{-t}$ as function of $\epsilon = t/s$ for the Skyrme and $\epsilon = s/t$ for the non-Skyrme branch, respectively (see Fig. 1).

3. Phenomenological Estimates of Soliton Masses

We now discuss constraints on the conventional baryon mass using constraints on the parameters of the low-energy effective chiral Lagrangian for QCD, and constraints on the masses of possible 'electroweak baryons' given by the constraints on higher-order electroweak Lagrangian parameters from electroweak theory, current phenomenology and the potential sensitivity from the LHC.[1,23]

Phenomenological constraints on higher-order Lagrangian parameters are often given in terms of the coefficients $\alpha_{4,5}$ that are related to the parameters s and t discussed in Section 2 in the following way:[1]

$$s = \frac{\alpha_4 - \alpha_5}{4}, \quad t = \frac{\alpha_4 + \alpha_5}{4}. \tag{14}$$

Although limits on electroweak baryon masses are our principal interest, we first discuss the limits that can be derived for ordinary baryons in QCD, comparing the non-Skyrmion case with $t \neq 0$ with the conventional Skyrmion case $t = 0$.

3.1. *QCD baryons*

Higher-order coefficients in the low-energy effective Lagrangian of QCD have been the subject of many studies. Here we use the following ranges extracted from low-energy strong-interaction data:[1,21,22]

$$11 \times 10^{-4} < \alpha_4 < 17 \times 10^{-4}, \qquad 14 \times 10^{-4} < \alpha_4 - \alpha_5 < 40 \times 10^{-4}. \tag{15}$$

We can compare these values with the predictions of the large-N_c approximation within the framework of chiral SU(3) × SU(3) → SU(3):[22]

$$\alpha_4 = \quad 18 \times 10^{-4}, \qquad \alpha_5 = -16 \times 10^{-4}. \tag{16}$$

The bounds (15) are displayed in Fig. 2, where the large-N_c prediction is also indicated, as a red spot [b].

Multiplying entries of Tab. I by $4\pi v\sqrt{s}$ or $4\pi v\sqrt{-t}$ one can easily convince oneself that the contours of constant $M_{\rm sol}$ are almost parallel to the line $t = 2s$.[2] Therefore the maximal mass corresponds to the right-most corner of the region allowed by the QCD bounds, called Point A below. For completeness, we also include two pure Skyrme points (B and C) and the point corresponding to minimal mass, located right below the stability line $t/s = 0.29$ (Point D):

- A : $(s, t) = (10, -4.5) \times 10^{-4}$ (maximal mass point overall): $M_{\rm A} \simeq 1354$ MeV;
- B : $(s, t) = (8.5, 0.0) \times 10^{-4}$ (maximal mass Skyrme point): $M_{\rm B} \simeq 1118$ MeV;
- C : $(s, t) = (5.5, 0.0) \times 10^{-4}$ (minimal mass Skyrme point): $M_{\rm C} \simeq 900$ MeV;
- D : $(s, t) = (4.3, 1.2) \times 10^{-4}$ (minimal mass point overall): $M_{\rm D} \simeq 728$ MeV.

We see that the range of classical masses allowed in the Skyrme case $t = 0$ includes the physical value of the nucleon mass. In terms of the conventional representation $s = 1/32e^2$, this range corresponds to

$$6 < e < 7.5. \tag{17}$$

[b]We represent the constraints (15), (18), (19) as parallelograms in the (s, t) plane, whereas they should be ellipses. However, the correlations between the errors in s and t are not available, so these ellipses cannot be drawn accurately.

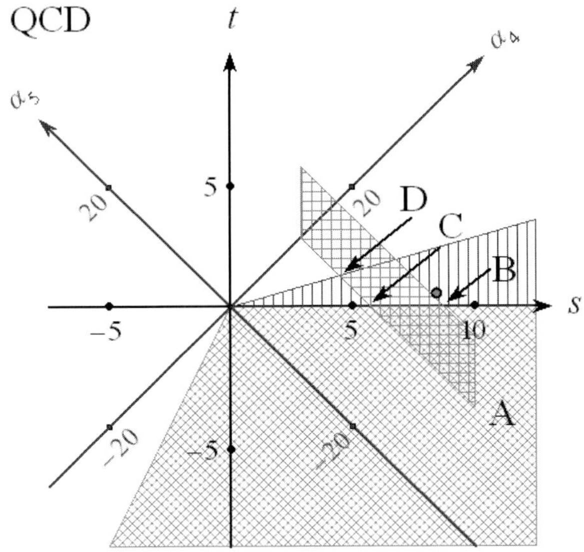

Fig. 2. Comparison of the parameter ranges allowed for the low-energy chiral perturbation theory coefficients in QCD,[21,22] eq. (15), with the range of parameters where stable baryonic solitons exist. The parameters α_4, α_5, s and t are in units of 10^{-4}. The red spot corresponds to the large-N_c values given in eq. (16). Point A has the maximal mass overall, Points B and C bracket the range of parameters allowed in the pure Skyrme limit: $t = 0$, and Point D lies on the stability boundary $t/s = 0.29$ and has the minimal mass overall.

We also note that the range of possible QCD baryon masses is extended significantly in the presence of a non-Skyrme term, by $\mathcal{O}(200)$ GeV in either direction. However, detailed QCD baryon phenomenology including the evaluation of corrections to the classical soliton mass due to the Casimir energy, inclusion of the current-quark mass terms and semiclassical quantization of the rotational modes lies beyond the scope of this work.

3.2. *Current bounds on electroweak baryon masses*

The higher-order coefficients in an effective electroweak Lagrangian are currently poorly constrained, namely by the following bounds:[23]

$$-3.5 \times 10^{-1} < \alpha_4 < 0.6 \times 10^{-1}, \quad -8.7 \times 10^{-1} < \alpha_5 < 1.5 \times 10^{-1}. \tag{18}$$

These constraints are superimposed on the (s, t) plane in Fig. 3. As in the QCD case, we superpose various points that illustrate the range of possible electroweak baryon masses currently allowed, as obtained using the values of the soliton mass of Tab. I. We note that there is no minimal electroweak baryon mass, since the constraints (18) include the possibility that $M_{\rm sol} = 0$ for $t = 2s < 0$. The points displayed include the maximal mass point denoted by A, the maximal-mass Skyrme case (Point B), and the maximal-mass case with vanishing Skyrme term $s = 0$ (Point C):

- $\mathrm{A} : (s,t) = (0.23, -0.20)$ (maximal mass point overall): $M_\mathrm{A} \simeq 59$ TeV;
- $\mathrm{B} : (s,t) = (0.03, \ 0.0\ )$ (maximal mass Skyrme point): $M_\mathrm{B} \simeq 18$ TeV;
- $\mathrm{C} : (s,t) = (0.0, \ -0.175)$ (maximal mass with $s = 0$): $M_\mathrm{C} \simeq 31$ TeV.

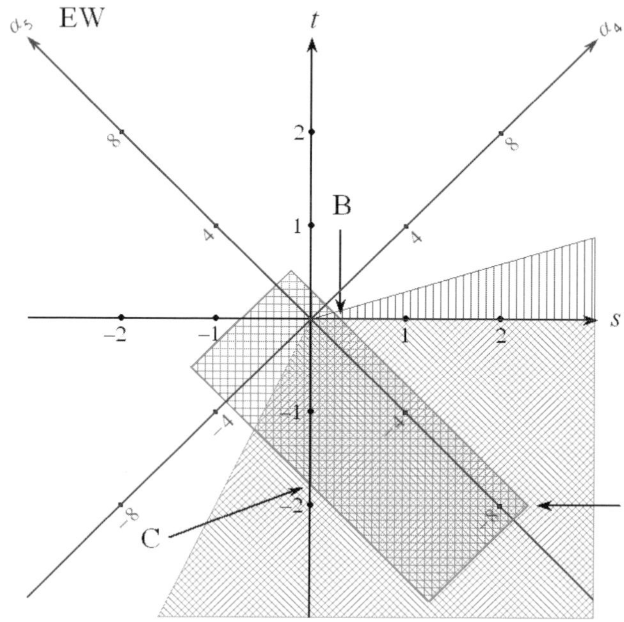

Fig. 3. Comparison of the parameter ranges currently allowed[23] for the effective low-energy electroweak Lagrangian with the range of parameters where stable electroweak baryonic solitons exist. The parameters α_4, α_5, s and t are in units of 10^{-1}. Point A has the maximal mass overall, Point B has the maximal mass allowed in the pure Skyrme limit: $t = 0$, and Point C has the maximal mass allowed in the limit where the Skyrme term vanishes: $s = 0$.

This analysis indicates that the upper bound on the possible electroweak baryon mass over the range of parameter space currently allowed is approximately 60 TeV [c]. This upper limit is somewhat higher than in [1], because we have extended the analysis to the case of a non-zero non-Skyrme term. If the conjectural strongly-interacting electroweak theory were to resemble QCD, i.e., the parameter t were small, the upper limit on the possible electroweak baryon mass limit would be significantly reduced to $\simeq 18$ TeV.

[c] Assuming that the model-dependent corrections to the classical mass calculation are not very large.

3.3. *Prospective LHC bounds on electroweak baryon masses*

A study has been made of the prospective LHC sensitivity to the higher-order electroweak Lagrangian terms. It was estimated that, in the absence of a signal, the LHC could yield the following allowed ranges[23]

$$-7.7 \times 10^{-3} < \alpha_4 < 15 \times 10^{-3}, \qquad -12 \times 10^{-3} < \alpha_5 < 10 \times 10^{-3}. \tag{19}$$

These ranges are superimposed on the (s, t) plane in Fig. 4. As in the previous examples, we list below the classical masses calculated for three illustrative points that are also displayed in Fig. 4:

- A : $(s, t) = (6.75, 0.75) \times 10^{-3}$ (maximal mass point): $M_{\mathrm{A}} \simeq 8.1$ TeV;
- B : $(s, t) = (6.0, \ 0.0\ ) \times 10^{-3}$ (Skyrme point): $M_{\mathrm{B}} \simeq 7.9$ TeV;
- C : $(s, t) = (0.0, -3.85) \times 10^{-3}$ (non-Skyrme point): $M_{\mathrm{C}} \simeq 4.6$ TeV.

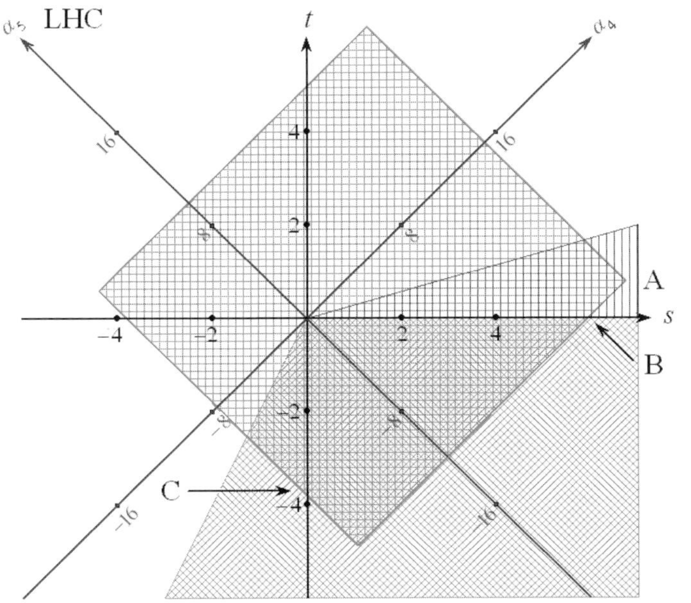

Fig. 4. Comparison of the estimated LHC sensitivity range[23] for higher-order electroweak Lagrangian parameters with the the range of parameters where stable electroweak baryonic solitons exist. The parameters α_4, α_5, s and t are in units of 10^{-3}. Point A has the maximal mass overall, Point B has the maximal mass allowed in the pure Skyrme limit: $t = 0$, and Point C has the maximal mass allowed in the limit where the Skyrme term vanishes: $s = 0$.

The upper limit on the possible electroweak baryon mass in the event that the LHC does not find non-zero higher-order effective Lagrangian parameters is $\simeq 8$ TeV, very similar to the bound quoted in [1]. This reflects the fact that the point with maximal mass overall is quite close to the Skyrme limit $t = 0$. In this case, unlike the previous examples, the upper limit on the soliton mass is not relaxed by allowing a non-Skyrme term with $t < 0$.

4. Electroweak Baryons as Dark Matter?

If they exist, electroweak baryons should be present in the Universe today, and could provide cold dark matter.[25] Their possible relic density today depends on the presence and magnitude of a primordial electroweak baryon asymmetry. If this was small, and the electroweak baryon density was completely equilibrated in the early Universe, the present abundance of electroweak baryons would be insufficient to provide the present density of cold dark matter. In this case, one or more other sources of cold dark matter would be required, and the constraints on dark matter scattering that we now discuss would be irrelevant.

Assuming that they make up the bulk of the cold dark matter density, it is clear that electroweak baryons should have no electric charge. Moreover, it was pointed out in [29] that they could not be fermions, as these would have an unacceptably large cross section for spin-independent scattering through magnetic moment couplings. Some scenarios for the effective low-energy electroweak Lagrangian in which the lightest electroweak baryon is a neutral boson have been enumerated in [30], in an analysis based on the Wess-Zumino-Witten term and the standard relation $Q_{\rm em} = I_3 + Y$. The neutral boson scenarios found *do not* include SU(3) $\times$ SU(3) $\to$ SU(3) (which could yield a neutral fermion if $N_c = 3$ or a bosonic baryon if N_c is even), but *do* include SU(N) $\times$ SU(N) $\to$ SU(N) with $N > 3$ and N_c even, *and* SO(N) $\times$ SO(N) $\to$ SO(N). However, SU(N) $\to$ SO(N) would yield a boson with charge N_c, and the other coset structures considered in [30] have trivial homotopy and hence no electroweak baryons.

It should be noted that dark matter scattering is a non-trivial constraint, even if the lightest electroweak baryon is a neutral boson. The scattering cross section for this case was estimated in [26], with results indicating that models containing a pseudo-dilaton identified with the recently-discovered boson X with mass $\simeq$ 125 GeV would be consistent with the (updated) XENON100 limit[28] if the pseudo-dilaton couplings were scaled by $\geq$ 1 TeV [d]. and the electroweak baryon weighed $\leq$ 1 TeV, as seen in Fig. 4 of [26]. The XENON100 Collaboration does not report results for larger dark matter particle masses, but naive extrapolation of the theoretical calculations and experimental sensitivity would suggest some tension for masses above 1 TeV.

Our results suggest that scenarios which, according to the analysis of [30], might yield either a charged bosonic baryon or a fermionic baryon *should not necessarily be abandoned.* By the same token, one might not need to worry about the prospective tension with XENON100. This is because the parameters of the fourth-order

[d] This is in the ball-park calculated in holographic dilaton models, and is reported to be consistent with the available data on $X(125)$,[31] once the contributions of the strongly-interacting electroweak sector the Xgg and $X\gamma\gamma$ couplings are taken into account.

effective Lagrangian might be in the range where no *stable* solitonic baryon exists, i.e., $t > 0$ or $t < 0$ and $t/s > 2$. That said, it could be that stable electroweak baryons nevertheless exist for such parameter choices, but cannot be found using the solitonic Ansatz discussed here.

5. Concluding Remarks

We have discussed the existence, stability and masses of classical solitons in effective Langrangians describing low-energy QCD and EWSB through a possible strongly interacting underlying theory. We focused on the consequences of a possible non-Skyrme quartic term.[13–15,17] Stability constraints in the (s, t) are displayed in Fig. 1 and masses in Table I.

We found that in the QCD case current data[21,22] allow values of classical soliton masses somewhat outside the range given by Skyrme term only. In the electroweak sector, the presently available bounds on quartic terms[23] are consistent with a somewhat wider range of t/s than in QCD and the ensuing range of electroweak baryon masses is proportionally larger than in the pure Skyrme case. On the other hand, given the precision with which LHC is expected to measure or put bounds on the quartic terms[23] we do not expect masses substantially larger than in the Skyrme case.

If the LHC reaches a higher sensitivity than indicated in (19) it should either find non-zero higher-order Lagrangian parameters outside the soliton stability region or discover electroweak baryons. The range of electroweak baryon masses that LHC could detect is likely to be narrower than the range of theoretical estimates quoted above.

As discussed earlier, a stable electroweak solitonic baryon should be a neutral boson on order to be a valid dark matter candidate. Models in which this occurs are quite constrained, as indicated by by Fig. 4 of [26], under the assumption that the Higgs is a pseudo-dilaton. The spin-independent cold dark matter scattering cross section rises with the Skyrmion mass,[26] increasing the discovery potential of heavier electroweak baryons by terrestrial experiments directly searching for dark matter such as XENON100.[28] Clearly, dark matter scattering experiments hold a significant promise as probes of electroweak models with an underlying strongly interacting sector with solitonic baryons.

Acknowledgements

The work of J.E. was supported partly by the London Centre for Terauniverse Studies (LCTS), using funding from the European Research Council via the Advanced Investigator Grant 267352. The work of M.P. was partly supported by the Polish NCN grant 2011/01/B/ST2/00492.

References

1. J. Ellis and M. Karliner, Phys. Lett. B **713** (2012) 233 [arXiv:1204.6642 [hep-ph]].
2. J. Ellis, M. Karliner and M. Praszalowicz JHEP, in print, arXiv:1209.6430 [hep-ph].
3. E. Witten, Nucl. Phys. B **223** (1983) 433.
4. G. S. Adkins, C. R. Nappi and E. Witten, Nucl. Phys. B **228** (1983) 552.
5. For a rigorous example, see N. Seiberg and E. Witten, Nucl. Phys. B **426** (1994) 19 [Erratum-ibid. B **430** (1994) 485] [hep-th/9407087] and N. Seiberg and E. Witten, Nucl. Phys. B **431** (1994) 484 [hep-th/9408099].
6. T.H.R. Skyrme, Proc. Roy. Soc. Lond. A**260** (1961) 127 and Nucl. Phys. **31** (1962) 556.
7. G. S. Adkins and C. R. Nappi, Nucl. Phys. B **233** (1984) 109 and Nucl. Phys. B **249** (1985) 507; M. Karliner and M. P. Mattis, Phys. Rev. Lett. **56** (1986) 428 and Phys. Rev. D **31** (1985) 2833 and Phys. Rev. D **34** (1986) 1991.
8. H. Weigel, Lect. Notes Phys. **743** (2008) 1.
9. S. Weinberg, Phys. Rev. D **13** (1976) 974; L. Susskind, Phys. Rev. D **20** (1979) 2619.
10. G. Aad *et al.* [ATLAS Collaboration], Phys. Lett. B**716** (2012) 1 [arXiv:1207.7214[hep-ex]]; S. Chatrchyan *et al.* [CMS Collaboration], Phys. Lett. B **716** (2012) 30 [arXiv:1207.7235[hep-ex]].
11. G. F. Giudice, C. Grojean, A. Pomarol and R. Rattazzi, JHEP **0706** (2007) 045 [hep-ph/0703164]; R. Contino, C. Grojean, M. Moretti, F. Piccinini and R. Rattazzi, JHEP **1005** (2010) 089 [arXiv:1002.1011 [hep-ph]]; R. Contino, arXiv:1005.4269 [hep-ph]; R. Grober and M. Muhlleitner, JHEP **1106** (2011) 020 [arXiv:1012.1562 [hep-ph]].
12. B. Holdom, Phys. Lett. B **150** (1985) 301; K. Yamawaki, M. Bando and K. i. Matumoto, Phys. Rev. Lett. **56** (1986) 1335; T. W. Appelquist, D. Karabali and L. C. R. Wijewardhana, Phys. Rev. Lett. **57** (1986) 957; T. Akiba and T. Yanagida, Phys. Lett. B **169** (1986) 432; M. Bando, K. -i. Matumoto and K. Yamawaki, Phys. Lett. B **178**, 308 (1986); W. D. Goldberger, B. Grinstein and W. Skiba, Phys. Rev. Lett. **100** (2008) 111802 [arXiv:0708.1463 [hep-ph]]; J. Fan, W. D. Goldberger, A. Ross and W. Skiba, Phys. Rev. D **79** (2009) 035017 [arXiv:0803.2040 [hep-ph]]; L. Vecchi, Phys. Rev. D **82** (2010) 076009; B. Grinstein and P. Uttayarat, JHEP **1107** (2011) 038 [arXiv:1105.2370 [hep-ph]]; S. Matsuzaki and K. Yamawaki, Prog. Theor. Phys. **127**, 209 (2012) [arXiv:1109.5448 [hep-ph]]; V. Barger, M. Ishida and W. Keung, arXiv:1111.2580 [hep-ph]; B. Coleppa, T. Gregoire and H. E. Logan, arXiv::1111.3276 [hep-ph]; C. Coriano, L. Delle Rose, A. Quintavalle and M. Serino, arXiv:1206.0590 [hep-ph]; S. Matsuzaki and K. Yamawaki, Phys. Rev. D **86** (2012) 035025 [arXiv:1206.6703 [hep-ph]]; C. Coriano, L. Delle Rose, C. Marzo and M. Serino, arXiv:1207.2930 [hep-ph]; S. Matsuzaki and K. Yamawaki, arXiv:1207.5911 [hep-ph]; D. Elander and M. Piai, arXiv:1208.0546 [hep-ph]; L. Delle Rose and M. Serino, arXiv:1208.6432 [hep-ph]; S. Matsuzaki and K. Yamawaki, arXiv:1209.2017 [hep-ph]; Z. Chacko and R. K. Mishra, arXiv:1209.3022 [hep-ph]; Z. Chacko, R. Franceschini and R. K. Mishra, arXiv:1209.3259 [hep-ph]; B. Bellazzini *et al.*,, arXiv:1209.3299 [hep-ph].
13. D. E. L. Pottinger and E. Rathske, Phys. Rev. D **33** (1986) 2448; E. Rathske, *Research*

on phenomenological models with chiral symmetry, BONN-IR-91-70.
14. K. Fujii, S. Otsuki and F. Toyoda, Prog. Theor. Phys. **73** (1985) 1287.
15. M. Lacombe, B. Loiseau, R. Vinh Mau and W. N. Cottingham, Phys. Lett. B **161** (1985) 31.
16. J. F. Donoghue, E. Golowich and B. R. Holstein, Phys. Rev. Lett. **53** (1984) 747.
17. A. A. Andrianov, V. A. Andrianov and V. Y. .Novozhilov, Phys. Rev. Lett. **56** (1986) 1882.
18. I. Aitchison, C. Fraser, E. Tudor and J. Zuk, Phys. Lett. B **165** (1985) 162.
19. B. Moussallam, Conf. Proc. C **9209271** (1992) 269 [hep-ph/9211229].
20. G. Holzwarth and H. Walliser, Nucl. Phys. A **587** (1995) 721.
21. G. Ecker, J. Gasser, A. Pich and E. de Rafael, Nucl. Phys. B **321** (1989) 311.
22. A. Pich, PoS Confinement **8** (2008) 026 [arXiv:0812.2631 [hep-ph]].
23. O. J. P. Eboli, M. C. Gonzalez-Garcia and J. K. Mizukoshi, Phys. Rev. D **74** (2006) 073005 [arXiv:hep-ph/0606118].
24. J. Distler, B. Grinstein, R. A. Porto and I. Z. Rothstein, Phys. Rev. Lett. **98** (2007) 041601 [arXiv:hep-ph/0604255].
25. S. Nussinov, Phys. Lett. B **165** (1985) 55.
26. B. A. Campbell, J. Ellis and K. A. Olive, JHEP **1203** (2012) 026 [arXiv:1111.4495 [hep-ph]].
27. T. N. Pham and T. N. Truong, Phys. Rev. D **31** (1985) 3027.
28. E. Aprile *et al.* [XENON100 Collaboration], arXiv:1207.5988 [astro-ph.CO].
29. J. Bagnasco, M. Dine and S. D. Thomas, Phys. Lett. B **320** (1994) 99 [arXiv:hep-ph/9310290].
30. M. Gillioz, JHEP **1202** (2012) 121 [arXiv:1111.2047 [hep-ph]].
31. See the last paper by S. Matsuzaki and K. Yamawaki in [12].

Quark-Gluon Plasma: from RHIC to LHC

Berndt Mueller

Department of Physics, Duke University, Durham, NC 27708, USA

This lecture presents an overview of the status of the investigation of the properties of the quark-gluon plasma using relativistic heavy ion collisions at the Rleativistic Heavy Ion Collider (RHIC) and the Large Hadron Collider (LHC). It focuses on the insights that have been obtained by the comparison between experimental data and theoretical calculations.

Keywords: Quark-gluon plasma; quark matter; Relativistic Heavy Ion Collider; Large Hadron Collider; relativistic heavy ions

1. Introduction

The quark-gluon plasma produced in nuclear collisions at LHC and RHIC is a new form of matter with unique properties:

- It is relativistic, yet strongly coupled;
- it is a liquid that cools into a gas;
- it is a nearly "perfect" liquid near the quantum limit of shear viscosity;
- it thermalizes as fast as causality permits;
- it creates its own new vacuum state to exist.

How is this possible?

Today we do not yet have a complete answer to this question, but theory has made impressive progress towards developing frameworks in which it can be meaningfully addressed and experimental data can be used to clarify those properties of hot QCD matter that theory cannot yet reliably predict from QCD. The theory toolkit in relativistic heavy ion physics is quite diverse. It includes QCD perturbation theory in the vacuum and in a thermal medium (especially for the description of jets and heavy quarkonia); semiclassical gauge theory (for the description of the initial conditions reached in the unclear collision); lattice gauge theory (for static thermodynamic properties of QCD matter, such as its equation of state and color screening); holographic methods mapping strongly coupled gauge theories on their gravity duals (for transport properties and the dynamics of thermalization); and transport theory, especially viscous hydrodynamics (for the evolution of the bulk matter).

2. Properties of QCD Matter

Lattice gauge theory has made impressive progress on the calculation of static thermodynamic properties of baryon symmetric QCD matter. The equation of state at $\mu_B = 0$ for physical quark masses is now known with a precision that exceeds that required in viscous hydrodynamics calculations. The quasi-critical temperature where susceptibilities related to chiral symmetry peak has been determined to lie at $T_c \approx 155$ MeV. While resumed thermal perturbation describes most properties of the quark-gluon plasma well at temperatures above $2 - -3T_c$, its failure at lower temperatures indicates that QCD matter in the range $T_c \leq T \leq 2T_c$ is highly non-perturbative and strongly coupled, making its description theoretically challenging and interesting.

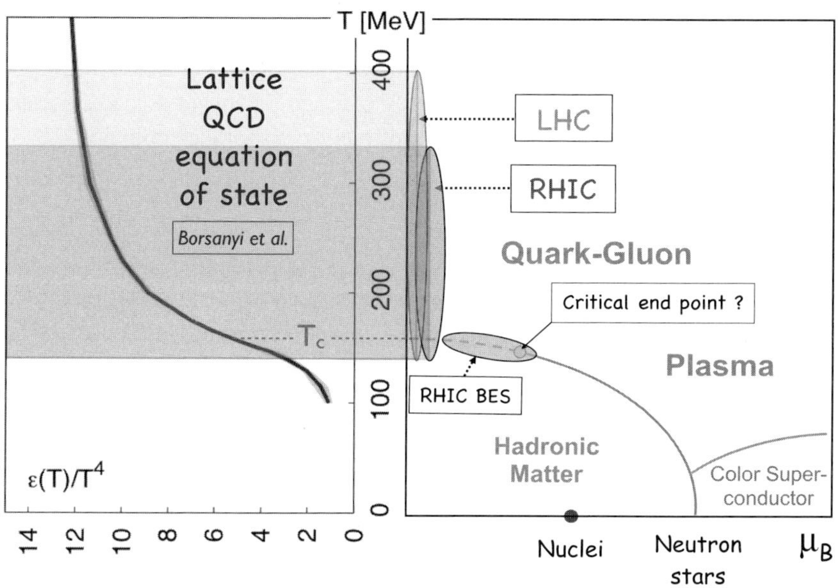

Fig. 1. Phase diagram of QCD matter (right panel) overlaid with regions covered by LHC and RHIC. The experimentally covered ranges are projected onto the energy density versus temperature at $\mu_B = 0$ curve calculated by lattice QCD (left panel).

While RHIC can explore the temperature range only up to $2T_c$, the higher energy range of the LHC makes temperatures up to nearly $3T_c$ accessible (see Fig. 2). Experimental information about the temperatures reached in heavy ion collisions come from the measurement of the spectrum of radiated photons in the energy range $1 - 3$ GeV, where direct photon emission is dominated by thermal radiation. The measurement at a fixed beam energy only yields a temporal average of the temperature of the medium, but hydrodynamical simulations of the time evolution allow us to deduce a lower limit to the initial temperature. In Au+Au collisions at

RHIC ($\sqrt{s_{\rm NN}} = 200$ GeV), this initial temperature exceeds 300 MeV; in Pb+Pb collisions at the LHC ($\sqrt{s_{\rm NN}} = 2.76$ TeV) the initial temperature is at least 37% higher. This increase is in good agreement with the observed scaling of the particle multiplicity from RHIC to LHC, which indicates an increase of at least a factor 3 in the initial energy density.

There are many similarities between the evolution of the matter produced in a relativistic heavy ion collision (the "little bang") and the expansion of the matter filled early universe (the "Big Bang"). In both expansions the initially imprinted quantum fluctuations propagate into macroscopic fluctuations in the final state via the acoustic and hydrodynamic response of the medium. In the Big Bang, the final temperature fluctuations probe the bulk dynamics; photons provide for penetrating probes, and light nuclei serve as chemical probes. In the little bang, the fluctuations in the final flow profile probe the expansion dynamics , photons and jets provide for the penetrating probes, and the various hadron species serve as the chemical probes. In each collision event the information that can be gathered from these probes is limited by the finite particle number; the advantage of the heavy ion experiments is that data can be gathered from many millions or billions of collisions.

3. Probing QCD Matter

It is worthwhile asking which intrinsic properties of the quark-gluon plasma we can hope to determine experimentally and from which observables. A not exhaustive list incudes:

- The equation of state of the matter, given by relations among the components of the energy-momentum tensor $T_{\mu\nu}$ at equilibrium and their temperature dependence are reflected in the spectra of emitted particles. Lattice QCD is able to compute these quantities reliably.
- Transport coefficients of the quark-gluon plasma, especially the shear viscosity η, the coefficient $\hat{q}$ governing the transverse momentum diffusion of a fast parton (often called the *jet quenching parameter*), the coefficient of linear energy loss $\hat{e}$, and the diffusion coefficient κ of a heavy quark, are related to the final-state flow pattern and the energy loss of fast partons that initiate jets. Lattice gauge theory presently cannot reliably calculate these dynamical quantities.
- The static color screening length λ_D (the inverse Debye mass m_D) governs the dissolution of bound states of heavy quarks in the quark-gluon plasma. This static quantity can be reliably calculated on the lattice.
- The electromagnetic response function of the quark-gluon plasma is reflected in the emission of thermal photons and lepton pairs. This dynamical quantity is difficult to calculate on the lattice, but moderate progress has been made recently.

All but the last of these properties is microscopically related to correlation functions of the gauge field. This implies that the associated experimental observables are mostly sensitive to the gluon structure of the quark-gluon plasma. On the other hand, much more is known theoretically from lattice simulations about the quark structure of hot QCD matter, because it is much easier to construct operators from quark fields that can be reliably calculated. In this respect, lattice calculations and heavy ion experiments are to a certain degree complementary. The presence of jets in heavy ion collisions at LHC and RHIC tell us that at high virtuality Q^2 or high momenta p, the QGP is weakly coupled and has quasiparticle structure. On the other hand, the collective flow properties of the matter produced in the collisions tells us that at thermal momentum scales the quark-gluon plasma is strongly coupled. At which Q^2 or p does the does the transition between strong and weak coupling occur? Does the quark-gluon plasma still contain quasiparticles at the thermal scale? Which observables (jets?) can help us pinpoint where the transition occurs?

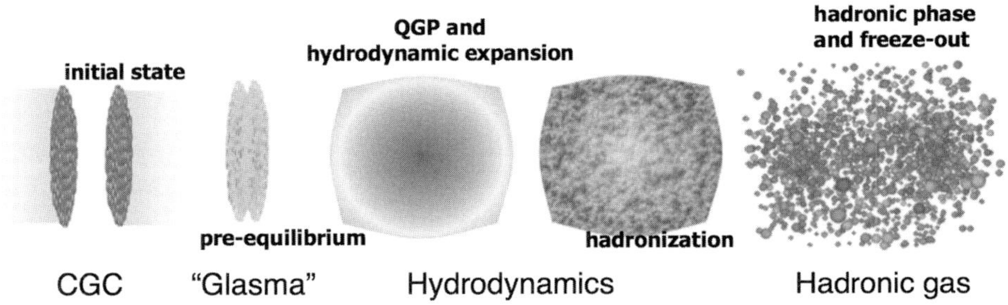

Fig. 2. Stages of a relativistic heavy ion collision in the "standard model" currently in use. The central stage of core interest is the quark-gluon plasma phase, which can be described by viscous hydrodynamics.

Maybe the most important theoretical achievement in the past decade is that a "standard model" of the dynamics of a relativistic heavy ion collision has emerged (see Fig. 3). After a very brief period of equilibration – most likely less than 1 fm/c – the space-time evolution of the quark-gluon plasma can be described by relativistic viscous hydrodynamics with an exceptionally low shear viscosity. After cooling below T_c the matter hadronizes. The subsequent expansion can be described by microscopic Boltzmann dynamics of a multi-component hadron gas until the gas becomes so dilute that all interactions cease and the hadron ensemble freezes out kinetically. The color glass condensate model of gluon saturation in the nuclear wave function at small x has proven to be remarkably successful in predicting the initial energy and entropy deposition in the nuclear collision, and thus the initial conditions of the hydrodynamic evolution. This has made it possible to predict the bulk behavior over a wide range of collision energies and for a wide range of collision systems without arbitrary fit parameters. This remarkable achievement has been possible by the success of hydrodynamics during the dense phase of the collision where no microscopic description could be reliable.

4. The Perfect Liquid

Hydrodynamics is the effective theory of the transport of energy and momentum in matter on long distance and time scales. In order to be applicable to the description of the quark-gluon plasma created in relativistic heavy ion collisions, which forms tiny, short-lived droplets of femtometer size, the hydrodynamic equations must be relativistic and include the effects of (shear) viscosity. The causal relativistic theory of viscous fluid has been worked out in full detail over the past few years. It is based on the framework of the Müller-Israel-Stewart formulation of second-order hydrodynamics, which includes relaxation effects for the dissipative part of the stress tensor. Schematically, the equations have the form

$$\partial_\mu T^{\mu\nu} = 0 \qquad \text{with} \qquad T^{\mu\nu} = (\varepsilon + P)u^\mu u^\nu + \Pi^{\mu\nu} \tag{1}$$

$$\tau_\Pi (d\Pi^{\mu\nu}/d\tau) + \Pi^{\mu\nu} = \eta(\partial^\mu u^\nu + \partial^\nu u^\mu - \text{trace}). \tag{2}$$

It turns out that the quantity that most directly controls the behavior of the fluid, in addition to its equation of state, is the ratio of the shear viscosity η to the entropy density s. The quantity η/s is the relativistic generalization of the well known kinematic viscosity. Since in kinetic theory η is proportional to the mean free path of particles in the fluid, which is inversely proportional to the transport cross section, unitarity limits how small η can become under given conditions. An interesting consequence of this observation is that the quantity eta/s has a lower bound of the order of 0.08 (in units of $\hbar$). The existence of such a bound was conjectured almost 30 years ago, but it was quantitatively derived only recently using the technique of holographic gravity duals, the AdS/CFT duality. It is now believed that $\eta/s \geq .(4\pi)^{-1}$ for any sensible quantum field theory.

This so-called KSS bound is one aspect of a deep and fruitful relationship between the formation of black holes and thermalization in strongly coupled quantum field theories, which associates thermalization on the formation of an event horizon and maps viscous hydrodynamiccs on the information absorbing dynamics of black hole horizons. Although a gravity dual for QCD is still unknown, the AdS/CFT duality has made it possible to explore thermalization and the approach to hydrodynamic behavior rigorously in strongly coupled gauge theories with some similarity to QCD. This has put the models used to describe the dynamics of the quark-gluon plasma in relativistic heavy ion collisions on a much firmer basis.

The experimental handle for the determination of η/s is the azimuthal anisotropy of the flow of final-state particles in off-central heavy-ion collisions, where the nuclear overlap region is elongated in the direction perpendicular to the reaction plane, as shown in Fig. 4. Hydrodynamics converts the anisotropy of the pressure gradient into a flow anisotropy, which sensitively depends on the value of η/s. The average geometric shape of the overlap region in symmetric nuclear collisions is dominated

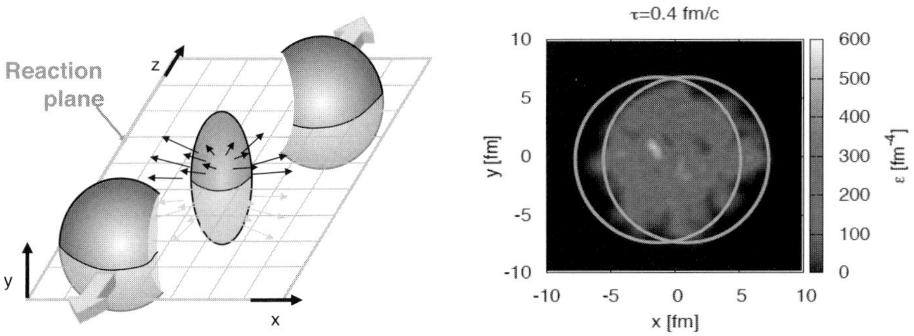

Fig. 3. Geometry of a non-central heavy ion collision (left panel). Density fluctuations in the transverse plane in a sample collision event (right panel).

by the elliptic eccentricity, resulting an elliptic flow anisotropy characterized by the second Fourier coefficient v_2. Event-by-event fluctuations of the density distribution within the overlap region generate higher Fourier coefficients for the initial geometry and final flow, encoded in higher Fourier coefficients v_3, v_4, etc. Their measurement is analogous to the mapping of the amplitudes of multipoles in the thermal fluctuations of the cosmic background radiation.

The precise results of such an analysis of event-by-event fluctuations of the flow distribution depends somewhat on the structure of the initial-state density fluctuations, especially their radial profile and spatial scale. The most complete study of this kind to date, by Schenke et al., starts from the fluctuations of the gluon distribution in the colliding nuclei, evolves them for a brief period using classical Yang-Mills equations, and then inserts the fluctuating energy density distribution into vicious hydrodynamics. The conclusion of this study is that the average value of η/s (averaged over the thermal history of the expansion) in Au+Au collisions at the top RHIC energy is 0.12; whereas the value for Pb+Pb collisions at LHC is 0.20 (see Fig. 4). While each of these values has systematic uncertainties of at least 50%, the ratio of these two values is probably rather stable against changes in the assumptions for the initial state.

This suggests analysis that the average value of η/s at the higher LHC energy is approximately 60% higher than at RHIC, indicating a strong temperature dependence of this quantity. It also indicates that the quark-gluon plasma at the lower temperature reached at RHIC is more strongly coupled and a more "perfect" liquid, making this energy domain especially interesting. Obviously, it would be of interest to make measurements of the flow fluctuations at energies between the top RHIC energy ($\sqrt{s_{\rm NN}} = 0.2$ TeV) and the present LHC energy ($\sqrt{s_{\rm NN}} = 2.76$ TeV). Future challenges include: Can we use p+A or d+A collisions to reduce the uncertainty of the initial state fluctuations in A+A collisions? Is the value for η/s independent of the collision system (Cu+Cu, Cu+Au, U+U)? Do coherent color fields affect the early generation of flow, e.g. in the form of an anomalous viscosity?

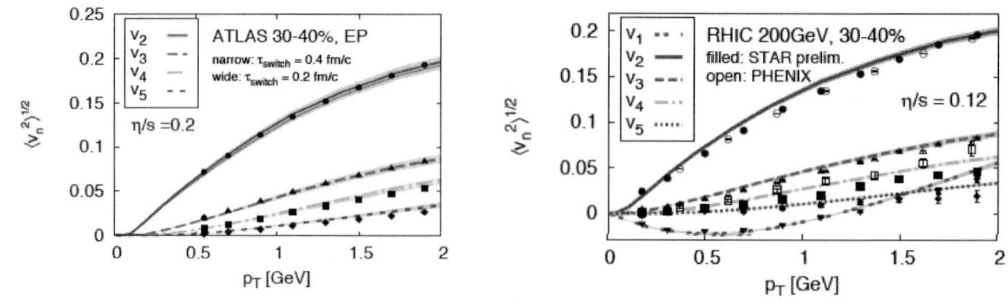

Fig. 4. Fourier components of collective flow, $v_n(p_T)$, for Au+Au collisions at RHIC (left panel) and Pb+Pb collisions at LHC (right panel), in compareison with viscous hydrodynamics calculations. The deduced average value of the ratio η/s is 60% larger at LHC (0.20) than at RHIC (0.12).

5. Jet Quenching

Energetic partons, the precursors of jets, lose energy while traversing the quark gluon plasma either by elastic collisions with the medium constituents or by gluon radiation. At high energies, radiation should dominate; collisional energy loss is expected to be important for intermediate energy partons and for heavy quarks. Each mechanism is encoded in a transport coefficient, $\hat{e}$ for collisional energy loss and $\hat{q}$ for radiative energy loss:

$$(dE/dx)_{\rm coll} = -C_2\hat{e}, \qquad (dE/dx)_{\rm rad} = -C_2\hat{q}L, \tag{3}$$

where L denotes the path length traversed in matter and C_2 is the quadratic Casimir of the fast parton. The value of $\hat{q}$ is given by the transverse momentum broadening of a fast light parton per unit path length.

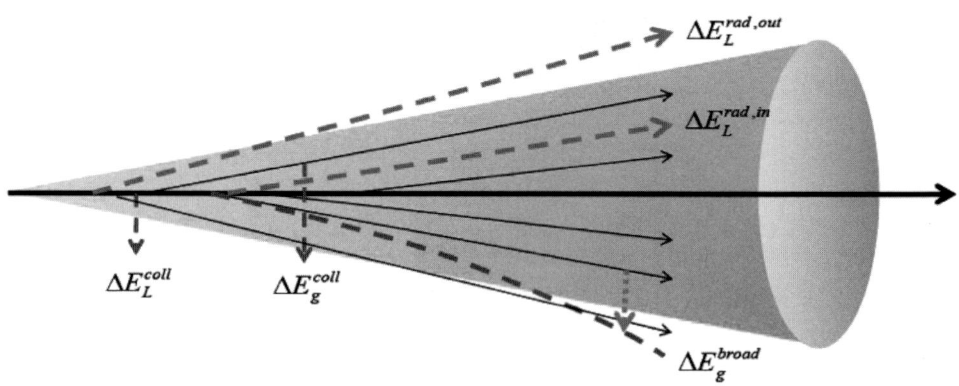

Fig. 5. Matter induced processes that contribute to jet quenching.

The evolution of a jet in the medium, shown schematically in Fig. 5, is characterized by several scales: The initial virtuality $Q_{\rm in}$ associated with the hard scattering process; the transverse scale at which the medium appears opaque, also called the saturation scale Q_s; and the transverse geometric extension of the jet, $r_\perp$. Those components of the jet, for which $r_\perp > Q_s^{-1}$, will be strongly modified by the medium. This means that the core of the jet will remain rather inert, except for an overall energy attenuation of the primary parton, but strong modifications are expected at larger angles and soft components of the jet. Figure 5 presents a schematic sketch of these features of jet modification. They can be encoded in a transport equation for the accompanying gluon radiation:

$$\frac{d}{dt} f(\omega, k_\perp^2, t) = \hat{e}\frac{\partial f}{\partial \omega} + \frac{\hat{q}}{2}\frac{\partial f}{\partial k_\perp^2} + \frac{dN_{\rm rad}}{d\omega dk_\perp^2 dt}, \tag{4}$$

where the last term denotes the gluon radiation induced by the medium.

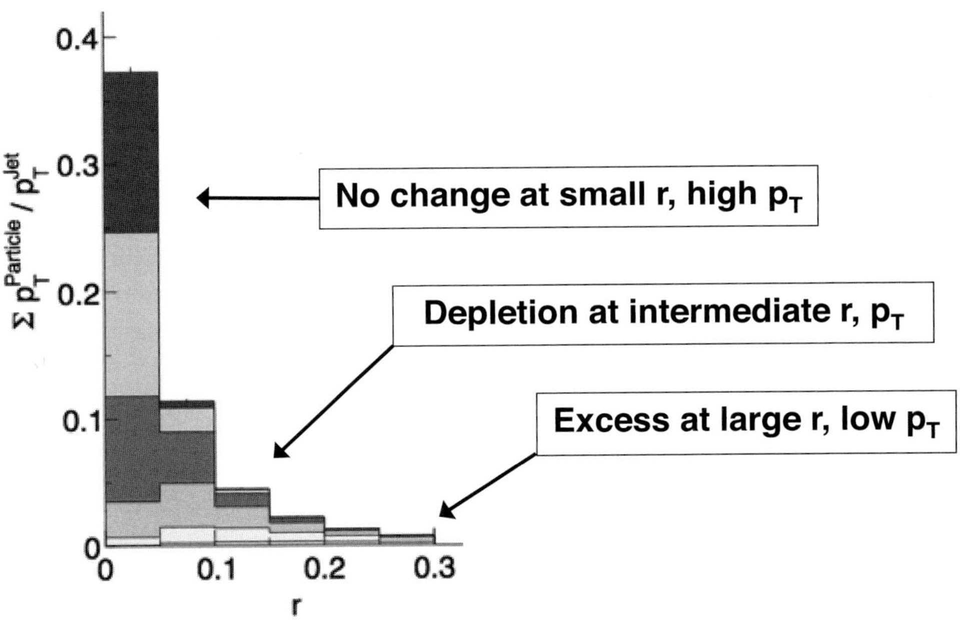

Fig. 6. Schematic representation of jet modifications in relativistic heavy ion collisions.

The "jet quenching parameter" $\hat{q}$ can be determined by analyzing the suppression of leading hadrons in A+A collisions, compared with the scaled p+p data, usually given by a suppression factor R_{AA}, which is of the order of 0.2 for hadrons of transverse momenta in the range of 10 MeV/c in Au+Au at RHIC and Pb+Pb at LHC. Analysis of RHIC and LHC data show that *hatq* grows less than linear with the matter density between RHIC and LHC. This again indicates that the quark-gluon plasma formed at higher temperatures is less strongly coupled. Considering that $T^3/\hat{q}$ should scale like η/s, one finds roughly the same factor – a 60% increase from RHIC to LHC energies.

Using the values of $\hat{q}$ and $\hat{e}$ determined by comparison with R_{AA} data, one can rather well explain the strong increase of the di-jet asymmetry observed in central Pb+Pb collisions at LHC. This gives confidence that the basic mechanisms of jet modification and parton energy loss are reasonably well understood. The phenomenology of jet quenching at the LHC, exemplified by the CMS data from Pb+Pb collisions shown in Fig. 5, agrees qualitatively well with the expectation that modifications are concentrated at large cone angles and soft momentum fractions within the jet. A quantitatively reliable analysis awaits the creation of complete jet Monte-Carlo programs, which include the full panoply of radiative and collisional processes.

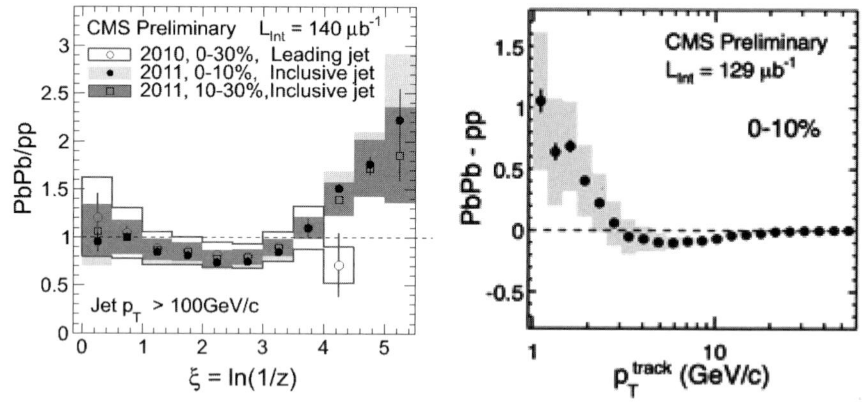

Fig. 7. Jet modifications observed in Pb+Pb collisions at LHC by the CMS collaboration.

The LHC jet quenching data have clearly ruled out speculations that the jet could be strongly coupled to the medium. The perturbative theory of jet evolution works, with appropriate medium modifications, confirming that partons with energies of 10 GeV or more behave like quasiparticles in the quark-gluon plasma. The central remaining question is at which momentum scale partons become strongly coupled and lose their quasiparticle nature. This must be the case at thermal momenta (below 1 GeV) as evidenced by the "perfect" liquid behavior. Improved theoretical tools and precise resolved jet measurements at RHIC will be needed to identify the transition scale between weak and strong coupling.

6. Quarkonium Melting

Bound states of heavy quarks, especially quarkonia (J/ψ, ψ', the Υ states), are sensitive to the distance at which the color force is screened in the quark-gluon plasma. Several mechanisms contribute to nuclear modification of the quarkonium yield (see Fig. 6). At sufficiently high temperatures the screening length becomes shorter than the size of the quarkonium radius and the $Q\overline{Q}$ bound state "melts".

Since the radii of the quarkonium states vary widely – from approximately 0.1fm for the Υ ground state to almost 1 fm for ψ' – the sequential melting of these states could enable at least a semi-quantitative determination of the color screening length.

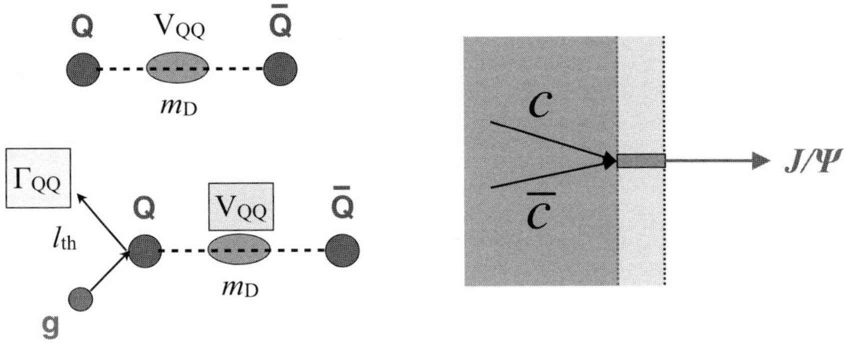

Fig. 8. Mechanisms contributing to matter induced changes in the yield of quarkonia: Color screening (upper left); ionization by thermal gluons (lower left); and recombination (right).

The static screening length, which is relevant for heavy quarks, can be calculated on the lattice. However, it has become well understood in recent years that static color screening is only part of the picture of quarkonium melting, and that quarkonium yields can not only be suppressed by the action of the medium, but also enhanced by recombination, if the density of heavy quarks and antiquarks is large enough. An important loss mechanism is ionization by absorption of thermal gluons. This mechanism gains in importance as the binding energy of a quarkonium state is lowered by color screening. The absorption channel can be included in the dynamical evolution of the amplitude as an imaginary part of the potential with a corresponding noise term ensuring ultimate approach to the equilibrium distribution:

$$i\hbar\frac{\partial}{\partial t}\Psi_{Q\overline{Q}} = \left[\frac{p_Q^2 + p_{\overline{Q}}^2}{2M_Q} + V_{Q\overline{Q}} - \frac{i}{2}\Gamma_{Q\overline{Q}} + \xi_{Q\overline{Q}}\right]\Psi_{Q\overline{Q}}. \tag{5}$$

Recombination of a heavy $Q\overline{Q}$-pair can occur at or near hadronization, similar to the sudden recombination mechanism that is thought to be responsible for the valence quark scaling of the identified particle elliptic flow. The yield of quarkonia formed in this manner grows quadratically with the heavy quark yield. Recombination of charm quark pairs into J/ψ and ψ' is thus expected to be much more frequent at LHC than at RHIC. This expectation seems to be borne out by a comparison of the centrality dependence of J/ψ suppression observed by PHENIX in Au+Au at RHIC and by ALICE in Pb+Pb at LHC (see Fig. 6). The LHC data show

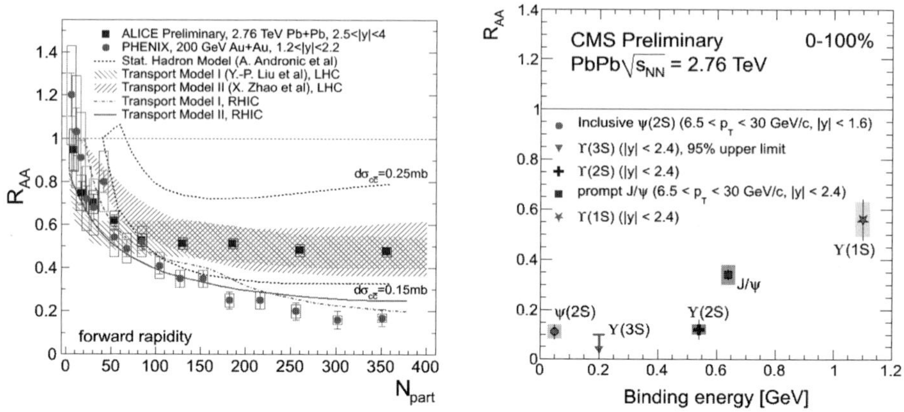

Fig. 9. Left panel: Nuclear suppression factor R_{AA} for J/ψ as function of collision centrality in Au+Au at RHIC (red) and Pb+Pb at LHC (blue). Right panel: R_{AA} in Pb+Pb at LHC for different quarkonium states. The surprisingly small amount of suppression for J/ψ at LHC is evidence for recombination.

less suppression in central collisions than the RHIC data, although the significantly hotter matter proceed at LHC energy must surely be more effective in melting the J/ψ state. Whether it is possible to measure enough observables in order to not only disentangle the action of these different mechanisms, but also determine the color screening length, will need to be seen. On the positive side, the theory of quarkonium transport in hot QCD matter has now reached a state of sophistication where this seems possible.

7. Summary

The insights and open questions generated by the experimental from LHC and RHIC can be summarized as follows:

- The quark-gluon plasma produced at the higher LHC energy is less strongly coupled than that produced at RHIC: The average value of η/s at LHC is larger than at RHIC and the quark-gluon plasma at LHC appears less opaque than at RHIC.
- Event-by-event fluctuations can be used as a versatile probe for the collective response of the quark-gluon plasma to differences in the initial conditions. The beam energy dependence can be used to vary the sensitivity to event-by-event fluctuations. There is reason for hope that details of the initial state structure can be separated from viscous effects, and both can be separately extracted from the data.
- Jet physics opens new avenues of probing the quark-gluon plasma at different scales, because the matter effect on the jet structure creates probes of scales. An important goal is to find the kinematic threshold between

quasiparticle and liquid domains in the quark-gluon plasma. The structure of the matter is reflected in the energy loss mechanisms and thus becomes amenable to investigation.

- Quarkonium spectroscopy is blossoming with the advent of the LHC and its superbly suited detectors. It is now understood thatu quarkonium melting is not just a static screening phenomenon but involve other mechanisms, such as thermal ionization by gluon bombardment. The first data from the LHC suggest that recombination likely dominates in central Pb+Pb collisions for the $c\bar{c}$ states.

Instead of listing individual references to the original literature, which is vast, I list a number of recent review articles where the reader can find extensive lists of the pertinent references.[1–8]

References

1. B. V. Jacak and B. Muller, *Science* **337**, 310 (2012).
2. B. Muller, J. Schukraft and B. Wyslouch, *Ann.Rev.Nucl.Part.Sci.* **62**, 361 (2012).
3. T. Hirano, P. Huovinen, K. Murase and Y. Nara, *Prog.Part.Nucl.Phys.* **70**, 108 (2013).
4. R. Fries and C. Nonaka, *Prog.Part.Nucl.Phys.* **66**, 607 (2011).
5. J. Casalderrey-Solana, H. Liu, D. Mateos, K. Rajagopal and U. A. Wiedemann arXiv:1101.0618 (2011).
6. A. Majumder and M. Van Leeuwen, *Prog.Part.Nucl.Phys.* **A66**, 41 (2011).
7. R. Rapp, D. Blaschke and P. Crochet, *Prog.Part.Nucl.Phys.* **65**, 209 (2010).
8. A. Mocsy, P. Petreczky and M. Strickland, *Int.J.Mod.Phys.* **A28**, p. 1340012 (2013).

Dynamical Model Based on Hydrodynamics for Relativistic Heavy-Ion Collisions

Chiho Nonaka*[AB], Yukinao Akamatsu[A], Shu-ichiro Inutsuka[B]

[A] *Kobayashi-Maskawa Institute for the Origin of Particles and the Universe (KMI), Nagoya University, Nagoya 464-8602, Japan,*

[B] *Department of Physics, Nagoya University, Nagoya 464-8602, Japan*

* *E-mail: nonaka@hken.phys.nagoya-u.ac.jp*

Makoto Takamoto

Max-Planck-Institut für Kernphysik, Postfach 103980, 69029 Heidelberg, Germany

We present a state-of-the-art algorithm for solving the relativistic viscous hydrodynamic equation with QCD equation of state. We apply the algorithm to several numerical test problems such as sound wave propagation and shock tube problem. In the sound wave propagation, the intrinsic *numerical* viscosity is measured and its explicit expression is shown. We apply our numerical algorithm to investigation of higher harmonics in relativistic heavy ion collisions.

1. Introduction

Since the Relativistic Heavy Ion Collider (RHIC) at Brookhaven National Laboratory (BNL) started its operation in 2000, a lot of discoveries have been made and gave us insight of quantum chromodynamics (QCD) phase transition and the QGP.[1] One of the most interesting and surprising outcomes at RHIC was the production of the strongly interacting QGP (sQGP), which was confirmed by both experimental and theory sides. For the achievement, relativistic hydrodynamic models have made a lot of contribution. For example, at that time only hydrodynamic models could explain the strong elliptic flow at RHIC, which was considered as a direct evidence for production of sQGP at RHIC. Because of the success of relativistic hydrodynamic model at RHIC, at present hydrodynamic analysis is a useful and powerful tool for understanding dynamics of hot and dense matter in high energy heavy-ion collisions. Furthermore in 2010 at LHC where collision energy is 10 times larger than that at RHIC heavy-ion collision experiments have been carried.

Currently, main purpose of the phenomenological study for relativistic heavy-ion collisions at RHIC and LHC is sifted to obtain detailed information of bulk property of QGP such as its transport coefficients. Besides, recent high statistical experimental data at RHIC and LHC impose more rigorous numerical treatment on the hydrodynamical models. Recently both at RHIC and LHC the higher harmonic anisotropic flow which is the Fourier coefficient of particle yield as a function of azimuthal angle is reported. One of important origins of the higher harmonics is considered as event-by-event fluctuations. To obtain the precise value of transport coefficients with relativistic viscous hydrodynamics, we need to choose an algorithm with small numerical dissipation and artifact for inviscid part carefully. Usually each algorithm has advantage or disadvantage in terms of coding, computational time, precision and stability. Up to now, unfortunately, only little attention has been paid to numerical aspects in hydrodynamic models for high energy heavy-ion collisions.

2. Hydrodynamic Model

The starting point of hydrodynamic models is the relativistic hydrodynamic equation which is given by the conservation laws of energy, momentum, and baryon number:

$$\partial_\mu T^{\mu\nu}(x) = 0\,, \tag{1}$$

$$\partial_\mu J^\mu_B(x) = 0, \tag{2}$$

where $T^{\mu\nu}(x)$ is the energy-momentum tensor and $J^\mu_B(x)$ is the baryon current. In the case of relativistic ideal fluid, $T^{\mu\nu}(x)$ and $J^\mu_B(x)$ are given uniquely. On the other hand, when one starts to include the effects of dissipation into relativistic hydrodynamics, one is confronted with a rather complicated situation. One of the difficulties is that a naive introduction of viscosities, first order theory (i.e. first order in gradients) suffers from acausality. In order to avoid this problem, several procedures are proposed, which means that several relativistic viscous hydrodynamic equations exist. The problem is that for the application to relativistic heavy-ion collisions we do not obtain a conclusive relativistic viscous hydrodynamic equation. Currently the Israel-Stewart equation,[2] which is derived phenomenologically is widely used for description of QGP dynamics.

Our physics application requires a numerical scheme that can treat a shock wave appropriately and has less numerical dissipation in order to gain comprehensive understanding of recent high energy heavy-ion collision physics. The Riemann solver is a method to calculate numerical flux by using the exact solution of the Riemann problems at the interfaces separating numerical grid cells, and can be used to describe the flows with strong shocks and sharp discontinuity stably and highly accurately. These advantages can be achieved by implementing a Riemann solver for the relativistic ideal hydrodynamics. In particular, we proposed a new Riemann solver for the QCD EoS at low baryon density (For details, please see Ref.[3]), which has not been considered in astrophysical application where baryon density is usually much higher. We derive our Riemann solver by analytically solving the relativistic Riemann problem in low baryon density within the approximation scheme proposed in Ref.[4] We show that from several numerical tests our new algorithm with the Riemann solver has an advantage over other prevailing algorithms such as KT, NT and SHASTA from the point of view of analyses for current relativistic heavy-ion collisions. By implementing our new Riemann solver for relativistic ideal hydrodynamics in a numerical scheme for causal viscous hydrodynamics recently proposed in Ref.,[5] we can also construct a new algorithm for causal viscous hydrodynamics for QGP.

The shock tube problem is analytically solvable for perfect fluid with the free gas EoS. It provides an important test to know performance and accuracy of different numerical schemes. To compare our numerical algorithm to other numerical schemes (SHASTA, NT and KT schemes) and the analytical solution, we start the test calculation with the same initial conditions as those of Ref.[6] The initial temperature on the left is $T_0 = 0.4$ GeV and that on the right is $T_0 = 0.2$ GeV. In the calculation we employ the free gas EoS. The spatial cell size and the Courant number are set to be $\Delta x = 0.1$ and $\lambda = 0.4$, respectively. Because a numerical calculation with enough fine grid and time step should converge on the analytical solution, the same discretization for spacial grid size and time step is important for accuracy test of numerical methods. Figure 1 shows the energy density distribution, the velocity and the invariant expansion rate with our algorithm, KT, NT and SHASTA together with the analytical solution. For these values, KT, NT, and SHASTA algorithms reproduce the analytical solution with almost the same accuracy and numerical artifacts. The difference between the analytical solution and numerical calculations indicates existence of artificial viscosity in numerical schemes. On the other hand, our numerical results are closer to analytical solution especially at $x = 3$ fm compared to KT, NT and SHASTA algorithms which suggests that our numerical algorithm contains less numerical dissipation. This tendency appears clearly in the invariant expansion rate θ in Fig. 1. Only our numerical scheme follows the shape of the analytical solution from $x = 3$ fm to $x = 5$ fm. The numerical dissipation is indispensable for stability of numerical calculation of the relativistic hydrodynamical equation. However too much artificial viscosity smears numerical results and leads a wrong solution which is far from the analytical solution.

From calculation of sound wave propagation we can evaluate the convergence speed of the numerical scheme to grid size and numerical dissipation of the numerical algorithm. The system length is $L_x = \lambda = 2$ fm and is discretized with $N_{\rm cell} = 50, 100, 200$, and 400 cells. We set an initial condition

$$\boldsymbol{V}(x, t=0) = \left(0, \frac{\delta p}{c_{s0}(e_0 + p_0)} \sin\left(2\pi x/\lambda\right), 0, 0, p_0 + \delta p \sin\left(2\pi x/\lambda\right)\right) \equiv \boldsymbol{V}_{\rm init}(x), \quad (3)$$

and impose a periodic boundary condition $\boldsymbol{V}(-\lambda/2, t) = \boldsymbol{V}(\lambda/2, t)$. Here $e_0 \equiv e(p_0, n_B = 0)$, $c_{s0} \equiv c_s(p_0, n_B = 0)$ and $p_0 = 10^3$ fm^{-4} and $\delta p = 10^{-1}$ fm^{-4}. Since the amplitude of the wave is small $\delta p/p_0 = 10^{-4} \ll 1$, the nonlinear hydrodynamic equation is approximated by linearized hydrodynamic equation, which possesses a sound wave mode $\boldsymbol{V}_s(x, t) = \boldsymbol{V}_{\rm init}(x - c_{s0} t)$ as its solution. We analyze the precision of our numerical scheme and its dependence on $N_{\rm cell}$ by calculating the L1 norm

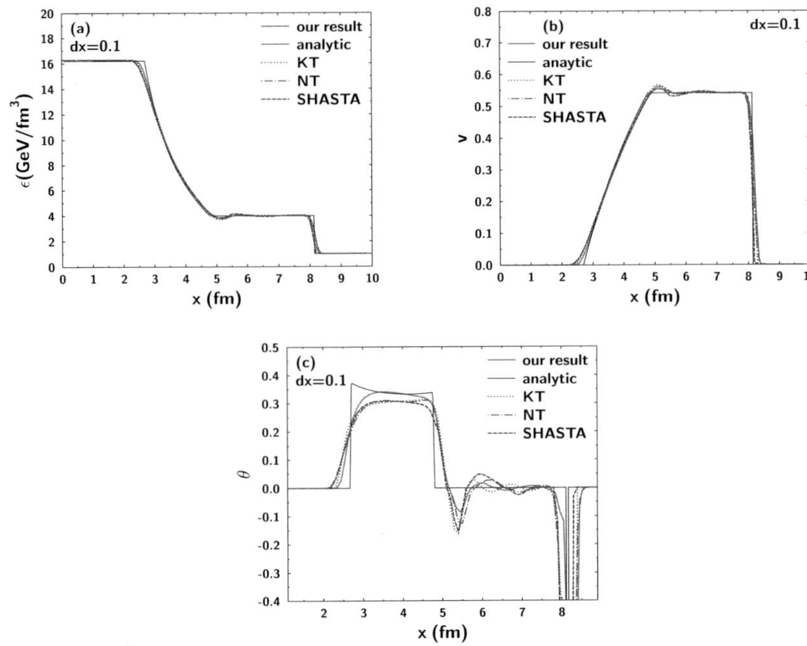

Fig. 1. The analytic (thin line) and numerical solutions of the relativistic Riemann problem on a grid with $N_x = 100$ cells with $\Delta x = 0.1$ fm, after $N_t = 100$ time steps at $t = 4$ fm/c. (a) the energy density distribution e (b) the velocity v (c) the invariant expansion rate θ with our algorithm (solid line) KT (dotted line), NT (dashed-dotted line) and SHASTA (dashed line).

for pressure after one cycle $t = \lambda/c_{s0}$:

$$L(p, p_s; N_{\text{cell}}) = \sum_{i=1}^{N_{\text{cell}}} \mid p(x_i, \lambda/c_{s0}) - p_s(x_i, \lambda/c_{s0}) \mid \frac{\lambda}{N_{\text{cell}}}. \tag{4}$$

We expect a scaling $L(p, p_s; N_{\text{cell}}) \propto (\delta p/N_{\text{cell}}^2) \cdot N_{\text{cell}} \cdot (\lambda/N_{\text{cell}}) = \lambda\delta p/N_{\text{cell}}^2$ after one cycle since our numerical scheme is of second order accuracy with respect to space and time discretization. The result of the L1 norm for the free gas EoS is shown in Fig. 2(a). We indeed find a scaling $L(p, p_s; N_{\text{cell}}) \propto 1/N_{\text{cell}}^2$ for both equations of states, which is consistent with the theoretical expectation.

The simulation of sound wave propagation can also be utilized to estimate the numerical dissipation of the scheme. Since any numerical scheme introduces tiny numerical dissipation, the sound wave in the simulation is attenuated even without physical viscosity. The value of numerical dissipation is evaluated by the value of physical shear viscosity which gives the same amount of sound wave attenuation in the linearized region. In Fig. 2 (b), we show the numerical dissipation of our scheme for the free gas EoS. The numerical dissipation can be approximated by $\eta_{\text{num}} \approx 1000(\Delta x)^2$, where $\Delta x = \lambda/N_{\text{cell}}$. From the dispersion relation and the second order accuracy of our numerical scheme $L(p, p_s; N_{\text{cell}}) \propto \lambda\delta p/N_{\text{cell}}^2 = (\delta p/\lambda) \cdot (\Delta x)^2$, the numerical dissipation is expected to scale with $\eta_{\rm num} \propto [c_{s0}(e_0+p_0)/\lambda]\cdot(\Delta x)^2$. Using the values of p_0, e_0, c_{s0}, and λ, we find

$$\eta_{\rm num} \approx 1\cdot \frac{c_{s0}(e_0+p_0)}{\lambda}(\Delta x)^2. \tag{5}$$

In Fig. 2 (c), we show our numerical result of the sound wave attenuation due to both physical and numerical viscosities by calculating

$$L(p(\eta), p_s; N_{\rm cell}) = \sum_{i=1}^{N_{\rm cell}} \mid p(x_i, \lambda/c_{s0};\eta) - p_s(x_i,\lambda/c_{s0}) \mid \frac{\lambda}{N_{\rm cell}}. \tag{6}$$

It shows that $L(p(\eta), p_s; N_{\rm cell})$ converges to $L_{\rm lin}(\eta)$ with larger $N_{\rm cell}$ at fixed η and the convergence is faster at larger η. This tendency is due to the numerical dissipation $\eta_{\rm num}(\Delta x)$; when $\eta_{\rm num}(\Delta x) \ll \eta$, the discretization effect is expected to be overwhelmed by the physical viscosity. In order to disentangle the physical and numerical viscosities, we calculate the following L1 norm:

$$L(p(\eta), p_s(\eta); N_{\rm cell}) = \sum_{i=1}^{N_{\rm cell}} \mid p(x_i, \lambda/c_{s0};\eta) - p_s(x_i,\lambda/c_{s0};\eta) \mid \frac{\lambda}{N_{\rm cell}}, \tag{7}$$

which eliminates the contribution from sound wave attenuation due to the physical viscosity. The result is shown in the right panel. We find that $L(p(\eta), p_s(\eta); N_{\rm cell})$ does not depend much on the physical viscosity η. This indicates that the numerical dissipation $\eta_{\rm num} \approx [c_{s0}(e_0+p_0)/\lambda]\cdot(\Delta x)^2$ gives a universal estimate of the numerical dissipation of our scheme in the presence of the physical viscosity.

In our numerical scheme, the numerical dissipation is $\eta_{\rm num} \approx [c_{\rm s}(e+p)/\lambda]\cdot(\Delta x)^2 = (c_{\rm s}sT/\lambda)\cdot(\Delta x)^2$, where s denotes the entropy density. Since we are interested in physical viscosity of $\eta \approx (0.1\text{-}1)s$, the condition $(\eta_{\rm num}/\eta) \approx [c_{\rm s}T/(0.1\text{-}1)\lambda]\cdot(\Delta x)^2 \ll 1$ gives $\Delta x \ll 0.8\text{-}2.6$ fm at $T = 500$ MeV and with $\lambda = 10$ fm, which are the typical temperature and system length scale at the relativistic heavy-ion collisions. This condition becomes more severe at higher temperature or when finer structure is of interest. For example, if fluctuations in initial conditions are included, the typical system length becomes $\lambda \sim 1$ fm. It requires the simulation with enough small grid cell size, $\Delta x \ll 0.25\text{-}0.82$ fm. The grid cell size and numerical dissipation depend on numerical scheme. Before we investigate the physical viscosity in relativistic heavy-ion collisions, we need to evaluate the artificial dissipation of the numerical algorithm. Otherwise we will reach the wrong physical viscosity which is smaller than actual physical viscosity of QGP which is produced at RHIC and LHC. We emphasize that appropriate fine grid size calculation is indispensable for any physical observables in heavy-ion collisions, to discuss the value of physical viscosity from comparison with experimental data.

3. Application to Relativistic Heavy-Ion Collisions

We apply our numerical method for the relativistic hydrodynamic equation to investigation of the higher harmonics. We prepare two initial conditions: smooth almond-

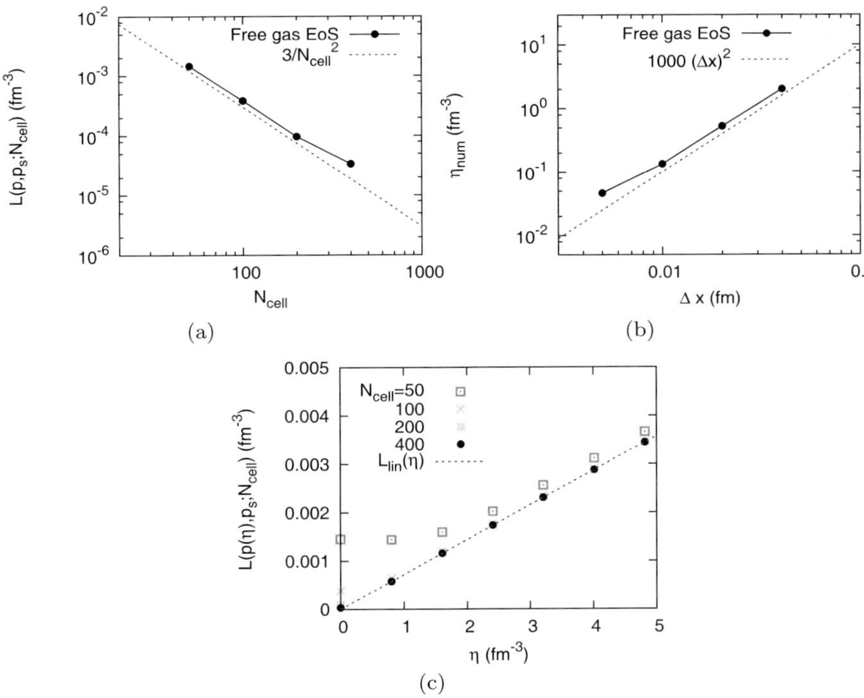

Fig. 2. (a) L1 norm for pressure at $t = \lambda/c_{s0}$ for the free gas EoS. The dotted line indicates a scaling $L(p; N_{\rm cell}) \propto 1/N_{\rm cell}^2$. (b) Numerical dissipation as a function of cell size Δx for the free gas EoS. The dotted line indicates $\eta_{\rm num} \approx 1000(\Delta x)^2$. (c) L1 norm $L(p(\eta), p_s; N_{\rm cell})$ as a function of physical viscosity η.

shaped and fluctuating initial conditions based on the Glauber model. Figure 3 shows time evolution of $v_n = \langle \cos(n(\phi - \Psi)) \rangle$ ($n = 1 \sim 5$) which is summed up with entropy density weight for smooth (left panel) and fluctuating initial conditions (right panel). In the case of smooth initial condition without fluctuations, v_2 is dominant over other higher harmonics, which is a consequence of smooth almond-shaped initial conditions. On the other hand, in the case of fluctuating initial condition not only even n higher harmonics but also odd n higher harmonics become finite, which is induced by event-by-event fluctuations in initial conditions. The energy distribution at final state from fluctuating initial conditions (right panel) shows a more complex structure which is produced by superposition of higher harmonics, compared to the final energy distribution (left panel) in the smooth initial condition.

4. Summary

We developed a state-of-the-art numerical scheme for the relativistic hydrodynamic equation which includes viscosity effects and a shock-wave capturing scheme. We showed that for the inviscid part of our algorithm has less artificial diffusion compared to other algorithms such as SHASTA, NT and KT, which is crucial for physical

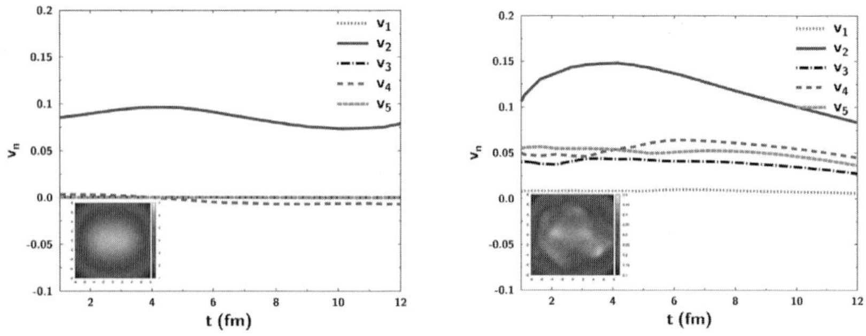

Fig. 3. Time evolution of v_n for smooth initial conditions (a) and that for fluctuating initial conditions (b) together with energy density distributions at final state ($t = 10$ fm).

viscosity analyses from comparison with experimental data. Because a relativistic hydrodynamic model is considered as one of promising dynamical models for understanding experimental data of heavy-ion collision, we need to pay more attention to numerical schemes for solving the relativistic hydrodynamic equation and choose the most suitable numerical scheme for analyses of experimental data. For the viscid part using the PES method we achieved a fast numerical calculation. We employed our numerical algorithm to analyses of the higher harmonics from which we can extract detail features of QGP property. This work is the first step for a realistic description of dynamics of relativistic heavy-ion collisions. Towards construction of realistic dynamical model for relativistic heavy-ion collisions, exploration of realistic initial conditions and freezeout processes with after burner should be taken into account, which is now working in progress.

References

1. R. J. Fries and C. Nonaka, Prog. Part. Nucl. Phys. **66**, 607 (2011) [arXiv:1012.1881 [nucl-th]], C. Nonaka and M. Asakawa, PTEP **2012**, 01A208 (2012) [arXiv:1204.4795 [nucl-th]].
2. W. Israel, Annals Phys. **100**, 310 (1976); W. Israel and J. M. Stewart, Phys. Lett.**A58**, 213 (1976); Annals Phys. **118**, 341 (1979).
3. Y. Akamatsu, S. -i. Inutsuka, C. Nonaka and M. Takamoto, arXiv:1302.1665 [nucl-th].
4. A. Mignone, T. Plewa and G. Bodo, Astrophys. J. S**160**, 199 (2005). [astro-ph/0505200].
5. M. Takamoto and S. Inutsuka, J. Comput. Phys. **230**, 7002 (2011) [arXiv:1106.1732 [astro-ph.HE]].
6. E. Molnar, H. Niemi and D. H. Rischke, Eur. Phys. J. C **65**, 615 (2010) [arXiv:0907.2583 [nucl-th]].

In-Medium QCD Forces at High Temperature

Yukinao Akamatsu

Kobayashi-Maskawa Institute for the Origin of Particles and the Universe (KMI), Nagoya University,
Nagoya 464-8602, Japan
**E-mail: akamatsu@kmi.nagoya-u.ac.jp*

We derive the quantum dynamics of heavy quark systems in quark-gluon plasma as open quantum systems. The scatterings of a heavy quark with the hard medium particles give rise to Debye screened QCD force and drag force and its fluctuation. We present a unified quantum description of these in-medium QCD forces at high temperature in the leading-order perturbation on the basis of the influence functional formalism.

Keywords: Quarkonium; Deconfinement; Quark-gluon plasma; Open quantum system.

1. Introduction

Heavy quarks play important roles in probing properties of the quark-gluon plasma (QGP). A single heavy quark in the QGP interacts with the medium particles and eventually it will be kinetically thermalized as is described by the Langevin equation:[1]

$$\frac{d\vec{p}}{dt} = -\frac{\gamma}{2MT}\vec{p} + \vec{f}(t), \quad \langle f_i(t) f_j(t') \rangle = \gamma \delta_{ij} \delta(t - t'). \tag{1}$$

If one can observe the degree of thermalization in a given time scale, one can derive the drag force of the heavy quark. When put in the QGP, a quarkonium, which is a bound state of a heavy quark-antiquark pair, gets unstable at sufficiently high temperature because the plasma constituents screen the color charges and turn the confining potential into the Debye screened potential:[2]

$$i\frac{\partial \Psi(r,t)}{\partial t} = \left(-\frac{\nabla_r^2}{M} + V(r;T)\right)\Psi(r,t), \quad V(r;T) = -C_{\rm F}\frac{g^2}{4\pi r}e^{-\omega_{\rm D} r}. \tag{2}$$

The Debye screening in the QGP and the related issues are still under active theoretical and conceptual developments.[3–7] If one can observe the "melting" of quarkonium, one can infer the temperature provided that there is a reliable theory for the melting.

These heavy quark probes are fundamentally different from each other: The drag force is a classical description for the non-potential irreversible force while the melting is a quantum phenomenon of bound-states in the screened potential. Moreover it is natural to expect that each heavy quark in a quarkonium experiences the irreversible drag process. In a theoretical point of view, it is desirable to have a unified quantum description for both of them, whose classical limit yields the Langevin dynamics in the single heavy quark case. In other words, we pursue the unified quantum description for in-medium forces in the QGP. The theoretical framework needed for this purpose is that of open quantum systems.

The basics of the open quantum systems[8] can be summarized as follows. When Hilbert space consists of direct product of two Hilbert spaces $\mathcal{H}_{\rm tot} = \mathcal{H}_{\rm sys}\otimes\mathcal{H}_{\rm env}$, the Hamiltonian for the total system is in general $\boldsymbol{H}_{\rm tot} = \boldsymbol{H}_{\rm sys}\otimes\boldsymbol{I}_{\rm env}+\boldsymbol{I}_{\rm sys}\otimes\boldsymbol{H}_{\rm env}+\boldsymbol{H}_{\rm int}$ and the density matrix of the total system evolves according to the von-Neumann equation:

$$\frac{d}{dt}\boldsymbol{\rho}_{\rm tot}(t) = -i[\boldsymbol{H}_{\rm tot}, \boldsymbol{\rho}_{\rm tot}(t)]. \tag{3}$$

The dynamics of the open quantum system is described by the master equation for the reduced density matrix:

$$\frac{d}{dt}\boldsymbol{\rho}_{\rm red}(t) = \mathcal{L}\boldsymbol{\rho}_{\rm red}(t), \quad \boldsymbol{\rho}_{\rm red}(t) \equiv {\rm Tr}_{\rm env}[\boldsymbol{\rho}_{\rm tot}(t)]. \tag{4}$$

In the case of the heavy quarks in the QGP, the system degrees of freedom are the heavy quarks and the environmental degrees of freedom are the light quarks and the gluons. In this article, we review the perturbative derivation of the master equation for the heavy quark systems at high temperature.[9]

2. Influence Functional of Heavy Quarks

The description of heavy quark systems as open quantum systems can be derived on the basis of the closed-time path formalism of non-equilibrium quantum field theory.[10] Applied to QCD, the partition function is

$$\begin{aligned} Z[\eta,\overline{\eta}] = &\int \mathcal{D}[\overline{\psi},\psi,\overline{q},q,A]\langle\overline{\psi},\overline{q},A\ (s=-\infty)|\boldsymbol{\rho}_{\rm tot}|\psi,q,A\ (s=\infty)\rangle \\ &\times \exp\left[i\int_C ds\int d^3x\left\{\mathcal{L}_{\rm QCD}(\overline{\psi},\psi,\overline{q},q,A)+\overline{\eta}\psi+\overline{\psi}\eta\right\}\right], \end{aligned} \tag{5}$$

where ψ and q are the heavy and light quark Dirac fields, A is the gluon field, and gauge fixing is implied in the QCD Lagrangian $\mathcal{L}_{\rm QCD}$. Here C denotes the closed-time path contour which starts from $t=-\infty$, extends to $t=\infty$, and then returns to the original $t=-\infty$. The initial density matrix (in the Schrödinger picture) is defined as $\boldsymbol{\rho}_{\rm tot} = \boldsymbol{\rho}^{\rm th}_{\rm env}\otimes\boldsymbol{\rho}_{\rm sys}$: At $t=-\infty$, a system in arbitrary state $\boldsymbol{\rho}_{\rm sys}$ starts to interact with the environmental degrees of freedom in thermal equilibrium. States $|\psi,q,A\rangle$ and $\langle\overline{\psi},\overline{q},A|$ are the coherent states, satisfying for example $\boldsymbol{\psi}(x)|\psi,q,A\rangle = \psi(x)|\psi,q,A\rangle$.

The dynamics of the open quantum system is described by the influence functional[11] obtained after integrating out the environmental degrees of freedom. The influence functional $S_{\rm FV}[j_1^{a\mu}, j_2^{a\mu}]$ is expressed as a functional of heavy quark color currents $j_1^{a\mu}$ on the forward time axis C_1 and $j_2^{a\mu}$ on the backward time axis C_2

where $\mathcal{C} = \mathcal{C}_1 + \mathcal{C}_2$, and is expanded in terms of the currents with coefficients being the correlation functions of the gluons in thermal equilibrium. In the leading-order perturbation and in the non-relativistic limit, $S_{\rm FV}$ (or $S_{\rm FV}^{\rm LONR}$) consists only of four-Fermi terms whose coefficients are the leading-order gluon two-point function $G^{\rm F}_{ab,00}$. The long-time dynamics of heavy quarks is summarized in the following effective action S_{1+2}:

$$S_{1+2} \equiv S_{\rm kin}^{\rm NR} + S_{\rm FV}^{\rm LONR} + \cdots , \tag{6}$$

$$S_{\rm kin}^{\rm NR} = \int d^4x \left[\mathcal{L}_{\rm kin}^{\rm NR}(\overline{\psi}_1, \psi_1) - \mathcal{L}_{\rm kin}^{\rm NR}(\overline{\psi}_2, \psi_2)\right], \tag{7}$$

$$\begin{aligned} S_{\rm FV}^{\rm LONR} &= -\frac{1}{2}\int dt d^3x d^3y (j_1^{a0}(t,\vec{x}), j_2^{a0}(t,\vec{x})) \begin{pmatrix} V(\vec{x}-\vec{y}) & -iD(\vec{x}-\vec{y}) \\ -iD(\vec{x}-\vec{y}) & -V^*(\vec{x}-\vec{y}) \end{pmatrix} \begin{pmatrix} j_1^{a0}(t,\vec{y}) \\ j_2^{a0}(t,\vec{y}) \end{pmatrix} \\ &\quad - \int dt d^3x d^3y \frac{\vec{\nabla}_x D(\vec{x}-\vec{y})}{4T} \cdot \left(\vec{j}^{a}_{1,\rm NR}(t,\vec{x}) j_2^{a0}(t,\vec{y}) + j_1^{a0}(t,\vec{x}) \vec{j}^{a}_{2,\rm NR}(t,\vec{y})\right), \end{aligned} \tag{8}$$

where $\delta_{ab}V(\vec{x}-\vec{y}) \equiv -ig^2 G^{\rm F}_{ab,00}(\omega = 0, \vec{x}-\vec{y})$, $D(\vec{x}-\vec{y}) = {\rm Im}V(\vec{x}-\vec{y})$, and $\vec{j}^{a}_{\rm NR}$ denotes the non-relativistic limit of the heavy quark color current. With this effective action S_{1+2}, the partition function is given by

$$\begin{aligned} Z[\eta_1, \overline{\eta}_1, \eta_2, \overline{\eta}_2] = &\int \mathcal{D}[\overline{\psi}_1, \psi_1, \overline{\psi}_2, \psi_2] \langle \psi_1(-\infty)|\boldsymbol{\rho}_{\rm S}|\psi_2(-\infty)\rangle \\ &\times \exp\left[iS_{1+2} + i\int d^4x \left(\overline{\eta}_1\psi_1 + \overline{\psi}_1\eta_1 - \overline{\eta}_2\psi_2 - \overline{\psi}_2\eta_2\right)\right]. \end{aligned} \tag{9}$$

From Eq. (9), the time evolution of $\langle\psi_1^\dagger|\boldsymbol{\rho}_{\rm S}(t)|\psi_2\rangle$ is obtained as a "functional Schrödinger equation" with an effective Hamiltonian $\boldsymbol{H}_{1+2}$ derived from S_{1+2}. However, the physically relevant matrix element of the density matrix is not $\langle\psi_1^\dagger|\boldsymbol{\rho}_{\rm S}(t)|\psi_2\rangle$ but $\rho_{\rm S}\left[t, Q^*_{1(c)}, Q_{2(c)}\right] \equiv \langle Q^*_{1(c)}|\boldsymbol{\rho}_{\rm S}(t)|Q_{2(c)}\rangle$, where $Q_{(c)}$s are the non-relativistic Pauli spinors for the heavy (anti)quarks. It needs to take following technical procedures to obtain the functional Schrödinger equation for $\rho_{\rm S}\left[t, Q^*_{1(c)}, Q_{2(c)}\right]$. In these procedures, introduction of conjugate fields $\tilde{\psi}_2 \equiv \psi_2^*$ on $\mathcal{C}_2$ allows a symmetric notation between ψ_1 and $\tilde{\psi}_2$.

(i) The operator ordering in $\boldsymbol{H}_{1+2}$ is determined according to the time ordering of the fields in the path integral formula.

(ii) The functional Schrödinger equation for $\rho_{\rm S}\left[t, Q^*_{1(c)}, \tilde{Q}^*_{2(c)}\right]$ is obtained from $\boldsymbol{H}_{1+2}$ with substitution

$$(\boldsymbol{Q}^\dagger_{1(c)}, \boldsymbol{Q}_{1(c)}) \to \left(Q^*_{1(c)}, \frac{\delta}{\delta Q^*_{1(c)}}\right), \quad (\tilde{\boldsymbol{Q}}^\dagger_{2(c)}, \tilde{\boldsymbol{Q}}_{2(c)}) \to \left(\tilde{Q}^*_{2(c)}, -\frac{\delta}{\delta \tilde{Q}^*_{2(c)}}\right). \tag{10}$$

(iii) The above substitution gives ultraviolet divergent contributions in the four-

Fermi terms arising from the Coulomb potential at the origin, e.g.

$$\int d^3x d^3y V(\vec{x}-\vec{y}) \left[Q^*_{1(c)}(\vec{x}) t^a \frac{\delta}{\delta Q^*_{1(c)}(\vec{x})} \right] \left[Q^*_{1(c)}(\vec{y}) t^a \frac{\delta}{\delta Q^*_{1(c)}(\vec{y})} \right]$$
$$= \int d^3x C_{\mathrm{F}} V(\vec{0}) \left[Q^*_{1(c)}(\vec{x}) \frac{\delta}{\delta Q^*_{1(c)}(\vec{x})} \right] + \cdots , \tag{11}$$

which needs to be renormalized.

(iv) Instead of (ii) and (iii), it is equivalent to derive first the renormalized effective Hamiltonian $\boldsymbol{H}^{\mathrm{ren}}_{1+2}$, in which the operators are reordered in the normal order and the ultraviolet divergences are renormalized, and then do the substitution (ii) to obtain a functional differential operator $H^{\mathrm{ren}}_{1+2}\left[Q^*_{1(c)}, \tilde{Q}^*_{2(c)}\right]$.

We adopt (iv) and obtain the following renormalized effective Hamiltonian $\boldsymbol{H}^{\mathrm{ren}}_{1+2}$:

$$\begin{aligned}
\boldsymbol{H}^{\mathrm{ren}}_{1+2} &= \int d^3x \left[aM \left(\boldsymbol{Q}^\dagger_{1(c)} \boldsymbol{Q}_{1(c)} \right) + \boldsymbol{Q}^\dagger_{1(c)} \left(-\frac{\nabla^2}{2M} \right) \boldsymbol{Q}_{1(c)} \right] \\
&+ \int d^3x \left[a^* M \left(\tilde{\boldsymbol{Q}}^\dagger_{2(c)} \tilde{\boldsymbol{Q}}_{2(c)} \right) + \tilde{\boldsymbol{Q}}^\dagger_{2(c)} \left(-\frac{\nabla^2}{2M} \right) \tilde{\boldsymbol{Q}}_{2(c)} \right] \\
&+ \frac{1}{2} \int d^3x d^3y \, \mathrm{N} \left\{ \left(\boldsymbol{j}^{a0}_1(\vec{x}), \boldsymbol{j}^{a0}_2(\vec{x}) \right) \begin{pmatrix} V(\vec{x}-\vec{y}) & -iD(\vec{x}-\vec{y}) \\ -iD(\vec{x}-\vec{y}) & -V^*(\vec{x}-\vec{y}) \end{pmatrix} \begin{pmatrix} \boldsymbol{j}^{a0}_1(\vec{y}) \\ \boldsymbol{j}^{a0}_2(\vec{y}) \end{pmatrix} \right\} \\
&+ \int d^3x d^3y \frac{\vec{\nabla}_x D(\vec{x}-\vec{y})}{4T} \cdot \mathrm{N} \left\{ \vec{\boldsymbol{j}}^a_{1,\mathrm{NR}}(\vec{x}) \boldsymbol{j}^{a0}_2(\vec{y}) + \boldsymbol{j}^{a0}_1(\vec{x}) \vec{\boldsymbol{j}}^a_{2,\mathrm{NR}}(\vec{y}) \right\} + \cdots ,
\end{aligned} \tag{12}$$

where $a \equiv 1 + C_{\mathrm{F}} V_{\mathrm{med}}/2M$, $V_{\mathrm{med}} \equiv V(\vec{0};T) - V(\vec{0};T=0)$, and N denotes taking the normal-ordered product. From $\boldsymbol{H}^{\mathrm{ren}}_{1+2}$, the functional Schrödinger equation is derived as

$$i\frac{\partial}{\partial t}\rho_{\mathrm{S}}\left[t, Q^*_{1(c)}, \tilde{Q}^*_{2(c)}\right] = H^{\mathrm{ren}}_{1+2}[Q^*_{1(c)}, \tilde{Q}^*_{2(c)}]\rho_{\mathrm{S}}\left[t, Q^*_{1(c)}, \tilde{Q}^*_{2(c)}\right]. \tag{13}$$

The calculation of the gluon two-point functions $G^{\mathrm{F}}_{00,ab}(\omega, \vec{k})$ at leading order in the hard thermal loop resummed perturbation theory yields

$$V(\vec{r}) = \frac{g^2}{4\pi} \left\{ \frac{e^{-\omega_{\mathrm{D}} r}}{r} - 2iT\omega_{\mathrm{D}}^2 \int_0^\infty dk \frac{\sin kr}{r(k^2+\omega_{\mathrm{D}}^2)^2} \right\}, \tag{14}$$

$$D(\vec{r}) = \mathrm{Im}\,[V(\vec{r})] = -\frac{g^2\omega_{\mathrm{D}}^2 T}{2\pi} \int_0^\infty dk \frac{\sin kr}{r(k^2+\omega_{\mathrm{D}}^2)^2}, \tag{15}$$

$$a = 1 + \frac{C_{\mathrm{F}}}{2M}\frac{g^2}{4\pi}(-\omega_{\mathrm{D}} - iT), \ \omega_{\mathrm{D}}^2 \equiv g^2 T^2 \left(\frac{N_{\mathrm{c}}}{3} + \frac{N_{\mathrm{f}}}{6} \right). \tag{16}$$

At this order of perturbation, the effect of t-channel scattering between a heavy quark and hard medium particles is taken into account. This interaction contributes to the diffusion process of the heavy quarks as well as the Debye screening of the color charges. The former is given by the imaginary part of $V(\vec{r})$ and $D(\vec{r})$ while the latter is by the real part of $V(\vec{r})$, and thus these in-medium forces are both contained in the renormalized effective Hamiltonian $\boldsymbol{H}^{\mathrm{ren}}_{1+2}$.

3. Real-time Dynamics of Heavy Quarks

The master equation for the heavy quark systems is obtained from the functional Schrödinger equation Eq. (13). The matrix element of $\boldsymbol{\rho}_{\mathrm{S}}(t)$ between the heavy quark coherent states $\langle Q^*_{1(c)}|$ and $|\tilde{Q}^*_{2(c)}\rangle$ is given as a functional of the heavy quark fields by

$$\rho_{\mathrm{S}}\left[t, Q^*_{1(c)}, \tilde{Q}^*_{2(c)}\right] = \langle Q^*_{1(c)}|\boldsymbol{\rho}_{\mathrm{S}}(t)|\tilde{Q}^*_{2(c)}\rangle, \tag{17}$$

$$\langle Q^*_{1(c)}| \equiv \langle\Omega| \exp\left[-\int d^3x \left\{\boldsymbol{Q}_{(c)}(\vec{x})Q^*_{1(c)}(\vec{x})\right\}\right], \tag{18}$$

$$|\tilde{Q}^*_{2(c)}\rangle \equiv \exp\left[-\int d^3x \left\{\tilde{Q}^*_{2(c)}(\vec{x})\boldsymbol{Q}^\dagger_{(c)}(\vec{x})\right\}\right]|\Omega\rangle. \tag{19}$$

Here $|\Omega\rangle$ is the heavy quark vacuum state. By functionally differentiating $\rho_{\mathrm{S}}\left[t, Q^*_{1(c)}, \tilde{Q}^*_{2(c)}\right]$, we obtain, for example, the reduced density matrix for a single heavy quark system:

$$\begin{aligned} \rho^{ij}_Q(t,\vec{x},\vec{y}) &= \langle \vec{x}, i|\boldsymbol{\rho}_Q(t)|\vec{y}, j\rangle \propto \langle\Omega|\boldsymbol{Q}^i(\vec{x})\boldsymbol{\rho}_{\mathrm{S}}(t)\boldsymbol{Q}^{j\dagger}(\vec{y})|\Omega\rangle \\ &= -\frac{\delta}{\delta Q^{i*}_1(\vec{x})}\frac{\delta}{\delta\tilde{Q}^{j*}_2(\vec{y})}\rho_{\mathrm{S}}\left[t, Q^*_{1(c)}, \tilde{Q}^*_{2(c)}\right]\Big|_{Q^*_{1,2(c)}=0}, \end{aligned} \tag{20}$$

and that for a heavy quark-antiquark system:

$$\begin{aligned} \rho^{ijkl}_{QQ_c}(t,\vec{x}_1,\vec{x}_2,\vec{y}_1,\vec{y}_2) &= \langle \vec{x}_1, i; \vec{x}_2, j|\boldsymbol{\rho}_Q(t)|\vec{y}_1, k; \vec{y}_2, l\rangle \\ &\propto \langle\Omega|\boldsymbol{Q}^i(\vec{x}_1)\boldsymbol{Q}^j_c(\vec{x}_2)\boldsymbol{\rho}_{\mathrm{S}}(t)\boldsymbol{Q}^{l\dagger}_c(\vec{y}_2)\boldsymbol{Q}^{k\dagger}(\vec{y}_1)|\Omega\rangle \\ &= \frac{\delta}{\delta Q^{i*}_1(\vec{x}_1)}\frac{\delta}{\delta Q^{j*}_{1c}(\vec{x}_2)}\frac{\delta}{\delta\tilde{Q}^{l*}_{2c}(\vec{y}_2)}\frac{\delta}{\delta\tilde{Q}^{k*}_2(\vec{y}_1)}\rho_{\mathrm{S}} \\ &\quad\times\left[t, Q^*_{1(c)}, \tilde{Q}^*_{2(c)}\right]\Big|_{Q^*_{1(c)}=\tilde{Q}^*_{2(c)}=0}. \end{aligned} \tag{21}$$

The master equation for each of these reduced density matrices is therefore obtained from the functional Schrödinger equation for $\rho_{\mathrm{S}}\left[t, Q^*_{1(c)}, \tilde{Q}^*_{2(c)}\right]$ in Eq. (13).

For example, let us examine the real-time dynamics of a single heavy quark system at high temperature. The master equation for a single heavy quark system reads

$$\begin{aligned} i\frac{\partial}{\partial t}\rho^{ij}_Q(t,\vec{x},\vec{y}) &= \left\{(a-a^*)M + \left(-\frac{\nabla^2_x - \nabla^2_y}{2M}\right)\right\}\rho^{ij}_Q(t,\vec{x},\vec{y}) \\ &\quad+\frac{1}{2}\left(\delta_{ij}\delta_{kl} - \frac{\delta_{ik}\delta_{jl}}{N_{\mathrm{c}}}\right)\left\{-id(\vec{x}-\vec{y}) + \frac{\vec{\nabla}_x d(\vec{x}-\vec{y})}{4T}\cdot\frac{\vec{\nabla}_x - \vec{\nabla}_y}{iM}\right\} \\ &\quad\times\rho^{kl}_Q(t,\vec{x},\vec{y}), \end{aligned} \tag{22}$$

and that for the color-traced reduced density matrix $\rho_Q(t,\vec{x},\vec{y}) \equiv \rho^{ii}_Q(t,\vec{x},\vec{y})$ is given by

$$\begin{aligned} i\frac{\partial}{\partial t}\rho_Q(t,\vec{x},\vec{y}) &= \left\{(a-a^*)M + \left(-\frac{\nabla^2_x - \nabla^2_y}{2M}\right)\right\}\rho_Q(t,\vec{x},\vec{y}) \\ &\quad+C_{\mathrm{F}}\left\{-id(\vec{x}-\vec{y}) + \frac{\vec{\nabla}_x d(\vec{x}-\vec{y})}{4T}\cdot\frac{\vec{\nabla}_x - \vec{\nabla}_y}{iM}\right\}\rho_Q(t,\vec{x},\vec{y}). \end{aligned} \tag{23}$$

Interestingly, taking short distance limit $d(\vec{x}) \approx d(\vec{0}) + d^{(2)}(\vec{0})\vec{x}^2/2$ in the master equation Eq. (23) reduces it to the well-known master equation of the Caldeira-Leggett model for the quantum Brownian motion.[12] Using the master equation Eq. (23), the following Ehrenfest relations for averaged expectation values $\langle \boldsymbol{O} \rangle(t) \equiv \mathrm{Tr}_x \{\boldsymbol{\rho}_Q(t)\boldsymbol{O}\} = \int d^3x d^3y \langle \vec{x}|\boldsymbol{\rho}_Q(t)|\vec{y}\rangle\langle\vec{y}|\boldsymbol{O}|\vec{x}\rangle$ are derived:

$$\frac{d}{dt}\langle \vec{\boldsymbol{x}} \rangle = \frac{\langle \vec{\boldsymbol{p}} \rangle}{M}, \quad \frac{d}{dt}\langle \vec{\boldsymbol{p}} \rangle = -\frac{\gamma}{2MT}\langle \vec{\boldsymbol{p}} \rangle, \tag{24}$$

$$\frac{d}{dt}\left\langle \frac{\boldsymbol{p}^2}{2M} \right\rangle = -\frac{\gamma}{MT}\left(\left\langle \frac{\boldsymbol{p}^2}{2M} \right\rangle - \frac{3T}{2} \right). \tag{25}$$

To be explicit,

$$\begin{aligned}\gamma &= \frac{C_{\mathrm{F}}}{3}\nabla^2 d(\vec{x})|_{x=0} = -\frac{C_{\mathrm{F}} g(T)^2}{9}\nabla^2 \overline{G}^{>}_{aa,00}(\vec{x})|_{x=0} \\ &= \frac{C_{\mathrm{F}} g(T)^2}{9}\int \frac{d^3k}{(2\pi)^3} k^2 G^{>}_{aa,00}(\omega = 0, \vec{k}),\end{aligned} \tag{26}$$

which is consistent with the leading-order perturbative calculation of the drag force.[13] In this way, we can derive the consequences of the classical Langevin dynamics through the quantum Ehrenfest relations.

4. Summary

In this article, we reviewed the derivation of the master equation for the reduced density matrix of heavy quark systems at high temperature. The derivation is based on the influence functional formalism applied to the heavy quark systems as the open quantum systems and is explicitly shown in the leading-order perturbation. The essential information is contained in the renormalized effective Hamiltonian $\boldsymbol{H}^{\mathrm{ren}}_{1+2}$, from which the master equation is derived.

Note that this method is applicable to systems with any finite number of heavy quarks and is not limited to the leading-order perturbation; once the renormalized effective Hamiltonian is derived, it is straight forward to obtain the master equation. As explained in Ref.,[9] the forward propagator of heavy quarks can also be derived from the renormalized effective Hamiltonian. The complex potential between heavy quark-antiquark pair is defined in terms of the forward propagator and thus can also be derived from our approach to yield consistent result with Refs.[3,4] in the leading-order perturbation.

Here we list several next steps of this approach:

- Derivation of the renormalized effective Hamiltonian in the next-to-leading order of perturbation. We expect the next-to-leading order contribution will introduce new processes, namely the so-called gluo-dissociation processes.[14,15]
- Consideration of how to extend this formalism to the general non-perturbative situation. One of the challenges is to understand how to describe heavy-light

meson bound states solely in terms of heavy quark fields, or to examine what is the correct choice of the system variables as the open quantum system in such cases.

- Phenomenological application to the dimuon spectra from bottomonia decays at the LHC.[16] If one has a reliable theory of melting, one can study the physics of deconfinement by comparing with the experimental data.

Acknowledgments

This work was supported by the Sasakawa Scientific Research Grant from the Japan Science Society.

References

1. B. Svetitsky, Phys. Rev. D **37**, 2484 (1988).
2. T. Matsui and H. Satz, Phys. Lett. B **178**, 416 (1986).
3. M. Laine, O. Philipsen, P. Romatschke and M. Tassler, JHEP **0703**, 054 (2007) [hep-ph/0611300]; A. Beraudo, J. -P. Blaizot and C. Ratti, Nucl. Phys. A **806**, 312 (2008) [arXiv:0712.4394 [nucl-th]].
4. N. Brambilla, J. Ghiglieri, A. Vairo and P. Petreczky, Phys. Rev. D **78**, 014017 (2008); N. Brambilla, M. A. Escobedo, J. Ghiglieri, J. Soto and A. Vairo, JHEP **1009**, 038 (2010) [arXiv:1007.4156 [hep-ph]].
5. A. Rothkopf, T. Hatsuda and S. Sasaki, Phys. Rev. Lett. **108**, 162001 (2012) [arXiv:1108.1579 [hep-lat]].
6. J. L. Albacete, Y. V. Kovchegov and A. Taliotis, Phys. Rev. D **78**, 115007 (2008) [arXiv:0807.4747 [hep-th]]; J. Noronha and A. Dumitru, Phys. Rev. Lett. **103**, 152304 (2009) [arXiv:0907.3062 [hep-ph]]; T. Hayata, K. Nawa and T. Hatsuda, arXiv:1211.4942 [hep-ph].
7. Y. Akamatsu and A. Rothkopf, Phys. Rev. D **85**, 105011 (2012) [arXiv:1110.1203 [hep-ph]].
8. H.-P. Breuer and F. Petruccione, *The Theory of Open Quantum Systems* (Oxford University, New York, 2002).
9. Y. Akamatsu, Phys. Rev. D **87**, 045016 (2013) [arXiv:1209.5068 [hep-ph]].
10. J. S. Schwinger, J. Math. Phys. **2**, 407 (1961); L. V. Keldysh, Zh. Eksp. Teor. Fiz. **47**, 1515 (1964) [Sov. Phys. JETP **20**, 1018 (1965)]; P. M. Bakshi and K. T. Mahanthappa, J. Math. Phys. **4**, 1 (1963); J. Math. Phys. **4**, 12 (1963).
11. R. P. Feynman and F. L. Vernon, Jr., Annals Phys. **24**, 118 (1963).
12. A. O. Caldeira and A. J. Leggett, Physica **121A**, 587 (1983).
13. G. D. Moore and D. Teaney, Phys. Rev. C **71**, 064904 (2005) [hep-ph/0412346]; S. Caron-Huot and G. D. Moore, Phys. Rev. Lett. **100**, 052301 (2008) [arXiv:0708.4232 [hep-ph]]; JHEP **0802**, 081 (2008) [arXiv:0801.2173 [hep-ph]].
14. D. Kharzeev and H. Satz, Phys. Lett. B **334**, 155 (1994) [hep-ph/9405414]; X. -M. Xu, D. Kharzeev, H. Satz and X. -N. Wang, Phys. Rev. C **53**, 3051 (1996) [hep-ph/9511331].
15. N. Borghini and C. Gombeaud, arXiv:1103.2945 [hep-ph]; Eur. Phys. J. C **72**, 2000 (2012) [arXiv:1109.4271 [nucl-th]].
16. S. Chatrchyan *et al.* [CMS Collaboration], Phys. Rev. Lett. **107**, 052302 (2011) [arXiv:1105.4894 [nucl-ex]]; JHEP **1205**, 063 (2012) [arXiv:1201.5069 [nucl-ex]]; arXiv:1208.2826 [nucl-ex].

Holographic Mean-Field Theory for Baryon Many-Body Systems *

Masayasu Harada[(a)†], Shin Nakamura[(b)‡] and Shinpei Takemoto[(a)§]

[(a)] *Department of Physics, Nagoya University, Nagoya 464-8602, Japan*
[(b)] *Department of Physics, Kyoto University, Kyoto 606-8502, Japan*

We report the recent work done in Ref. 1, where we propose a mean-field approach to analyze many-body systems of fermions in the gauge/gravity duality. We introduce a non-vanishing classical fermionic field in the gravity dual, which we call the holographic mean field for fermions. The holographic mean field takes account of the many-body dynamics of the fermions in the bulk. We predict the equation of state between the chemical potential and the baryon number density.

Keywords: dense matter, holographic QCD

1. Introduction

Many-body physics of strongly-correlated fermions is one of the central subjects in modern physics. Strongly-correlated fermions appear in various places, such as the systems of cold atoms, high-T_c superconductors, neutron stars and the quark-gluon plasma. However, the theoretical description of many-body systems of strongly-correlated fermions is still a challenge owing to their non-perturbative natures. One of the successful computational frameworks for strongly-interacting fermions is the lattice gauge theory, but the Monte Carlo simulation at finite baryon density suffers from the sign problem.

Recently, a great deal of attention has been paid to the AdS/CFT correspondence[2–4] because of its ability for non-perturbative computations and the absence of the sign problem. However, the current framework of the AdS/CFT correspondence for finite baryon density still has room for improvement in the confinement phase. It can be summarized as follows: 1) dynamical baryons at finite density have not been satisfactorily incorporated, and 2) the Gubser-Klebanov-Polyakov-Witten (GKP-Witten) prescription[3,4] to relate the baryon chemical potential to the baryon density has an ambiguity.[5]

In Ref. 1, we proposed a new approach, a holographic mean-field theory for

*Talk given by M. Harada.
†E-mail: harada@hken.phys.nagoya-u.ac.jp
‡E-mail: nakamura@hken.phys.nagoya-u.ac.jp
Address after March 1, 2013: Department of Physics, Nagoya University, Nagoya 464-8602, Japan
§E-mail: takemoto@hken.phys.nagoya-u.ac.jp

fermions, to resolve these problems. The self-consistent incorporation of dynamical baryons in our approach provides an eigenvalue equation which fixes the relationship between the baryon chemical potential and the baryon density unambiguously. In this write-up, we summarize main points of the work.

Let us consider a five-dimensional (5D) dual geometry without an event horizon whose metric depends only on the radial direction, for example. It is convenient, although not necessary, to make a coordinate transformation after which the geometry is given by the conformally flat metric,

$$ds_{5d}^2 = H(w)(\eta_{\mu\nu}dx^\mu dx^\nu - dw^2)\,, \tag{1}$$

where $\eta_{\mu\nu} = \mathrm{diag}(+,-,-,-)$ and w is the radial direction. An example of $H(w)$ is given in Ref. 6. We introduce a 5D Dirac spinor Ψ which interacts with a bulk $U(1)$ gauge field A_M on this geometry. The actions for them are

$$\begin{aligned} S_\Psi &= \int d^4x dw \left[i\overline{\Psi}\Gamma^M(\partial_M - iqA_M)\Psi - m_5(w)\overline{\Psi}\Psi \right] , \\ S_A &= \int d^4x dw \mathcal{L}_A \,, \end{aligned} \tag{2}$$

where q and $m_5(w)$ denote the charge and the 5D mass of the fermion, respectively. Γ^M are written in terms of the four dimensional (4D) Dirac matrices as $\Gamma^\nu = \gamma^\nu$ $(\nu = 0,1,2,3)$ and $i\Gamma^w = \gamma^5$. The possible boundary terms and the counter terms are omitted here. Note that the Dirac field is on the curved spacetime (1), but the vielbein and the spin connection are absorbed by a suitable field redefinition of the fermion field.[6] The w dependence of $m_5(w)$ reflects the curvature of the geometry. Typically, $m_5(w)$ diverges at the boundary, and its explicit example is given in.[6] We assume that S_A is a functional of $\partial_N A_M$, which may be given by a Dirac-Born-Infeld (DBI) action on the flavor brane, for example. However, its explicit form is not important at this stage. We employ the probe approximation such that the fermion field and the gauge field do not modify the geometry: we regard $S = S_\Psi + S_A$ as the total action up to the constant action of the gravity sector.

We now explain how we formulate the holographic mean-field approach. We work in the $A_w(x,w) = 0$ gauge, and replace the gauge field with the mean field as

$$A_0(x,w) \to A_0(w)\,, \quad A_i(x,w) \to A_i(w)\,, \tag{3}$$

where $i = (1,2,3)$. We impose

$$A_0(w)\big|_{\mathrm{boundary}} = \mu\,, \quad A_i(w)\big|_{\mathrm{boundary}} = 0\,, \tag{4}$$

where μ is the chemical potential associated with the fermion charge. We also introduce the mean field for the fermion as

$$\Psi(x,w) \to \Psi(w)\,. \tag{5}$$

The non-vanishing mean field realizes the dynamical distribution of the charge in w direction, which is given by

$$\rho(w) = q\Psi^\dagger(w)\Psi(w)\,. \tag{6}$$

The integration of $\rho(w)$ over w gives the 4D charge density n. In the present analysis, we switch off the fermionic source:

$$\Psi(w)\big|_{\text{boundary}} = 0\,. \tag{7}$$

We also require that the mean fields are regular.

Now $A'_0 \equiv \partial_w A_0$ is given by the Gauss-law constraint,

$$\partial_w[\partial\mathcal{L}_{\text{A}}/\partial A'_0(w)] - q\Psi^\dagger(w)\Psi(w) = 0\,, \tag{8}$$

and $A'_i \equiv \partial_w A_i$ is given by

$$\partial_w[\partial\mathcal{L}_{\text{A}}/\partial A'_i(w)] - q\Psi^\dagger(w)\Gamma^0\Gamma^i\Psi(w) = 0\,. \tag{9}$$

The scalar potential $A_0(w)$ is given by using $A'_0(w)$:

$$A_0(w) = \mu + \int_{\text{boundary}}^{w} dw' A'_0(w') \equiv \mu + \tilde{A}_0(w)\,. \tag{10}$$

However, the boundary value μ, that is the constant of integration, is arbitrary at this stage. In other words, whatever value of μ can give the same charge distribution. This has been a problem in holographic models at finite density. This problem will be resolved by solving the equation of motion for $\Psi(w)$:

$$\left[i\Gamma^w\partial_w + q\Gamma^0 A_0(w) + q\Gamma^i A_i(w) - m_5(w)\right]\Psi(w) = 0 \tag{11}$$

that depends on the boundary value of $A_0(w)$ explicitly.

Let us see how it works. We write $\Psi(w)$ in terms of the two-component spinors $\Psi_\pm$ as $\Psi(w) = (\Psi_+, \Psi_-)^T$ with the Dirac matrices given by

$$i\Gamma^w = \begin{pmatrix} 0 & 1 \\ 1 & 0 \end{pmatrix}\,,\ \Gamma^0 = \begin{pmatrix} 1 & 0 \\ 0 & -1 \end{pmatrix}\,,\ \Gamma^i = \begin{pmatrix} 0 & \sigma^i \\ -\sigma^i & 0 \end{pmatrix}\,. \tag{12}$$

We can take $A_1(w) = A_2(w) = 0$, and hence $\sigma^3\Psi_\pm = \Psi_\pm$ without loss of generality by virtue of the rotational invariance. From (11), we obtain

$$\begin{aligned} &[\partial_w - qA_3(w)]\Psi_+ - [m_5(w) + q\tilde{A}_0(w)]\Psi_- = q\mu\Psi_-\,, \\ &[\partial_w + qA_3(w)]\Psi_- - [m_5(w) - q\tilde{A}_0(w)]\Psi_+ = -q\mu\Psi_+\,. \end{aligned} \tag{13}$$

Solving (13) coupled with (8) and (9), $q\mu$ is obtained as the eigenvalue for a given value of $n = \int dw\, q\Psi^\dagger(w)\Psi(w)$. The point here is that the eigenvalue equations (13) are nonlinear equations after rewriting the gauge fields in terms of Ψ_+ and Ψ_-, and the amplitudes of Ψ_+ and Ψ_- are not arbitrary if we specify $q\mu$. If we start with a given $q\mu$, the density n is determined as a function of $q\mu$. In this way, the holographic mean-field theory provides the equation of state (EOS): the relation between the 4D charge density n and the chemical potential μ is determined dynamically. The new bulk condition to fix the ambiguity of μ in the GKP-Witten prescription is the regularity of the fermionic mean field under (7).

Once the mean fields are obtained, the energy spectrum of the physical excitations can be computed by introducing the fluctuations $\psi(x, w), a_\nu(x, w)$ on top of the mean fields as $\Psi(x,w) = \Psi(w) + \psi(x,w)$ and $A_\nu(x,w) = A_\nu(w) + a_\nu(x,w)$, and solving the coupled equations of motion for the fluctuations. One of the equations of motion is given by

$$[i\Gamma^w\partial_w + \Gamma^0(p_0 + qA_0(w)) + \Gamma^3 qA_3(w) + \vec{\Gamma}\cdot\vec{p} \\ - m_5(w)]\psi(p_0,\vec{p},w) + \Gamma^\nu\Psi(w)qa_\nu(p_0,\vec{p},w) = 0\,. \tag{14}$$

Let us discuss the zero density limit, $n \to +0$. (8) and (9) tell us that $\tilde{A}_0(w) \to 0, A_3(w) \to 0$, and (13) becomes

$$\partial_w\Psi_+ - m_5(w)\Psi_- = q\mu\Psi_-\,, \\ \partial_w\Psi_- - m_5(w)\Psi_+ = -q\mu\Psi_+\,. \tag{15}$$

On the other hand, at $\mu = 0$, all the mean fields vanish, then (14) is reduced to be

$$\partial_w\psi_+ - m_5(w)\psi_- = m_f\psi_-\,, \\ \partial_w\psi_- - m_5(w)\psi_+ = -m_f\psi_+\,, \tag{16}$$

where $\psi = (\psi_+,\psi_-)^T$ and we used $\Gamma^\nu p_\nu\psi = m_f\psi$.

The eigenvalue m_f of (16) for the non-vanishing eigenfunction under the boundary condition $\psi|_{\text{boundary}} = 0$ is the mass of the fermion. Therefore, the threshold of the chemical potential for the non-vanishing density given by (15) coincides with the fermion mass, i.e., $q\mu \to m_f$ at the limit $n \to +0$. We do not find any nontrivial solutions for $\Psi(w)$ for $|q\mu| < m_f$, where there is no formation of the Fermi surface.

Now we demonstrate how to obtain the EOS for baryon many-body systems. We apply our method to the model given in Ref. 6 where the baryon field is introduced into the Sakai-Sugimoto model.[7] The boundaries in this model are located at $w = \pm w_{\max}$, thus the boundary conditions for the mean fields are $A_0(\pm w_{\max}) = \mu$, $A_3(\pm w_{\max}) = 0$, $\Psi(\pm w_{\max}) = 0$. Here μ is the baryon chemical potential and $q = 1$. We assume that the theory is invariant under the 5D parity $(t,\mathbf{x},w) \to (t,-\mathbf{x},-w)$.[7] Therefore, $A_0(w)$ ($A_3(w)$) is an even (odd) function of w, and the regularity condition leads to $A_0'(0) = 0$ ($A_3(0) = 0$). From the parity invariance, Ψ_+ (Ψ_-) is either an even (odd) function of w or an odd (even) function. Since (13) is invariant under $(\Psi_+,\Psi_-,q) \to (\Psi_-,\Psi_+,-q)$, the assignment of the parity is a convention up to the sign of q. In this demonstration, we assign even parity to Ψ_+, and the regularity condition we employ is $\Psi_+'(0) = 0$.

We use the DBI action of the D8-branes as S_A:

$$S_A = -C\int d^4x dw U^{1/4}(w) \\ \times\sqrt{H^3(w)\{H^2(w) - (2\pi l_s^2/N_c)^2[(A_0'(w))^2 - (A_3'(w))^2]\}}\,, \tag{17}$$

where $C = (2\pi)^{-7}(4\pi^2/3)l_s^{-11/2}\lambda^{1/4}(2M_{\text{KK}})^{-1/4}N_f N_c$, and the $A_\nu(w)$ here is N_c times $A_\nu(w)$ in the standard convention. Here, M_{KK} is the energy scale of the theory, λ is the 't Hooft coupling, N_f is the number of flavors, and l_s is the string length which does not show up in the end of the calculation. The definitions of $U(w)$ and $H(w)$ are given in.[6]

We fix $M_{\rm KK}$ and λ in the same way as in:[7] $M_{\rm KK} = 949\,{\rm MeV}$ and $\lambda N_c = 50$. We consider $N_f = 2$ case, and use the normal nuclear matter density $n_0 = 0.16\,{\rm fm}^{-3}$ to scale the baryon number density n. Figure 1(a) shows the resultant EOS compared with the one for the free baryons. The chemical potential μ increases as the density grows. The increase of the chemical potential of free baryons is very tiny on the figure, whereas that of our dynamical baryons is considerably larger: this can be understood as an effect of repulsive interactions among the baryons.

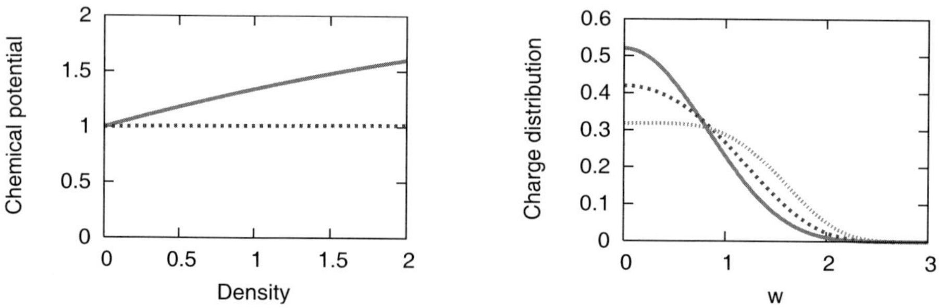

Fig. 1. (a) [Left Panel] The EOS of the model (solid) compared with the one for the free baryons (dashed) given by $\mu = \sqrt{m_n^2 + p_F^2}$, where m_n and p_F are the mass and the Fermi momentum of the baryons, respectively. The horizontal axis is n/n_0, while the vertical axis is μ/m_n.
(b) [Right Panel] The baryon charge distribution $\rho(w)/n$ for $n/n_0 = 0.1$ (solid), 1 (dashed) and 2 (dotted). The boundary is located at $w = w_{\rm max} \simeq 3.64/M_{\rm KK}$. The unit of w on the figure is $1/M_{\rm KK}$.

Figure 1(b) shows the density dependence of the baryon charge distribution $\rho(w)$, which is indeed regular. The distribution shifts to the boundary side as the density grows. This implies that taking account of the distribution of the baryon charge along the radial direction becomes more important in the higher density region.

Finally, we give some discussions on future directions. It is interesting to study how the equation of state is changed when higher excited states of baryons are included.[8] It is also important to study the effect of the scalar meson which gives an atractive force.[9] We also expect that our approach is applicable to holographic descriptions of various systems of fermions.[10]

Acknowledgments

We thank D. K. Hong, Y. Kim, M. Rho, T. Sakai, H. Suganuma and P. Yi for useful discussions and S. Hartnoll for the comment on the first version. This work was initiated at the workshop Dense11 at YITP. This work was supported in part by the JSPS Grant-in-Aid (GIA) ♯ 22224003 and ♯ 24540266 (M.H.) and ♯ 23654132 (S.N.), the GIA for Scientific Research on Innovative Areas No. 2104 (M.H. and S.N.), Nagoya Univ. GCOE Program QFPU (M.H. and S.T.) from MEXT.

References

1. M. Harada, S. Nakamura and S. Takemoto, Phys. Rev. D **86**, 021901 (2012) [arXiv:1112.2114 [hep-th]].
2. J. M. Maldacena, Adv. Theor. Math. Phys. **2**, 231 (1998) [Int. J. Theor. Phys. **38**, 1113 (1999)].
3. S. S. Gubser, I. R. Klebanov and A. M. Polyakov, Phys. Lett. B **428**, 105 (1998).
4. E. Witten, Adv. Theor. Math. Phys. **2**, 253 (1998).
5. See, e.g., S. Nakamura, Prog. Theor. Phys. **119**, 839 (2008).
6. D. K. Hong, M. Rho, H.-U. Yee and P. Yi, Phys. Rev. D **76**, 061901 (2007); JHEP **0709**, 063 (2007).
7. T. Sakai and S. Sugimoto, Prog. Theor. Phys. **113**, 843 (2005); T. Sakai and S. Sugimoto, Prog. Theor. Phys. **114**, 1083 (2005).
8. M. Harada, H. Hoshino and S. Nakamura, work in progress.
9. M. Harada and B.-R. He, work in progress.
10. See, e.g., S. A. Hartnoll, arXiv:1106.4324 [hep-th], and references therein.

Dilatons in Dense Baryonic Matter

Hyun Kyu Lee* and Mannque Rho†

*Department of Physics, Hanyang University, Seoul 133-791, Korea
E-mail: hyunkyu@hanyang.ac.kr

† Institut de Physique Théorique, CEA Saclay, 91191 Gif-sur-Yvette cédex, France
Department of Physics, Hanyang University, Seoul 133-791, Korea
E-mail:mannque.rho@cea.fr

We discuss the role of dilaton, which is supposed to be representing a special feature of scale symmetry of QCD, trace anomaly, in dense baryonic matter. The idea that the scale symmetry breaking of QCD is responsible for the spontaneous breaking of chiral symmetry is presented along the similar spirit of Freund-Nambu model. The incorporation of dilaton field in the hidden local symmetric parity doublet model is briefly sketched with the possible role of dilaton at high density baryonic matter, the emergence of linear sigma model in dilaton limit.

Keywords: dilaton, QCD, scale anomaly, hidden local symmetry, baryonic matter

1. Introduction

The recent discovery of Higgs-like particles at LHC activates many interesting research not only on phenomenology but also on the origin of Higs particle.[1,2] In the framework of Standard Model the masses of leptons and quarks are explained by Higgs mechanism. But the question on how masses of hadrons are generated has not been satisfactorily answered yet.[3] Since the light quark masses are too small to explain the hadron mass, the proton mass, for example, it is believed that the masses of hadrons are generated dynamically in the framework of quantum chromodynamics (QCD).

If we ignore quark masses, a very good approximation for the hadrons consisting of light quarks), QCD has no scale at the classical level, but a scale emerges at the quantum level in the form of trace anomaly, an explicit breaking of scale symmetry. On the other hand, the spontaneous symmetry breaking of chiral symmetry(χSB) gives rise to another important scale determined by the quark condensate, $\langle\bar{q}q\rangle$, which is supposed to generate most of the hadron masses. Thus it is natural to ask whether the chiral symmetry breaking and the scale symmetry breaking (SSB) are linked to each other and if so, how intricately.

The idea as to how χSB is tied to SSB has been discussed recently in connection to the discovery of Higgs-like particles in LHC. In the technicolor scheme, the *explicit* χSB in heavy quark sector triggers the spontaneous scale-symmetry breaking (SSB), of which a Nambu-Goldstone boson is identified as a Higgs-like particle. It turns out that, in this walking technicolor scheme,[2] the phenomenology of Goldston boson of SSB can be differentiated from Higgs in the standard model in more detailed analysis of LHC data. And this development may shed new light on new mechanism of mass generation for hadrons.

In hadronic sector, the idea that the scale symmetry and chiral symmetry are tied to each other has been explored in the line of thought that the explicit breaking of scale symmetry due to the QCD anomaly triggers the spontaneous χSB.[4] Assuming that the QCD anomaly can be decomposed into two parts, "hard" and "soft," the explicit symmetry breaking due to the soft part can be treated as small enough so that there is an approximate scale symmetry at low energy scale. Then the dilaton emerges as a Nambu-Goldstone boson as argued in the work of Freund and Nambu.[5]

We can construct a phenomenological Lagrangian which implements the above idea by introducing dilaton field into the Lagrangian of pions and vector mesons constructed under the hidden local gauge symmetry principle. One of the interesting consequences is the scaling of Lagrangian parameters: masses and coupling constants are dialed by the change of vacuum expectation value of dilaton field, i.e., BR scaling.[6] The possible forms of scaling laws have been explored in various contexts. One of them is the observation of a half-skyrmion phase when skyrmions are put on the lattice which can be implemented into a new scaling law giving a stiffer equation of state.[7,8] Another interesting feature of dilaton field is the dilaton limit, that leads to the condition that a particular combination of pions and dilaton lead to the linear sigma model giving rise to a nontrivial scaling behavior.[9,10]

In section 2, the work of Freund and Nambu is briefly sketched to develop a model in which pions and dilaton are introduced to take care of spontaneous breaking of both chiral symmetry and approximate scale symmetry. We introduce the phenomenological Lagrangian, denoted dHLS, which implements the above idea by introducing dilaton field into the Lagrangian constructed under the hidden local gauge symmetry principle[11] . In section 3, we elaborate on how a dilaton field can be applied to the entangled transformation of pions in nonlinear realization of χSB at lower scale such that the chiral symmetry is linearly realized in the form of linear sigma model.[9] The summary is given in section 4.

2. Dilaton and Chiral Symmetry Breaking

The first example which demonstrates the possibility of spontaneous symmetry breaking of the scale symmetry and its physical application of the dilaton field is the Freund-Nambu model.[5,12] It has two real scalar fields, ψ and ϕ, with the potential

$$V(\psi,\phi) = V_a + V_b, \tag{1}$$

where

$$V_a = \frac{1}{2}f^2\psi^2\phi^2, \quad V_b = \frac{\tau}{4}[\frac{\phi^2}{g^2} - \frac{1}{2}\phi^4 - \frac{1}{2g^4}] = \frac{\tau}{8g^4}(g^2\phi^2 - 1)^2. \tag{2}$$

V_a is scale invariant but V_b includes scale symmetry breaking terms $\frac{\phi^2}{g^2}$ and $-\frac{1}{2g^4}$. One can introduce the new field, χ, defined by

$$\chi = (g^2\phi^2 - 1)/(2g), \tag{3}$$

of which the scale transformation

$$\delta\chi = \epsilon(x \cdot \partial + 2)\chi + \frac{\epsilon}{g}, \tag{4}$$

manifests one of the characteristics of Nambu-Goldstone bosons. In terms of the condensate ϕ_0 given by the minimum of the potential, $\phi_0 = 1/g$, the masses take the form

$$m_\psi^2 = f^2\phi_0^2, \quad m_\chi^2 = \tau\phi_0^2. \tag{5}$$

It should be noted that the mass of ψ comes from the scale invariant term V_a, so is independent of τ. Thus it can have an arbitrary value. However m_χ depends on τ, going to zero linearly as $\tau \to 0$. This is known to be a characteristic of an approximate spontaneously-broken scale symmetry.[12]

Let us consider the Lagrangian of linear sigma model with $\Phi = \sigma + i\tau \cdot \vec{\phi}$ with a potential of the following form

$$V(\phi) = \lambda(\frac{1}{2}Tr\Phi\Phi^\dagger - v^2)^2 = \lambda(\sigma^2 + \Sigma_i\phi_i^2 - v^2)^2. \tag{6}$$

The scale invariance is broken due to nonvanishing v^2. But it is also this term which breaks chiral symmetry spontaneously. When we assign the vacuum expectation only to σ, $\phi_i, i = 1, 2, 3$ become Nambu-Goldstone bosons related to the spontaneous chiral symmetry breaking and we call them pions, π. For $\lambda = 0$, there is no explicit breaking of scale symmetry regardless of the value of v^2. We introduce a scalar field, χ, and reformulate the above potential in the following form,

$$V(\phi, \chi) = \lambda(\frac{1}{2}Tr\Phi\Phi^\dagger - \chi^2)^2 + V(\chi) \tag{7}$$

The first term is analogous to V_a in eq.(2), which is scale invariant. The second term is to break scale symmetry. In general, V_χ gives nonvavinshing vacuum expectation value for χ, which can be of the following forms

$$V_\chi = \frac{\tau}{8}(\chi^2 - v^2)^2, \quad or \quad V_\chi = \tau\chi^4 \ln\frac{\chi}{e^{1/4}v}. \tag{8}$$

One can see that the scale symmetry breaking in eq.(6) is transferred to the second term of eq.(7), which plays a similar role as eq.(2). If the scale symmetry breaking term with τ is small enough such that there is an *approximate* scale symmetry, then it can be considered as spontaneously broken and associate a Nambu-Goldstone boson of the form,[12,13]

$$\eta = (\chi^2 - v^2)/2v, \quad or \quad \chi = ve^{\eta/v}. \tag{9}$$

The origin of scale symmetry breaking in hadron physics can be traced back to the QCD trace anomaly and a dilaton field can be associated with the process. Then eq.(7) reduces to eq.(6). It is now clear in this formulation that the spontaneous chiral symmetry breaking is not possible without scale symmetry breaking. Or put differently, *scale symmetry breaking is responsible for the chiral symmetry breaking.* This indicates one of the possible ways that the trace anomaly of QCD can be linked to chiral symmetry breaking.

The chiral symmetry, which is broken spontaneously, can be realized non-linearly by introducing $U(\pi) = e^{i\,\vec{\tau}\cdot\vec{\pi}/F_\pi}$ instead of four-component scalar fields, Φ. The pions $\pi_i,\ \ i = 1, 2, 3$ transform nonlinearly under the broken sector of chiral transformation, with the pions derivatively coupled. The lowest derivative term of pion dynamics (namely the current algebra term) can be written

$$\mathcal{L} = \frac{F_\pi^2}{4}\mathrm{tr}\left[\partial_\mu U \partial^\mu U^\dagger\right]. \tag{10}$$

This term breaks scale symmetry explicitly because of the explicit scale F_π. This scale can be traced back to the vacuum expectation value of the scalar field χ as in eq.(7) with $F_\pi = v$ and the scale symmetry breaking can be traded in the potential $V(\chi)$

$$\mathcal{L}(U,\chi) = \frac{\chi^2}{4}\mathrm{tr}\left[\partial_\mu U \partial^\mu U^\dagger\right] + V(\chi) + \frac{1}{2}\partial_\mu \chi \partial^\mu \chi. \tag{11}$$

Defining a new field

$$\Sigma \equiv \chi U(\pi), \tag{12}$$

we can see that

$$\mathcal{L}(U,\chi) \to \mathcal{L}(\Sigma) = \frac{1}{4}\mathrm{tr}\left[\partial_\mu \Sigma \partial^\mu \Sigma^\dagger\right] + V(\Sigma\Sigma^\dagger), \tag{13}$$

where we have used $\chi = \frac{1}{2}Tr\Sigma\Sigma^\dagger$. If we can identify Σ as

$$\Sigma = \sigma + i\vec{\pi}\cdot\vec{\tau}, \tag{14}$$

then it is nothing but the linear sigma model. However it is not clear how the linear representation of nonlinear field can be achieved by adding a dilaton, which is chiral singlet. Apart from the theoretical justification, we may expect this can be achieved physically in a certain limit, for example in baryonic matter with higher density. This is the issue that concerns the "dilaton limit" discussed in.[9,10]

3. Dilton Limit of dHLS

It has been observed that the dilaton limit leads to a very interesting consequence when nucleons are included in the chiral Lagrangian. The vector coupling constant is found to have a limiting value $g \to 1$ in the dilaton limit. This idea has been extended[14] to the effective theory where dilaton is incorporated into the hidden local symmetry theory, where vector mesons are present as explicit degrees of freedom. In this section the results are briefly sketched.

The dilaton field which is responsible for the spontaneous symmetry breaking of chiral symmetry, thus generating pions, can be naturally included in the effective Lagrangian at low baryon density. The redundant symmetry, i.e., hidden local symmetry(HLS), appearing in the formalism of nonlinear realization of spontaneous chiral symmetry breaking has been used to incorporate the vector mesons systematically. However the Lagrangian of HLS is, by construction, noninvariant under scale symmetry. The pattern of symmetry breaking is thus not consistent with the scale symmetry breaking due to QCD trace anomaly, which is believed to be the only reason why the effective theory of QCD should have scale symmetry breaking term modulo the light quark mass. The dilated HLS is constructed such that all terms in the Lagrangian are scale invariant except for the potential of dilaton, that reproduces the relevant scale symmetry breaking for QCD trace anomaly. One of the simplest ways to construct the dilated HLS is to follow the standard trick of inserting the dilaton field χ as "conformal compensator" into the Lagrangian to obtain scale symmetric Lagrangian, given by

$$\mathcal{L} = \mathcal{L}_N + \mathcal{L}_M + \mathcal{L}_\chi \,, \tag{15}$$

$$\begin{aligned}\mathcal{L}_N = {} & \overline{Q} i\gamma^\mu D_\mu Q - g_1 F_\pi \frac{\chi}{F_\chi}\overline{Q}Q + g_2 F_\pi \frac{\chi}{F_\chi}\overline{Q}\rho_3 Q - i m_0 \overline{Q}\rho_2\gamma_5 Q \\ & + g_{V\rho}\overline{Q}\gamma^\mu \hat{\alpha}_{\parallel\mu} Q + g_{V0}\overline{Q}\gamma^\mu \mathrm{tr}\left[\hat{\alpha}_{\parallel\mu}\right] Q + g_A \overline{Q}\rho_3\gamma^\mu \hat{\alpha}_{\perp\mu}\gamma_5 Q \,,\end{aligned} \tag{16}$$

$$\begin{aligned}\mathcal{L}_M = {} & \frac{F_\pi^2}{F_\chi^2}\chi^2 \mathrm{tr}\left[\hat{\alpha}_{\perp\mu}\hat{\alpha}_\perp^\mu\right] + \frac{F_{\sigma\rho}^2}{F_\chi^2}\chi^2 \mathrm{tr}\left[\hat{\alpha}_{\parallel\mu}\hat{\alpha}_\parallel^\mu\right] + \frac{F_{\sigma\omega}^2 - F_{\sigma\rho}^2}{2F_\chi^2}\chi^2 \mathrm{tr}\left[\hat{\alpha}_{\parallel\mu}\right] \mathrm{tr}\left[\hat{\alpha}_\parallel^\mu\right] \\ & - \frac{1}{2}\mathrm{tr}\left[\rho_{\mu\nu}\rho^{\mu\nu}\right] - \frac{1}{2}\mathrm{tr}\left[\omega_{\mu\nu}\omega^{\mu\nu}\right] \,,\end{aligned} \tag{17}$$

$$\mathcal{L}_\chi = \frac{1}{2}\partial_\mu\chi \cdot \partial^\mu\chi - V(\chi) \tag{18}$$

where $V(\chi)$ is the dilaton potential that breaks scale symmetry spontaneously and F_χ is the vacuum expectation value of χ at zero temperature and density. The detailed expressions needed in this section can be found in ref.[14] . As in the previous section, we do the field re-parameterizations $\Sigma_s = U\chi\frac{F_\pi}{F_\chi} = s + i\vec{\tau}\cdot\vec{\pi}$. And using the nonlinearly transforming nucleon field of parity eigenstates, one finds a complicated expression for (15) composed of a part that is regular, $\mathcal{L}_{\rm reg}$, and a part that is singular, $\mathcal{L}_{\rm sing}$, as $\mathrm{tr}(\Sigma_s\Sigma_s^\dagger) = 2\left(s^2 + \pi^{a\,2}\right) \to 0$, where a is iso-spin index. The singular part that arises solely from the scale invariant part of the original Lagrangian (15) has the form

$$\mathcal{L}_{\rm sing} = (g_{V\rho} - g_A)\,\mathcal{A}\left(1/\mathrm{tr}\left[\Sigma_s\Sigma_s^\dagger\right]\right) + (\alpha - 1)\,\mathcal{B}\left(1/\mathrm{tr}\left[\Sigma_s\Sigma_s^\dagger\right]\right) \,, \tag{19}$$

One can see that the condition $\alpha \equiv \frac{F_\pi^2}{F_\chi^2} = 1$ has been already used in getting Eq.(13). That $\mathcal{L}_{\rm sing}$ be absent also leads to the condition that

$$g_{V\rho} - g_A \to 0 \,. \tag{20}$$

Using large N_c sum-rule arguments[9] and the RGE,[10] we infer

$$g_A - 1 \to 0 \,. \tag{21}$$

It follows then that

$$g_{\rho NN} = g_\rho(g_{V\rho} - 1) \to 0. \tag{22}$$

This set of limits defines what is referred to as "dilaton limit." We thus find that in the dilaton limit, the ρ meson decouples from the nucleon. In contrast, the limiting $\mathrm{tr}(\Sigma_s \Sigma_s^\dagger) \to 0$ *does not* give any constraint on $(g_{V\omega} - 1)$. The ω-nucleon coupling remains non-vanishing in the Lagrangian of fluctuations $\tilde{s}$ and $\tilde{\pi}$ around their expectation values, which in terms of the mass-diagonalizing field $\mathcal{N}$, takes the form,

$$\begin{aligned} \mathcal{L}_N = \quad & \overline{\mathcal{N}} i \partial\!\!\!/ \mathcal{N} - \overline{\mathcal{N}} \hat{M} \mathcal{N} - g_1 \overline{\mathcal{N}} \left(\hat{G} \tilde{s} + \rho_3 \gamma_5 i \vec{\tau} \cdot \vec{\tilde{\pi}} \right) \mathcal{N} \\ & + g_2 \overline{\mathcal{N}} \left(\rho_3 \tilde{s} + \hat{G} \gamma_5 i \vec{\tau} \cdot \vec{\tilde{\pi}} \right) \mathcal{N} + (1 - g_{V\omega}) g_\omega \mathcal{N} \frac{\omega\!\!\!/}{2} \mathcal{N}. \end{aligned} \tag{23}$$

This is just the nucleon part of the linear sigma model in which the ω is minimally coupled to the nucleon.

4. Summary

The origin of scale symmetry breaking in hadron physics can be traced back to the QCD trace anomaly, and it can be implemented into an effective theory by introducing a dilaton field. The idea that the scale symmetry and chiral symmetry are tied to each other has been explored in a line of thought that the explicit breaking of scale symmetry due to the QCD anomaly triggers the spontaneous χSB.[4] We demonstrate that using a model similar to that of Freund-Nambu the spontaneous chiral symmetry breaking is locked to the scale symmetry breaking through dilaton field. This provides one of the possible ways to relate the trace anomaly of QCD to chiral symmetry breaking.

It is possible to construct a phenomenological Lagrangian[14] that encapsulates the above idea by introducing dilaton field into the Lagrangian of pions and vector mesons constructed under the hidden local gauge symmetry principle with parity doubled baryons. One of the interesting consequences is the scaling of parameters of the Lagrangian: masses and coupling constants are dialed by the change of vacuum expectation value of dilaton field as in BR scaling.[6,8] Another interesting feature of dilaton field is the dilaton limit that gives the condition that a particular combination of pions and dilaton lead to the linear sigma model, giving rise to a nontrivial scaling behavior.[9,10] Most of the conventional effective actions constructed for low-density phenomena without dilaton available in the literature do not reveal the above feature. To realize the linear realization feature, the coupling constants should satisfy the particular condition. A highly non-trivial support for the dilaton limit properties comes from a renormalization group analysis of hidden local symmetry implemented with baryons.[10] An intriguing consequence of the analysis is that the ρ meson nucleon coupling vanishes at critical density corresponding to the dilaton limit *before* reaching chiral restoration.

Acknowledgments

HKL would like to thank Koichi Yamawaki for kind invitation to SCGT 12, Nagoya and for hospitalities during the conference. He also would like to thank Mannque Rho, Masayasu Harada, Tom Kuo, Won-Gi Paeng, Byung-Yoon Park and Chihiro Sasaki for helpful discussions. This work is supported by WCU (World Class University) program: Hadronic Matter under Extreme Conditions through the National Research Foundation of Korea funded by the Ministry of Education, Science and Technology (R33-2008-000-10087-0).

References

1. M. E. Peskin, arXiv:1208.5152 [hep-ph]
2. S. Matsuzaki and K. Yamawaki, arXiv:1207.5911 [hep-ph] and references there in.
3. M. Rho, "Proton mass, topology change and tensor forces in compressed baryonic matter," arXiv:1301.0066 [nucl-th].
4. H. K. Lee and M. Rho, "Dilatons in hidden local symmetry for hadrons in dense matter," Nucl. Phys. A **829**, 76 (2009) [arXiv:0902.3361 [hep-ph]]; "Dilatons for Dense Hadronic Matter," Nucl. Phys. A **844**, 80C (2010) [arXiv:0910.0679 [hep-ph]]; "Flavor symmetry and topology change in nuclear symmetry energy for compact stars," arXiv:1201.6486 [nucl-th].
5. P.G.O. Freund and Y. Nambu, "Scalar field coupled to the trace of the energy-momentum tensor," Phys. Rev. **174** (1968) 1741.
6. G.E. Brown and M. Rho, "Scaling effective Lagrangians in a dense medium," Phys. Rev. Lett. **66** (1991) 2720.
7. H. K. Lee, B. Y. Park and M. Rho, "Half-skyrmions, tensor forces and symmetry energy in cold dense matter," Phys. Rev. C **83**, 025206 (2011).
8. H. Dong, T. T. S. Kuo, H. K. Lee, R. Machleidt and M. Rho, "Half-skyrmions and the equation of state for compact-star matter," arXiv:1207.0429 [nucl-th].
9. S. R. Beane and U. van Kolck, "The Dilated chiral quark model," Phys. Lett. B **328**, 137 (1994) [hep-ph/9401218].
10. W. -G. Paeng, H. K. Lee, M. Rho and C. Sasaki, "Dilaton-Limit Fixed Point in Hidden Local Symmetric Parity Doublet Model," Phys. Rev. D **85**, 054022 (2012) [arXiv:1109.5431 [hep-ph]].
11. M. Harada and K. Yamawaki, "Vector manifestation of the chiral symmetry," Phys. Rev. Lett. **86** (2001) 757 [arXiv:hep-ph/0010207]; "Hidden local symmetry at loop: A new perspective of composite gauge bosons and chiral phase transition," Phys. Rept. **381** (2003) 1.
12. B. Zumino, "Effective Lagrangians and broken symmetries," in *1970 Brandeis lectures on 'Elementary particles and quantum field theory*, Vol. 2, Cambridge, Mass, 1970, pp. 437-500.
13. M Shaposhnikov and D. Zenhausern, "Scale invariance, unimodular gravity and dark energy," [arXiv:0809.3395[hep-ph]]
14. W.-G. Paeng, H.K. Lee, M. Rho and C. Sasaki, "ω-Nucleon Interaction and Nucleon Mass in Dense Baryonic Matter", arXiv:1303.2898[nucl-th].

Baryonic Matter in the Hidden Local Symmetry Induced from Holographic QCD Models

Yong-Liang Ma[†], Ghil-Seok Yang[‡], Yongseok Oh[‡], Masayasu Harada[†], Hyun Kyu Lee[††], Byung-Yoon Park[*] and Mannque Rho[††,§]

[†]*Department of Physics, Nagoya University, Nagoya, 464-8602, Japan*
[‡]*Department of Physics, Kyungpook National University, Daegu 702-701, Korea*
[††]*Department of Physics, Hanyang University, Seoul 133-791, Korea*
[*]*Department of Physics, Chungnam National University, Daejeon 305-764, Korea*
[§]*Institut de Physique Théorique, CEA Saclay, 91191 Gif-sur-Yvette cédex, France*

Baryonic matter is studied in the Skyrme model by taking into account the roles of π, ρ, and ω mesons through the hidden local symmetry up to $\mathcal{O}(p^4)$ terms including the homogeneous Wess-Zumino (hWZ) terms. Using the master formulas for the low energy constants derived from holographic QCD models the skyrmion matter properties can be quantitatively calculated with the input values of the pion decay constant f_π and the vector meson mass m_ρ. We find that the hWZ terms are responsible for the repulsive interactions of the ω meson. In addition, the self-consistently included $\mathcal{O}(p^4)$ terms with the hWZ terms is found to increase the half skyrmion phase transition point above the normal nucleon density.

Keywords: Baryonic matter; Skyrme model; Hidden local symmetry

1. Introduction

Understanding the QCD phase structure is an extremely important problem in particle and nucleon physics, but is highly nontrivial because of the nonperturbative nature of strong interactions. Therefore, any theoretical investigation of the QCD phase structure requires effective models for strong interactions at low energies. In this talk, we discuss the QCD phase structure for zero temperature baryonic matter in the Skyrme model based on the hidden local symmetry (HLS) up to $\mathcal{O}(p^4)$ terms including the homogeneous Wess-Zumino (hWZ) terms.[1–3] To uniformly describe the baryon, baryonic matter and also the change of hadron properties in baryonic matter background, we apply the Skyrme's conjecture[4] and describe baryons as topological solitons in a mesonic theory. We also put the skyrmions into the face-centered-cubic (FCC) crystal[5] for studying baryonic matter.

Although the original Skyrme model is quite simple as it is just the nonlinear sigma model stabilized by the Skyrme term, it can reasonably describe the properties of a single nucleon in vacuum in the sense of large N_c expansion.[6] Since the original Skyrme model contains only pion degree of freedom, it is natural to improve the model predictions by including more dynamical degrees of freedom. When we increase the energy scale, the first resonance entering is the ground state ρ and ω vector mesons. In the present work, we adopt the Hidden Local Symmetry (HLS) approach which is gauge equivalent to the nonlinear sigma model for including vector mesons.[7]

Once extra degrees of freedom are added to an effective theory, the dynamics generally becomes more complicated by introducing more independent low energy constants (LECs) which are to be determined by empirical data. However, it is not possible to determine all the LECs phenomenologically without ambiguity as the number of LECs increases, and this hinders the systematic investigation of the baryon and baryonic matter properties. In Ref. 2 we derived general master formulas for the LECs of HLS from a kind of holographic QCD (hQCD) models. This allows us to investigate the skyrmion and hadronic matter properties with only two input values of f_π and m_ρ.

In Refs. 1,2 we applied the HLS up to $\mathcal{O}(p^4)$ terms including the hWZ terms to investigate the single-skyrmion properties. It was found that the hWZ terms are responsible for the role of the ω meson and that the ω and ρ mesons have very different roles indicated by the attractive nature of the ρ meson and the repulsive interaction from the ω meson. As the next step for studying the role of vector mesons, we apply the above model to baryonic matter[3] to study the skyrmion matter properties and the half skyrmion phase transition point.

This paper is organized as follows. In Sec. 2 we investigate the single skyrmion properties from the HLS Lagrangian after discussing the implications of the infinite towers of vector mesons in the skyrmion properties. We then study the skyrmion matter properties from HLS in Sec. 3, where we also compare our results with other model calculations. Section 4 contains a summary and perspective of our work.

2. Skyrmion properties from the hidden local symmetry

We start with the study of the single skyrmion properties based on the HLS up to $\mathcal{O}(p^4)$ terms including the hWZ terms. Therefore, this model includes the π, ρ and ω mesons. The LECs of the HLS Lagrangian can be determined from the master formulas derived from hQCD models which allows us to control the LECs so that the skyrmion properties can be calculated in a self-consistent way.

The role of the infinite tower structure in five-dimension hQCD models in the skyrmion properties have been addressed in the large 't Hooft coupling λ limit.[8] It was found that, when only the pion was kept in the truncated model, the skyrmion has an excessive energy of about 25% of the Bogomol'nyi bound, while the excess energy is reduced to 7% when the ρ meson is included. This observation indicates that high-lying vector mesons make the theory flow to a conformal theory. However, in the large λ limit, the theory does not include the repulsive interactions due to the ω meson which is the corrections of the next order in λ, where the Chern-Simons (CS) term brings in a U(1) degree of freedom responsible for the ω meson and its tower.

To include the ω meson, we consider the hWZ terms in the HLS Lagrangian, which allows us to study the role of the ω meson in the baryon properties. In addition, since the validity of the Skyrme model is supported by the large N_c limit, for consistency, we need to take into account the higher order terms such as the $\mathcal{O}(p^4)$ terms in addition to the $\mathcal{O}(p^2)$ terms. Our Lagrangian is then summarized as

$$\mathcal{L}_{\rm HLS} = \mathcal{L}_{(2)} + \mathcal{L}_{(4)} + \mathcal{L}_{\rm anom}, \tag{1}$$

where $\mathcal{L}_{(2)}, \mathcal{L}_{(4)}$ and $\mathcal{L}_{\rm anom}$ stand for the $\mathcal{O}(p^2)$, $\mathcal{O}(p^4)$ and the hWZ terms, respectively. The explicit expressions for the Lagrangian (1) responsible for the calculation of the skyrmion properties can be found in Ref. 1.

The Lagrangian (1) has seventeen LECs, namely, f_π, g, a, $y_1 \sim y_9$, z_4, z_5 and $c_1 \sim c_3$. These LECs can be determined from the general master formula[1,2] derived from a class of hQCD models which is defined by the following general 5D action:

$$S_5 = S_5^{\rm DBI} + S_5^{\rm CS}, \tag{2}$$

where the Dirac-Born-Infeld part $S_5^{\rm DBI}$ and the CS part $S_5^{\rm CS}$ are expressed as

$$S_5^{\rm DBI} = N_c G_{\rm YM} \int d^4x dz \Big\{ -\frac{1}{2} K_1(z) {\rm Tr}\left[\mathcal{F}_{\mu\nu}\mathcal{F}^{\mu\nu}\right] + K_2(z) M_{KK}^2 {\rm Tr}\left[\mathcal{F}_{\mu z}\mathcal{F}^{\mu z}\right]\Big\}, \tag{3}$$

$$S_5^{\rm CS} = \frac{N_c}{24\pi^2} \int_{M^4\times R} w_5(A), \tag{4}$$

where $G_{\rm YM} \equiv \lambda/(108\pi^3)$ and $\mathcal{F}_{MN} = \partial_M \mathcal{A}_N - \partial_N \mathcal{A}_M - i[\mathcal{A}_M, \mathcal{A}_N]$ is the field strength of the 5D gauge field $\mathcal{A}_M$. Here, we use the index $M = (\mu, z)$ with the convention $\mu = 0, 1, 2, 3$ and $K_{1,2}(z)$ are the metric functions constrained by the gauge/gravity duality. In Eq. (4), M^4 and R stand for the four-dimensional Minkowski space-time and z-coordinate space, respectively, and $w_5(A)$ is the CS 5-form. Note that the DBI part is of $\mathcal{O}(\lambda^1)$ while the CS term is of $\mathcal{O}(\lambda^0)$. From the action (2), after integrating out all the modes except the zero and the lowest-lying vector modes, the LECs of HLS are expressed as

$$\begin{aligned}
& f_\pi^2 = N_c G_{\rm YM} M_{KK}^2 \int dz K_2(z) \left[\dot\psi_0(z)\right]^2, \quad f_\pi^2 = N_c G_{\rm YM} M_{KK}^2 \lambda_1 \langle \psi_1^2 \rangle, \\
& \frac{1}{g^2} = N_c G_{\rm YM} \langle \psi_1^2 \rangle, \quad y_1 = -y_2 = -N_c G_{\rm YM} \left\langle \left(1 + \psi_1 - \psi_0^2\right)^2 \right\rangle, \\
& y_3 = -y_4 = -N_c G_{\rm YM} \left\langle \psi_1^2 \left(1 + \psi_1\right)^2 \right\rangle, \quad y_5 = 2y_8 = -y_9 = -2N_c G_{\rm YM} \left\langle \psi_1^2 \psi_0^2 \right\rangle, \\
& y_6 = -\left(y_5 + y_7\right), \quad y_7 = 2N_c G_{\rm YM} \left\langle \psi_1 \left(1 + \psi_1\right)\left(1 + \psi_1 - \psi_0^2\right) \right\rangle, \\
& z_4 = 2N_c G_{\rm YM} \left\langle \psi_1 \left(1 + \psi_1 - \psi_0^2\right) \right\rangle, \quad z_5 = -2N_c G_{\rm YM} \left\langle \psi_1^2 \left(1 + \psi_1\right) \right\rangle, \\
& c_1 = \left\langle\!\!\left\langle \dot\psi_0 \psi_1 \left(\frac{1}{2}\psi_0^2 + \frac{1}{6}\psi_1^2 - \frac{1}{2} \right) \right\rangle\!\!\right\rangle, \\
& c_2 = \left\langle\!\!\left\langle \dot\psi_0 \psi_1 \left(-\frac{1}{2}\psi_0^2 + \frac{1}{6}\psi_1^2 + \frac{1}{2}\psi_1 + \frac{1}{2} \right) \right\rangle\!\!\right\rangle, \quad c_3 = \left\langle\!\!\left\langle \frac{1}{2}\dot\psi_0 \psi_1^2 \right\rangle\!\!\right\rangle,
\end{aligned} \tag{5}$$

where λ_1 is the smallest (non-zero) eigenvalue of the eigenvalue equation,

$$-K_1^{-1}(z)\partial_z\left[K_2(z)\partial_z\psi_n(z)\right] = \lambda_n \psi_n(z), \tag{6}$$

and

$$\langle A \rangle \equiv \int_{-\infty}^{\infty} dz K_1(z) A(z), \qquad \langle\langle A \rangle\rangle \equiv \int_{-\infty}^{\infty} dz A(z). \tag{7}$$

With the help of the formulas in Eq. (5), sixteen of the seventeen parameters can be fixed from the two parameters $G_{\rm YM}$ and $M_{\rm KK}$ which, in turn, can be determined by the pion decay constant f_π and the vector meson mass m_ρ. The only undetermined parameter is the HLS-parameter a or equivalently the coupling constant g as they are connected by the mass relation $m_\rho^2 = a g^2 f_\pi^2$. From the point view of the 5D holographic model, this ambiguity is attributed to the normalization of the 5D wave function $\psi_1(z)$ that cannot be determined from the *homogeneous* eigenvalue equation (6). In Refs. 1,2 we have shown that the parameter g disappears from the expressions for the skyrmion mass if we express the LECs in terms of the normalized $\psi_1(z)$, which leads to the independence of the skyrmion properties on the value of the HLS parameter a. This conclusion was numerically verified in Ref. 2.

For numerical calculations, we use the empirical values, $m_\rho = 775.5$ MeV and $f_\pi = 92.4$ MeV as input values. And for the hQCD model, we take the Sakai-Sugimoto model,[9] where $K_1(z) = (1+z^2)^{-1/3}$ and $K_2(z) = 1+z^2$. Here, we consider three versions of the HLS model: HLS(π, ρ, ω), HLS(π, ρ) and HLS(π). HLS(π, ρ, ω) is the full model including the pion, ρ, and ω mesons and HLS(π, ρ) is the model without the hWZ terms. When the vector fields are integrated out, we have the original Skyrme model but with the Skyrme parameter determined by the hQCD model. This is called HLS(π). Our numerical results are summarized in Table 1 that shows the skyrmion mass $M_{\rm sol}$, the Δ-N mass splitting Δ_M, the winding number rms radius $\sqrt{\langle r^2 \rangle_W}$ and the energy rms radius $\sqrt{\langle r^2 \rangle_E}$. For a comparison, we also show the results of "the minimal model" which includes the ω meson in a minimal way[11] in the column of HLS$_{\rm min}(\pi, \rho, \omega)$.

Table 1. Skyrmion properties calculated from the HLS models. The soliton mass $M_{\rm sol}$ and the Δ-N mass difference Δ_M are in units of MeV, while $\sqrt{\langle r^2 \rangle_W}$ and $\sqrt{\langle r^2 \rangle_E}$ are in units of fm.

	HLS(π, ρ, ω)	HLS(π, ρ)	HLS(π)	HLS$_{\rm min}(\pi, \rho, \omega)$
$M_{\rm sol}$	1184	834	922	1407
Δ_M	448	1707	1014	259
$\sqrt{\langle r^2 \rangle_W}$	0.433	0.247	0.309	0.540
$\sqrt{\langle r^2 \rangle_E}$	0.608	0.371	0.417	0.725

Close inspection of the results given in Table 1 leads to the following observations. Namely, the results from HLS(π) and HLS(π, ρ) explicitly show the attractive interaction nature arising from the ρ meson, while comparison of the results of HLS(π, ρ) and HLS(π, ρ, ω) explicitly illustrates the repulsive nature due to the ω meson. In addition, the differences between the results from HLS(π, ρ, ω) and HLS$_{\rm min}(\pi, \rho, \omega)$ come from the $\mathcal{O}(p^4)$ terms and the extra terms in the hWZ terms.

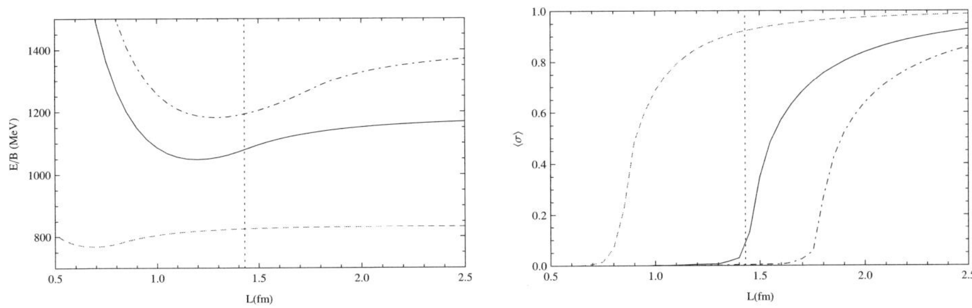

Fig. 1. The energy per baryon number E/B and the expectation value $\langle\sigma\rangle$ of the minimum energy configuration for a given value of L.

3. Skyrmion matter from the hidden local symmetry

In this Section, we apply the HLS model of the previous Section for the study of the skyrmion matter properties[3] using the procedure explained in Ref. 5, i.e., by putting skyrmions into the FCC crystal. Shown in Fig. 1 is the numerical results for E/B and $\langle\sigma\rangle$ as functions of the FCC size parameter L. Here, $\langle\sigma\rangle$ is the average of the chiral field U over the space where a single skyrmion occupies, which vanishes in the half-skyrmion phase. Here, we present the numerical results for HLS(π,ρ,ω), HLS(π,ρ) and HLS$_{\rm min}(\pi,\rho,\omega)$ by the solid, dahed, and dash-dotted lines, respectively. The position of the normal nuclear matter density is denoted by the vertical dotted line.

One can see that, in HLS(π,ρ,ω), the half-skyrmion phase transition happens almost at the normal nuclear matter density n_0. Comparing to that of HLS$_{\rm min}(\pi,\rho,\omega)$, we see that the inclusion of the $\mathcal{O}(p^4)$ interaction terms and the other π-ρ-ω interactions through the hWZ terms is crucial to make $n_{1/2}$ larger by the factor of 1.7. This is quite a noticeable improvement from the previous work,[10] where the position of the phase transition could not be explicitly pinned down. The higher value of $n_{1/2}$ of HLS(π,ρ,ω) compared to the HLS$_{\rm min}(\pi,\rho,\omega)$ result can be understood from the fact that the size of the single skyrmion is smaller in the former and the additional interactions in HLS(π,ρ,ω) weaken the repulsive interactions from the ω meson. This analysis is supported by the results from HLS(π,ρ).

In addition, we find the weak dependence of E/B on the density in HLS(π,ρ). This again emphasizes the role of the vector mesons. As we can see from Table 1 as well as from the conclusion of Ref. 8, when only the pions are considered, the skyrmion carries a lot of excessive energy, while the inclusion of the ρ meson reduces the soliton mass a lot and almost saturates the Bogomol'nyi bound. Therefore, only very little interaction energy remains in the skyrmion made of pion and the ρ, and consequently it yields the (nearly) flat dashed curve for E/B.

In Fig. 2, we present the skyrmion energy from each term in Eq. (1) as a function of L. The plot shows that $(E/B)_{\mathcal{O}(p^4)}$ is negative for all the values of density, which explicitly shows that the inclusion of the higher order terms and vector mesons provide attractions and lead to the weak density dependence. On

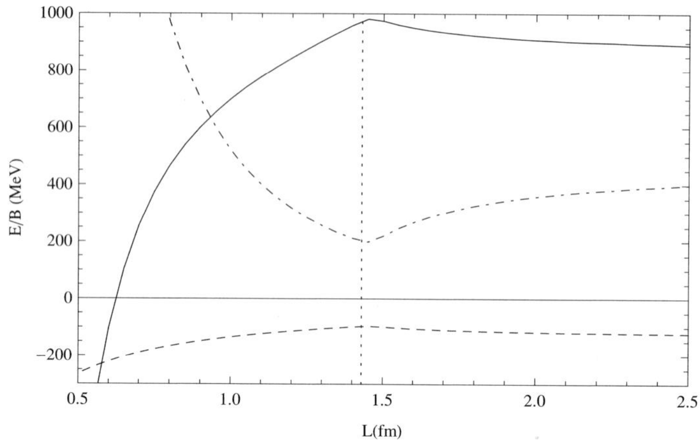

Fig. 2. The energy per baryon number is plotted as a function of L. The solid, dashed and dash-dotted lines are $(E/B)_{\mathcal{O}(p^2)}$, $(E/B)_{\mathcal{O}(p^4)}$ and $(E/B)_{\text{hWZ}}$, respectively.

the contrary, in the single skyrmion phase, $(E/B)_{\mathcal{O}(p^2)}$ is dominant and the ratio $(E/B)_{\mathcal{O}(p^2)}/(E/B)_{\text{hWZ}}$ even increases from ~ 2 to ~ 3 as the density increases. However, after the half-skyrmion phase transition, $(E/B)_{\mathcal{O}(p^2)}$ starts to decrease, while $(E/B)_{\text{hWZ}}$ starts to increase. Then, at some density the repulsive interaction due to the ω meson becomes the dominant interaction.

4. Remarks and perspectives

In the present work, we discuss the skyrmion and baryonic matter properties using the hidden local symmetry Lagrangian up to $\mathcal{O}(p^4)$ terms including the hWZ terms. Thanks to the holographic QCD models, the LECs can be determined by the master formulas and the skyrmion properties can be calculated self-consistently using two input parameters m_ρ and f_π. This self-consistent approach was also recently applied to study the bound state picture of heavy baryons.[12]

Due to the quantitative predictive power of the present framework, we apply this model to investigate the skyrmion matter properties by putting the skyrmions into the FCC crystal. For the crystal size dependence of the skyrmion energy, we found that, in $\text{HLS}_{\text{min}}(\pi, \rho, \omega)$ and $\text{HLS}(\pi, \rho, \omega)$, the minimum of the skyrmion energy n_{min} appears at the same density and n_{min} in $\text{HLS}(\pi, \rho)$ appears at a very deep region. A noticeable result is that, in contrast to $\text{HLS}_{\text{min}}(\pi, \rho, \omega)$, $n_{1/2}$ of $\text{HLS}(\pi, \rho, \omega)$ and $\text{HLS}(\pi, \rho)$ appears at density higher than the normal nuclear density n_0, which is consistent with our expectation.

It will be also interesting to study the role of the dilaton field in our HLS model and the change of hadron properties in baryonic matter. Work in this direction is in progress and will be reported elsewhere.

Acknowledgments

The work reported here was partially supported by the WCU project of Korean Ministry of Education, Science and Technology (R33-2008-000-10087-0). The work of M.H. and Y.-L.M. was supported in part by Grant-in-Aid for Scientific Research on Innovative Areas (No. 2104) "Quest on New Hadrons with Variety of Flavors" from MEXT. Y.-L.M. was supported in part by the National Science Foundation of China (NNSFC) under Grant No. 10905060. The work of M.H. was supported in part by the Grant-in-Aid for Nagoya University Global COE Program "Quest for Fundamental Principles in the Universe: from Particles to the Solar System and the Cosmos" from MEXT, the JSPS Grant-in-Aid for Scientific Research (S) ♯ 22224003, (c) ♯ 24540266. The work of Y.O. and G.-S.Y. was supported in part by the Basic Science Research Program through the National Research Foundation of Korea (NRF) funded by the Ministry of Education, Science and Technology (Grant No. 2010-0009381).

References

1. Y.-L. Ma, Y. Oh, G.-S. Yang, M. Harada, H. K. Lee, B.-Y. Park and M. Rho, Phys. Rev. D **86**, 074025 (2012).
2. Y.-L. Ma, G.-S. Yang, Y. Oh, M. Harada, Phys. Rev. D **87**, 034023 (2013).
3. Y.-L. Ma *et al.*, to appear.
4. T. H. R. Skyrme, Proc. Roy. Soc. Lond. A **260**, 127 (1961).
5. H.-J. Lee, B.-Y. Park, D.-P. Min, M. Rho and V. Vento, Nucl. Phys. A **723**, 427 (2003).
6. I. Zahed and G. E. Brown, Phys. Rept. **142**, 1 (1986).
7. M. Harada and K. Yamawaki, Phys. Rept. **381**, 1 (2003).
8. P. Sutcliffe, JHEP **1104**, 045 (2011).
9. T. Sakai and S. Sugimoto, Prog. Theor. Phys. **113**, 843 (2005).
10. B.-Y. Park, M. Rho and V. Vento, Nucl. Phys. A **736**, 129 (2004).
11. U. G. Meissner, N. Kaiser and W. Weise, Nucl. Phys. A **466**, 685 (1987).
12. M. Harada and Y.-L. Ma, Phys. Rev. D **87**, 056007 (2013).

Twisted Reduction in Large N QCD with Adjoint Wilson Fermions

Antonio González-Arroyo[ab]

[a]*Instituto de Física Teórica UAM/CSIC*
[b]*Departamento de Física Teórica, C-15*
Universidad Autónoma de Madrid, E-28049–Madrid, Spain
E-mail: antonio.gonzalez-arroyo@uam.es

Masanori Okawa[c]

[c]*Graduate School of Science, Hiroshima University*
Higashi-Hiroshima, Hiroshima 739-8526, Japan
E-mail: okawa@sci.hiroshima-u.ac.jp

The twisted reduced model of large N QCD with two adjoint Wilson fermions is studied numerically using the Hybrid Monte Carlo method. This is the one-site model, whose large N limit (large volume limit) is expected to be conformal or nearly conformal. The string tension calculated at N=289 approaches zero as we decrease quark mass and the preliminary value of the mass anomalous dimension γ_* is close to one if we assume that the theory is governed by an infrared fixed point. We also discuss the twisted reduced model with single adjoint Wilson fermion. The string tension remains finite as the quark mass decreases to zero, supporting that this is the confining theory.

Keywords: large N QCD, twisted space-time reduction, adjoint fermion

1. Introduction

The standard model of the elementary particle has set the foundation to SU(N) gauge theories. Although SU(N) gauge theories generally have very complicated structures, significant simplifications could occur in the large N limit. In fact, Eguchi and Kawai considered the space-time reduced model of the lattice gauge theory having only one space-time point.[1] This model is now called the Eguchi-Kawai model (EK-model). The EK model possesses the Z(N) symmetry. If this symmetry is not spontaneously broken, the lattice gauge theory and the EK model are equivalent in the large N limit. This means that space-time degrees of freedom of the lattice gauge theory could be encoded into those of internal SU(N) group. Soon after the proposal of Eguchi and Kawai, however, the Z(N) symmetry was shown to be spontaneously broken in the weak coupling limit, thus invalidating the EK model.[2] To circumvent this difficulty, the present authors have proposed to consider the one-site model with twisted boundary conditions (the twisted Eguch-Kawai (TEK) model), thus avoiding the Z(N) symmetry being spontaneously broken.[3,4] Recently, gauge theories coupled with fermions in the adjoint representation received much attention in the large N limit in connection, for example, to the AdS/CFT correspondence. The idea of the twisted space-time reduction can also be applied to these models. The purpose of this talk is to present our recent progress in this direction. In the next section, we calculate the continuum string tension of the SU(N) pure gauge theory in the large N limit using the TEK model. In sect. 3, large N QCD with two adjoint Wilson fermions is studied with the twisted space-time reduction, and sect. 4 is devoted to the large N QCD with single adjoint Wilson fermion. Conclusions are given in sect. 5.

2. The TEK model

We consider the SU(N) group with $N = L^2$, L being some positive integer. Then the action of the TEK model is obtained from the action of the SU(N) lattice gauge theory

$$S = -bN \sum_{n} \sum_{\mu \neq \nu = 1}^{4} \mathrm{Tr}\left[U_{n,\mu} U_{n+\mu,\nu} U^{\dagger}_{n+\nu,\mu} U^{\dagger}_{n,\nu}\right] \tag{1}$$

by neglecting the space-time dependence of the link variable $U_{n,\mu} \to U_{\mu}$ and multiplying the twist tensor $Z_{\mu\nu}$ in front of the plaquette action as[3]

$$S_{TEK} = -bN \sum_{\mu \neq \nu = 1}^{4} \mathrm{Tr}\left[Z_{\mu\nu} U_{\mu} U_{\nu} U^{\dagger}_{\mu} U^{\dagger}_{\nu}\right] \tag{2}$$

As a result, there remain only four SU(N) matrices U_{μ}. The inverse 't Hooft coupling is labelled $b = 1/g^2 N$. The elements of the twist tensor belong to Z(L) and the explicit form is given by

$$Z_{\mu\nu} = \exp\left(k \frac{2\pi i}{L}\right), \quad Z_{\nu\mu} = Z^{*}_{\mu\nu}, \quad \mu > \nu \tag{3}$$

The action S_{TEK} is invariant under the Z(L) transformation $U_{\mu} \to z U_{\mu}$, $z \in$ Z(L). If this symmetry is not spontaneously broken, the TEK model and the corresponding lattice gauge theory are equivalent in the large N limit. k and L should be coprime, and a general prescription for choosing k and L to preserve the Z(L) symmetry has been given in ref. [4]. In this talk, we choose L=29 and k=11 for the TEK model.[5,6] For finite L, the TEK model is closely related to the lattice gauge theory on a finite space-time volume L^4, and we are able to calculate Wilson loops up to size 14×14 for L=29. In the TEK model, the Wilson loop $W(R,T)$ with size $R \times T$ is obtained from four link variables U_{μ} as

$$W(R,T) = Z^{RT}_{\mu\nu} < U^{R}_{\mu} U^{T}_{\nu} U^{R\dagger}_{\mu} U^{T\dagger}_{\nu} > . \tag{4}$$

Then the string tension σ is extracted from the large distance behavior of the Creutz ratio

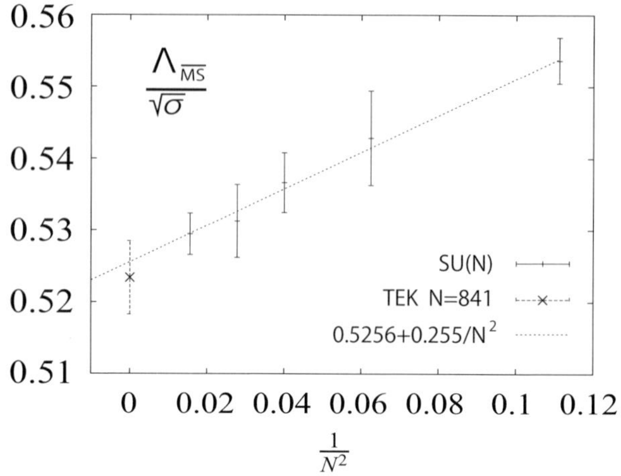

Fig. 1. N dependence of the continuum string tension.

$$\chi(R', T') = -\log \frac{W(R'+0.5, T'+0.5)W(R'-0.5, T'-0.5)}{W(R'+0.5, T'-0.5)W(R'-0.5, T'+0.5)} \tag{5}$$

as

$$\chi(R', R') = \sigma + \frac{2\gamma}{R'^2} + \frac{\eta}{R'^4} \tag{6}$$

with half-integer R' and T'. We calculate the continuum string tension by extrapolating the TEK data at six values of b. For comparison, we also calculate the continuum string tension of the usual SU(N) lattice gauge theory with $N = 3, 4, 5, 6, 8$ on a $V = 32^4$ lattice. Results are summarized in Fig. 1. The $\times$ point is $\Lambda_{\overline{MS}}/\sqrt{\sigma}$ for the TEK model with N=841. The + symbols are results for the lattice gauge theory, and dotted line is a linear fit of these data as a function of $1/N^2$. The large N extrapolated value agrees remarkably well with the result of the TEK model, demonstrating the correctness of the twisted space-time reduction idea.

3. Large N QCD with two adjoint fermions

SU(N) gauge theories with two adjoint fermions are thought to be conformal or nearly conformal for any value of N, since the first two coefficients of the beta function expressed in term of 't Hooft coupling are independent of N, namely, $b_0 = (4N_f - 11)/24\pi^2$ and $b_1 = (16N_f - 17)/192\pi^4$. For $N_f = 2$, $b_0 < 0$ and $b_1 > 0$, then we naturally expect that there is a infrared fixed point for finite value of 't Hooft coupling. In fact, for N=2 (minimal walking technicolor), there are now many lattice simulations indicating that the theory is conformal at vanishing fermion mass.[7] The mass anomalous dimension, however, is shown to be rather small $\gamma_* \sim 0.3$, to explain the large values of the observed quark masses, within the walking technicolor scenario.

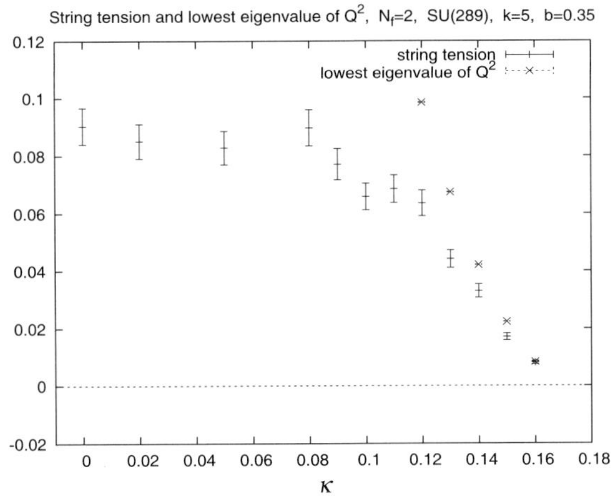

Fig. 2. String tension and lowest eigenvalue of Q^2 at b=0.35 with $N_f = 2$.

We can also consider the twisted space-time reduced model of large N QCD with two adjoint fermions. [a] The action reads

$$S = -bN \sum_{\mu\neq\nu=1}^{4} \mathrm{Tr}\left[Z_{\mu\nu}U_\mu U_\nu U_\mu^\dagger U_\nu^\dagger\right] + \sum_{j=1}^{N_f} \bar{\Psi}_j D_W \Psi_j \tag{7}$$

with N_f=2 and the Wilson-Dirac operator D_W given by

$$D_W = 1 - \kappa \sum_{\mu=1}^{4} \left[(1-\gamma_\mu)U_\mu^{adj} + (1+\gamma_\mu)U_\mu^{\dagger adj}\right] \tag{8}$$

Ψ_j is the fermion matrix in the color $(N, \bar{N})$ representation. Thus the link variable in the adjoint representation U_μ^{adj} actually acts on Ψ_j as $U_\mu^{adj}\Psi_j = U_\mu \Psi_j U_\mu^\dagger$. The hopping parameter of the fermion field κ is related to the bare fermion mass as $m_f = (1/2)(1/\kappa - 1/\kappa_c)$ with κ_c the critical hopping parameter.

We calculate the string tension at two values of b, 0.35 and 0.36, and for various values of κ. We are using $N = 289 = 17^2$ and $k = 5$, which should compare with ordinary lattice theory with $V = 17^4$.[11] For $N_f = 2$, we can use the standard hybrid Monte Carlo method. Simulations have been done on Hitachi SR16000 supercomputer at KEK having peek speed of 980 GFlops per node. Thanks to the Hitachi system engineers, our code is highly optimized for the SR16000. The sustained speed of our code being 600 GFlops.

[a]For previous works of the space-time reduced model of large N QCD with adjoint fermions, see ref. [8–10].

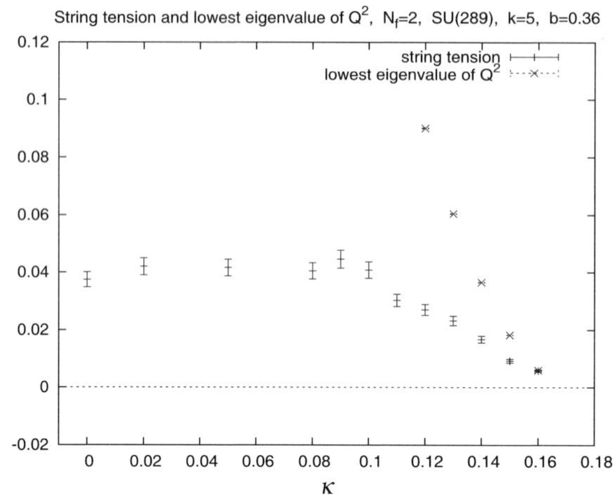

Fig. 3. String tension and lowest eigenvalue of Q^2 at b=0.36 with $N_f = 2$.

In Fig. 2, we show the string tension σ as a function of κ at b=0.35 with + symbols. The value of σ at κ=0 is obtained from the TEK model without fermions. We clearly see that as κ increases the string tension rapidly decreases and seems to vanish at $\kappa \sim 0.17$. So far, we have not calculated any hadronic spectrum. It is quite straightforward, however, to calculate the lowest eigenvalue of the positive hermitian Wilson-Dirac operator $Q^2 = (D_W\gamma_5)^2$, which should be related to the physical fermion mass square. In Fig. 2, we also show the lowest eigenvalue of Q^2 as $\times$ symbols. The string tension and the lowest eigenvalue of Q^2 vanish simultaneously, which strongly supports that the string tension is zero at critical κ_c where the fermion mass vanishes.

We can fit both the string tension σ and the lowest eigenvalue of Q^2 with the same fitting form $a(1/\kappa - 1/\kappa_c)^b$. Requiring that both quantities vanish at the same critical κ_c, we have $\kappa_c = 0.1694(7)$ and, for the string tension, $b = 1.10(4)$. If the massless theory is governed by an infrared fixed point, b is related to the mass anomalous dimension γ_* as $b = 2/(1+\gamma_*)$, then we have $\gamma_* = 0.81(8)$.

We have repeated the analysis at b=0.36, a little bit closer to the continuum limit. Results are shown in Fig. 3. It seems that the data at $\kappa = 0.16$ are largely affected by the finite -size effects. Excluding the data at $\kappa = 0.16$ from our analysis, we have $\kappa_c = 0.168(1)$, $b = 1.33(4)$, and thus $\gamma_* = 1.17(21)$. Although our data are still preliminary, large N QCD with two adjoint fermions seems to be a conformal field theory with large mass anomalous dimension $\gamma_* \sim 1$.

4. Large N QCD with single adjoint fermion

In the large N limit, $N_f = 1$ adjoint fermion is equivalent to the $N_f = 2$ fundamental fermions in the rank two anti-symmetric representation (the orientifold equivalence[8]). For $N = 3$, the latter theory is just the two flavor QCD. This means

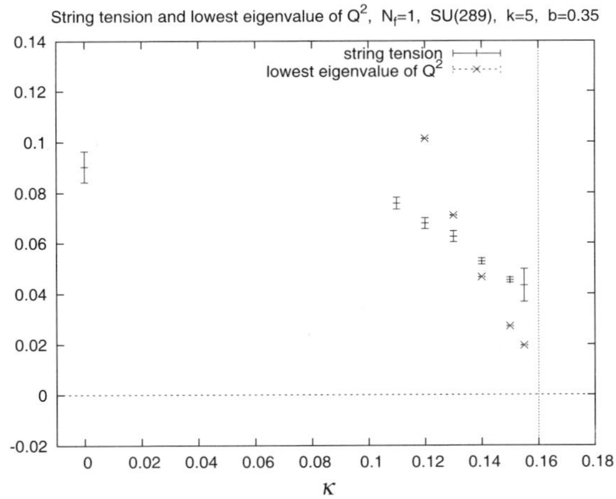

Fig. 4. String tension and lowest eigenvalue of Q^2 at b=0.35 with $N_f = 1$.

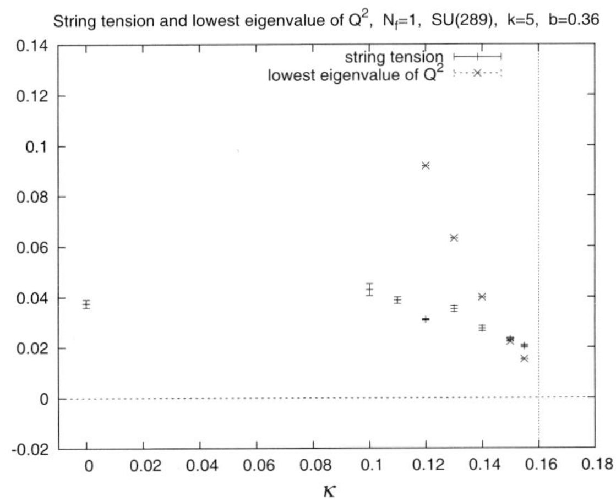

Fig. 5. String tension and lowest eigenvalue of Q^2 at b=0.36 with $N_f = 1$.

that the large N QCD with single adjoint fermion corresponds to the Corrigan-Ramond large N limit[12] of realistic QCD, and thus it is quite interesting to analyze this theory from the phenomenological point of view.

We use the rational hybrid Monte Carlo method to simulate the $N_f = 1$ theory. In Fig. 4, we plot the string tension σ as a function of κ at b=0.35 with $+$ symbols. Also shown is the lowest eigenvalue of Q^2 with $\times$ symbols. The lowest eigenvalue of Q^2 seems to vanish around $\kappa \sim 0.165$, while the string tension seems to remain finite at this point, indicating that the large N QCD with single adjoint fermion is a confining theory. This is quite natural since our theory should be related to the two flavor $N = 3$ QCD. We found that, for $\kappa \geq 0.16$, we cannot make simulations since the CG iteration does not converge, which is also the common phenomena encountered in the simulation of QCD.

In Fig. 5, we show the results at b=0.36. The absolute values of the string tension are smaller than those of $b = 0.35$ since we are approaching to the continuum limit. The string tension, however, remains finite around $\kappa \sim 0.165$ as in Fig. 4. We also confirm that the CG iteration does not converge for $\kappa \geq 0.16$. Hence, both Figs. 4 and 5 indicate that the continuum theory of the large N QCD with single adjoint fermion is a confining theory.

5. Conclusion

We have demonstrated that the twisted reduction works quite well for the description of large N QCD. For pure gauge theory, we have succeeded in calculating the continuum string tension using TEK model. Although our calculations with dynamical fermions are still preliminary, we have also shown that large N QCD with two adjoint fermions is a conformal theory with large mass anomalous dimension $\gamma_* \sim 1$, and that the theory with single adjoint fermion is a confining theory. Since the rational hybrid Monte Carlo method we used for the simulation with one flavor can directly applicable to any number of flavor, we are planning to extend our analysis to $N_f = 1/2$ super-symmetric theory and also $N_f = 3/2$ and $5/2$ theories if they are conformal or confining theories.

Acknowledgments

A.G-A is supported by grants FPA2009-08785, FPA2009-09017, CSD2007-00042, HEPHACOS S2009/ESP-1473 and ITN PITN-GA-2009-238353. M. O. is supported by the Japanese MEXT grant No 23540310. The main calculations have been done on SR16000 computer at KEK supported by the Large Scale Simulation Prog. No.12-01 (FY2011-12). The calculations have also been done on SR16000 computer at YITP in Kyoto University and on the INSAM cluster system at Hiroshima University.

References

1. T. Eguchi and H. Kawai, *Phys. Rev. Lett.* **48** (1982) 1004.
2. G. Bhanot, U. M. Heller and H. Neuberger, *Phys. Lett.* **B113** (1982) 47.
3. A. Gonzalez-Arroyo and M. Okawa, *Phys. Rev.* **D27** (1983) 2397.
4. A. Gonzalez-Arroyo and M. Okawa, *JHEP* **1007** (2010) 043.
5. A. Gonzalez-Arroyo and M. Okawa, *PoS Lattice 2012* (2012) 221.
6. A. Gonzalez-Arroyo and M. Okawa, *Phys. Lett.* **B718** (2013) 1524.
7. L. Del Debbio, *PoS Lattice 2010* (2010) 004.
8. P. Kovtun, M. Unsal, L. G. Yaffe, *JHEP* **0706** 019 (2007) 019.
9. T. Azeyanagi, M. Hanada, M. Unsal, R. Yacoby, *Phys. Rev.* **D82** (2010) 125013.
10. B. Bringoltz, M. Koren, S. R. Sharpe, *Phys. Rev.* **D85** (2012) 094504.
11. A. Gonzalez-Arroyo and M. Okawa, *PoS Lattice 2012* (2012) 046.
12. E. Corrigan and P. Ramond, *Phys. Lett.* **B87** (1979) 73.

Large-N_c Gauge Theory and Chiral Random Matrix Theory

Masanori Hanada* , Jong-Wan Lee† and Norikazu Yamada

KEK Theory Center, High Energy Accelerator Research Organization (KEK), Tsukuba 305-0801, Japan

Effective theory approaches and the large-N_c limit are useful for studying the strongly coupled gauge theories. In this talk we consider how the chiral random matrix theory (χRMT) can be used in the study of large-N_c gauge theories. It turns out the parameter regions, in which each of these two approaches are valid, are different. Still, however, we show that the breakdown of chiral symmetry can be detected by combining the large-N_c argument and the χRMT with some cares. As a demonstration, we numerically study the four dimensional $SU(N_c)$ gauge theory with $N_f = 2$ heavy adjoint fermions on a 2^4 lattice by using Monte-Carlo simulations, which is related to the infinite volume lattice through the Eguchi-Kawai equivalence.

Keywords: Large-N_c; effective field theory; lattice gauge theory.

1. Introduction

Strongly coupled gauge theories such as QCD are of great interest. Because it is difficult to study those theories analytically, effective theory approaches such as the χRMT and the large-N_c limit are often considered. Given their importance, it is natural to consider how they can be combined to study various QCD-like theories.

In this talk, based on our paper,[1] we discuss how to utilize the χRMT to study the large-N_c gauge theories, and provide the theoretical argument for the detection of SχSB. As a demonstration, we numerically study the four dimensional $SU(N_c)$ gauge theory with $N_f = 2$ heavy adjoint fermions on a 2^4 lattice by using Monte-Carlo simulations, which is related to the infinite volume lattice through the Eguchi-Kawai equivalence[2] [a]. It turns out that some cares are required because the valid parameter regions for these two techniques are different in general: in the χRMT limit (see Sec. 4 for the definition), the 't Hooft large-N_c expansion fails due to the non-trivial N_c- dependence of the expansion coefficients.

*Current affiliation: Yukawa Institute for Theoretical Physics, Kyoto University

†Current affiliation: The City College of New York, Department of Physics

[a]In Ref.[3] and,[4,5] the authors already used the χRMT techniques for numerical studies of chiral symmetry breaking of $SU(N_c)$ gauge theories in the quenched limit and the theories with adjoint fermions, respectively. However, there is a subtlety (the difference of the limiting procedures explained in Sec. 4), which makes the conclusion ambiguous.[5] In this paper we clarify this point in order to fully justify the method.

2. The 't Hooft large-N_c limit

Let us first consider the $SU(N_c)$ pure Yang-Mills theory,

$$S_{YM} = \frac{1}{4g_{YM}^2} \int d^4x \, \mathrm{Tr} F_{\mu\nu}^2. \tag{1}$$

The 't Hooft large-N_c limit[6] is the large-N_c limit in which the 't Hooft coupling constant $\lambda = g_{YM}^2 N_c$ and the space-time V are fixed. The energy scale under consideration (e.g. distance between operators, the size of the Wilson loop) is also fixed. In this limit the $1/N_c$ expansion has a natural topological structure,

$$\langle \hat{O} \rangle = \sum_{g=0}^{\infty} c_g(\lambda, V) N_c^{-2g}, \tag{2}$$

where $\hat{O}$ is a properly normalized single trace operator. In the perturbation theory, contribution of order N_c^{-2g} comes from the genus-g nonplanar diagrams, i.e. the Feynman diagrams which can be drawn on the two-dimensional surface with g handles. The connected correlation functions of more than one operators have the same structure.

When N_f fundamental fermions with the mass m_f are introduced, in addition to the 't Hooft coupling and the space-time volume, the fermion mass m_f must also be fixed. Because the fermions have $O(N_f N_c)$ degrees of freedom while the gluons have $O(N_c^2)$, every time we replace the gluon loop with the fermion loop we obtain the factor N_f/N_c. Hence the $1/N_c$-expansion becomes

$$\langle \hat{O} \rangle = \sum_{g,b=0}^{\infty} c_{g,b}(\lambda, V, m_f) N_c^{-2g} \cdot (N_f/N_c)^b, \tag{3}$$

where b is the number of fermion loops. Therefore, when N_f is fixed, diagrams with fermion loops are suppressed. (They are not suppressed if one considers the adjoint fermion instead.) In the 't Hooft large-N_c limit, the parameters λ, V and m_f are fixed, because otherwise a nontrivial N_c-dependence can appear through the coefficient $c_{g,b}(\lambda, V, m_f)$. For example, in QCD, if m_f decreases with N_c, N_c-dependent infrared divergence appears because the pion becomes lighter and the standard $1/N_c$-counting in the 't Hooft limit can be destroyed completely.

2.1. *The Eguchi-Kawai equivalence*

Let us consider Wilson's lattice gauge theory on the L^4 periodic lattice,

$$S_W = 2N_c^2 b \sum_{\mathbf{n}} \sum_{\mu<\nu} \left(1 - \frac{1}{N_c} \mathrm{ReTr}\left(U_{\mathbf{n},\mu} U_{\mathbf{n}+\mu,\nu} U_{\mathbf{n}+\nu,\mu}^{\dagger} U_{\mathbf{n},\nu}^{\dagger}\right)\right), \tag{4}$$

where μ, ν run from 1 to 4, $b = 1/\lambda = 1/(g_{YM}^2 N_c)$, and $\mathbf{n}$ runs through the L^4 lattice. This theory has the $(\mathbb{Z}_{N_c})^4$ center symmetry, which multiply a phase factor to the link variables:

$$U_{\mathbf{n},\mu} \to e^{2\pi i k_\mu / N_c} U_{\mathbf{n},\mu} \qquad (k_\mu \in \mathbb{Z}). \tag{5}$$

Let us consider the 't Hooft large-N_c limit, in which the coupling constant b and the lattice size L are fixed. As long as the $(\mathbb{Z}_{N_c})^4$ center symmetry is not broken spontaneously, various quantities like the expectation values of the Wilson loops do not depend on L. This is called the Eguchi-Kawai equivalence.[2] One can also introduce fermions with the periodic boundary condition; for example, for both the fundamental and adjoint fermions, the Dirac spectrum does not depend on L when N_f and the mass are fixed. Therefore one can study the large-N_c theory on the infinite lattice by using a small lattice, say 1^4 or 2^4.

In the pure glue theory, the $(\mathbb{Z}_{N_c})^4$ center symmetry is actually broken in the weak coupling limit $b \to \infty$ with fixed L.[7] However the situation changes drastically when one adds the adjoint fermions, because they provide the repulsive force between eigenvalues. If we add one or more massless Dirac adjoint fermions, the center symmetry is unbroken irrespectively of the detail of the lattice action.[8] Furthermore, somehow surprisingly at first glance, the center symmetry does not break spontaneously even with very heavy adjoint fermions, whose mass is as heavy as the ultraviolet cutoff scale;[9,10] therefore one can study the pure Yang-Mills theory by using the EK equivalence, by taking the fermion mass very large.

3. The chiral random matrix theory (χRMT)

In this section we provide a brief review of the χRMT. For more details, see e.g.[11,12]

We consider the ϵ-regime of QCD, where the space-time volume $V = L^4$ is taken such that L is much smaller than the pion Compton wavelength and is much larger than $1/\Lambda_{QCD}$,[13] $1/\Lambda_{QCD} \ll L \ll 1/m_\pi$. In the ϵ-regime, the only relevant degrees of freedom are zero momentum modes of pions. Then the system has a *universality*, that is, the dynamics is determined by the symmetry breaking pattern and the microscopic details of the theory do not matter. Therefore QCD can be replaced by the χRMT, which is a random matrix model with the same symmetry breaking pattern; in particular, the spectrum of the Dirac operator can be calculated from the χRMT. Note that the same argument holds for other QCD-like theories as long as chiral symmetry is broken spontaneously.

The partition function of the χRMT is given by

$$Z = \int d\Phi \prod_{f=1}^{N_f} \det \mathcal{D}_f \; e^{-\frac{N\beta}{2} G^2 \mathrm{tr}\Phi^\dagger\Phi}, \tag{6}$$

where Φ is a $N \times (N+\nu)$ matrix and G is a normalization parameter. N corresponds to the size of the system ($\simeq$ the space-time volume), and ν is the topological charge. Correspondingly to the thermodynamic limit of QCD, N is sent to infinity. In this limit, however, the fermion mass m_f must be scaled so that $m_f N$, which is (roughly speaking) identified with $m_f V$, is fixed.

Note that we can define the 't Hooft large-N limit (not the large-N_c limit) for the χRMT, in which m_f is fixed. The limit one has to take for the comparison to QCD ($m_f N$ fixed) is *not* this 't Hooft limit. This difference is crucial when we compare the large-N_c gauge theories and χRMT, as we will see in Sec. 4. In addition, it is important to notice that N is identified with the total degrees of freedom associated with the low-energy dynamics and the individual degrees of freedom, such as the volume and the number of colors, does not appear in the χRMT explicitly.

The ensemble and the Dirac operator $\mathcal{D}$ are chosen so that the Dirac operator has the same symmetries as the counterparts in QCD and QCD-like theories. Depending on the universality classes, there are three χRMTs, which are called as the Chiral Gaussian Sympletic Ensemble (ChGSE), Chiral Gaussian Orthogonal Ensemble (ChGOE) and Chiral Gaussian Unitary Ensemble (ChGUE).[14] $SU(N_c)$ with the adjoint fermions is described by the ChGSE, in which

$$\mathcal{D}_f = \begin{pmatrix} m_f \mathbf{1} & \Phi \\ -\Phi^\dagger & m_f \mathbf{1} \end{pmatrix}, \tag{7}$$

where Φ is a $2N \times 2(N+\nu)$ quaternion real matrix, which can be written by using four $N \times (N+\nu)$ real matrices ϕ_0, ϕ_1, ϕ_2 and ϕ_3 as

$$\Phi = \begin{pmatrix} \phi_0 + i\phi_3 & i\phi_1 + \phi_2 \\ i\phi_1 - \phi_2 & \phi_0 - i\phi_3. \end{pmatrix}. \tag{8}$$

4. Large-N_c versus χRMT

In this section, we establish the way to apply the χRMT techniques to large-N_c gauge theories. For concreteness, we consider the $SU(N_c)$ lattice theory with volume V and fermion mass m. Here the volume V is arbitrary, although we concentrate on a small fixed volume in Sec. 5.

4.1. *Difference of the limits*

When one compares the χRMT with the gauge theory, the matrix size N of the χRMT is identified with the degrees of freedom in the gauge theory which are important for the low-energy dynamics, $N \sim VN_c^\alpha$, where the constant α depends on the representation of the fermion in general. (As we will see, $\alpha = 1$ for the massless probe adjoint fermion of the 2^4-lattice model. Note that it is different from the usual counting in the 't Hooft limit, $\alpha = 2$.) In order for the universality to hold, $mN \sim mVN_c^\alpha$ must be fixed as we take the large-N_c limit. Let us call it the 'χRMT limit'.On the other hand, for the standard $1/N_c$ counting, the ordinary 't Hooft large-N_c limit, in which m and V are fixed, must be taken. Indeed we can explicitly confirm the failure of the 't Hooft counting in the χRMT limit.

4.2. *The comparison*

First let us recall how one can realize SχSB in numerical simulations of QCD in the ε-regime. The criterion for SχSB is whether the Dirac spectrum agrees with the χRMT prediction. Once they are found to agree, the value of the chiral condensate is determined. Notice that the value of the chiral condensate obtained in the ε-regime is the same as the one obtained by taking the chiral limit after the large-volume (or thermodynamic) limit. It means that, whatever the order of the limiting procedures is, we can determine the chiral condensate.

The same logic holds for large N_c gauge theories. But this time, the thermodynamic limit can be achieved not just by taking large-V but also by sending N_c to infinity. Therefore, the χRMT limit is achieved by taking the large N_c limit with mVN_c^α fixed. The constant α is determined by the *effective* degrees of freedom in the low mode region. As we will see, in the 2^4-lattice model with the massless probe adjoint fermion, $\alpha = 1$. Then, the criterion for SχSB is again the agreement of the Dirac spectrum in the χRMT.

Strictly speaking, there is a subtlety in the χRMT limit of the 2^4-lattice model: because the space-time volume of this lattice is very small, usual derivation of the χRMT from QCD via the chiral perturbation theory may not be applicable. In order to circumvent this subtlety, one can take another path as follows. First let us consider a sufficiently large volume, where the usual derivation of the χRMT is valid when chiral symmetry is spontaneously broken. Then, the distribution of the Dirac eigenvalues should agree with that from the χRMT after properly normalizing the eigenvalues. Therefore, as long as V is large enough to justify the chiral perturbation, V-dependence does not appear manifestly, rather a certain combination VN_c^α appears. Then we take the volume even smaller. Because there is no phase transition thanks to the heavy adjoint fermion, it is expected that the same expression for the eigenvalue distribution holds at very small volume as a function of VN_c^α, where the value of α could be different from that in the large-volume theory. Then, the agreement between the χRMT and the 2^4-lattice model in the RMT limit translates into the agreement between the χRMT and the large-volume lattice provided unbroken centery symmetry. At large volume and in the 't Hooft limit, one can conclude SχSB in the ordinary manner. By further using the EK equivalence, SχSB of the 2^4-lattice model in the 't Hooft limit can also be concluded.

5. Numerical results

The simplest way to achieve the χRMT limit without losing the generality is to set $m = 0$ and $N_c \to \infty$. [b] For this purpose, we calculate the low-lying spectrum of the overlap-Dirac operator $\mathcal{D}$ for a massless fermion in the adjoint representation. The eigenvalues $\tilde{\lambda}_k$ lie on a circle in the complex plane. To compare the Dirac spectrum with the χRMT, we consider the projection of the eigenvalues to the imaginary axis, $\lambda_k = \mathrm{Im}[\tilde{\lambda}_k]/(1 - \mathrm{Re}[\tilde{\lambda}_k]/m_0)$. Note that the Dirac eigenvalues for adjoint fermions appears as conjugate pairs with 2-fold degeneracy and we take the distinct eigenvalues on the upper-half plane for our numerical results.

For the configuration generation, we use the standard Wilson plaquette gauge action. In order to preserve the center symmetry, we add two-flavor of very heavy

[b] Although this limit looks like the 't Hooft limit (m is fixed), actually one should not regard it so because the infrared (IR) regulator is assumed in the usual 't Hooft counting.

adjoint fermions, for which we choose the plain Wilson fermion. For the detail of the simulations, see.[1]

5.1. *Chiral symmetry breaking*

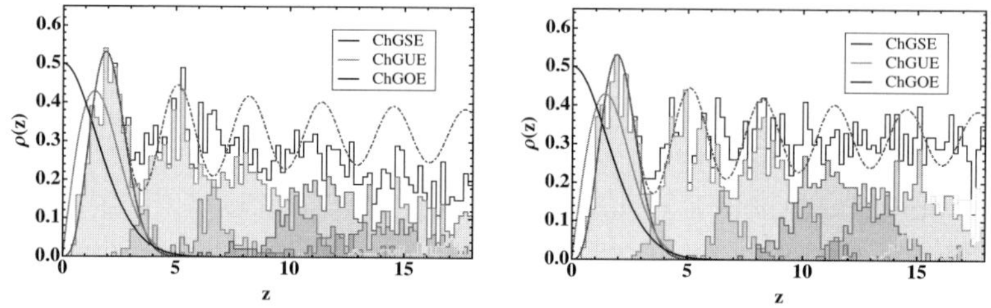

Fig. 1. Low-lying Dirac spectrum for $N_c = 8$ (left) and $N_c = 16$ (right). The Dirac eigenvalues are rescaled by matching the ensemble average of the lowest eigenvalue with the expectation value of the lowest eigenvalue in the χRMT for the ChGSE. The solid and dotted lines represent the χRMT predictions of the distributions of the lowest eigenvalue and the spectral density, respectively. The colored histograms are for individual eigenvalues, while the black histograms are for all eigenvalues.

We compare the simulation data with the χRMT prediction for the quenched theory. We consider only the zero-topological charge sector. The adjoint QCD with any number of flavors belongs to the universal class of the ChGSE. However, we also consider two other universal classes, ChGOE and ChGUE, in order to make the comparison manifest. We borrow the analytic expressions for the χRMT from Ref.[15]

Histograms of the small Dirac eigenvalues for $N_c = 8, 16$ at $b = 0.5$ are shown in Fig. 1. For the lowest eigenvalue, we found that its distribution is well described by the χRMT for the ChGSE (solid blue curve) after introducing a rescaled eigenvalue z by $z = \lambda V \Sigma$ to fit the data, where $V = 2^4$ and Σ is a free parameter. This result is a strong evidence that chiral symmetry of quenched large N_c gauge theory is spontaneously broken.

5.2. *N_c-scaling of the Dirac eigenvalues*

In Sec. 4, we argued that the χRMT limit of the 2^4-lattice model is analogous to that of ordinary QCD by replacing mV with mVN_c^α, where the exponent α can be determined so that N_c^α is the number of degrees of freedom important at low energy, which is proportional to the inverse of the eigenvalue density around the origin. Then, because the low-lying eigenvalues scale as $1/N_c$ as we will see below, we obtain $\alpha = 1$, which is different from a naive expectation in the 't Hooft limit.

In the left panel of Fig. 2, we plot the spacing between the adjacent low-lying Dirac eigenvalues $\delta\lambda_k = \lambda_k - \lambda_{k-1}$ multiplied by N_c for $N_c = 8, 12$ and 16. We see a nice agreement of the data points at up to $k = 2$ for $N_c = 8$ and at up to

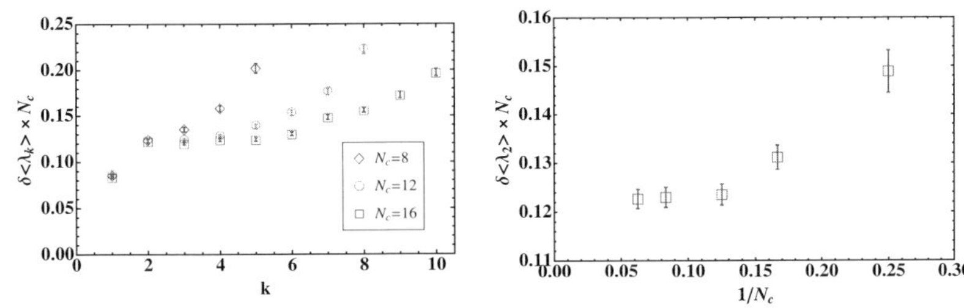

Fig. 2. (Left) Spacing between the adjacent Dirac eigenvalues multiplied by N_c for $N_c = 8, 12$ and 16. (Right) Spacing between the first and second Dirac eigenvalues multiplied by N_c for $N_c = 4, 6, 8, 12$ and 16.

$k = 4$ for $N_c = 12$. This result implies that the near-zero eigenvalues, which are expected to reproduce the χRMT prediction well, scale as $1/N_c$. (The same scaling had been reported in,[5] by using the single-site model with a very light dynamical overlap adjoint fermion.) The right figure of Fig. 2 shows $\delta\langle\lambda_2\rangle$ multiplied by N_c for various N_c; for given statistics, they agree to each other for $N_c \geq 8$. For more detailed argument about this scaling, see.[1]

Acknowledgements

The authors would like to thank S. Hashimoto, A. Hietanen, M. Kieburg, M. Koren, S. Sharpe, M. Tezuka, J. Verbaarschot and N. Yamamoto for stimulating discussions and comments. The numerical computations used in this work were carried out on cluster at KEK. This work is supported in part by the Grant-in-Aid for Scientific Research of the Japanese Ministry of Education, Culture, Sports, Science and Technology and JSPS (Nos. 20105002,20105005,and 22740183).

References

1. M. Hanada, J.-W. Lee and N. Yamada (2013).
2. T. Eguchi and H. Kawai, *Phys. Rev. Lett.* **48**, p. 1063 (1982).
3. R. Narayanan and H. Neuberger, *Nucl. Phys.* **B696**, 107 (2004).
4. A. Hietanen and R. Narayanan, *JHEP* **01**, p. 079 (2010).
5. A. Hietanen and R. Narayanan, *Phys.Rev.* **D86**, p. 085002 (2012).
6. G. 't Hooft, *Nucl.Phys.* **B72**, p. 461 (1974).
7. G. Bhanot, U. M. Heller and H. Neuberger, *Phys. Lett.* **B113**, p. 47 (1982).
8. P. Kovtun, M. Unsal and L. G. Yaffe, *JHEP* **06**, p. 019 (2007).
9. B. Bringoltz and S. R. Sharpe, *Phys. Rev.* **D80**, p. 065031 (2009).
10. T. Azeyanagi, M. Hanada, M. Unsal and R. Yacoby, *Phys. Rev.* **D82**, p. 125013 (2010).
11. J. Verbaarschot and T. Wettig, *Ann.Rev.Nucl.Part.Sci.* **50**, 343 (2000).
12. G. Akemann, *Int.J.Mod.Phys.* **A22**, 1077 (2007).
13. H. Leutwyler and A. V. Smilga, *Phys.Rev.* **D46**, 5607 (1992).
14. J. J. M. Verbaarschot, *Phys.Rev.Lett.* **72**, 2531 (1994).
15. D. Toublan and J. Verbaarschot, *Nucl.Phys.* **B560**, 259 (1999).

Non-Abelian Dual Superconductivity in SU(3) Yang-Mills Theory Due to Non-Abelian Magnetic Monopoles

Akihiro Shibata

Computing Research Center, High Energy Accelerator Research Organization (KEK), and Graduate Univ. for Advanced Studies (Sokendai), Tsukuba 305-0801, Japan
E-mail: akihiro.shibata@kek.jp

Kei-Ichi Kondo

Department of Physics, Graduate School of Science, Chiba University, Chiba 263-8522, Japan
E-mail: kondok@faculty.chiba-u.jp

Seikou Kato

Fukui National College of Technology, Sabae 916-8507, Japan E-mail: skato@fukui-nct.ac.jp

Toru Shinohara

Department of Physics, Graduate School of Science, Chiba University, Chiba 263-8522, Japan
E-mail: kondok@faculty.chiba-u.jp

The dual superconductivity is the promising mechanism for quark confinement. We have proposed the non-Abelian dual superconductivity picture in the $SU(3)$ Yang-Mills theory, and already presented numerical evidences for the restricted field dominance and the non-Abelian magnetic monopole dominance in the string tension, by applying our new formulation of Yang-Mills theory to a lattice. In this talk, we focus on the non-Abelian dual Meissner effect and the type of dual superconductivity. We find that the measured chromo-electric flux tube between a quark and antiquark pair strongly supports the non-Abelian dual Meissner effect due to non-Abelian magnetic monopoles. Moreover, we give a remarkable result that the type of the resulting dual superconductor is the type I in $SU(3)$ Yang-Mills, rather than the border between the type I and II, in marked contrast to the $SU(2)$ case.

Keywords: Quark connement; dual superconductivity; non-Abelian magnetic monopole

1. Introduction

The dual superconductivity is the promising mechanism for quark confinement.[3] There are many preceding works to investigate the dual superconductivity based on the Abelian projection. However, these results are obtained only in special gauges such as the maximal Abelian (MA) gauge and the Laplacian Abelian gauge within the Abelian projection, which breaks the gauge symmetry as well as color symmetry (global symmetry). In order to overcome the shortcomings of the Abelian projection, we have presented a new lattice formulation of $SU(N)$ Yang-Mills (YM) theory and presented non-Abelian dual superconductor picture in the previous papers (see Ref. 1 and Refs. therein).

By applying the minimal option of our new formulation to the $SU(3)$ YM theory on a lattice, in the static potential for a pair of quark and antiquark in the fundamental representation, we have demonstrated: (i) the restricted-field dominance or "Abelian" dominance (which is a gauge-independent (invariant) extension of the conventionally called Abelian dominance): the string tension $\sigma_{\rm V}$ obtained from the decomposed V-field (i.e., restricted field in the minimal option) reproduced the string tension $\sigma_{\rm full}$ of the original YM field, $\sigma_{\rm V}/\sigma_{\rm full} = 93 \pm 16\%$, (ii) the gauge-independent non-Abelian magnetic monopole dominance: the string tension $\sigma_{\rm V}$ extracted from the restricted field was reproduced by only the (non-Abelian) magnetic monopole part $\sigma_{\rm mon}$, $\sigma_{\rm mon}/\sigma_{\rm V} = 94 \pm 9\%$.

In this talk, we give further evidences for establishing the non-Abelian dual superconductivity picture for quark confinement in $SU(3)$ Yang-Mills theory. First, we study the dual Meissner effect by measuring the distribution of chromo-flux created by a pair of static quark and antiquark. Second, we measure the possible magnetic monopole current induced around the flux connecting a pair of static quark and antiquark. Third, we focus on the type of dual superconductivity, i.e., type I or type II.[2]

2. Lattice formulation

We give a quick review of a new formulation of the lattice $SU(3)$ YM theory.[1] We focus our attention on quarks belonging to the fundamental representation, and the minimal option of the formulation enables us to extract the dominant mode of quark confinement. For the original $SU(3)$ gauge link variable $U_{x,\mu}$, we wish to decompose it into new variables $X_{x,\mu} \in SU(3)$ and $V_{x,\mu} \in SU(3)$, i.e., $U_{x,\mu} = X_{x,\mu}V_{x,\mu}$, so that, $V_{x.\mu}$ could be the dominant mode for quark confinement, while $X_{x,\mu}$ is the remainder. In this decomposition, we require that $V_{x,\mu}$ is transformed in the same way as the original gauge link variable $U_{x,\mu}$ and $X_{x,\mu}$ as a site variable by the full $SU(3)$ gauge transformation.

To obtain the decomposition, we introduce the color field $\mathbf{h}_x$. In the minimal option, the color field is defined by $\mathbf{h}_x = \xi(\lambda^8/2)\xi^\dagger \in Lie[SU(3)/U(2)]$, with λ^8 being the Gell-Mann matrix and ξ the $SU(3)$ group element. Once the color field is introduced, the decomposition is obtained by solving the defining equation:

$$D^{\epsilon}_{\mu}[V]\mathbf{h}_x := \frac{1}{\epsilon}\left[V_{x,\mu}\mathbf{h}_{x+\mu} - \mathbf{h}_x V_{x,\mu}\right] = 0. \tag{1}$$

This defining equation can be solved exactly, and the solution is given by

$$X_{x,\mu} = \widehat{L}^{\dagger}_{x,\mu}\det(\widehat{L}_{x,\mu})^{1/3}g_x^{-1}, \qquad V_{x,\mu} = X^{\dagger}_{x,\mu}U_{x,\mu} = g_x\widehat{L}_{x,\mu}U_{x,\mu}, \tag{2a}$$

$$\widehat{L}_{x,\mu} = \left(L_{x,\mu}L^{\dagger}_{x,\mu}\right)^{-1/2}L_{x,\mu}, \tag{2b}$$

$$L_{x,\mu} = \frac{5}{3}\mathbf{1} + \sqrt{\frac{4}{3}}(\mathbf{h}_x + U_{x,\mu}\mathbf{h}_{x+\mu}U^{\dagger}_{x,\mu}) + 8\mathbf{h}_x U_{x,\mu}\mathbf{h}_{x+\mu}U^{\dagger}_{x,\mu}\,. \tag{2c}$$

Here the variable g_x is the $U(2)$ part undetermined from Eq.(1) alone. In what follows, we put $g_x = 1$, so that the defining equations correspond respectively to the continuum version.

Thus the decomposition is uniquely determined from Eqs.(2), if the color field $\mathbf{h}_x$ is specified. In order to determine the configuration $\{\mathbf{h}_x\}$ of color fields, we use the reduction condition which guarantees that the new theory written in terms of new variables $(X_{x,\mu}, V_{x,\mu})$ is equipollent to the original YM theory. Here, we use the reduction condition: for a given configuration of the original link variables $U_{x,\mu}$, color fields $\{\mathbf{h}_x\}$ are obtained by minimizing the functional:

$$F_{\rm red}[\{\mathbf{h}_x\}] = \sum\nolimits_{x,\mu} \mathrm{tr}\left\{(D_\mu^\epsilon[U_{x,\mu}]\mathbf{h}_x)^\dagger(D_\mu^\epsilon[U_{x,\mu}]\mathbf{h}_x)\right\}. \tag{3}$$

3. Lattice method and result

We generate configurations of the YM gauge link variable $\{U_{x,\mu}\}$ using the standard Wilson action on a 24^4 lattice at $\beta = 6.2$. The gauge link decomposition is obtained according to the framework given in the previous section: the color field configuration $\{h_x\}$ is obtained by solving the reduction condition of minimizing the functional Eq.(3) for each gauge configuration $\{U_{x,\mu}\}$, and then the decomposed variables $\{V_{x,\mu}\}$, $\{X_{x,\mu}\}$ are obtained by using the formula Eq.(2). In the measurement of the Wilson loop average, we apply the APE smearing technique to reduce noises.

3.1. *Dual Meissner effect*

To investigate the non-Abelian dual Meissner effect as the mechanism of quark confinement, we measure the chromofield strength connecting a quark and an antiquark for both the original YM field and the restricted field. In order to extract the chromo-field, we use a gauge-invariant correlation function proposed in Ref. 4. The chromo-field created by a quark-antiquark pair in $SU(N)$ Yang-Mills theory is measured by using a gauge-invariant connected correlator between a plaquette and the Wilson loop (see Fig.1):

$$\rho_W[U] := \frac{\langle \mathrm{tr}\left(U_P L^\dagger W[U] L\right)\rangle}{\langle \mathrm{tr}\left(W[U]\right)\rangle} - \frac{1}{N}\frac{\langle \mathrm{tr}\left(U_P\right)\mathrm{tr}\left(W[U]\right)\rangle}{\langle \mathrm{tr}\left(W[U]\right)\rangle}, \tag{4}$$

where $W[U]$ is the Wilson loop in Z-T plane representing a pair of quark and antiquark, U_P a plaquette variable as the probe operator to measure the chromo-field strength at the point P, and L the Wilson line connecting the source $W[U]$ and the probe U_P. Here L is necessary to guarantee the gauge invariance of the correlator ρ_W and hence the probe is identified with $LU_PL^\dagger$. The symbol $\langle \mathcal{O} \rangle$ denotes the average of the operator $\mathcal{O}$ in the space and the ensemble of the configurations. The gauge-invariant chromo-field strength $F_{\mu\nu}[U]$ is given by $F_{\mu\nu}[U]$:,where $\beta := 2N/g^2$ is the lattice gauge coupling constant. The chromo-field strength of the restricted

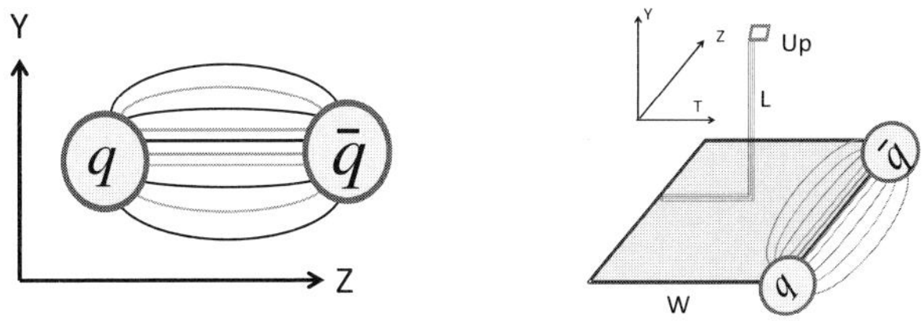

Fig. 1. (Left) The setup of measuring the chromo-flux produced by a quark–antiquark pair. (Right) The gauge-invariant connected correlator $(U_p L W L^\dagger)$ between a plaquette U and the Wilson loop W.

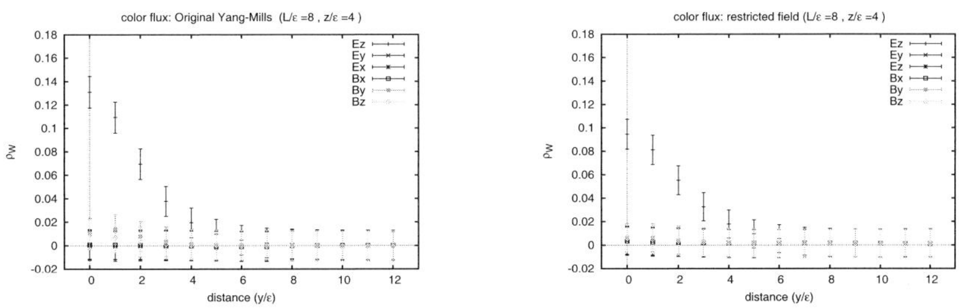

Fig. 2. Measurement of components of the chromoelectric field $\boldsymbol{E}$ and chromomagnetic field $\boldsymbol{B}$ as functions of the distance y from the z axis. (Left panel) the original $SU(3)$ YM field, (Right panel) the restricted field.

field $F_{\mu\nu}[V] := \epsilon^{-2}\sqrt{\frac{\beta}{2N}}\rho_W[U]$ is measured as well as the original YM field where the field strength is obtained by using $V_{x,\mu}$ in Eq.(4) instead of $U_{x,\mu}$.

Fig. 2 show the results of measurements for the chromoelectric and chromomagnetic fields $F_{\mu\nu}[U]$ for the original $SU(3)$ field U (left panel) and $F_{\mu\nu}[V]$ for the restricted filed (right panel). Here the quark and antiquark source is introduced as 8×8 Wilson loop ($W[U]$ or $W[V]$) in the Z-T plane, and the probe (U_p or V_p) is set at the center of the Wilson loop and moved along the Y-direction. We find that only the E_z component of the chromoelectric field $(E_x, E_y, E_z) = (F_{10}, F_{20}, F_{30})$ connecting q and $\bar{q}$ has non-zero value for both the restricted field V and the original YM field U. The other components are zero consistently within the numerical errors. This means that the chromomagnetic field $(B_x, B_y, B_z) = (F_{23}, F_{31}, F_{12})$ connecting q and $\bar{q}$ does not exist and that the chromoelectric field is parallel to the z axis on which quark and antiquark are located. The magnitude E_z quickly decreases in the distance y away from the Wilson loop.

To see the profile of the nonvanishing component E_z of the chromoelectric field in detail, we explore the distribution of chromoelectric field on the 2-dimensional plane. Fig. 3 shows the distribution of E_z component of the chromoelectric field, where the quark-antiquark source represented as 9×11 Wilson loop W is placed at $(Y, Z) = (0, 0), (0, 9)$, and the probe U is displaced on the Y-Z plane at the

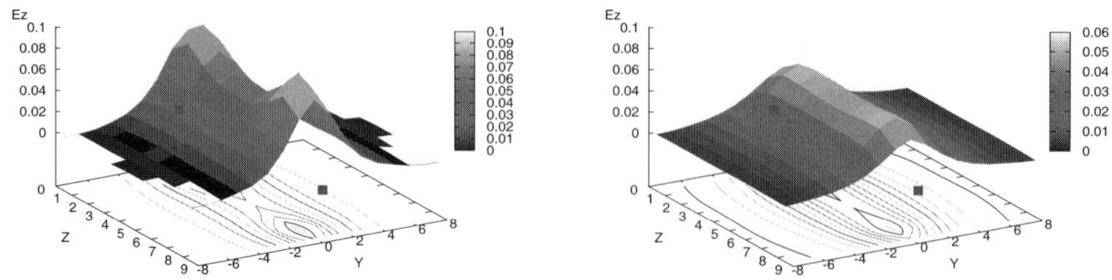

Fig. 3. The distribution in Y-Z plane of the chromoelectric field E_z connecting a pair of quark and antiquark: (Left panel) chromoelectric field produced from the original YM field, (Right panel) chromoelectric field produced from the restricted field.

midpoint of the T-direction. The position of a quark and an antiquark is marked by the solid (blue) box. The magnitude of E_z is shown by the height of the 3D plot and also the contour plot in the bottom plane. The left panel of Fig. 3 shows the plot of E_z for the $SU(3)$ YM field U, and the right panel of Fig. 3 for the restricted-field V. We find that the magnitude E_z is quite uniform for the restricted part V, while it is almost uniform for the original part U except for the neighborhoods of the locations of q, $\bar{q}$ source. This difference is due to the contributions from the remaining part X which affects only the short distance, as will be discussed in the next section.

3.2. *Magnetic current*

Next, We measure gauge-invariant magnetic-monopole current is induced around the chromoelectric field, which supports the dual picture of the ordinary superconductor exhibiting the electric current around the magnetic field:$\mathbf{k} = {}^*dF[\mathbf{V}]$, where $F[\mathbf{V}]$ is the field strength (2-form) of the restricted field (1-form) $\mathbf{V}$, d the exterior derivative and * denotes the Hodge dual operation. Note that non-vanishing magnetic current follows from violation of the Bianchi identity (If the field strength was given by the exterior derivative of $\mathbf{V}$ field (one-form), $F[\mathbf{V}] = d\mathbf{V}$,we would obtain $\mathbf{k} = {}^*d^2\mathbf{V} = 0$). Fig. 4 shows the magnetic current measured in X-Y plane at the midpoint of quark and antiquark pair in the Z-direction. The left panel of Fig. 4 shows the positional relationship between chromoelectric flux and magnetic current. The right panel of Fig. 4 shows the magnitude of the chromoelectric field E_z (left scale) and the magnetic current k (right scale). The existence of nonvanishing magnetic current k around the chromoelectric field E_z supports the dual picture of the ordinary superconductor exhibiting the electric current J around the magnetic field B.

3.3. *Type of dual superconductivity*

In order to examine QCD vacuum, i.e., the type of the dual superconductivity, we apply the formula for the magnetic field derived by Clem[5] in the ordinary superconductor based on the Ginzburg-Landau (GL) theory to the chromoelectric field in

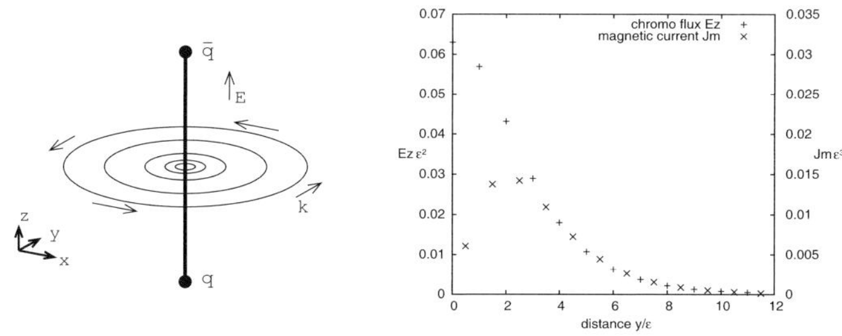

Fig. 4. The magnetic-monopole current $\mathbf{k}$ induced around the flux along the z axis connecting a quark-antiquark pair. (Left panel) The positional relationship between the chromoelectric field E_z and the magnetic current $\mathbf{k}$. (Right panel) The magnitude of the chromo-electronic current E_z and the magnetic current $J_m = |\mathbf{k}|$ as functions of the distance y from the z axis.

the dual superconductor, where an approximated form for the scalar field ϕ (given below in (6)) instead of obtaining the analytical solution of the GL equation explicitly. Therefore, we can obtain the formula which is valid for any distance (core radius) y from the axis connecting q and $\bar{q}$: the profile of chromoelectric field in the dual superconductor is obtained as

$$E_z(y) = \frac{\Phi}{2\pi}\frac{1}{\zeta\lambda}\frac{K_0(R/\lambda)}{K_1(\zeta/\lambda)}, \quad R = \sqrt{y^2+\zeta^2}, \tag{5}$$

$$\phi(y) = \frac{\Phi}{2\pi}\frac{1}{\sqrt{2}\lambda}\frac{y}{\sqrt{y^2+\zeta^2}}, \tag{6}$$

where K_ν is the modified Bessel function of the ν-th order, λ the parameter corresponding to the London penetration length, ζ a variational parameter for the core radius, and Φ external electric flux. The GL parameter κ as the ratio of the London penetration length λ and the coherence length ξ which measures the coherence of the magnetic monopole condensate (the dual version of the Cooper pair condensate):

$$\kappa := \frac{\lambda}{\xi} = \sqrt{2}\frac{\lambda}{\zeta}\sqrt{1 - K_0^2(\zeta/\lambda)/K_1^2(\zeta/\lambda)}. \tag{7}$$

The superconductor is type I if $\kappa < \kappa_c$, while type II if $\kappa > \kappa_c$, where the critical value of GL parameter dividing the type of the superconductor is given by $\kappa_c = 1/\sqrt{2} \simeq 0.707$. According to the formula Eqs.(5)(7), we estimate parameters and the GL parameter κ for the dual superconductor of $SU(3)$ YM theory, and the result of fitting is shown Fig. 5 in units of the string tension $\sigma_{\text{phys}} = (440\text{MeV})^2$. Our data clearly shows that the dual superconductor of $SU(3)$ Yang-Mills theory is type I with

$$\kappa = 0.45 \pm 0.01,\ \lambda = 0.1207(17)\text{fm},\ \ \xi = 0.2707(86)\text{fm}, \tag{8}$$

which is which should be compared with the border of type I and II for the SU(2) YM theory. Moreover, our result shows that the restricted part plays the dominant role in determining the type of the non-Abelian dual superconductivity of the $SU(3)$ Yang-Mills theory, reproducing the same result i.e., type I with

$$\kappa = 0.48 \pm 0.02, \lambda = 0.132(3)\text{fm}, \ \xi = 0.277(14)\text{fm}. \tag{9}$$

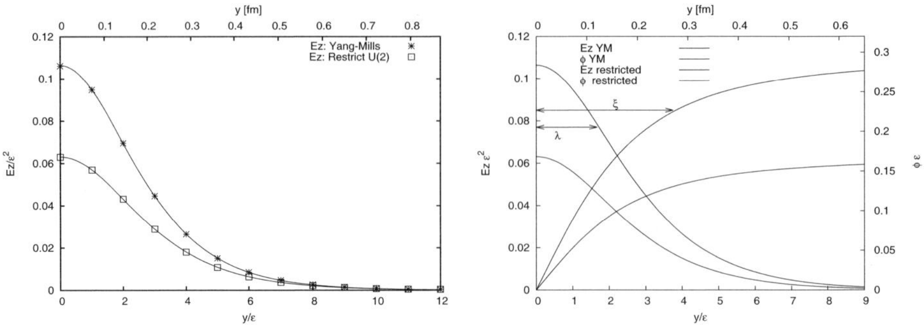

Fig. 5. (Left panel) The plot of the chromoelectric field E_z versus the distance y in units of the lattice spacing ϵ and the fitting as a function $E_z(y)$ of y according to Eq.(5). The red cross for the original $SU(3)$ field and the green square symbol for the restricted field. (Right panel) The order parameter ϕ reproduced as a function $\phi(y)$ of y according to Eq.(6), togather with the chromoelectric field $E_z(y)$.

4. Summary and outlook

These results confirm the non-Abelian dual superconductivity picture for quark confinement. In order to confirm the type of the non-Abelian dual superconductivity, it will be interesting to investigate if the interactions between two chromoelectric flux tubes are attractive (expected for the type I) or repulsive (expected for the type II). The interactions should reflect the internal non-Abelian nature which depends on the distance between two tubes. It will be also interesting to extend our result to the finite temperature case to see the fate of the dual superconductivity, and to see when and how the dual superconductivity disappears above the critical temperature.

Acknowledgement

This work is supported by Grant-in-Aid for Scientific Research (C) 24540252 from the Japan Society for the Promotion Science (JSPS), and also in part by JSPS Grant-in-Aid for Scientific Research (S) 22224003. The numerical calculations are supported by the Large Scale Simulation Program No.T11-15 (FY2011) and No.12-13 (FY2012) of High Energy Accelerator Research Organization (KEK).

References

1. K.-I. Kondo, A. Shibata, T. Shinohara, and S. Kato, arXiv:1007.2696 [hep-th], Phys. Rev. D83, 114016 (2011).
2. A. Shibata, K.-I. Kondo, S. Kato and T. Shinohara, arXiv:1212.6512 [hep-lat], Phys.Rev. D87, 054011 (2013).
3. Y. Nambu, Phys. Rev. D10, 4262 (1974); G. 't Hooft, in: High Energy Physics, edited by A.; Zichichi (Editorice Compositori, Bologna, 1975); S. Mandelstam, Phys. Report 23, 245 (1976); A.M. Polyakov, Nucl. Phys. B120, 429 (1977).
4. A. Di Giacomo, M. Maggiore and S. Olejnik, Nucl. Phys. B **347**, 441 (1990).; A. Di Giacomo, M. Maggiore and S. Olejnik, Phys. Lett. B **236**, 199 (1990).
5. J.R. Clem, J. Low. Temp. Phys. 18, 427 (1975).

Confinement via Strongly-Coupled Nonabelian Monopoles

K. Konishi*

Department of Physics, "E. Fermi", University of Pisa,
Largo Pntecorvo, 3, 56127, Pisa, Italy
**E-mail: ab_konishi@df.unipi.it*
http://www.df.unipi.it/ konishi/

New types of confinement phase emerge as singular SCFT's appearing as infrared-fixed-points of $\mathcal{N} = 2$ supersymmetric QCD (SQCD) are perturbed by an $\mathcal{N} = 1$ adjoint mass term. Based on a recent remarkable work on infrared-fixed-point SCFT of highest criticalities by Gaiotto, Seiberg and Tachikawa, we discuss physics of certain confining systems in $SU(N)$, $USp(2N)$ or $SO(N)$ gauge theories. These show features different from a straightforward dual superconductivity picture of confinement à la 't Hooft and Mandelstam, which might suggest a new venue in exploring the quark confinement mechanism in the real-world QCD.

1. Quark confinement versus chiral symmetry breaking

Often a theory at an UV conformal fixed point flows in the infrared to another, infrared-fixed-point conformal theory. If a small relevant operator is either introduced by hand or generated dynamically, the system can instead flow into one in confinement phase. If in some sense the relevant deformation is small, the knowledge about the infrared conformal theory (in the absence of such a deformation) is important, as the degrees of freedom in the latter describe also how confinement and dynamical symmetry breaking take place.

In Quantum Chromodynamics (QCD), the UV degrees of freedom are quarks and gluons, and their behavior at short distances is well understood because of asymptotic freedom, but the nature of their collective behavior in the infrared is still covered by mysteries. The familiar Nambu-'t Hooft-Mandelstam picture[1,2] of confinement assumes a dynamical gauge symmetry breaking,

$$SU(3) \to U(1) \times U(1) \to \mathbf{1} \,, \tag{1}$$

with the magnetic monopoles M_i $(i = 1, 2)$ generated by the first step of gauge symmetry breaking, acting as confinement order parameters. Indeed, their condensation $\langle M_i \rangle \neq 0$ would lead to a (dual) superconductor vacuum, so that the color electric flux is squeezed into narrow dual Abrikosov-Nielsen-Olesen (ANO) vortices: the quarks are confined, just as the magnetic monopoles of the usual electromagnetism would be in the standard type II superconductor.

This well-known picture of confinement leads however to difficulties. If the confinement and chiral symmetry breaking take place simultaneously, as suggested by $SU(3)$ lattice simulation data, it is natural to assume that one or both of the magnetic monopoles carry flavor $SU_L(N_f) \times SU_R(N_f)$ charges à la Jackiw-Rebbi;[3] their condensation leads both to confinement and chiral symmetry breaking. In softly broken $\mathcal{N} = 2$ SQCD with $SU(2)$ gauge group and with $N_f = 1, 2, 3$ flavors, exactly such a phenomenon is dynamically realized at low energies.[4–8] In the case of non-supersymmetric QCD, however, the pattern of the chiral symmetry breaking

$$SU_L(N_f) \times SU_R(N_f) \to SU_V(N_f) \tag{2}$$

would require one (or both) of the magnetic monopoles to carry both left and right flavor $SU(N_f)$ charges

$$M_b^a, \qquad a, b = 1, 2, \ldots N_f \ , \tag{3}$$

so that the condensation

$$\langle M_b^a \rangle = c \, \delta_b^a \, \Lambda \tag{4}$$

explains the symmetry breaking pattern Eq. (2). This scenario however implies an accidental global $SU(N_f^2)$ symmetry of the Abelian monopole theory and as a consequence, a proliferation of Nambu-Goldstone bosons (in the realistic case of $N_f = 2$, twelve NG bosons instead of three), of which of course there are no traces in nature.

Another, perhaps more serious, problem is the fact that the low-energy theory has

$$\Pi_1(U(1) \times U(1)) = \mathbb{Z} \times \mathbb{Z}, \tag{5}$$

implying a doubling of the meson spectrum (quarks confined by two distinct confining strings), while in the real world one observes unique, universal $q - \bar{q}$ meson trajectory for each flavor quantum number.

A possible way out of this conundrum is that the system does not completely Abelianize, i.e., that the gauge symmetry dynamically breaks as

$$SU(3) \to SU(2) \times U(1) \to \mathbf{1} \ . \tag{6}$$

In this case magnetic monopoles generated are of nonAbelian variety. As

$$\Pi_1(SU(2) \times U(1)) = \mathbb{Z}, \tag{7}$$

there would be a unique confining string in this case. We shall not discuss here the well-known "existence problems" for the nonAbelian monopoles, but it is believed that the light flavors play a crucial role for the quantum mechanical behavior of the nonAbelian monopoles[9] solving the difficulties arising for the monopoles in pure Yang-Mills theories. It is to be seen whether and how the problem of the proliferation of the Nambu-Goldstone bosons can be avoided in this case.

At the same time, however, this introduces a new (apparent) difficulty. In contrast to what happens in the r vacua of the softly broken $\mathcal{N} = 2$ supersymmetric QCD[8] (see below), the nonAbelian monopoles of the standard QCD are expected to be strongly coupled at low energies. One might wonder whether it makes sense to assume that the strongly coupled quarks and gluons get replaced in the infrared by another strongly coupled (e.g., magnetic) system. Actually, the idea is not new: there are many known examples of systems possessing similar properties (tumbling gauge theories,[10] duality cascades,[11] etc.) and, as will be seen below, some supersymmetric version of QCD exhibit exactly this kind of behavior.

It is possible that ultimately one must accept the idea that the color magnetic degrees of freedom of QCD are strongly coupled, and that confinement and dynamical chiral symmetry breaking are phenomena subtler than expected from a straightforward dual superconductivity idea.

2. What softly broken $\mathcal{N} = 2$ SQCD teaches us

As many exact quantum results are known in the softly broken $\mathcal{N} = 2$ SQCD, it is natural to ask what these theories can tell us about confinement, dynamical symmetry breaking and about the mechanism underlying them. Confining vacua in these theories can be classified as follows.

Table 1. The massless non-Abelian and Abelian monopoles and their charges at the root of a "non-baryonic" r-th Higgs branch.

	$SU(r)$	$U(1)_0$	$U(1)_1$	$\ldots$	$U(1)_{N-r-1}$	$U(1)_B$
$N_f \times \mathcal{M}$	$\underline{\mathbf{r}}$	1	0	$\ldots$	0	0
M_1	$\underline{\mathbf{1}}$	0	1	$\ldots$	0	0
$\vdots$	$\vdots$	$\vdots$	$\vdots$	$\ddots$	$\vdots$	$\vdots$
M_{N-r-1}	$\underline{\mathbf{1}}$	0	0	$\ldots$	1	0

(i) Abelian dual superconductor vacuum is realized in certain class of theories, as in the $SU(2)$ gauge theories, or in all pure $\mathcal{N} = 2$ supersymmetric theories with any gauge group G. The system dynamically Abelianizes, and in low energies becomes a $U(1)^R$ theory, where R is the rank of the gauge group G. In the case of $SU(2)$ theory with $N_f = 1, 2, 3$ flavors, the monopole condensation induces confinement and dynamical flavor symmetry breaking. This is beautiful, but does not look like what happens in the real-world QCD, as discussed above.

(ii) Massless nonAbelian monopoles, carrying flavor quantum numbers appear in the so-called r-vacua of SQCD. (See Fig. 1). In the $SU(N)$ theory with N_f flavors, the low-energy action is[6,8] an $SU(r) \times U(1) \times \ldots U(1)$ theory, with light monopoles carrying the quantum numbers shown in Table 1. The low-enrgy effective $SU(r) \times U(1)$ theory is a local, infrared-free theory: the monopoles are weakly coupled [a]. Upon $\mu\Phi^2$ perturbation the monopoles condense in a dual-color-flavor locked vacuum,

$$\langle M_i^{\alpha} \rangle = \delta_i^{\alpha} \sqrt{\mu\Lambda} \,, \tag{8}$$

breaking the symmetry as

$$SU(N_f) \times U(1) \to U(r) \times U(N_f - r) \,. \tag{9}$$

Again, in spite of the beautiful aspects - nonAbelian monopoles appear quantum mechanically as low-energy degrees of freedom, and they act as order parameters of confinement and dynamical symmetry breaking - these systems do not look like a good model for QCD.

(iii) Still another possibility, actually realized in softly broken $\mathcal{N} = 2$ supersymmetric theories, is that strongly interacting nonAbelian monopoles appear in a nontrivial infrared-fixed point SCFT. The low-energy effective theory is a non-Langrangian theory involving monopoles and dyons. This occurs in $USp(2N)$ and $SO(N)$ theories with vanishing bare quark masses[8] (see Fig. 2), and as recently realized by Di Pietro and Giacomelli,[12] for some critical value of the quark masses, in $SU(N)$ theories as well. Our discussions below will be mainly concerned with these singular vacua.

3. Recent developments

A development of considerable interest for our purpose is the recent elucidation of the nature of certain infrared-fixed-point SCFT of highest criticalities[15,18] in the context of $\mathcal{N} = 2$ supersymmetric QCD. These SCFT's occur at particular points of the vacuum moduli space and/or for special values of the bare quark mass parameters. A straightforward interpretation of these points would involve monopoles and dyons in an infinite-coupling regime, where it is not easy draw any clear physics picture.

3.1. *Argyres-Seiberg S duality*

The first key step forward has been the discovery by Argyres and Seiberg[13] of an elegant S-dual description of some "infinitely-strongly-coupled SCFT". For example consider the $\mathcal{N} = 2$, $SU(3)$ gauge theory with $N_f = 6$ flavors of hypermultiplets (quarks). This theory is superconformal, with an exactly marginal coupling constant g. By studying the behavior of the Seiberg-Witten curve in the limit $g \to \infty$, Argyres and Seiberg were able to show that the system in this limit admits a weakly coupled dual description. The dual is two separate SCFT sectors coupled weakly by $SU(2)$ gauge interactions. In this particular $SU(3)$ theory, one sector is a free doublet of

---

[a]The nonAbelian monopoles acquire flavor multiplicities through the fermion zeromodes:[3] this is essential for the sign flip of the beta function. Indeed the quantum r vacua occur only for $r \leq N_f/2$.

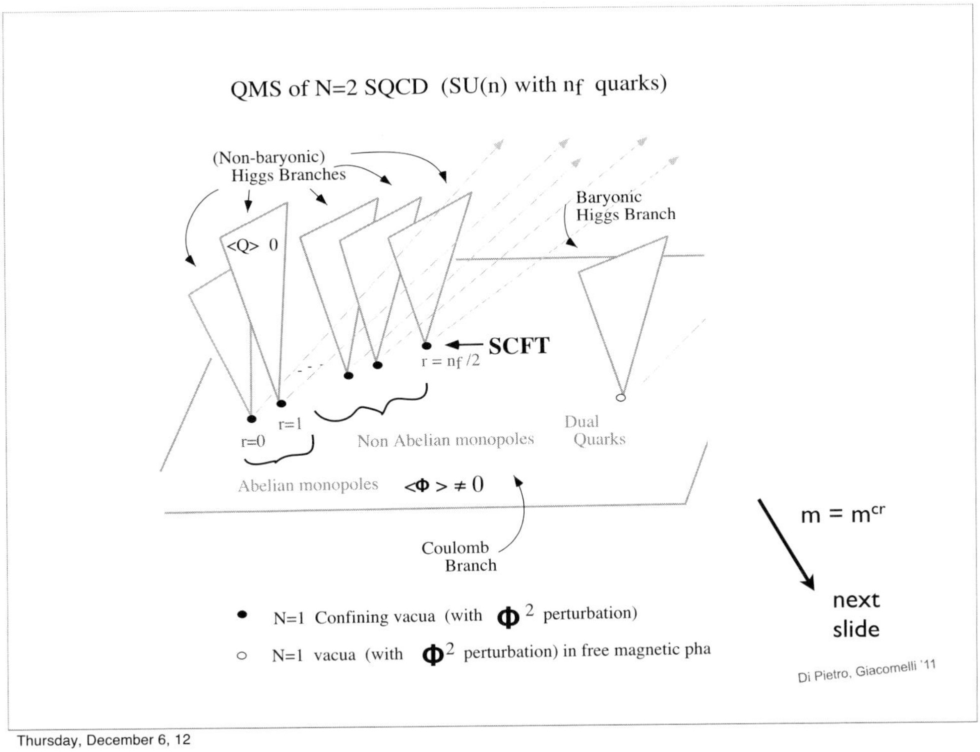

Fig. 1. A schematic view of the quantum moduli space of vacua in softly broken $\mathcal{N} = 2$ SQCD with $SU(N)$ gauge group and N_f flavors of quarks. Upon $\mu\Phi^2$ perturbation only the interceptions where various Higgs branches ($\langle Q \rangle \neq 0$) and the Coulomb branch ($\Phi \neq 0$) meet, survive (the black and while points). The black points are the r vacua; the white point represents the baryonic Higgs branch root with the vacua which are not confining. In the limit $m \to m^{cr}$ the r-vacua collapse to a singular SCFT as in the $USp(2N)$ case (the next Figure).

hypermultiplet, whereas the other sector is a non-Lagrangian SCFT with E_6 global symmetry,[14] whose $SU(2)$ subgroup ($SU(2) \times SU(6) \subset E_6$) is weakly gauged. The commutant $SU(6)$ and the $U(1)$ associated wit the free doublet hypermultiplet make up the $U(6)$ global symmetry of the underlying $SU(3)$, $N_f = 6$, theory.

In another example of $USp(4)$ theory with $N_f = 6$ (with a global $SO(12)$ symmetry), the infinite strong coupling limit is dual to an SCFT with E_7 global symmetry, whose $SU(2)$ subgroup ($SU(2) \times SO(10) \subset E_7$) is weakly gauged.

3.2. *Gaiotto-Seiberg-Tachikawa duals*

The next, crucial step was made by Gaiotto, Seiberg and Tachikawa (GST),[15] who applied the Argyres-Seiberg S-dual description to those SCFT's appearing as infrared fixed points of $\mathcal{N} = 2$ $SU(N)$ SQCD. This way they have been able to solve certain puzzles which plagued earlier studies on the IFPT conformal theories of highest criticalities.

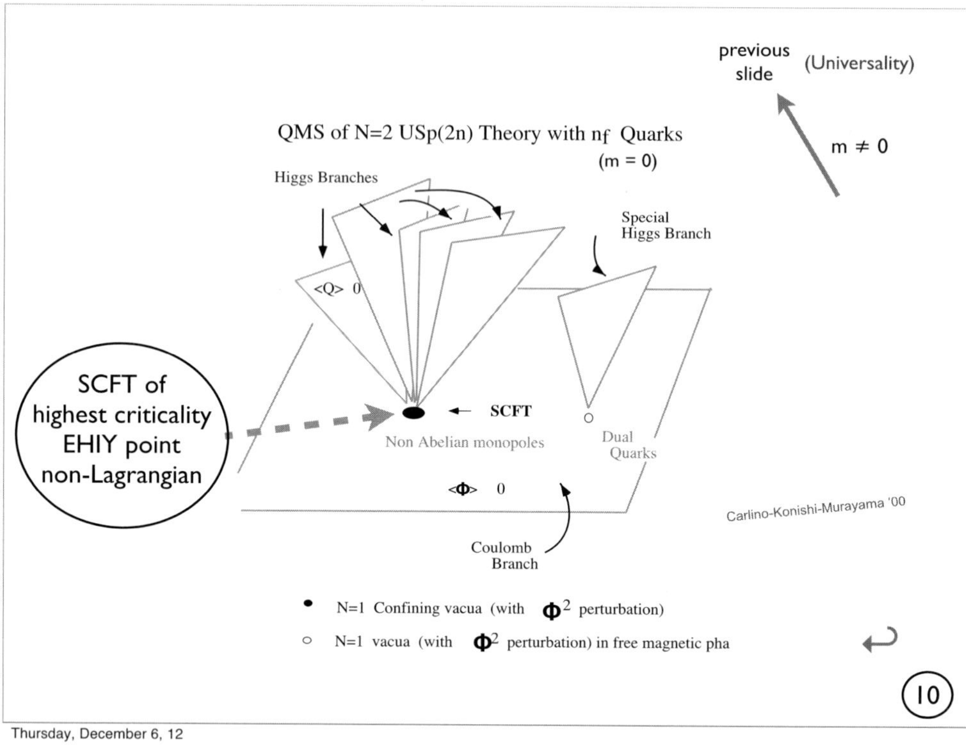

Fig. 2. The quantum moduli space of vacua in $USp(2N)$ theory with N_f flavors of quarks. For vanishing bare quark masses all confining vacua are in the Chebyshev points (the black point) which are SCFT of the highest criticality. Upon turning on the equal bare quark masses $m \neq 0$, they split into various r vacua which are identical to those in the $SU(N)$ theory (the previous Figure).

In the case of $SU(N)$ gauge theory with $N_f = 2n$ flavors, the Seiberg-Witten (SW) curve is given by

$$y^2 = (x^N + u_1 x^{N-1} + u_2 x^{N-2} + \ldots + u_N)^2 - \Lambda^{2N-2n} \prod_{i=1}^{2n} (x + m_i) \qquad (10)$$

Setting $u_j = m_i = 0$ (except for u_{N-n} which is chosen at $u_{N-n} = \Lambda^{N-n}$) one is at the EHIY point[19] where

$$y^2 \sim x^{N+n} . \qquad (11)$$

A straightforward scaling argument on the fluctuation around this point, by requiring the meromorphic (SW) differential

$$\lambda \sim \frac{y\, dx}{x^n} \qquad (12)$$

to have the unit canonical dimension, would assign the dimensions $[x] = \frac{1}{N+1}$; $[y] = \frac{N+n}{2(N+1)}$. Assumption that the curve scales uniformly would then determine how various u_i's and masses scale. But this leads to mass parameters associated with nonAbelian flavor symmetries to possess anomalous scaling dimensions, which they should not.[17] This is the same, well known argument that nonAbelian flavor charges do not get renormalized, as they satisfy inhomogeous current algebra commutation relations. From the point of view of finding infrared SCFT as points of quantum moduli space of vacua where the SW curves exhibit certain singular behaviors, there is an *a priori* ambiguity how to define (i.e., how to approach) the conformal limit, $u_i \to 0$, $m_i \to 0$.

It was found[15] that in order to have an SCFT with the correct mass dimensions it is necessary to scale various u_i's towards 0 nontrivially, as

$$u_{N-n+2} \sim O(\epsilon_A^2), \quad u_{N-n+3} \sim O(\epsilon_A^3), \quad \ldots \quad u_N \sim O(\epsilon_A^n), \tag{13}$$

$$u_1 \sim O(\epsilon_B), \quad u_2 \sim O(\epsilon_B^2), \quad \ldots \quad u_{N-n+2} \sim O(\epsilon_B^{N-n+2}), \tag{14}$$

where

$$\epsilon_A^2 \sim \epsilon_B^{N-2+2}, \qquad \epsilon_A \ll \epsilon_B \tag{15}$$

and

$$m_{i_1} \cdots m_{i_k} \sim \epsilon_A^k \,. \tag{16}$$

Accordingly, the branch points of the SW curve (10) get separated into two groups of different orders of magnitudes

$$x \sim \epsilon_A, \qquad \text{and} \qquad x \sim \epsilon_B \tag{17}$$

and consequenty the system splits into different sectors. The result is the system composed of

(i) Some decoupled $U(1)^{N-n-1}$ gauge multiplets;
(ii) An SU(2) gauge multiplet (infrared free) coupled to the $SU(2)$ flavor symmetries of the two SCFT's, A and B;
(iii) The A sector is a SCFT entering the Argyres-Seiberg dual of a $SU(n)$, $N_f = 2n$ theory, having $SU(2) \times SU(2n)$ flavor symmetry;
(iv) The B sector is a maximally singular SCFT of the $SU(N-n+1)$ theory with two flavors,

which may be schematically represented as in Fig. 3.

Remarks:

(i) In the case $n = 2$, the A theory describes simply three free hypermultiplets; for $n = 3$, it is a non-Lagrangian SCFT with E_6 global symmetry first found by Minahan and Nemeschansky;[14]
(ii) For $N = 3$, $n = 2$, the B sector corresponds to the most singular SCFT of the $SU(2)$, $N_f = 2$ theory.[17]
(iii) The Gaiotto-Seiberg-Tachikawa analysis has been generalized to the cases of $USp(2N)$ and $SO(N)$ theories by Giacomelli.[18]

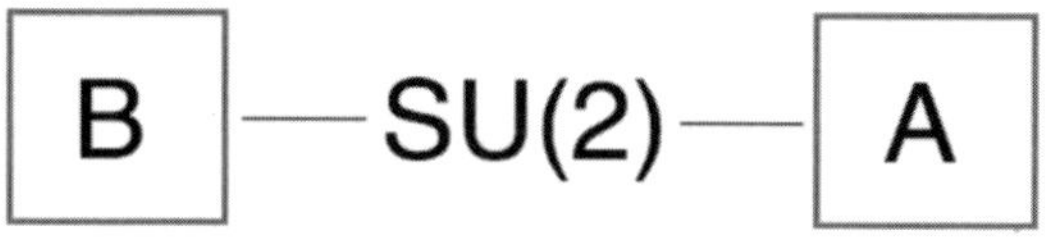

Fig. 3.

These developments enable us to study new types of confining systems arising as deformation of these strongly critical SCFT's, to which we now turn our attention.

4. Confinement near the singular vacua

4.1. $USp(2N)$ theory with $N_f = 2n$

In the $\mathcal{N} = 2$ $USp(2N)$ theory, the relevant most singular vacua are[8] two "Chebyshev" vacua, at $\phi_1 = \phi_2 = \ldots = \phi_{n-1} = 0$, $\phi_n^2 = \pm\Lambda^2$ (other ϕ's are determined à la Douglas-Shenker by using a Chebyshev polynomial):

$$x\, y^2 \sim [\, x^n (x - \phi_n^2)\,]^2 - 4\, \Lambda^4\, x^{2n} = x^{2n}\, (x - \phi_n^2 - 2\, \Lambda^2)\, (x - \phi_n^2 + 2\, \Lambda^2) \tag{18}$$

that is,

$$y^2 \sim x^{2n} \tag{19}$$

(see Fig. 2). Such a singular behavior implies that strongly-coupled massless monopoles and dyons, relatively nonlocal to each other, appear simultaneously in the low-energy effective action.

The strategy adopted in Carlino et. al.[8] was to try to "resolve" this vacuum, by introducing generic, nearly equal quark masses m_i alongside the adjoint scalar mass μ. By requiring the factorization property of the SW curve to be of maximally Abelian type (the criterion for $\mathcal{N} = 1$ supersymmetric vacua), this point was found to split into various r vacua which are local $SU(r) \times U(1)^{N-r}$ gauge theories, identical to those appearing in the infrared limit of $SU(N)$ SQCD (the universality of the infrared fixed points). One type of the Chebyshev vacua ($\phi_n^2 = +2\Lambda^2$) yields

$$\binom{N_f}{0} + \binom{N_f}{2} + \ldots \binom{N_f}{N_f} = 2^{N_f - 1} \tag{20}$$

vacua, whereas the other vacua ($\phi_n^2 = -2\Lambda^2$) split into odd r vacua, with the total multiplicity

$$\binom{N_f}{1} + \binom{N_f}{3} + \ldots \binom{N_f}{N_f - 1} = 2^{N_f - 1}\,. \tag{21}$$

Now the GST dual description of the point (18) was found by Giacomelli:[18]

(i) Some decoupled $U(1)^{N-n}$ gauge multiplets;
(ii) An SU(2) gauge multiplet (infrared free) coupled to the $SU(2)$ flavor symmetries of the two SCFT's, A and B;
(iii) The A sector is a non-Lagrangian SCFT having $SU(2) \times SO(4n)$ flavor symmetry;
(iv) The B sector is a free doublet coupled to a $U(1)$ gauge field.

For a particular choice of the number of flavors, $N_f = 4$, even the A sector becomes a trivial SCFT: four free doublets, so let us concentrate on this particular case.

We wish to verify that the SCFT vacua, *deformed by the $\mu\Phi^2$ perturbation*, are correctly described by the GST variables. The superpotential is given by

$$\sqrt{2}\, Q_0 A_D \tilde{Q}^0 + \sqrt{2}\, Q_0 \phi \tilde{Q}^0 + \sum_{i=1}^{4} \sqrt{2}\, Q_i \phi \tilde{Q}^i + \mu A_D \Lambda + \mu \, \mathrm{Tr}\phi^2 + \sum_{i=1}^{4} m_i \, Q_i \tilde{Q}^i \, . \quad (22)$$

For equal and nonvanishing masses the system has $SU(4) \times U(1)$ flavor symmetry. In the massless limit the symmetry gets enhanced to $SO(8)$, in accordance with the symmetry of the underlying $USp(2N)$ theory.

The vacuum equations are:

$$\sqrt{2}\, Q_0 \tilde{Q}_0 + \mu\Lambda = 0 \, ; \quad (23)$$

$$(\sqrt{2}\, \phi + A_D) \tilde{Q}_0 = Q_0 \, (\sqrt{2}\, \phi + A_D) = 0 \, ; \quad (24)$$

$$\sqrt{2} \, \left[\frac{1}{2} \sum_{i=1}^{4} Q_i^a \tilde{Q}_b^i - \frac{1}{4} \delta_b^a Q_i \tilde{Q}^i + \frac{1}{2} Q_0^a \tilde{Q}_b^0 - \frac{1}{4} \delta_b^a Q_0 \tilde{Q}^0 \right] + \mu \, \phi_b^a = 0 \, ; \quad (25)$$

$$(\sqrt{2}\, \phi + m_i) \, \tilde{Q}^i = Q_i \, (\sqrt{2}\, \phi + m_i) = 0, \qquad \forall i \, . \quad (26)$$

The first tells that $Q_0 \neq 0$. By gauge choice

$$Q_0 = \tilde{Q}_0 = \begin{pmatrix} 2^{-1/4} \sqrt{-\mu\Lambda} \\ 0 \end{pmatrix} \quad (27)$$

so that

$$\frac{1}{2} Q_0^a \tilde{Q}_b^0 - \frac{1}{4} (Q_0 \tilde{Q}^0) \, \delta_b^a = \frac{(-\mu\Lambda)}{4\sqrt{2}} \, \tau^3 \, . \quad (28)$$

The second equation can be satisfied by adjusting A_D.

The solutions can be found by having one of Q_i's canceling the contributions of Q_0 and ϕ in Eq (25). Which of Q_i is nonvanishing is related to the value of ϕ through Eq (26). For instance, four solutions can be found by choosing ($i = 1, 2, 3, 4$)

$$a = -\frac{m_i}{\sqrt{2}}, \qquad Q_i = \tilde{Q}_i = \begin{pmatrix} f_i \\ 0 \end{pmatrix} ; \qquad Q_j = \tilde{Q}_j = 0, \quad j \neq i, \quad (29)$$

such that

$$f_i^2 = \frac{\mu\Lambda - 4\, a}{\sqrt{2}} = \mu (\frac{\Lambda}{\sqrt{2}} + 2 m_i) \, . \quad (30)$$

There are four more solutions with the similar form as above but with $a = +\frac{m_i}{\sqrt{2}}$.[21] These are unrelated to each other and to (29) by any gauge transformation, so that we find $2^3 = 8$ solutions in all, consistently with Eq. (20).

Approaching the equal mass limit these eight solutions group into two set of four nearby vacua, clearly related by the $SU(4)$. So they are the $4+4=8$, two $r=1$ vacua, from one of the Chebyshev vacua, see Eq. (21). The other Chebyshev vacuum should give $1+6+1=8$ vacua, corresponding to $r=0,2$ vacua. Where are they?

A possible solution is that in the other Chebyshev vacuum the superpotential has a similar form as (22) but with Q_i's carrying different flavor charges. The $SU(4)$ symmetry of the equal mass theory may be represented as $SO(6)$:

$$\sqrt{2}\, Q_0 A_D \tilde{Q}^0 + \sqrt{2}\, Q_0 \phi \tilde{Q}^0 + \sum_{i=1}^{4} \sqrt{2}\, Q_i \phi \tilde{Q}^i + \mu A_D \Lambda + \mu\, \mathrm{Tr}\phi^2 + \sum_{i=1}^{4} \tilde{m}_i\, Q_i \tilde{Q}^i\ , \tag{31}$$

where

$$\begin{aligned} \tilde{m}_1 &= \frac{1}{4}(m_1 + m_2 - m_3 - m_4)\ ; \\ \tilde{m}_2 &= \frac{1}{4}(m_1 - m_2 + m_3 - m_4)\ ; \\ \tilde{m}_3 &= \frac{1}{4}(m_1 - m_2 - m_3 + m_4)\ ; \\ \tilde{m}_4 &= \frac{1}{4}(m_1 + m_2 + m_3 + m_4)\ . \end{aligned} \tag{32}$$

The correct realization of the underlying symmetry in various cases is not obvious, so let us check them:

(i) In the equal mass limit, $m_i = m_0$,

$$\tilde{m}_4 = m_0, \qquad \tilde{m}_2 = \tilde{m}_3 = \tilde{m}_4 = 0\ , \tag{33}$$

so the symmetry is

$$U(1) \times SO(6) = U(1) \times SU(4)\ . \tag{34}$$

Clearly in the $m_i = 0$ limit the symmetry is enhanced to $SO(8)$.

(ii) $m_1 = m_2$, m_3, m_4 generic. In this case $\tilde{m}_2 = -\tilde{m}_3$ and $\tilde{m}_4$ and $\tilde{m}_1$ are generic, so the symmetry is $U(1) \times U(1) \times U(2)$, as in the underlying theory;

(iii) $m_1 = m_2 \neq 0$, $m_3 = m_4 = 0$. In this case, $\tilde{m}_4 = \tilde{m}_1 \neq 0$ and $\tilde{m}_2 = \tilde{m}_3 = 0$, so obviously the symmetry is $U(2) \times SO(4)$ both in the UV and in (31).

(iv) $m_1 = m_2 = m_3 \neq 0$, m_4 generic. In this case, $\tilde{m}_1 = \tilde{m}_2 = -\tilde{m}_3 \neq 0$, $\tilde{m}_4$ generic. Again the symmetry is $U(3) \times U(1)$ both at the UV and IR.

(v) $m_1 = m_2 \neq 0$ and $m_3 = m_4 \neq 0$ but $m_1 \neq m_3$. In this case $\tilde{m}_2 = \tilde{m}_3 = 0$ and $\tilde{m}_4$ and $\tilde{m}_1$ generic. The flavor symmetry is

$$SO(4) \times U(1) \times U(1) = SU(2) \times SU(2) \times U(1) \times U(1)\ ; \tag{35}$$

this is equal to the symmetry

$$(SU(2) \times U(1)) \times (SU(2) \times U(1)) \tag{36}$$

of the underlying theory.

(vi) $m_1 \neq 0$, $m_2 = m_3 = m_4 = 0$. In this case $\tilde{m}_1 = \tilde{m}_2 = \tilde{m}_3 = \tilde{m}_4 \neq 0$. The symmetry is $U(1) \times SO(6)$ in the UV, and $U(4)$ in the infrared.

(vii) $m_1 \neq 0$, $m_2 \neq 0$, $m_1 \neq m_2$, $m_3 = m_4 = 0$. In this case $\tilde{m}_1 = \tilde{m}_4$ $\tilde{m}_2 = \tilde{m}_3 \neq \tilde{m}_1$. The symmetry is $U(1)^2 \times SO(4)$ in the UV, and $U(2) \times U(2)$ in the infrared.

The cases of masses equal except sign, e.g., $m_1 = -m_2$, are similar.

Thus in all cases Eq.(31) has the correct symmetry properties as the underlying theory. The vacuum solutions which follow from it are similar to those found from Eq.(22), with simple replacement,

$$m_i \to \tilde{m}_i \tag{37}$$

so there are 8 of them. The interpretation and their positions in the quantum moduli space (QMS) are different, however. In the equal mass limit, $m_i \to m_0$, The two solutions with

$$a = -\frac{\tilde{m}_4}{\sqrt{2}}, \qquad \text{or} \qquad a = \frac{\tilde{m}_4}{\sqrt{2}}, \tag{38}$$

can be regarded as two $r = 0$ vacua. Note that as $|f_1| \neq |g_1|$ they are in distinct points of the moduli space. On the other hand, in the other six vacua $a = 0$ always and $|f_i| = |g_i|$, these six solutions are at the same point of the moduli space: they can be associated with the $r = 2$ (sextet) vacua.

Remarks: The flavor charges (32) suggest that Q's are really non-Abelian magnetic monopoles, as semiclassically magnetic monopoles appear in the spinor representations of $SO(2N_f)$.

4.2. *Confinement and dynamical symmetry breaking*

In the massless limit, in the original SW description we had singular Chebyshev vacua. Having relatively non-local, strongly coupled monopoles and dyons, physics there was not very obvious, especially as to the effect of the $\mu\Phi^2$ perturbation and the types of confinement and dynamical symmetry breaking which might ensue.

In the GST description, once can take smoothly the $m_i \to 0$ limit, keeping $\mu \neq 0$, to get a system,

$$\sqrt{2}\, Q_0 A_D \tilde{Q}^0 + \sqrt{2}\, Q_0 \phi \tilde{Q}^0 + \sum_{i=1}^{4} \sqrt{2}\, Q_i \phi \tilde{Q}^i + \mu A_D \Lambda + \mu\, \mathrm{Tr}\phi^2 \,. \tag{39}$$

The vacuum of this system can be easily found:[20]

$$Q_0 = \tilde{Q}_0 = \begin{pmatrix} 2^{-1/4}\sqrt{-\mu\Lambda} \\ 0 \end{pmatrix} \tag{40}$$

$$\phi = 0, \quad A_D = 0\,. \tag{41}$$

The contribution from Q_i's must then cancel that of Q_0 in Eq. (25). By flavor rotation the nonzero VEV can be attributed to $Q_1, \tilde{Q}^1$, i.e., either of the form

$$(Q_1)^1 = (\tilde{Q}^1)_1 = 2^{-1/4}\sqrt{\mu\Lambda}\,, \qquad Q_i = \tilde{Q}_i = 0, \quad i = 2,3,4. \tag{42}$$

or

$$(Q_1)^2 = (\tilde{Q}^1)_2 = 2^{-1/4}\sqrt{-\mu\Lambda}\,, \qquad Q_i = \tilde{Q}_i = 0, \quad i = 2,3,4. \tag{43}$$

The $U(1)$ gauge symmetry is broken by the Q_0 condensation: an ANO vortex is formed. As the gauge group of the underlying theory is simply connected, such a low-energy vortex must end. The quarks are confined. The flavor symmetry breaking

$$SO(8) \to U(1) \times SO(6) = U(1) \times SU(4) = U(4), \tag{44}$$

is induced by the condensation of Q_1, which does not carry the $U(1)$ gauge charge. We note that the pattern of the symmetry breaking above agrees with that found at large μ.[8]

The vortex is made of the Q_0 field and the effective Abelian gauge field. The most interesting feature of this system is that there is no dynamical Abelianization, i.e., the effective low-energy gauge group is $SU(2) \times U(1)$. The confining string is unique and does not leads to the doubling of the meson spectrum. The global symmetry breaking of the low-energy effective theory is the right one (44), but the vacuum is not color-flavor locked. The confining string is of Abelian type, and is not a non-Abelian vortex as the one appearing in an r vacuum.[20] These facts clearly distinguish the confining system found here both from the standard Abelian dual superconductor type systems and from the non-Abelian dual Higgs system found in the r-vacua of SQCD. The dynamical symmetry breaking and confinement are linked to each other (the former is induced by the Q condensates, which in turn, is triggered by the Q_0 condensation which is the order parameter of confinement), but are not described by one and the same type of condensates.

5. Conclusion

We have studied several other theories having a similar strongly critical SCFT in the infrared.

- **(i)** Colliding r-vacua [b] of $SU(3)$ theory with $N_f = 4$: with the $U(4)$ flavor symmetry unbroken;
- **(ii)** Singular $r = 2$ vacua[20] of $SU(4)$, $N_f = 4$ theory, with symmetry breaking $U(4) \to U(2) \times U(2)$;
- **(iii)** $SO(2N{+}1)$, $N_f = 1$ theory with symmetry breaking $USp(2) = SU(2) \to U(1)$;
- **(iv)** $SO(2N)$, $N_f = 2$ theory, with symmetry breaking $USp(4) \to U(2)$.

---

[b] This occurs at the critical mass $m = \pm 2^{\frac{6N-2N_f}{4N-2N_f}} \frac{2N-N_f}{2N} \Lambda$ for general $SU(N)$, N_f theory.[22]

In all these cases confinement and flavor symmetry breaking have been found to be appropriately described by the GST dual variables. As in the case of the $USp(4)$ theory discussed in some detail above, the picture of confinement and dynamical symmetry breaking in these systems seems be somewhat subtler than is expected in a straightforward dual superconductivity mechanism. This might suggest a new direction in improving our understanding of quark confinement in the real-world QCD.

For the moment our analyses are restricted to special cases where the GST description simplifies particularly, i.e., either to local ones or to those which can be replaced by another simple system, containing a local but asymptotically free sector (to generate effectively[21] the structure of the GST system, Fig. 3). It is to be seen whether such a procedure can be applied to more general systems with arbitrary gauge group and for general values of N_f.

Acknowledgement

This talk is based mainly on the two papers with Simone Giacomelli.[20,21] The author thanks the organizer of the Workshop SCGT12 (Nagoya), K. Yamawaki, for inviting him to present these results in a nice stimulating meeting.

References

1. G. 't Hooft, Nucl. Phys. **B190** (1981) 455.
2. Y. Nambu, Phys. Rev. **D10** (1974) 4262, S. Mandelstam, Phys. Lett. **53B** (1975) 476, Phys. Rep. **23C** (1976) 245.
3. R. Jackiw and C. Rebbi, Phys. Rev. **D13** (1976) 3398.
4. N. Seiberg and E. Witten, Nucl.Phys. **B426** (1994) 19; Erratum *ibid.* **B430** (1994) 485, hep-th/9407087.
5. N. Seiberg and E. Witten, Nucl. Phys. **B431** (1994) 484, hep-th/9408099.
6. P.C. Argyres, M.R. Plesser and N. Seiberg, Nucl. Phys. B **471**, 159 (1996), hep-th/9603042.
7. P.C. Argyres, M.R. Plesser and A.D. Shapere, Nucl. Phys. B **483**, 172 (1997), hep-th/9608129; P.C. Argyres, M.R. Plesser, and A.D. Shapere, Nucl. Phys. **B483** (1997) 172, hep-th/9608129; K. Hori, H. Ooguri and Y. Oz, Adv. Theor. Math. Phys. **1**, 1 (1998), hep-th/9706082.
8. G. Carlino, K. Konishi and H. Murayama, Nucl. Phys. **B590** (2000) 37, hep-th/0005076.
9. K. Konishi, Lect. Notes in Phys. **737**, 471 (2008), hep-th/0702102 [hep-th].
10. S. Raby, S. Dimopoulos and L. Susskind, Nucl. Phys. B **169**, 373 (1980).
11. I. R. Klebanov and M. J. Strassler, JHEP **0008**, 052 (2000) [hep-th/0007191].
12. L. Di Pietro and S. Giacomelli, JHEP 1202 (2012) 087, arXiv:1108.6049 [hep-th].
13. P. C. Argyres and N. Seiberg, JHEP 0712:088 (2007), arXiv:0711.0054 [hep-th].
14. J. A. Minahan and D. Nemeschansky, Nucl. Phys. B **482**, 142 (1996) [hep-th/9608047].
15. D. Gaiotto, N. Seiberg and Y. Tachikawa, JHEP **1101**, 078 (2011), arXiv:1011.4568 [hep-th].
16. P. C. Argyres, M. R. Plesser, N. Seiberg and E. Witten, Nucl. Phys. B **461**, 71 (1996) [hep-th/9511154].

17. P. C. Argyres, M. R. Plesser, N. Seiberg and E. Witten, Nucl. Phys. B **461**, 71 (1996) [hep-th/9511154].
18. S. Giacomelli, JHEP 1209 (2012) 040, arXiv:1207.4037 [hep-th].
19. T. Eguchi, K. Hori, K. Ito and S. -K. Yang, Nucl. Phys. B **471**, 430 (1996) [hep-th/9603002].
20. S. Giacomelli and K. Konishi, JHEP 1212 (2012) 083, arXiv:1209.5328 [hep-th].
21. S. Giacomelli and K. Konishi, arXiv:1301.0420 [hep-th].
22. L. Di Pietro and S. Giacomelli, JHEP **1202**, 087 (2012) [arXiv:1108.6049 [hep-th]].

Non-Abelian Strings in Supersymmetric Yang-Mills and Beyond

M. Shifman

William I. Fine Theoretical Physics Institute, University of Minnesota, Minneapolis, MN 55455, USA
shifman@umn.edu

I give a broad review of novel phenomena discovered in certain Yang-Mills theories: non-Abelian strings and confined monopoles.Then I explain how these phenomena allow one to study strong dynamics of gauge theories in four dimensions from two-dimensional models emerging on the string world sheet.

Keywords: Supersymmetry; Non-Abelian strings; World-sheet Theories.

1. Introduction

Gauge dynamics at strong coupling is an intriguing mystery challenging theoretical physicists for four decades. If before QCD only two phases were known – the Coulomb phase in QED and the Higgs regime in the Standard Model – the advent of QCD changed that in a dramatic way. The hypothesis of color confinement (quark confinement in a more narrow sense) formulated simultaneously with the discovery of asymptotic freedom, was a brilliant conjecture that allowed one to explain basic regularities of the hadronic world. Theoretical understanding (let alone derivation) of the underlying mechanisms of confinement turned out to be a much harder problem.

Color confinement means that if we consider Yang–Mills theory with probe (heavy) color charges, and try to separate them, asymptotic separation is impossible since the energy between the probe quark and antiquark grows linearly with separation. This is a very peculiar behavior since usually interaction energy between interacting bodies falls off with distance. Has the linear growth mentioned above any analogs in nature at all?

The only phenomenon known previously that was similar to linear confinement could occur in a rather exotic environment: superconductors of the second type. Assume that we have a large superconducting sample. Assume that we have a magnetic monopole and an antimonopole (very long and thin solenoids can simulate them), and we want to immerse them inside the superconductor and study their interaction as a function of separation.

The superconductor is a condensate of Cooper pairs (electric charges). These electric charges cannot screen the magnetic charges of the probe monopoles: the magnetic flux emanating from the monopole must find its way to the antimonopole which can then absorb it. However, this is in contradiction with the Meissner effect which tells us that the magnetic field cannot exist inside the superconductor since it suppresses the Cooper pair condensation and destroys superconductivity.

The clash between these two requirements is solved through a compromise. Instead of being dispersed *á la* Coulomb's law, the magnetic flux inside the superconductor squeezes itself in a tube of a fixed cross section. This ensures that, on the one hand, the magnetic flux is transmitted from the monopole to antimonopole with no loss, while on the other hand, superconductivity is destroyed in a minimal volume. Since the tension of the flux tube is a fixed number, the energy of the monopole-antimonopole probe pair grows linearly with their separation inside the superconducting sample. Such flux tubes (strings) were predicted by Abrikosov, simultaneously with the very existence of the type-II superconductors. In high-energy physics they are known as Abrikosov–Nielsen–Olesen (ANO) strings.

In the mid-1970s Nambu, 't Hooft, and Mandelstam (independently) suggested[1] that quark confinement is in a sense similar in nature to the Abrikosov phenomenon described above. They even invented an appropriate name: the dual Meissner effect. Why dual?

In Yang–Mills theories confined are quarks, i.e. color charges. Therefore, the flux to be transmitted through a tube is chromoelectric rather than chromomagnetic. Correspondingly, the objects to be condensed are not Cooper pairs (i.e. charges), but, rather, act like magnetic monopoles.

In the 1970s the Nambu–'t Hooft–Mandelstam conjecture looked more like a dream rather than a physical scenario, let alone a quantitative side since none of the objects above could be described or even identified, because of the strong-coupling nature of the phenomenon. Despite vigorous efforts and many published papers, not much happened in the 20 years that elapsed after the Nambu–'t Hooft–Mandelstam conjecture. A breakthrough came in 1994, from an unexpected side.

Seiberg and Witten thought up[2] to use supersymmetry as a tool which had proved to be powerful in yielding exact results at strong coupling in some previous instances. The price they had to pay is a certain complexification of the theory under consideration. Instead of pure Yang–Mills they had to deal with its supersymmetric version; to be more exact, with the $\mathcal{N}=2$ extension (8 supercharges). This theory can be viewed as a cousin of QCD. In addition to color gluons, it has a Dirac spinor field in the adjoint representation of the gauge group (for simplicity, let us take SU(2)), and a complex scalar field, a^a, also in the adjoint. Supersymmetry rigidly fixes self-interaction term of the scalar fields, $V_a \propto |\varepsilon^{abc}a^b a^c|^2$.

If we limit ourselves to the bosonic sector of this theory, we will immediately see that it includes the Georgi–Glashow model.[3] This is the model in which the 't Hooft–Polyakov monopoles were originally discovered[4] (in the quasi-classical regime). The self-interaction term $V_a \propto |\varepsilon^{abc}a^b a^c|^2$ has a remarkable feature: it vanishes along a (one-dimensional) manifold, say, $a^{1,2}=0$ with arbitrary a^3. Thus, the theory at hand has a flat direction: classically every point $a^{1,2}=0,\ a^3\neq 0$ is a vacuum in which the SU(2) gauge symmetry is spontaneously broken down to U(1). As a result, two out of three gluons act as W bosons while the third one becomes a photon. If, in addition, $|a^3| \gg \Lambda$, the theory is at weak coupling, and quasi-classical methods are applicable. (Here Λ is the dynamical scale parameter.)

Thus, $\mathcal{N}=2$ Yang–Mills theory naturally supports monopoles. Of course, in the limit $|a^3| \gg \Lambda$ they are very heavy and play little role, if at all, in dynamics.

So far, special features of supersymmetry were not used at all. The power of supersymmetric tools becomes evident as we move (in the complex plane of $u \equiv \mathrm{Tr}\, a^2$) towards the strong coupling domain $|u| \sim \Lambda^2$, inaccessible otherwise, since semiclassical treatment fails here. However, exact results based on holomorphic dependencies of certain protected quantities[a] on u remain applicable for all values of the parameter u. In this way Seiberg and Witten determined that there exists two special points (the Seiberg–Witten points), $u = \pm\Lambda^2$, in which either the monopole or the dyon mass vanish. Moreover, we learn that the pattern of the symmetry breaking

$$\mathrm{SU}(2) \to \mathrm{U}(1) \tag{1}$$

is valid not only in the quasi classical domain, but everywhere on the complex plane of u, including the point $u = 0$.

Now, let us focus on the vicinity of one of the two Seiberg–Witten (SW) points. In this domain monopoles are light, the photon is massless while the W bosons are heavy; their mass is of order of Λ. In the low-energy limit we can limit ourselves to monopoles and photons. In fact, after the existence of the Seiberg–Witten points is established, the non-Abelian nature of the theory under consideration no longer plays any role.

Given that the monopole-photon coupling is strong and highly nonlocal a straightforward theoretical analysis is impossible. However, as was realized by Seiberg and Witten, one cam dualize electromagnetic interaction through the transition

$$\mathrm{U}(1)_E \to \mathrm{U}(1)_M\,, \tag{2}$$

where the subscripts E and M mark electric and magnetic descriptions, respectively. The magnetic photon is coupled to the magnetic monopole locally and weakly, just in the same way as the electrons are coupled to the electric photons.

Now the last step is rather easy because when one has light charged objects (massless in the SW points) one can readily force them to condense using a small deformation of the theory. The deformation that does the job is a small mass term m (where $|m| \ll \Lambda$) for the a superfield. It explicitly breaks $\mathcal{N}=2$ down to $\mathcal{N}=1$. The monopole condensation implies that the residual U(1) gauge symmetry is broken down to nothing.

[a]They are referred to as Bogomol'nyi–Prasad–Sommerfield saturated, or BPS saturated for short.

What was achieved in this way? It was proved – analytically, for the first time ever – that the dual Meissner effect, which had been conjectured by Nambu, 't Hooft and Mandelstam as a mechanism of color confinement, does take place at least in some Yang–Mills theories. The magnetic monopoles condense in the vacuum triggering formation of the dual analogs of the ANO strings. The latter carry electric fluxes and, thus, confine electric charges. Note that everything is well-defined in the Seiberg–Witten solution due to the possibility of analytic continuation to the quasiclassical domain, which is ensured, in turn, by supersymmetry.

This was a breakthrough of enormous proportions. When the initial euphoria subsided, people realized, however, that the strings in the SW solution are not quite the strings we want to obtain in QCD. They are Abelian (although occur in a non-Abelian theory). Correspondingly, the spectrum of the states bound by such strings is much richer than we expect in QCD.[5]

2. Non-Abelian Strings

What is the intrinsic defect of the SW solution? The culprit is the two-scale structure of the spontaneous symmetry breaking. First, at a scale Λ the gauge SU(2) is Higgsed to U(1), see (1). Then at a much lower scale, proportional to $\sqrt{m\Lambda}$ the residual gauge symmetry is Higgsed. Correspondingly, the electric flux inside the SW string lies in the Cartan sub algebra of S(2), instead of fluctuating freely over the entire gauge group. In QCD-like theories the gauge symmetry is not Higgsed at all, there is only one scale Λ and all gauge boson play an equally important role in the (conjectured) string formation.

Therefore, our first task will be to find a model in which the latter situation can be realized.

3. Basic Model

Such a model was developed in.[6] It is $\mathcal{N} = 2$ super-Yang–Mills theory with the number of colors equal to the number of flavors (i.e. if the gauge group is SU(2) we have two (s)quark flavors . Moreover, we add a U(1) factor to the gauge group, so that, in fact, it is U(2), and endow this U(1) factor with the Fayet–Iliopoulos term ξ. The latter is needed to make non-Abelian strings BPS saturated.

The bosonic part of our SU(N)×U(1) theory has the form[6]

$$S = \int d^4x \left[\frac{1}{4g_2^2}\left(F^a_{\mu\nu}\right)^2 + \frac{1}{4g_1^2}\left(F_{\mu\nu}\right)^2 + \frac{1}{g_2^2}\left|D_\mu a^a\right|^2 + \frac{1}{g_1^2}\left|\partial_\mu a\right|^2 \right.$$

$$\left. + \left|\nabla_\mu q^A\right|^2 + \left|\nabla_\mu \bar{\tilde{q}}^A\right|^2 + V(q^A, \tilde{q}_A, a^a, a)\right] . \qquad (3)$$

Here D_μ is the covariant derivative in the adjoint representation of SU(2), and

$$\nabla_\mu = \partial_\mu - \frac{i}{2}\, A_\mu - i A^a_\mu\, T^a\,, \qquad T^a = \frac{1}{2}\tau^a\,. \qquad (4)$$

The coupling constants g_1 and g_2 correspond to the U(1) and SU(N) sectors, respectively. With our conventions, the U(1) charges of the fundamental matter fields are $\pm 1/2$.

The potential $V(q^A, \tilde{q}_A, a^a, a)$ in the action (3) is a sum of D and F terms,

$$V(q^A, \tilde{q}_A, a^a, a) = \frac{g_2^2}{2}\left(\frac{i}{g_2^2}\,\varepsilon^{abc}\bar{a}^b a^c + \bar{q}_A\,T^a q^A - \tilde{q}_A T^a\,\bar{\tilde{q}}^A\right)^2$$

$$+\frac{g_1^2}{8}\left(\bar{q}_A q^A - \tilde{q}_A \bar{\tilde{q}}^A\right)^2 + 2g_2^2\left|\tilde{q}_A T^a q^A\right|^2 + \frac{g_1^2}{2}\left|\tilde{q}_A q^A - \xi\right|^2$$

$$+\frac{1}{2}\sum_{A=1}^{N}\left\{\left|(a+\sqrt{2}m_A + 2T^a a^a)q^A\right|^2 + \left|(a+\sqrt{2}m_A + 2T^a a^a)\bar{\tilde{q}}^A\right|^2\right\}. \tag{5}$$

Here the sum over the repeated flavor indices A is implied.

Let us discuss the vacuum structure of this model. We will limit ourselves to isolated vacua, with the maximal possible value of condensed quarks – two. The vacua of the theory (3) are determined by the zeros of the potential (5). The adjoint fields develop the following VEV's:

$$\langle\Phi\rangle = -\frac{1}{\sqrt{2}}\begin{pmatrix} m_1 & 0 \\ 0 & m_2 \end{pmatrix}, \tag{6}$$

where we defined the scalar adjoint matrix as

$$\Phi = \frac{1}{2}\,a + T^a\,a^a. \tag{7}$$

If

$$m_1 = m_2 \tag{8}$$

the SU(2)×U(1) gauge group remains classically unbroken, unless we take into account the squark VEVs. In fact, in this case m can be absorbed in a, or, alternatively set $m = 0$.

The squark VEVs take the color-flavor locked form

$$\langle q^{kA}\rangle = \langle\bar{\tilde{q}}^{kA}\rangle = \sqrt{\frac{\xi}{2}}\begin{pmatrix} 1 & 0 \\ 0 & 1 \end{pmatrix}, \qquad k = 1,2, \qquad A = 1,2\,, \tag{9}$$

where we write down the quark fields as an 2×2 matrix in the color and flavor indices. The Fayet–Iliopoulos parameter ξ is assumed to be large $\xi \gg \Lambda^2$.

The vacuum field (9) results in the spontaneous breaking of both gauge and flavor SU(2)'s. A diagonal global SU(2) survives, however,

$$\mathrm{U}(2)_{\rm gauge}\times\mathrm{SU}(2)_{\rm flavor}\to\mathrm{SU}(2)_{C+F}\,. \tag{10}$$

Thus, a color-flavor locking takes place in the vacuum. The above condensates imply that the basic theory under consideration is fully Higgsed.

4. Non-Abelian Strings

Why does the model described above support a novel type of strings, non-Abelian? The ANO string corresponds to a U(1) winding of the phase of the squark fields in the plane, perpendicular to the string axis,

$$q^{kA} \longrightarrow \sqrt{\frac{\xi}{2}}\, e^{i\alpha} \begin{pmatrix} 1 & 0 \\ 0 & 1 \end{pmatrix}, \tag{11}$$

where α is the polar angle in the perpendicular plane. Its topological stability is due to $\pi_1(\mathrm{U}(1)) = Z$. Now we have more options. Usually people say that $\pi_1(\mathrm{SU}(2))$ is trivial and, therefore there are no other topologically stable strings.

This is not the case. Observe that the center of the SU(2) group, Z_2, belongs to the U(1) factor too. This means that we can split the 2π windings in two halves: the first (from 1 to -1) is carried out in U(1), while the second, from -1 to 1 in SU(2) (e.g. around the third axis). This is clearly, a topologically stable configuration, albeit the stability is of the Z_2 type. Correspondingly, the winding ansatz takes the form

$$q^{kA} \longrightarrow \sqrt{\frac{\xi}{2}} \begin{pmatrix} e^{i\alpha} & 0 \\ 0 & 1 \end{pmatrix} \quad \text{or} \quad q^{kA} \longrightarrow \sqrt{\frac{\xi}{2}} \begin{pmatrix} 1 & 0 \\ 0 & e^{i\alpha} \end{pmatrix}, \tag{12}$$

depending on whether we use the combination of generators $T_{\mathrm{U}(1)} + T^3_{\mathrm{SU}(2)}$ or $T_{\mathrm{U}(1)} - T^3_{\mathrm{SU}(2)}$.

It is clear, that the ansatz (12) breaks the color-flavor locked SU(2) down to U(1). The particular way of embedding is unimportant. Instead of T^3 we could have chosen any other generator of SU(2). In other words, the existence of the string (12) implies the existence of the whole family of strings parametrized by SU(2)/U(1) moduli. The theory of the moduli fields on the string world sheet is the sigma model on the SU(2)/U(1) coset space. This is the celebrated CP(1) model. It is asymptotically free in the UV and strongly coupled in the IR. Since the string is 1/2 BPS saturated, the worldsheet model has four supercharges. The existence of the orientational moduli means that the flux through the string does not have a preferred orientation inside SU(2). This is a genuinely non-Abelian string.

If the tension of the ANO string is $4\pi\xi$, the tension of the non-Abelian string is $2\pi\xi$.

5. Kinks as Confined Monopoles

$\mathcal{N} = 2$ CP(1) model has two degenerate vacua. This is dictated by the Witten index. From the bulk standpoint this means that there exist two distinct strings, with the tension $2\pi\xi$. If so, there should exist a junction of these two strings. Returning to two dimensions we can say that this junction is a kink interpolating between two distinct vacua of the supersymmetric CP(1) model. The kink itself is BPS saturated. $\mathcal{N} = (2,2)$ algebra in CP(1) has a central extension. Therefore, the mass of the kink is exactly calculable.

Moreover, using supersymmetry one can prove that this two-dimensional kink is a reincarnation of a confined monopole, i.e. a monopole with inalienable strings attached to it. To this end we must use exact analytic dependencies protected by supersymmetry.

Since so far we set $m_1 = m_2 = m = 0$, the theory under consideration has two parameters of dimension of mass: the string tension ξ and the dynamical scale Λ (it is one and the same in the bulk and on the string world sheet). In the limit $m_1 = m_2$ the kink is practically unrecognizable as monopole. However, of we introduce another parameter of dimension of mass and allow it to vary continuously, we will see that we start from the classical 't Hooft–Polyakov monopole, and then analytically continue (smoothly, with no phase transitions) to the kink on the string. The crucial parameter is the squark mass difference $m_1 = -m_2 \neq 0$. In this case we cannot absorb the squark mass terms in a. In fact, the difference $m_1 - m_2 \equiv 2m$ is an observable (albeit complex) physical parameter. If $|m| \gg \xi$ (or, alternatively, $\xi \to 0$), we arrive at the Georgi–Glashow model, with the classical 't Hooft–Polyakov monopoles. Making ξ nonvanishing but small we attach strings to the 't Hooft–Polyakov monopole, although these strings start far away from its core, which remains easily recognizable. Evolving towards smaller values of $|m|$ and larger values of ξ we arrive, at the end of the evolution process, at the kink described above. Thus, indeed, this kink is the monopole's apparent heir.

Figure 1 illustrates this analytic continuation.

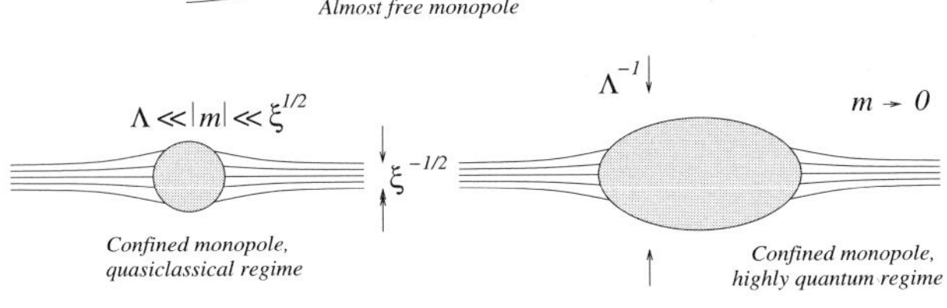

Fig. 1. Evolution of the confined monopoles.

Four-dimensional expression for the (confined) monopole mass matches an exact two-dimensional expression for the kink mass.[7,8] This is a manifestation of the $4D - 2D$ correspondence. For a thorough review see.[9]

6. Other Occurrences of Non-Abelean Moduli on Strings

Non-Abelian strings with interesting applications appear not only in supersymmetric theories. Below I will describe a nonsupersymmetric model (which may or may not apply to some condensed-matter systems) in which n-Abelian strings (vortices) are supported.

Recently some models supporting the Abelian Abrikosov-Nielsen-Olesen (ANO)[10] or similar strings (vortices) were considered[11–13] with "extra" non-Abelian moduli (or quasimoduli) on the string world sheet. Such extra moduli fields can appear in bulk models in which there are order parameters carrying spatial indices, such as those relevant for superfluidity in ^{3}He (see e.g.[14]). This particular example was studied in.[12]

Let us discuss consider the simplest model[11] of this type,

$$\mathcal{L} = \mathcal{L}_0 + \mathcal{L}_\chi \tag{13}$$

where

$$\mathcal{L}_0 = -\frac{1}{4e^2} F_{\mu\nu}^2 + |\mathcal{D}^\mu \phi|^2 - V(\phi)\,, \qquad V = \lambda\left(|\phi|^2 - v^2\right)^2\,, \tag{14}$$

and

$$\mathcal{L}_\chi = \partial_\mu \chi^i\, \partial^\mu \chi^i - U(\chi, \phi)\,, \tag{15}$$

$$U = \gamma\left[\left(-\mu^2 + |\phi|^2\right)\chi^i\chi^i + \beta\left(\chi^i\chi^i\right)^2\right], \tag{16}$$

with self-evident definitions of the fields involved, the covariant derivative, and the kinetic and potential terms. The parameters e, λ, β, μ, and v can be chosen at will, with some mild constraints (e.g. $v > \mu$) discussed in.[12]

In the vacuum the complex field ϕ develops a vacuum expectation value $|\phi_{\rm vac}| = v$ while its phase is eaten up by the Higgs mechanism. The masses of the (Higgsed) photon and the Higgs excitation are

$$m_A^2 = 2e^2 v^2\,, \quad m_\phi^2 = 4\lambda v^2\,. \tag{17}$$

Denote the ratio of the masses

$$x = m_A^2/m_\phi^2 \equiv \frac{e^2}{2\lambda}\,. \tag{18}$$

Moreover, in the vacuum the field χ^i does *not* condense. Its mass is

$$m_\chi^2 = \gamma\left(v^2 - \mu^2\right). \tag{19}$$

For what follows it is convenient to introduce two extra dimensionless parameters:

$$y = m_\chi^2/m_\phi^2 \equiv \frac{\gamma}{4\lambda}\,\frac{z-1}{z}\,, \qquad z = v^2/\mu^2\,. \tag{20}$$

The first measures the ratio of the χ to ϕ masses in the bulk. The second parameter is constrained, $z > 1$. Both of them are assumed to be of the order of unity. As for the spatial orientation, the string will be assumed to lie along the z axis. We introduce a dimensionelss radius in the perpendicular $\{x, y\}$ plane,

$$\rho = m_\phi \sqrt{x^2 + y^2}\,. \tag{21}$$

Let us start from the conventional ANO string (see e.g.[16]). The ϕ field winds ensuring topological stability, which entails in turn its vanishing at the origin. This implies the following ansätze:

$$A_0 = 0\,, \quad A_i = -\varepsilon_{ij}\frac{x_j}{r^2}\left(1 - f(r)\right)\,, \quad \phi = v\varphi(\rho)e^{i\alpha}\,, \tag{22}$$

where α is the polar angle in the perpendicular plane, and we assume for simplicity the minimal (unit) winding. The boundary conditions supplementing (22) are

$$f(\infty) = 0\,, \qquad f(0) = 1\,; \qquad \varphi(\infty) = 1\,, \quad \varphi(0) = 0\,. \tag{23}$$

In the core of such a tube the ϕ field tends to zero, see (23). The vanishing of the ϕ field results in the χ^i field destabilization in the core of the string (as follows from Eq. (16)). Hence, inside the core, the χ^i field no longer vanishes,

$$(\chi^i\chi^i)_{\rm core} \approx \frac{\mu^2}{2\beta}\,. \tag{24}$$

Choosing the value of λ judiciously, we can make $\mu^2/\beta \gg m_\chi^2$, implying that the O(3) symmetry is broken in the core. The appropriate ansatz is

$$\chi^i = \frac{\mu}{\sqrt{2\beta}}\,\chi(\rho)\,S^i\,, \quad i = 1,2,3\,, \quad S^iS^i = 1\,. \tag{25}$$

with the boundary conditions

$$\chi(\infty) = 0\,, \quad \chi(0) \approx 1\,. \tag{26}$$

Thus, we have three profile functions, f, φ, and χ, depending on ρ. Minimizing the energy functional we derive the system of equations for the profile functions

$$\begin{aligned}
&\left(\frac{f'}{\rho}\right)' = x\frac{\varphi^2 f}{\rho}\,,\\
&\left(\phi'\rho\right)' = \frac{f^2\varphi}{\rho} + \frac{\rho\varphi\left(\varphi^2-1\right)}{2} + \frac{\rho\,\varphi\,\chi^2}{2\beta}\,\frac{y}{z-1}\,,\\
&\left(\chi'\rho\right)' = \frac{y}{z-1}\rho\chi\left(z\varphi^2 + \chi^2 - 1\right)\,,
\end{aligned} \tag{27}$$

where the primes denote differentiation with respect to ρ. In the numerical solution to be presented below we will assume for simplicity that

$$x = 1\,, \quad \text{i.e. } m_\phi = m_A\,. \tag{28}$$

In the absence of the χ filed this would imply the Bogomol'nyi-Prasad-Sommerfield (BPS) limit[17] with the tension[b]

$$T_0 = 2\pi v^2\,. \tag{29}$$

[b] Alternatively, this is the boundary between type-I and type-II superconductors.

Below we will see how the presence of the χ filed changes the tension, using T_0 as a reference point.

It is not difficult to integrated Eqs. (27) numerically for various choices of $\{y, z, \beta\}$ keeping $x = 1$. Then the parameter λ appears only as an overall factor, with the analytically known dependence.

7. The World-Sheet Theory

S^i are rotational moduli. They describe the O(3) model on the string world sheet. In addition, two translational moduli are obvious.

To derive the theory on the string world sheet we, as usual, introduce t, z dependence converting the S^i moduli into the moduli fields $S^i(t,z)$. Substituting this in the Lagrangian (3) and (4) we obtain the low-energy effective action

$$S = \frac{1}{2g^2} \int dt\, dz\, \left(\partial_a S^i\right)^2 , \tag{30}$$

where

$$\frac{1}{2g^2} = \frac{1}{8z\beta\lambda} \int_0^\infty 2\pi\rho\, \chi^2(\rho)\, d\rho\, . \tag{31}$$

One can rewrite this as

$$\frac{g^2}{2\pi} = \lambda\, \frac{\beta}{\pi^2}\, \frac{z}{I_1}\, , \tag{32}$$

where

$$I_1 = \int_0^\infty \rho\, \chi^2(\rho)\, d\rho\, . \tag{33}$$

If $z \approx 1.25$ and $\beta \approx 10$, the integral I_1 is close to unity. Hence,

$$\frac{g^2}{2\pi} \approx \lambda\, . \tag{34}$$

If the bulk theory is weakly coupled (i.e. $\lambda \ll 1$), so is the O(3) sigma model on the string world sheet. The O(3) sigma model is asymptotically free.[18] More exactly, at the scale m_ϕ the coupling constant is small according to (34). As we descend down into the infrared domain, this constant runs and eventually becomes strong, entailing the mass gap Λ generation. The ratio Λ^2/m_ϕ^2 is exponentially small, of the order of $\exp(-1/\lambda)$.

As for the tension T it is very close to T_0, only slightly lower,

$$\frac{T}{T_0} \approx 0.95\, . \tag{35}$$

The imact of the χ^i field on the string tension is small and negative. The positive contribution of its kinetic energy is compensated by the negative potential energy.

8. Spin-Orbit Interaction

The "two-component" ϕ-χ string solution presented above spontaneously breaks two translational symmetries, in the perpendicular x, y plane, and O(3) rotations. The latter are spontaneously broken by the string orientation along the z axis (more exactly, O(3)→O(2)), and by the orientation of the spin field χ^i inside the core of the flux tube introduced through S^i.

Now, we deform Eq. (15) by adding a "spin-orbit" interation,[13]

$$\mathcal{L}_\chi = \partial_\mu \chi^i \, \partial^\mu \chi^i - \varepsilon (\partial_i \chi^i)^2 - U(\chi, \phi) \,, \tag{36}$$

where ε is to be treated as a perturbation parameter.

If $\varepsilon = 0$ (i.e. Eq. (15) is valid) the breaking O(3)→O(2) produces no extra zero modes (other than translational) in the ϕ-A_μ sector.[19] Due to the fact that $\chi \neq 0$ in the core, we obtain two extra moduli S^i on the world sheet. This is due to the fact that at $\varepsilon = 0$ the rotational O(3) symmetry is enhanced[12,13] because of the O(3) rotations of the "spin" field χ^i, independent of the coordinate spacial rotations.

What happens at $\varepsilon \neq 0$, see Eq. (36)? If ε is small, to the leading order in this parameter, we can determine the effective world-sheet action using the solution calculated above at $\varepsilon = 0$. Two distinct O(3) rotations mentioned above become entangled: O(3)×O(3) is no longer the exact symmetry of the model, but, rather, an approximate symmetry. The low-energy effective action on the string world sheet takes the form

$$\mathcal{L}_{\mathrm{O}(3)} = \left\{ \frac{1}{2g^2} \left[\left(\partial_a S^i\right)^2 - \varepsilon \left(\partial_z S^3\right)^2 \right] \right\} - M^2 \left(1 - (S^3)^2\right) . \tag{37}$$

The induced mass term M^2 is

$$M^2 = \varepsilon \, v^2 \, \frac{I_2}{4z\beta} \,, \tag{38}$$

where

$$I_2 = \int_0^\infty \rho \, (\chi'(\rho))^2 \, d\rho \approx 0.4 \,. \tag{39}$$

It is seen that $M^2/T \sim \varepsilon/\beta$ and is small for sufficiently small ratio ε/β. This justifies the above calculation.

Acknowledgments

This work was supported by the DOE grant DE-FG02-94ER40823.

References

1. Y. Nambu, Phys. Rev. D **10**, 4262 (1974);
G. 't Hooft, *Gauge theories with unified weak, electromagnetic and strong interactions,* in Proc. of the E.P.S. Int. Conf. on High Energy Physics, Palermo, 23-28 June, 1975 ed. A. Zichichi (Editrice Compositori, Bologna, 1976); S. Mandelstam, Phys. Rept. **23**, 245 (1976).

2. N. Seiberg and E. Witten, Nucl. Phys. **B426**, 19 (1994), (E) **B430**, 485 (1994) [hep-th/9407087]; Nucl. Phys. **B431**, 484 (1994) [hep-th/9408099].
3. H. Georgi and S. L. Glashow, Phys. Rev. Lett. **28**, 1494 (1972).
4. G. 't Hooft, Nucl. Phys. B **79**, 276 (1974).;
   A. M. Polyakov, JETP Lett. **20**, 194 (1974).
5. M. R. Douglas and S. H. Shenker, Nucl. Phys. B **447**, 271 (1995) [hep-th/9503163]; A. Hanany, M. J. Strassler and A. Zaffaroni, Nucl. Phys. B **513**, 87 (1998) [hep-th/9707244].
6. A. Hanany and D. Tong, JHEP **0307**, 037 (2003) [hep-th/0306150]; R. Auzzi, S. Bolognesi, J. Evslin, K. Konishi and A. Yung, Nucl. Phys. B **673**, 187 (2003) [hep-th/0307287]; M. Shifman, A. Yung, Phys. Rev. **D70**, 045004 (2004). [hep-th/0403149]; A. Hanany and D. Tong, JHEP **0404**, 066 (2004) [hep-th/0403158].
7. N. Dorey, JHEP **9811**, 005 (1998) [hep-th/9806056].
8. P. A. Bolokhov, M. Shifman and A. Yung, Phys. Rev. D **85**, 085028 (2012) [arXiv:1202.5612 [hep-th]].
9. M. Shifman and A. Yung, *Supersymmetric Solitons*, (Cambridge University Press, 2009).
10. A. Abrikosov, Sov. Phys. JETP **32** 1442 (1957); H. Nielsen and P. Olesen, Nucl. Phys. **B61** 45 (1973).
11. M. Shifman, Phys. Rev. D **87**, 025025 (2013) [arXiv:1212.4823 [hep-th]].
12. M. Nitta, M. Shifman, and W. Vinci, arXiv:1301.3544 [cond-mat.other].
13. M. Shifman, and A. Yung, *Abrikosov-Nielsen-Olesen string with Non-Abelian Moduli and "Spin-Orbit" Interaction*, arXiv:1303.7010 [hep-th].
14. G. Volovik, *The Universe in a Helium Droplet*, (Oxford University Press, 2003); A. J. Leggett, *Quantum Liquids* (Oxford University Press, 2006).
15. E. Witten, Nucl. Phys. B **249**, 557 (1985).
16. M. Shifman, *Advanced Topics in Quantum Field Theory*, (Cambridge University Press, 2012).
17. E. B. Bogomol'nyi, *Sov. J. Nucl. Phys.* **24**, 449 (1976), reprinted in *Solitons and Particles*, eds. C. Rebbi and G. Soliani (World Scientific, Singapore, 1984) p. 389. M. K. Prasad and C. M. Sommerfield, *Phys. Rev. Lett.* **35**, 760 (1975), reprinted in *Solitons and Particles*, Eds. C. Rebbi and G. Soliani (World Scientific, Singapore, 1984) p. 530.
18. A. M. Polyakov, Phys. Lett. B **59**, 79 (1975).
19. I. Low and A. V. Manohar, Phys. Rev. Lett. **88**, 101602 (2002) [hep-th/0110285].

Finite-Time Future Singularities and Related Cosmologies in Modified Gravitational Theories

Kazuharu Bamba *[(a)], Ratbay Myrzakulov†[(b)], Shin'ichi Nojiri‡[(a),(c)], Sergei D. Odintsov§[(b),(c),(d),(e),(f)] and Misao Sasaki¶[(g),(h)]

[(a)] *Kobayashi-Maskawa Institute for the Origin of Particles and the Universe, Nagoya University, Nagoya 464-8602, Japan*
[(b)] *Eurasian International Center for Theoretical Physics, Eurasian National University, Astana 010008, Kazakhstan*
[(c)] *Department of Physics, Nagoya University, Nagoya 464-8602, Japan*
[(d)] *Instituciò Catalana de Recerca i Estudis Avançats (ICREA), Barcelona, Spain*
[(e)] *Institut de Ciencies de l'Espai (CSIC-IEEC), Campus UAB, Facultat de Ciencies, Torre C5-Par-2a pl, E-08193 Bellaterra (Barcelona), Spain*
[(f)] *Tomsk State Pedagogical University, Tomsk, Russia* [(g)] *Yukawa Institute for Theoretical Physics, Kyoto University, Kyoto 606-8502, Japan*
[(h)] *Korea Institute for Advanced Study 207-43 Cheongnyangni 2-dong, Dongdaemun-gu, Seoul 130-722, Republic of Korea*

We investigate finite-time future singularities in non-local gravity as well as $f(T)$ gravity with T being the torsion scalar. It is found that three types of the finite-time future singularities can occur in non-local gravity. Furthermore, it is explicitly demonstrated that there appear finite-time future singularities in $f(T)$ gravity and an $f(T)$ gravity model with realizing those singularities is reconstructed.

Keywords: Modified theories of gravity; Dark energy; Cosmology.

1. Introduction

Observations such as Supernovae Ia (SNe Ia), cosmic microwave background (CMB) radiation, large scale structure (LSS), baryon acoustic oscillations (BAO), and weak lensing imply the accelerated expansion of the universe at the present time. Two representative approaches to explain the current cosmic acceleration exist. One is to introduce the so-called "dark energy" (for a very recent review, see, e.g., Ref. 1). The other is to modify the gravitational theory such as $F(R)$ gravity (for reviews, see, for instance, Ref. 2). As one example of the latter approach, there have proposed non-local gravity produced by quantum effects.[3] In addition, there is the so-called

*Email address: bamba@kmi.nagoya-u.ac.jp
†Email address: rmyrzakulov@gmail.com
‡Email address: nojiri@phys.nagoya-u.ac.jp
§Email address: odintsov@ieec.uab.es
¶Email address: misao@yukawa.kyoto-u.ac.jp

teleparallelism in which the Weitzenböck connection is used. This is constructed by not the scalar curvature R defined with the Levi-Civita connection but the torsion scalar T.[4] In $f(T)$ gravity, not only inflation[5] but also the late time cosmic acceleration[6–8] can occur. In this paper, we review our results in Refs. 9,10. We show that there can appear the finite-time future singularities in non-local gravity[9] and $f(T)$ gravity.[10] The procedure used here has also been extended to Loop quantum cosmology.[11] We use units of $k_{\mathrm{B}} = c = \hbar = 1$ and denote the gravitational constant $8\pi G$ by $\kappa^2 \equiv 8\pi/{M_{\mathrm{Pl}}}^2$ with the Planck mass of $M_{\mathrm{Pl}} = G^{-1/2} = 1.2 \times 10^{19}\mathrm{GeV}$.

2. Finite-time future singularities in non-local gravity

The action of non-local gravity is expressed as $S = \int d^4x\sqrt{-g}\ \left\{ \left(2\kappa^2\right)^{-1} \left[R\left(1 + f(\Box^{-1}R)\right) - 2\Lambda\right] + \mathcal{L}_{\mathrm{M}}\right\}$. Here, g is the determinant of the metric tensor $g_{\mu\nu}$, f is some function, $\Box \equiv g^{\mu\nu}\nabla_\mu\nabla_\nu$ with ∇_μ being the covariant derivative is the covariant d'Almbertian for a scalar field, Λ is a cosmological constant, and $\mathcal{L}_{\mathrm{M}}$ is the matter Lagrangian. This action can be rewritten by introducing two scalar fields η and ξ in the following form: $S = \int d^4x\sqrt{-g}\left\{\left(2\kappa^2\right)^{-1}\left[R\left(1+f(\eta)\right) - \partial_\mu\xi\partial^\mu\eta - \xi R - 2\Lambda\right] + \mathcal{L}_{\mathrm{matter}}\right\}$. We take the flat Friedmann-Lemaître-Robertson-Walker (FLRW) metric $ds^2 = dt^2 - a^2(t)\sum_{i=1,2,3}\left(dx^i\right)^2$ with $a(t)$ the scale factor and $H = \dot{a}/a$ the Hubble parameter, where the dot denotes the time derivative, $\partial/\partial t$. We suppose that the scalar fields η and ξ only depend on time and examine whether there exist the finite-time future singularities in non-local gravity. We analyze an asymptotic solution of the gravitational field equations in the limit of the time when the finite-time future singularities appear. We consider the case that the Hubble parameter H is expressed as $H \sim h_{\mathrm{s}}\left(t_{\mathrm{s}} - t\right)^{-q}$ with h_{s} a positive constant, q a non-zero constant larger than -1 $(q > -1, q \neq 0)$, t_{s} the time when the finite-time future singularity appears, and $0 < t < t_{\mathrm{s}}$. We take the form of $\eta = f_{\mathrm{s}}\eta^\sigma$ with $f_{\mathrm{s}}(\neq 0)$ and $\eta_{\mathrm{c}}(\neq 0)$ non-zero constants. Four types of the finite-time future singularities are known.[12] Type I ("Big Rip"[13]): For the limit $t \to t_{\mathrm{s}}$, $a \to \infty$, $\rho_{\mathrm{eff}} \to \infty$ and $|P_{\mathrm{eff}}| \to \infty$. The case that ρ_{eff} and P_{eff} are finite values at $t = t_{\mathrm{s}}$[14] is included. (ii) Type II ("sudden"[15]): $a \to a_{\mathrm{s}}$, $\rho_{\mathrm{eff}} \to \rho_{\mathrm{s}}$ and $|P_{\mathrm{eff}}| \to \infty$. (iii) Type III: $a \to a_{\mathrm{s}}$, $\rho_{\mathrm{eff}} \to \infty$ and $|P_{\mathrm{eff}}| \to \infty$. (iv) Type IV: $a \to a_{\mathrm{s}}$, $\rho_{\mathrm{eff}} \to 0$, $|P_{\mathrm{eff}}| \to 0$, and higher derivatives of H diverge.

Table 1. Range and conditions for the value of parameters of $f(\eta)$, H, and integration constants η_{c} and ξ_{c} of the expressions for η and ξ, respectively, in order that the finite-time future singularities can exist.

$f(\eta) = f_{\mathrm{s}}\eta^\sigma$	$H \sim h_{\mathrm{s}}\left(t_{\mathrm{s}} - t\right)^{-q}$	η_{c} , ξ_{c}
$f_{\mathrm{s}} \neq 0$	$h_{\mathrm{s}} > 0$	$\eta_{\mathrm{c}} \neq 0$
$\sigma \neq 0$	$q > -1$, $q \neq 0$	
$\sigma < 0$	$q > 1$ [Type I ("Big Rip") singularity]	$\xi_{\mathrm{c}} = 1$
$f_{\mathrm{s}}\eta_{\mathrm{c}}^{\sigma-1}\left(6\sigma - \eta_{\mathrm{c}}\right) + \xi_{\mathrm{c}} - 1 = 0$	$0 < q < 1$ [Type III singularity]	
	$-1 < q < 0$ [Type II ("sudden") singularity]	

The case that $\rho_{\rm eff}$ and/or $|P_{\rm eff}|$ become finite values at $t = t_{\rm s}$ is also included. Here, $a_{\rm s}(\neq 0)$ and $\rho_{\rm s}$ are constants. The results are described in Table 1.

3. Finite-time future singularities in $f(T)$ gravity

In the teleparallelism, the metric $g^{\mu\nu}$ is written $g_{\mu\nu} = \eta_{AB} e^A_\mu e^B_\nu$ with $e_A(x^\mu)$ orthonormal tetrad components, where $A = 0, 1, 2, 3$ for the tangent space at each point x^μ of the manifold, μ, $\nu = 0, 1, 2, 3$ are coordinate indices on the manifold, and e^μ_A is the tangent vector of the manifold. The teleparallel Lagrangian density is expressed by using the torsion scalar T, although in general relativity the Lagrangian density is described by the Ricci scalar R. The torsion scalar T is given by $T \equiv S_\rho{}^{\mu\nu} T^\rho{}_{\mu\nu}$ with $T^\rho{}_{\mu\nu} \equiv e^\rho_A \left(\partial_\mu e^A_\nu - \partial_\nu e^A_\mu\right)$ the torsion tensor and $S_\rho{}^{\mu\nu} \equiv (1/2)\left(K^{\mu\nu}{}_\rho + \delta^\mu_\rho\, T^{\alpha\nu}{}_\alpha - \delta^\nu_\rho\, T^{\alpha\mu}{}_\alpha\right)$, where $K^{\mu\nu}{}_\rho \equiv -(1/2)\left(T^{\mu\nu}{}_\rho - T^{\nu\mu}{}_\rho - T_\rho{}^{\mu\nu}\right)$. The modified teleparallel action is given by[7] $I = \int d^4x |e| \left[f(T)/\left(2\kappa^2\right) + \mathcal{L}_{\rm M}\right]$, where $|e| = \det\left(e^A_\mu\right) = \sqrt{-g}$. We assume the flat FLRW space-time with the metric $ds^2 = dt^2 - a^2(t) \sum_{i=1,2,3} \left(dx^i\right)^2$. In this background, we have $g_{\mu\nu} = \mathrm{diag}(1, -a^2, -a^2, -a^2)$, the tetrad components $e^A_\mu = (1, a, a, a)$, and the torsion scalar $T = -6H^2$. The gravitational field equations read $H^2 = \left(\kappa^2/3\right)(\rho_{\rm M} + \rho_{\rm DE})$ and $\dot{H} = -\left(\kappa^2/2\right)(\rho_{\rm M} + P_{\rm M} + \rho_{\rm DE} + P_{\rm DE})$ with where $\rho_{\rm M}$ and $P_{\rm M}$ the energy density and pressure of all perfect fluids of generic matter, respectively. In addition, the energy density and pressure of dark components become $\rho_{\rm DE} = \left[1/\left(2\kappa^2\right)\right] J_1$ and $P_{\rm DE} = -\left[1/\left(2\kappa^2\right)\right](4J_2 + J_1)$, respectively, where $J_1 \equiv -T - f + 2TF$ and $J_2 \equiv (1 - F - 2TF')\,\dot{H}$ with $F \equiv df/dT$ and $F' = dF/dT$. The standard continuity equations for the dark energy component as well as the perfect fluid are satisfied. We examine the case that H is expressed as[16] $H \sim h_{\rm s}\left(t_{\rm s} - t\right)^{-q}$ for $q > 0$, $H \sim H_{\rm s} + h_{\rm s}\left(t_{\rm s} - t\right)^{-q}$ for $q < -1$, $-1 < q < 0$ with $h_{\rm s}(> 0)$, $H_{\rm s}(> 0)$ and $q(\neq 0, -1)$ constants. For the expression of H shown above, we reconstruct mod-

Table 2. Necessary conditions on the model parameters of a power-law model of $f(T)$ in order to realize the finite-time future singularities and those of the correction term $f_{\rm c}(T) = BT^\beta$ with removing the finite-time future singularities. Only Type I and IV finally emerge.

$q(\neq 0, -1)$	$f(T) = AT^\alpha$ $(A \neq 0,\ \alpha \neq 0)$	$f_{\rm c}(T) = BT^\beta$ $(B \neq 0,\ \beta \neq 0)$
$q \geq 1$ [Type I ("Big Rip") singularity]	$\alpha < 0$	$\beta > 1$
$0 < q < 1$ [Type III singularity]	$\alpha < 0$	$\beta > 1$
$-1 < q < 0$ [Type II ("sudden") singularity]	$\alpha = 1/2$	$\beta \neq 1/2$
$q < -1$, but q is not any integer [Type IV singularity]	$\alpha = 1/2$	$\beta \neq 1/2$

els of $f(T)$ gravity where the finite-time future singularities occur. We note that there have been proposed reconstruction method for modified gravity such as an $F(R)$ gravity in Ref. 17. As a result, we find that for a power-law form of $f(T)$ as $f(T) = AT^\alpha$ with $A(\neq 0)$ and $\alpha(\neq 0)$ non-zero constants, in the flat FLRW background both the two gravitational filed equations can be satisfied. In addition, we investigate a correction term to remove the finite-time future singularities.

As an example of a correction term $f_c(T)$, we take $f_c(T) = BT^\beta$ where $B(\neq 0)$ and $\beta(\neq 0)$ are non-zero constants. If we take into account this correction term as $f(T) = AT^\alpha + BT^\beta$, the two gravitational filed equations cannot be met simultaneously. This implies that such a power-low correction term can remove the finite-time future singularities. In $F(R)$ gravity, a T^2 correction term (i.e., $\beta = 2$) can cure the finite-time future singularities.[2] In Table 2, we depict necessary conditions for the model parameters of a power-law form of $f(T)$ where the finite-time future singularities appear and those of a power-low the correction term which can remove the finite-time future singularities. Only Type I and IV eventually appear.

4. Summary

We have demonstrated that Type I, II and III finite-time future singularities can occur in non-local gravity, and that that Type I and IV finite-time future singularities can appear in $f(T)$ gravity. We have derived explicit $f(T)$ gravity models with those singularities and the forms of the correction term curing these singularities.

References

1. K. Bamba, S. Capozziello, S. Nojiri and S. D. Odintsov, Astrophys. Space Sci. **342**, 155 (2012).
2. S. Nojiri and S. D. Odintsov, Phys. Rept. **505**, 59 (2011); eConf **C0602061**, 06 (2006) [Int. J. Geom. Meth. Mod. Phys. **4**, 115 (2007)] [arXiv:hep-th/0601213].
3. S. Deser and R. P. Woodard, Phys. Rev. Lett. **99**, 111301 (2007); N. Arkani-Hamed, S. Dimopoulos, G. Dvali and G. Gabadadze, arXiv:hep-th/0209227; S. Nojiri, S. D. Odintsov, M. Sasaki and Y. l. Zhang, Phys. Lett. B **696**, 278 (2011).
4. F. W. Hehl, P. Von Der Heyde, G. D. Kerlick and J. M. Nester, Rev. Mod. Phys. **48**, 393 (1976); K. Hayashi and T. Shirafuji, Phys. Rev. D **19**, 3524 (1979) [Addendum-ibid. D **24**, 3312 (1981)]; E. E. Flanagan and E. Rosenthal, *ibid.* **75**, 124016 (2007).
5. R. Ferraro and F. Fiorini, Phys. Rev. D **75**, 084031 (2007); *ibid.* **78**, 124019 (2008).
6. G. R. Bengochea and R. Ferraro, Phys. Rev. D **79**, 124019 (2009).
7. E. V. Linder, Phys. Rev. D **81**, 127301 (2010) [Erratum-ibid. D **82**, 109902 (2010)].
8. K. Bamba, C. Q. Geng and C. C. Lee, arXiv:1008.4036 [astro-ph.CO]; K. Bamba, C. Q. Geng, C. C. Lee and L. W. Luo, JCAP **1101**, 021 (2011).
9. K. Bamba, S. Nojiri, S. D. Odintsov and M. Sasaki, Gen. Rel. Grav. **44**, 1321 (2012).
10. K. Bamba, R. Myrzakulov, S. Nojiri and S. D. Odintsov, Phys. Rev. D **85**, 104036 (2012).
11. K. Bamba, J. de Haro and S. D. Odintsov, arXiv:1211.2968 [gr-qc], to appear in JCAP.
12. S. Nojiri, S. D. Odintsov and S. Tsujikawa, Phys. Rev. D **71**, 063004 (2005).
13. R. R. Caldwell, M. Kamionkowski and N. N. Weinberg, Phys. Rev. Lett. **91**, 071301 (2003); B. McInnes, JHEP **0208** (2002) 029.
14. Y. Shtanov and V. Sahni, Class. Quant. Grav. **19**, L101 (2002).
15. J. D. Barrow, Class. Quant. Grav. **21**, L79 (2004); S. Nojiri and S. D. Odintsov, Phys. Lett. B **595**, 1 (2004).
16. S. Nojiri and S. D. Odintsov, Phys. Rev. D **78**, 046006 (2008).
17. S. Nojiri and S. D. Odintsov, Phys. Rev. D **74**, 086005 (2006); J. Phys. Conf. Ser. **66**, 012005 (2007) [arXiv:hep-th/0611071].

Neutron Electric Dipole Moment Induced by the Strangeness Revisited

Kaori Fuyuto[1], Junji Hisano[1,2], Natsumi Nagata[1,3]
[1] *Department of Physics, Nagoya University, Nagoya 464-8602, Japan*
[2] *Kavli IPMU, University of Tokyo, Kashiwa 277-8584, Japan*
[3] *Department of Physics, University of Tokyo, Tokyo 113-0033, Japan*

We evaluate the neutron electric dipole moment in the presence of the CP-violating operators up to dimension five based on the chiral perturbation theory. In the calculation, we apply the nucleon matrix elements of scalar-type quark operators obtained by the lattice QCD simulations. On the other hand, those of the dipole-type quark-gluon operators are evaluated by the QCD sum rules. The obtained resultant indicates that, though the quantity of strange quark in nucleon is small, the contribution of the chromoelectric dipole moment of that quark may be sizable.

1. Introduction

As one of the physical quantities which are sensitive to the CP-violation in the high-energy theories beyond the Standard Model, the neutron electric dipole moment (nEDM) exists. The Standard Model contains a CP phase in the Cabibbo-Kobayashi-Maskawa matrices. If we estimate the value of nEDM with this phase, it will be much smaller than a sensitivity which some experiment groups of nEDM currently aim. Therefore the existence of nEDM indicates the CP-violation in the high energy physics beyond the Standard Model. In the theoretical side, in regard to the experimental limit $|d_n| < 2.9 \times 10^{-26} e \cdot \mathrm{cm}$,[1] we need to intereprete this limit to it of the CP-violating parameters at parton level. Therefore, in this study, we would like to estimate nEDM with the chiral perturbation theory.

In the calculation, the nucleon matrix elements of both the quark scalar operators and the dipole-type quark-gluon operators are needed. In the case of the former, the previous study extracted them from the baryon mass splittings and the nucleon sigma terms evaluated by the chiral perturbation theory.[2] On the other hand, the recent lattice simulations indicate the nucleon matrix element of the strange quark scalar operator is actually quite small compared to it used in the previous study. Meanwhile, in the case of the dipole-type quark-gluon operators, we use the QCD sum rules in order to obtain them. Therefore, we need to reevaluate the nEDM based on the chiral perturbation theory with the up-to-date results for the nucleon matrix elements.

2. CP-odd meson-nucleon couplings

In this study, we focus on the effective CP-violating interactions at the scale of 1 GeV which consist of the flavor-diagonal operators of light quarks and gluon up to dimension five in QCD as follows,

$$\mathcal{L}_{\not{C}\not{P}} = - \sum_{q=u,d,s} m_q \bar{q} i\theta_q \gamma_5 q + \theta_G \frac{\alpha_s}{8\pi} G^A_{\mu\nu} \tilde{G}^{A\mu\nu} - \frac{i}{2} \sum_{q=u,d,s} \tilde{d}_q \bar{q} g_s (G\cdot\sigma)\gamma_5 q. \tag{1}$$

In the above lagrangian, m_q are the quark masses, $G^A_{\mu\nu}$ is the gluon field strength tensor, $G\cdot\sigma \equiv G^A_{\mu\nu}\sigma^{\mu\nu}T^A$, and $\tilde{G}^A_{\mu\nu} \equiv= \frac{1}{2}\epsilon_{\mu\nu\rho\sigma}G^{A\mu\nu}$ with $\epsilon^{0123} = 1$. T^A and g_s are the generators and the coupling constant ($\alpha_s = g_s^2/4\pi$) of the $\mathrm{SU(3)_C}$. The first term is the mass term which contains the γ_5 matrix and the second terms is called the effective QCD θ term. Finally, the third term expresses the chromoelectric dipole moments (cEDM) for light quarks. Here the QCD θ term can be rotated out into the first term with the use of the chiral $\mathrm{U(1)_A}$ transformation. After this transformation, the effective lagrangian becomes as follows,

$$\mathcal{L}_{\not{C}\not{P}} = - \sum_{q=u,d,s} m_q \bar{q} i\theta'_q \gamma_5 q - \frac{i}{2} \sum_{q=u,d,s} \tilde{d}_q \bar{q} g_s (G\cdot\sigma)\gamma_5 q, \tag{2}$$

where $\sum_q \theta'_q = \bar{\theta} \equiv \theta_G + \sum_q \theta_q$. From here, we omit the prime, *i.e.*, $\sum_q \theta_q = \bar{\theta}$.

Moreover, in this paper, we would like to focus on the case of the presence of the Peccei-Quinn (PQ) symmetry.[3] When the cEDM exists, $\bar{\theta}$ doesn't vanish but has the following relation,

$$\bar{\theta}_{\mathrm{PQ}} = \frac{m_0^2}{2} \sum_{q=u,d,s} \frac{\tilde{d}_q}{m_q}, \tag{3}$$

where we parametrize $\langle \bar{q} g_s (G\cdot\sigma) q\rangle = -m_0^2 \langle \bar{q} q\rangle$.[4] With this notation, the effective lagrangian at baryon level can be written down as follows,

$$\mathcal{L}^{\mathrm{baryon}}_{\not{C}\not{P}} = \langle \Phi^A B^B | \mathcal{L}^{\mathrm{parton}}_{\not{C}\not{P}} | B^C \rangle \Phi^A \overline{B}^B B^C, \tag{4}$$

where B^A and Φ^A respectively mean the baryon and meson octets. And here, $\mathcal{L}^{\mathrm{parton}}_{\not{C}\not{P}}$ indicates Eq. (2). Then, we can extract the r.h.s in the above equation with the PCAC relation and group-theoretical arguments. Finally, they are given as

$$\mathcal{L}^{(n)}_{\not{C}\not{P}} = \bar{g}_{pn\pi} \bar{p} n \pi^+ + \bar{g}_{\Sigma n K} \overline{\Sigma^-} n K^- + \mathrm{h.c.}, \tag{5}$$

with

$$\bar{g}_{pn\pi}|_{\mathrm{PQ}} = -\frac{1}{\sqrt{2} f_\pi} \left[\frac{m_0^2}{2}(S_u - S_d) + \frac{1}{2}(D_u - D_d) \right] (\tilde{d}_u + \tilde{d}_d) \tag{6}$$

$$\bar{g}_{|\Sigma n K}|_{\mathrm{PQ}} = -\frac{1}{\sqrt{2} f_\pi} \left[\frac{m_0^2}{2}(S_s - S_d) + \frac{1}{2}(D_s - D_d) \right] (\tilde{d}_u + \tilde{d}_s) \tag{7}$$

where we define $S_q \equiv \langle p|\bar{q}q|p\rangle$ as the nucleon matrix elements of scalar-type quark operators and $D_q \equiv \langle p|\bar{q} g_s G\cdot\sigma q|p\rangle$ as them of the dipole-type quark-gluon operators. Here, it is important to recognize that, if the S_s equals 0, the effect of the cEDM of strange quark doesn't vanish.

3. Formula of nEDM and the nucleon matrix elements

In order to obtain a formula of nEDM, we need the CP-even interaction terms. In this study, we would like to introduce them from the chiral perturbation theory with baryons. Then we can calculate the one-loop diagrams which contribute the nEDM with these interaction terms and finally obtain the following formula,[5]

$$d_n = -\frac{e}{4\sqrt{2}\pi^2 f_\pi}\left[(D+F)\bar{g}_{pn\pi}\log\left(\frac{\Lambda}{m_\pi}\right) - (D-F)\bar{g}_{\Sigma nK}\log\left(\frac{\Lambda}{m_K}\right)\right]. \tag{8}$$

In the above formula, f_π is the pion decay constant. And both D and F are the low-energy constants determined by the semi-leptonic decays $B \to B' + e^- + \bar{\nu}_e$: $D = 0.80(1), F = 0.47(1)$.[6] Moreover, the reason why this formula contains the cut off scale Λ is to include higher order contribution, which act as counter terms, at tree level in the chiral lagrangian.[5]

It only remains for us to extract the nucleon matrix elements for the estimate of nEDM. As it was previously mentioned, we use the nucleon matrix elements of the scale-type quark operators obtained by the lattice QCD simulations. With the results of the recent lattice simulations,[7,8] they are extracted as follows,

$$S_u = 5.8, \qquad S_d = 4.0, \qquad S_s = 0.18. \tag{9}$$

Note that the values of S_u and S_d are almost as twice as those in the previous work.[2] However, S_s is much smaller, since the recent lattice simulations predict relatively small values for σ_s, which is defined as $\sigma_s = m_s \times S_s$.

Furthermore, we have to extract the case of the dipole-type quark-gluon operators. Therefore, we introduce the method of QCD Sum rules. QCD sum rules is to evaluate the correlation functions of hadron field in terms of two different ways and connect each result by using the Borel transformation. In the two different ways, one is to calculate it with the phenomenological description as a hadron picture and the other is to calculate it with the operator product expansion as a quark picture. In particular, in this study, we use the following correlation function of proton field in the presence of the external field J,

$$\Pi(k)|_J \equiv i\int d^4x e^{ikx}\langle\Omega|T\left\{\eta_p(x), \bar{\eta}_p(0)\right\}|\Omega\rangle_J. \tag{10}$$

Here, η_p means the interpolating field of proton. J exists as the extra term $\mathcal{L}_{\text{ext}} = J_q\bar{q}g_sG\cdot\sigma q$, in the ordinary QCD lagrangian. Calculating this correlation function with two different ways, we can connect these with the Borel transformation. This transformation has a role which can exponentially suppress the contributions from the excited states and continuum states.

In this study, we estimated D_q up to the leading order and finally obtained the following results,

$$D_u \simeq \frac{3m_p^3}{64\pi^2}\frac{\langle\bar{q}q\rangle}{\lambda_p^2}m_0^2\cdot\exp[2], \qquad D_d \simeq \frac{m_p^3}{32\pi^2}\frac{\langle\bar{q}q\rangle}{\lambda_p^2}m_0^2\cdot\exp[2], \tag{11}$$

and $D_s = 0$ up to the leading order calculation. In the above results, D_u and D_d contain λ_p which is the field renotmalization constant. In this work, we use the result from the lattice QCD simulations for λ_p.[9]

4. Results and conclusion

As a numerical result, the following values are obtained,

$$d_n = e(3.0\tilde{d}_u + 2.5\tilde{d}_d + 0.5\tilde{d}_s), \tag{12}$$

where we regard Λ as M_p. These values are about as twice as those in the previous work. Moreover, in this study, we focus on the contribution from the strange quark to the nEDM. In order to examine this, we normalize the cEDMs by the quark masses as follows,

$$d_n|_{\mathrm{PQ}} = \sum_q eC_q \frac{\tilde{d}_q}{m_q}. \tag{13}$$

Since in most cases the quark cEDMs are proportional to the quark masses, $\tilde{d}_q \propto m_q$, we normalize it like above equation. Moreover, when we examine the ratio of C_s/C_q with the cut off scale changing from 600 MeV to 1200 MeV, the ratio may change by $\mathcal{O}(1)$ along the cut off scale. However, it seems that the cEDM of the strange quark tends to remain important. To summarize this point, although the recent lattice QCD results show that the nucleon matrix element of strange quark is small, the cEDM of strange quark may have an important role.

As a conclusion, we reevaluate the neutron electric dipole moment induced by the CP-violating operators up to dimension five based on the chiral perturbation theory. In the results, nevertheless the recent lattice results indicate that S_s is smaller, the contribution of the cEDM of the strange quark to nEDM ma be sizable. Moreover, as a prospect, we need to calculate the CP-odd meson baryon couplings which contain the contribution from higher order terms with QCD sum rules.

References

1. C. A. Baker, D. D. Doyle, P. Geltenbort, K. Green, M. G. D. van der Grinten, P. G. Harris, P. Iaydjiev and S. N. Ivanov *et al.*, Phys. Rev. Lett. **97**, 131801 (2006).
2. J. Hisano and Y. Shimizu, Phys. Rev. D **70**, 093001 (2004).
3. R. D. Peccei and H. R. Quinn, Phys. Rev. Lett. **38**, 1440 (1977).
4. V. M. Belyaev and B. L. Ioffe, Sov. Phys. JETP **56**, 493 (1982) [Zh. Eksp. Teor. Fiz. **83**, 876 (1982)].
5. A. Pich and E. de Rafael, Nucl. Phys. B **367**, 313 (1991).
6. S. Y. Hsueh, D. Muller, J. Tang, R. Winston, G. Zapalac, E. C. Swallow, J. P. Berge and A. E. Brenner *et al.*, Phys. Rev. D **38**, 2056 (1988).
7. S. Aoki *et al.* [PACS-CS Collaboration], Phys. Rev. D **79**, 034503 (2009).
8. P. E. Shanahan, A. W. Thomas and R. D. Young, arXiv:1205.5365 [nucl-th].
9. Y. Aoki *et al.* [RBC-UKQCD Collaboration], Phys. Rev. D **78**, 054505 (2008).

Enhanced diphoton Higgs decay rate and isospin symmetric Higgs boson

Michio Hashimoto*
Chubu University,
1200 Matsumoto-cho, Kasugai-shi,
Aichi, 487-8501, JAPAN
**E-mail: michioh@isc.chubu.ac.jp*

Based on Ref. 1, we introduce a model with an isospin symmetric Higgs boson and study the properties of this particle, including the enhancement of its diphoton decay rate. The predictions of the model relevant for future experiments are also discussed.

Keywords: Composite Higgs boson

1. Introduction

Recently, the ATLAS and the CMS Collaborations discovered a new boson h in the mass range 125–126 GeV.[2] The observed diphoton decay rate is about 1.6 times larger than that of the Standard model (SM) Higgs boson H, while the other properties seem rather similar to the SM. This deviation in the diphoton channel, if established, would be an indication of a new physics beyond the SM.

Based on Ref. 1, we show that the ATLAS and CMS data for the enhanced Higgs diphoton branching ratio can be explained in the class of models with isospin symmetric (IS) electroweak Higgs boson. We also discuss the predictions of these models, which can be checked at the LHC in the near future.

2. IS Higgs model and Higgs decay rates

The main characteristics of the IS Higgs boson models are the following.[1,3] a) It is assumed that the dynamics primarily responsible for the electroweak symmetry breaking (EWSB) leads to the mass spectrum of quarks with no (or weak) isospin violation. *Moreover, it is assumed that the values of these masses are of the order of the observed masses of the down-type quarks.* b) The second (central) assumption is introducing the horizontal interactions for the quarks in the three families. As a first step, a *subcritical* (although nearcritical, i.e., strong) diagonal horizontal interactions for the top quark is utilized which lead to the observed ratio $\frac{m_t}{m_b} \simeq 41.5$.[4] The second step is introducing *equal* strength horizontal flavor-changing-neutral (FCN) interactions between the t and c quarks and the b and s ones. As was shown in Ref. 3, these interactions naturally provide the observed ratio $m_c/m_s \simeq 13.4$ in the second family.[4] As to the mild isospin violation in the first family, it was studied together with the effects of the family mixing, reflected in the Cabibbo-Kobayashi-Maskawa (CKM) matrix.[3]

In this scenario, the main source of the isospin violation is only the strong top quark interactions. However, because these interactions are subcritical, the top quark plays a minor role in EWSB. This distinguishes the IS Higgs scenario from the top quark condensate model.[5–9]

One of the signatures of this scenario is the appearance of a composite top-Higgs boson h_t composed of the quarks and antiquarks of the third family.[3] Note that unlike the topcolor assisted technicolor model (TC2),[10] this class of models utilizes subcritical dynamics for the top quark, so that the top-Higgs h_t is heavy in general. Here we also emphasize that while the top-Higgs boson h_t has a large top-Yukawa coupling, the IS Higgs boson h does not, $y_t \simeq y_b \sim 10^{-2}$. On the other hand, the hWW^* and hZZ^* coupling constants are close to those in the SM. Also, the mixing between h and much heavier h_t should be small.

Let us now describe the decay processes of the IS Higgs h.

It is well known that the W-loop contribution to $H \to \gamma\gamma$ is dominant in the SM, while the top-loop effect is destructive against the W-loop. In the IS Higgs model, however, the Yukawa coupling between the top and the IS Higgs h is as small as the bottom Yukawa coupling, so that the top-loop contribution is strongly suppressed. The partial decay width of $h \to \gamma\gamma$ is thus enhanced without changing essentially $h \to ZZ^*$ and $h \to WW^*$. A rough estimate taking the isospin symmetric top and bottom Yukawa couplings $y_t \simeq y_b \approx 10^{-2}$ is as follows:

$$\frac{\Gamma^{\rm IS}(h \to \gamma\gamma)}{\Gamma^{\rm SM}(H \to \gamma\gamma)} \simeq 1.56, \quad \frac{\Gamma^{\rm IS}(h \to WW^*)}{\Gamma^{\rm SM}(H \to WW^*)} = \frac{\Gamma^{\rm IS}(h \to ZZ^*)}{\Gamma^{\rm SM}(H \to ZZ^*)} = \left(\frac{v_h}{v}\right)^2 \simeq 0.96. \tag{1}$$

Here using the Pagels-Stokar formula,[11] we estimated the vacuum expectation value (VEV) of the top-Higgs h_t as $v_t = 50$ GeV, and the VEV v_h of the IS Higgs h is given by the relation $v^2 = v_h^2 + v_t^2$ with $v = 246$ GeV. Note that the values of the ratios in Eq. (1) are not very sensitive to the value of v_t, e.g., for $v_t = 40$–100 GeV, the suppression factor in the pair decay modes to WW^* and ZZ^* is 0.97–0.84 and the enhancement factor in the diphoton channel is 1.58–1.37. For the decay mode of $h \to Z\gamma$, this model yields

$$\frac{\Gamma^{\rm IS}(h \to Z\gamma)}{\Gamma^{\rm SM}(H \to Z\gamma)} \simeq 1.07\,. \tag{2}$$

The values in Eq. (1) agree well with the data in the ATLAS and CMS experiments. However, obviously, the main production mechanism of the Higgs boson, the gluon fusion process $gg \to h$, is now in trouble. The presence of new chargeless colored particles, which considered by several authors[12] can help to resolve this problem. We pursue this possibility in the next section.

3. Benchmark model with colored scalar

As a benchmark model, we may introduce a real scalar field S in the adjoint representation of the color $SU(3)_c$:

$$\mathcal{L} \supset \mathcal{L}_S = \frac{1}{2}(D_\mu S)^2 - \frac{1}{2}m_{0,S}^2 S^2 - \frac{\lambda_S}{4}S^4 - \frac{\lambda_{hS}}{2}S^2\Phi_h^\dagger\Phi_h, \tag{3}$$

where Φ_h represents the IS Higgs doublet. The effective Lagrangian $\mathcal{L}$ also contains the IS Higgs quartic couplings λ_h, $\mathcal{L} \supset -\lambda_h|\Phi_h|^4$. The IS Higgs mass is $m_h = \sqrt{2\lambda_h}v_h$, and we will take it to be equal to 125 GeV. The mass-squared term for the scalar S is $M_S^2 = m_{0,S}^2 + \frac{\lambda_{hS}}{2}v_h^2$, and should be positive in order to avoid the color symmetry breaking. Typically, $M_S \sim 200$ GeV.

Taking into account the S contribution to $gg \to h$, we find appropriate values of the Higgs-portal coupling,

$$\lambda_{hS} \simeq 2.5\text{–}2.7 \times \frac{M_S^2}{vv_h}. \tag{4}$$

As a typical value, we may take $\lambda_{hS} = 1.8$ for $M_S = 200$ GeV and $v_t = 50$ GeV.

A comment concerning the IS Higgs quartic coupling λ_h is in order. In the SM, the Higgs mass 125 GeV suggests that the theory is perturbative up to an extremely high energy scale.[13] On the contrary, in the present model, when we take a large Higgs-portal coupling λ_{hS} that reproduces $gg \to h$ correctly, the quartic coupling λ_h will grow because the β-function for λ_h contains the λ_{hS}^2 term. Also, there is no large negative contribution to the β-function for λ_h from the top-Yukawa coupling $y_t \sim 10^{-2}$.

One can demonstrate such a behavior more explicitly by using the renormalization group equations. In Fig. 1, the running of the coupling λ_h is shown. Taking a large Higgs-portal coupling $\lambda_{hS} = 1.8$ and the S^4-coupling $\lambda_S = 1.5$, it turns out that the coupling λ_h rapidly grows. The blowup scale strongly depends on the initial values of λ_{hS} and λ_S. A detailed analysis will be performed elsewhere.

Last but not least, we would like to mention that other realizations of the enhancement of the h production are also possible.

4. Conclusion

We studied the properties of the IS Higgs boson. The IS Higgs model can explain the enhanced Higgs diphoton decay rate observed at the LHC, and also makes several predictions. The most important of them is that the value of the top-Yukawa coupling h-t-$\bar{t}$ should be close to the bottom-Yukawa one. Another prediction relates to the decay mode $h \to Z\gamma$, which is enhanced only slightly, $\Gamma^{\rm IS}(h \to Z\gamma) = 1.07 \times \Gamma^{\rm SM}(H \to Z\gamma)$, unlike $h \to \gamma\gamma$. Last but not least, the LHC might potentially discover the top-Higgs resonance h_t, if lucky. For details, see Ref. 1.

References

1. M. Hashimoto and V. A. Miransky, Phys. Rev. D **86**, 095018 (2012).

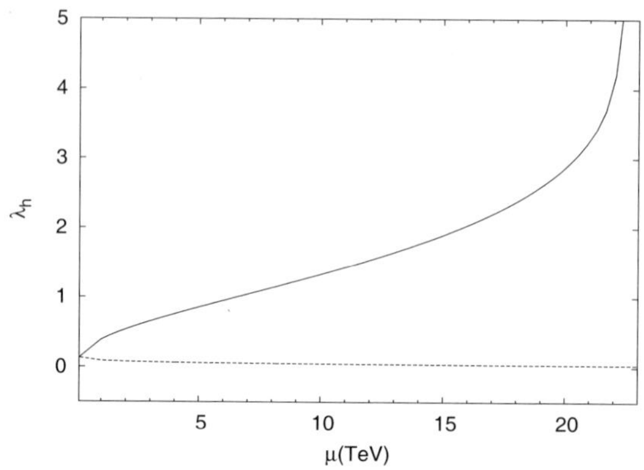

Fig. 1. The running behavior of the IS Higgs quartic coupling λ_h. The solid and dashed lines correspond to λ_h and the SM Higgs quartic coupling, respectively. We fixed the IS Higgs mass $m_h = \sqrt{2\lambda_h} v_h = 125$ GeV and took $\lambda_{hS} = 1.8$ and $\lambda_S = 1.5$. Unlike the SM, the IS Higgs quartic coupling grows up due to a large Higgs-portal coupling λ_{hS} and a small top-Yukawa coupling y_t.

2. ATLAS Collaboration, Phys. Lett. B **716**, 1 (2012); CMS Collaboration, Phys. Lett. B **716**, 30 (2012).
3. M. Hashimoto and V. A. Miransky, Phys. Rev. D **80**, 013004 (2009); Phys. Rev. D **81**, 055014 (2010).
4. J. Beringer *et al.* [Particle Data Group], Phys. Rev. D **86**, 010001 (2012).
5. V. A. Miransky, M. Tanabashi and K. Yamawaki, Phys. Lett. B **221**, 177 (1989); Mod. Phys. Lett. A **4**, 1043 (1989).
6. Y. Nambu, Enrico Fermi Institute Report No. 89-08, 1989; in *Proceedings of the 1988 Kazimierz Workshop*, eds. Z. Ajduk et al. (World Scientific Publishing Co., Singapore, 1989).
7. W. J. Marciano, Phys. Rev. Lett. **62**, 2793 (1989); Phys. Rev. D **41**, 219 (1990).
8. W. A. Bardeen, C. T. Hill and M. Lindner, Phys. Rev. D **41**, 1647 (1990).
9. For a dynamical model with extra dimensions, see, e.g., M. Hashimoto, M. Tanabashi and K. Yamawaki, Phys. Rev. D **64**, 056003 (2001); *ibid* D **69**, 076004 (2004); V. Gusynin, M. Hashimoto, M. Tanabashi and K. Yamawaki, Phys. Rev. D **65**, 116008 (2002); For models with the fourth generation, see, e.g., Ref.3, and M. Hashimoto, Phys. Rev. D **81**, 075023 (2010).
10. C. T. Hill, Phys. Lett. B **345**, 483 (1995); For a comprehensive review, see C. T. Hill and E. H. Simmons, Phys. Rept. **381**, 235 (2003); Erratum-ibid. **390**, 553 (2004).
11. H. Pagels and S. Stokar, Phys. Rev. D **20**, 2947 (1979).
12. R. Boughezal and F. Petriello, Phys. Rev. D **81**, 114033 (2010); B. Batell, S. Gori and L. -T. Wang, JHEP **1206**, 172 (2012); B. A. Dobrescu, G. D. Kribs and A. Martin, Phys. Rev. D **85**, 074031 (2012); G. D. Kribs and A. Martin, Phys. Rev. D **86**, 095023 (2012).
13. J. A. Casas, J. R. Espinosa and M. Quiros, Phys. Lett. B **342**, 171 (1995); *ibid* B **382**, 374 (1996); G. Isidori, G. Ridolfi and A. Strumia, Nucl. Phys. B **609**, 387 (2001); J. Elias-Miro, J. R. Espinosa, G. F. Giudice, G. Isidori, A. Riotto and A. Strumia, Phys. Lett. B **709**, 222 (2012); S. Alekhin, A. Djouadi and S. Moch, Phys. Lett. B **716**, 214 (2012).

Partial Mass-Degeneracy and Spontaneous CP Violation in the Lepton Sector *

Hiroyuki Ishida

Maskawa Institute, Kyoto Sangyo University,
Motoyama, Kamigamo, Kyoto 603-8555, Japan
E-mail: h_ishida@cc.kyoto-su.ac.jp

We have investigated a flavour model which inspired by small squared-mass difference measured in solar neutrino oscillation experiments and observability in neutrinoless double beta decay experiments. In our model, the $1^{\text{st.}}$ and $2^{\text{nd.}}$ generations of fermions have a common mass at the leading order. Such limit may be a good starting point from the points of view of understanding the mixing patterns and mass spectra. In this limit, the mass matrices are respected an $O(2)$ symmetry on flavor space of the first two generations. For simplicity, we propose a model for lepton sector based on the D_N group which is a discrete subgroup of $O(2)$. We show that our model can reproduce the experimental data without hierarchical couplings except for 5-10% tuning partially for the large neutrino mixing. Further, we show a novel relation between the tiny electron mass and the relatively large $\theta_{13}^{\text{PMNS}}$ via CP violation by the complex vacuum expectation values of the extra scalar fields.

Keywords: Flavour physics , neutrino mass

1. Introduction and Motivation

On $4^{\text{th.}}$ July 2013, we have been reported Higgs-like boson at long last. Although we don't know that whether this new boson is the Higgs boson or not, we would have found the last piece of the standard model (SM).

Although the SM of particle physics can predict various experiments very accuracy, there have been some problems and unsatisfactory points yet. In this paper, we would like to focus on the flavour structure of fermions among these problems. When we consider to solve this problem, there is no guiding principle at the present. For a long while, many scientists have employed discrete flavour symmetries (for a review see Ref.[2]) in the neutrino sector. Such symmetries derive tri-bi maximal type mixing[3] and this mixing have been able to describe neutrino oscillation data before growth of the reactor experiments.[4] These experiments have reported non-zero reactor angle, $\theta_{13}^{\text{PMNS}}$, at very high confidence level. At this point, we have to reconsider the origin of the flavour structure of fermions.

Searching for the clue of the guidepost for the flavour structure, it is very im-

*Poster presentation based on Ref. 1.

Table 1. Charge assignments.

	L_I	L_3	ℓ_i	H	S_I
D_N	$\mathbf{2}_2$	$\mathbf{1}$	$\mathbf{1}$	$\mathbf{1}$	$\mathbf{2}_1$

portant to seek a small parameter in the model. From this point of view, we show the following candidates as small parameters for examples :

(i) $\theta_{13}^{\rm PMNS} \ll \theta_{12}^{\rm PMNS},\ \theta_{23}^{\rm PMNS}$ (or $|V_{e3}^{\rm PMNS}| \ll$ the others),

(ii) $|\theta_{23}^{\rm PMNS} - 45^\circ| \ll \theta_{23}^{\rm PMNS}$ (or $|V_{\mu 3}^{\rm PMNS} - 1/\sqrt{2}| \ll |V_{\mu 3}^{\rm PMNS}|$),

(iii) $\Delta m_{12}^2 = (m_2^\nu)^2 - (m_1^\nu)^2 \ll \Delta m_{23}^2 = |(m_3^\nu)^2 - (m_2^\nu)^2|$,

(iv) $m_1^{u,d,\ell},\ m_2^{u,d,\ell} \ll m_3^{u,d,\ell}$,

(v) $\theta_{ij}^{\rm CKM} \ll \theta_{ij}^{\rm PMNS}$ (or $|V_{ij}^{\rm CKM}| \ll |V_{ij}^{\rm PMNS}|$),

where $\theta_{ij}^{\rm PMNS}$ and $\theta_{ij}^{\rm CKM}$ stand for the mixing angles in the Pontecorvo-Maki-Nakagawa-Sakata (PMNS), $V^{\rm PMNS}$, and Cabibbo-Kobayashi-Maskawa (CKM), $V^{\rm CKM}$, mixing matrices[a], respectively, and m_i^f with $f = u, d, \ell, \nu$ and $i = 1, 2, 3$ denote masses of the up-type quarks, down-type quarks, charged leptons and neutrinos. In this paper, we pay our attentions from (iii) to (v) and discuss the gifts of these viewpoints.

In the limit of $\Delta m_{12}^2 \to 0$, that is $m_1^\nu = m_2^\nu$, the $O(2)$ flavour symmetry appears in the Majorana neutrino mass matrix. If we do not necessary to make unnatural hierarchy between the masses for first two neutrinos and the third one, it can be naturally realized to obtain the degenerated neutrino masses. Such region of neutrino masses (roughly speaking $m_2^\nu = 0.05$–0.10eV) is very attractive from the neutrinoless double beta ($0\nu\beta\beta$) decay. The effective mass of $0\nu\beta\beta$ decay can be constrained on the promising region for the next generation experiments for example KamLAND-Zen and EXO experiments.

2. Model

In this section, we briefly explain about our model. At the beginning, we summarize the charge assignments of each fields in Table.1. Our model is constructed with respect to D_N flavour symmetry which is discrete subgroup of $O(2)$. D_N and $O(2)$ have same multiplication rules. We introduce a new SM gauge singlet scalar fields, S_I (I denotes the index of generations), to break this flavour symmetry. Furthermore, we impose this new scalar fields have complex vacuum expectation values (VEVs) and denote the CP violating phases as $\phi_{1,2}$. Here, we skip to show the complete Lagrangian but note that due to an extra charge of D_N which is described as a subscript in Table.1, we can construct ideal terms with Froggatt-Nielsen[6] like mechanism. For concluding of explanation of our model, we have to note that our

[a]In this paper, we adopt the standard parametrization defined by the Particle Data Group.[5]

model do not need to have any hierarchical couplings to reproduce the experimental data. 5-10% tuning between f_ν and f'_ν, however, is necessary in order to realize the large neutrino mixings. Although we do not show the results, we have checked that our model can reproduce the experimental data by numerical analysis. Actually, we have enough number of parameters to reproduce experimental data.

3. Discussion and Conclusion

We study a D_N flavour symmetric model inspired by the smallness of solar neutrino mass square difference comparing with atmospheric one and testability of the $0\nu\beta\beta$ decay. In this model, we introduce a new SM gauge singlet scalar with complex VEVs. This model has a non-trivial feature between electron mass and a mixing angle, $\theta_{13}^{\rm PMNS}$, via CP violating phases in these. Here, we assume that $\phi_2 = \phi_1 + \delta\phi$ with $\delta\phi \ll 1$ and from this assumption we can rewritten the mass matrix of charged lepton as

$$\begin{pmatrix} Y_1\left(s_1^2 - s_2^2\right) & Y_2\left(s_1^2 - s_2^2\right) & Y_3\left(s_1^2 - s_2^2\right) \\ Y_1 2s_1s_2 & Y_2 2s_1s_2 & Y_3 2s_1s_2 \\ 0 & 0 & 0 \end{pmatrix} + i\delta\phi \begin{pmatrix} -Y_1' 2s_2^2 & Y_2' 2s_2^2 & -Y_3' 2s_2^2 \\ Y_1' 2s_1s_2 & Y_2' 2s_1s_2 & Y_3' 2s_1s_2 \\ 0 & 0 & 0 \end{pmatrix}, \quad (1)$$

where $Y_i \equiv y_i e^{2i\phi_1} + y'_i e^{-2i\phi_1} + y''_i$ and $Y'_i \equiv y_i e^{2i\phi_1} - y'_i e^{-2i\phi_1}$. We can see that each term is rank $= 1$ matrix. The first term is mainly responsible for muon mass and the second one is for electron mass, respectively. We suppose that $\delta\phi \ll 1$ so that we can obtain the small electron mass. Furthermore, we need one more assumption. To avoid accidental contribution to 1-2 mixing from charged lepton mass matrix, we assume $s_1 \simeq s_2$.

On the other hand, the 1-3 element of neutrino mass matrix can similarly be rewritten as

$$(\mathcal{M})_{13} \propto \left[g_\nu e^{2i\phi_1} + g'_\nu e^{-2i\phi_1} + g''_\nu\right]\left(s_1^2 - s_2^2\right) - i\delta\phi\left[g_\nu e^{2i\phi_1} - g'_\nu e^{-2i\phi_1}\right] 2s_2^2. \quad (2)$$

Each term is suppressed by $s_1^2 - s_2^2$ and $\delta\phi$, therefore $\theta_{13}^{\rm PMNS}$ can be small because of this angle is determined by the element $(\mathcal{M})_{13}$ dominantly.

We can see that our model can reproduce the experimental data with some assumptions but we need to check whether such assumptions are valid or not. To do this, we discuss the possibility of spontaneous CP violation (SCPV). At first, we assume that the singlet scalar fields were completely decoupled from the theory at high energy. In this condition, the only to do is to investigate only the potential of the singlet scalar fields. Although the scalar potential is a little bit complicated then we do not write it here, the minimization conditions respect with the phases are given as

$$\begin{aligned}\frac{\partial V_S}{\partial \phi_1} &= -2s_1^2\left[\ \{\alpha'_S+\gamma_S\left(s_1^2+s_2^2\right)\}\sin 2\phi_1+2(\beta_S^b+\beta_S^c)s_2^2\sin\left[2\left(\phi_1-\phi_2\right)\right]\right.\\ &\quad \left.+2\beta'_S\left(s_1^2\sin 4\phi_1+s_2^2\sin\left[2\left(\phi_1+\phi_2\right)\right]\right)\ \right]=0\,,\\ \frac{\partial V_S}{\partial \phi_2} &= -2s_2^2\left[\ \{\alpha'_S+\gamma_S\left(s_1^2+s_2^2\right)\}\sin 2\phi_2-2(\beta_S^b+\beta_S^c)s_1^2\sin\left[2\left(\phi_1-\phi_2\right)\right]\right]\\ &\quad +2\beta'_S\left(s_2^2\sin 4\phi_2+s_1^2\sin\left[2\left(\phi_1+\phi_2\right)\right]\right)\ \Big]=0\,.\end{aligned} \tag{3}$$

For simplicity, we set $\alpha'_S=\beta'_S=\gamma_S=0$. From these condition, the minimization conditions turn to be

$$2(\beta_S^b+\beta_S^c)s_2^2\sin[2(\phi_1-\phi_2)]=0\,,\quad 2(\beta_S^b+\beta_S^c)s_1^2\sin[2(\phi_1-\phi_2)]=0\,. \tag{4}$$

This analysis is very naive one but we can see that the assumption $\phi_1\simeq\phi_2$ is very good perspective from potential analysis.

For summarizing this paper, we would like to see our future works at a glance. In the present work, we adopt the D_N flavour symmetry in our model in order to focus on only the flavour structure of fermions. It is, of course, very interesting and challenging that to construct the model with $O(2)$ and take into account new gauge bosons. Actually, there may exist gauge anomalies. Furthermore, the quark sector should be included. We have to check that whether the lightest up- and down-type quarks also obtain their masses through CP violation.

We do not mention about the specific generation mechanism of neutrino masses in this work. The neutrinos must have their masses from the experimental facts and its generation mechanism is discussed by much amount of frameworks. We have to decide it and discuss the implement of solving the problem for example the origin of the baryon asymmetry of the universe. These issues will be studied elsewhere.

Acknowledgments

The author thanks the organizers of SCGT12 for giving the chance to present our poster at this fruitful conference.

References

1. T. Araki, H. Ishida and , arXiv:1211.4452 [hep-ph].
2. H. Ishimori, T. Kobayashi, H. Ohki, Y. Shimizu, H. Okada and M. Tanimoto, Prog. Theor. Phys. Suppl. **183** (2010) 1
3. P. F. Harrison, D. H. Perkins and W. G. Scott, Phys. Lett. B **530** (2002) 167
4. KamLAND Collaboration, Phys. Rev. D**83**, 052002 (2011); DOUBLE-CHOOZ Collaboration, Phys. Rev. Lett. **108**, 131801 (2012); DAYA-BAY Collaboration, Phys. Rev. Lett. **108**, 171803 (2012); RENO collaboration, Phys. Rev. Lett. **108**, 191802 (2012).
5. J. Beringer *et al.* (Particle Data Group), Phys. Rev. D**86**, 010001 (2012).
6. C. D. Froggatt and H. B. Nielsen, Nucl. Phys. B**147**, 277 (1079).

One-Family Walking Techni-Pions

Junji Jia *

Kobayashi-Maskawa Institute for the Origin of Particles and the Universe (KMI), Nagoya University, Nagoya 464-8602, Japan.

Shinya Matsuzaki †

Maskawa Institute for Science and Culture, Kyoto Sangyo University, Motoyama, Kamigamo, Kita-Ku, Kyoto 603-8555, Japan.

Koichi Yamawaki

Kobayashi-Maskawa Institute for the Origin of Particles and the Universe (KMI), Nagoya University, Nagoya 464-8602, Japan.

We compute masses of the techni-pions in the one-family walking technicolor model by properly taking into account the characteristic walking features. We find the techni-pion masses are generally on the order of 300 – 500 GeV, to be accessible at the ongoing LHC experiments. The techni-pion cross sections are compared with the currently available LHC data.

1. Introduction

Technicolor (TC) achieves the electroweak (EW) symmetry breaking by the techni-fermion condensation. Of phenomenologically interest is walking TC[1] having the almost non-running gauge coupling, i.e., the approximate scale invariance which provides a Higgs-like composite scalar (techni-dilaton[1,2]) consistent with the 125 GeV LHC boson.[3] The walking gauge theory can be realized in a class of models having a large number of techni-flavors (a la large N_f QCD),[4] such as the one-family model.[5] The enhanced chiral $SU(8)_L \times SU(8)_R$ symmetry in the one-family model predicts 60 pseudo Nambu-Goldstone bosons, techni-pions, as listed in Table 1. Probing those techni-pions at collider experiments would thus be necessary for discovering the walking TC. In this talk we explicitly calculate the techni-pion masses including the essential walking features and evaluate the couplings to the SM particles to explore the techni-pion signatures at the LHC. The details for these computations can be found in the published paper.[6]

---

*J.J. is currently at School of Physical Sciences and Technology, Wuhan University, Wuhan, Hubei, 430072, China.

†S.M. is currently at Department of Physics, Nagoya University, Nagoya 464-8602, Japan.

Table 1. The one-family walking techni-pions, their associated currents constructed from techni-quarks (Q) and -leptons (L) and estimated masses including the enhancement of techni-fermion condensate in the case of walking TC with $\Lambda_{\rm TC} = 10^3\,(10^4)$ TeV. Here λ_a ($a = 1, \cdots, 8$) are the Gell-Mann matrices, σ^i ($i = 1, 2, 3$) the Pauli matrices, and the label c attached on the color-triplet techni-pion field T_c stands for QCD-three colors, $c = r, g, b$.

techni-pion	current	mass [GeV]
θ_a^i	$\frac{1}{2\sqrt{2}}\bar{Q}\gamma_\mu\gamma_5\lambda_a\sigma^i Q$	$449(537)\sqrt{\frac{3}{N_{\rm TC}}}$
θ_a^0	$\frac{1}{2\sqrt{2}}\bar{Q}\gamma_\mu\gamma_5\lambda_a Q$	$449(537)\sqrt{\frac{3}{N_{\rm TC}}}$
$T_c^i\ (\bar{T}_c^i)$	$\frac{1}{2\sqrt{2}}\bar{Q}_c\gamma_\mu\gamma_5\sigma^i L$ (h.c.)	$299(358)\sqrt{\frac{3}{N_{\rm TC}}}$
$T_c^0\ (\bar{T}_c^0)$	$\frac{1}{2\sqrt{2}}\bar{Q}_c\gamma_\mu\gamma_5 L$ (h.c.)	$299(358)\sqrt{\frac{3}{N_{\rm TC}}}$
P^i	$\frac{1}{4\sqrt{3}}(\bar{Q}\gamma_\mu\gamma_5\sigma^i Q - 3\bar{L}\gamma_\mu\gamma_5\sigma^i L)$	502 (ETC)
P^0	$\frac{1}{4\sqrt{3}}(\bar{Q}\gamma_\mu\gamma_5 Q - 3\bar{L}\gamma_\mu\gamma_5 L)$	397 (ETC)

2. Estimate of techni-pion masses

The SM gauge corrections give rise to masses for the charged ($\theta_a^\pm, T_c^\pm, P^\pm$ in Table 1) and QCD-colored ($\theta_a^{i,0}, T_c^{i,0}$) techni-pions through the one-gauge boson exchanges:

$$\Delta m_\pi^2|_{\rm SM-gauge} \sim \frac{\alpha_{\rm EM(s)}}{16\pi F_\pi^2}\int_0^{\Lambda_{\rm TC}^2\to\infty} dQ^2\,Q^2\,D_{\rm EM(s)}(Q^2)\,\Pi_{V-A}(Q^2)\,, \tag{1}$$

up to intrinsic group factors, where F_π denotes the techni-pion decay constant related to the EW scale $v_{\rm EW} \simeq 246$ GeV as $F_\pi = v_{\rm EW}/2 \simeq 123$ GeV; Π_{V-A} the difference of the vector and axial-vector current correlators; $\alpha_{\rm EM(s)}$ are the electromagnetic (EM) and QCD gauge couplings, respectively; the momentum integral is cutoff by a typical ultraviolet (UV) scale of TC, $\Lambda_{\rm TC} = 10^3 - 10^4$ TeV; $D_{\rm EM}(Q^2) = \left(\frac{1}{Q^2} - \frac{1}{Q^2+m_Z^2}\right)$ for the photon and Z boson exchange contributions and $D_s(Q^2) = \frac{1}{Q^2}$ for the QCD gluon exchange contribution. The momentum integral in Eq.(1) can be split into the infrared (IR) region $0 \le Q^2 \le \Lambda_\chi^2$ and the UV region $\Lambda_\chi^2 \le Q^2 \le \Lambda_{\rm TC}^2$, where $\Lambda_\chi \simeq 4\pi F_\pi/\sqrt{N_f}|_{N_f=8}$ above which the operator product expansion is assumed to be applicable to estimate of Π_{V-A}:

$$\Pi_{V-A}(Q^2) \overset{Q^2>\Lambda_\chi^2}{\simeq} \frac{4(N_{\rm TC}^2-1)}{N_{\rm TC}^2}\left(\frac{Q^2}{\mu^2}\right)^{\gamma_m}\frac{\alpha(\mu)\langle\bar{F}F\rangle_\mu^2}{Q^4}\,, \tag{2}$$

in which $\gamma_m \simeq 1$ in the case of walking TC and μ denotes the renormalization scale of the techni-fermion condensate $\langle\bar{F}F\rangle$. The integral kernel then turns out to be dominated by the UV region due to the large anomalous dimension $\gamma_m \simeq 1$ in Eq.(2): the one-gauge boson exchange contribution in the UV region goes like $\int_{\Lambda_\chi^2}^{\Lambda_{\rm TC}^2} dQ^2\,Q^2(1/Q^2)(1/Q^{4-2\gamma_m})|_{\gamma_m\simeq 1} \sim \log\Lambda_{\rm TC}^2/\Lambda_\chi^2$. Accordingly, the techni-pion masses are generically enhanced by the characteristic walking feature, compared to the case of a naive scale-up version of QCD with $\gamma_m \simeq 0$, to be on the order of a few hundred GeV as shown in Table 1. Note, however, the relative sign between the photon and Z boson exchange contributions in $D_{\rm EW}(Q^2)$, which leads to the dramatic cancellation for the enhanced UV term for the masses of techni-pions charged only under the EM gauge, such as $P^{\pm}$.

The masses of $P^{\pm}$ as well as $P^{3,0}$ can get larger corrections from extended TC (ETC)-induced four-fermion interactions, $\mathcal{L}_{4-\text{fermi}}^{\text{ETC}} \sim \Lambda_{\text{ETC}}^{-2}(\overline{Q}Q\overline{L}L - \overline{Q}\gamma_5\sigma^a Q\overline{L}\gamma_5\sigma^a L)$ with Λ_{ETC} being identified as Λ_{TC}. The masses are thus estimated to be

$$\Delta m_{Pi,0}^2|_{\text{ETC}} \sim \frac{\langle 0|(\overline{Q}Q\overline{L}L)_{\Lambda_{\text{ETC}}}|0\rangle}{F_\pi^2\Lambda_{\text{ETC}}^2}, \tag{3}$$

with the intrinsic group factors suppressed. Note that the characteristic walking feature comes in $\langle 0|(\overline{Q}Q\overline{L}L)_{\Lambda_{\text{ETC}}}|0\rangle \sim 9\langle 0|\overline{L}L|0\rangle_{\Lambda_{\text{ETC}}}^2$: using the scaling law of the chiral condensate $\langle \overline{L}L\rangle_{\Lambda_{\text{ETC}}} \simeq (\Lambda_{\text{ETC}}/\Lambda_\chi)^{\gamma_m\simeq 1}\langle \overline{L}L\rangle_{\Lambda_\chi}$ and noting $\langle \overline{L}L\rangle_{\Lambda_\chi} \sim -4\pi F_\pi^3$ up to some overall coefficient determined by certain nonperturbative calculation, we find the ETC-origin mass is enhanced by the walking feature with $\gamma_m \simeq 1$ to be of the order of a few hundred GeV.

The details for the explicit computations of all the techni-pion masses that have been discussed so far can be found in the paper[6] and the estimated size of the mases are summarized in Table 1.

3. Couplings to the SM gauge bosons and fermions

In order to study the techni-pion LHC phenomenologies, we shall next discuss the techni-pion couplings to the SM particles. The techni-pion couplings in the one-family model are read off by writing a nonlinear Lagrangian following the nonlinear realization of the chiral $SU(8)_L \times SU(8)_R$ symmetry with the SM gauge group partially gauged. To be concrete, we shall focus on the isospin singlet and neutral techni-pions, P^0 and θ_a^0

The P^0 and θ_a^0 couplings to two SM gauge bosons arise from the (covariantized) Wess-Zumino-Witten term related to the non-Abelian $SU(8)_L \times SU(8)_R$ anomaly. We then find that the P^0 is almost weak-gaugephobic, i.e., $g_{P^0WW} = 0$ and $g_{P^0ZZ} \simeq 0$, which is due to the cancellation between the techni-quark and -lepton loop contributions. The explicit expressions for other couplings such as those to $gg, g\gamma, gZ$ are given in the paper.[6]

The couplings to the SM fermions can be constructed from four-fermion interactions that is generally induced by ETC gauge boson exchanges and using low energy theorem based on the Ward-Takahashi identities for the axialvector current. The explicit forms of those couplings can be found in the paper.[6]

4. The LHC signatures of one-family walking techni-pions

The main decaying channels of P^0 are thus $P^0 \to \gamma\gamma$; ZZ; γZ; gg; $\overline{f}f$ and for θ_a^0 we have $\theta_a^0 \to gg$; Zg; γg; $\overline{q}q$. The gluon fusion (GF) cross section is enhanced by the extra QCD colored fermions in the one-family model, so the P^0 can be seen at the LHC through the decays to τ and top quark pairs. The left panel of Fig. 1

shows the $\sigma_{\rm GF}(pp \to P^0 \to \tau^+\tau^-)$ for $N_{\rm TC} = 3$ (dashed black) and 4 (dotted black) at the 7 TeV LHC in unit of pb. The red dotted line stands for the current 95% CL upper bound from the ATLAS experiments with 1.06 fb^{-1},[7] which places the lower limit on the P^0 mass to be $2m_t$, consistently with our estimate of the mass (See Table 1). Figure 1 (center and right) also shows the predicted cross sections $\sigma_{\rm GF}(pp \to \theta^a \to gZ/\gamma)$ which can be tested at a high luminosity LHC run in the future.

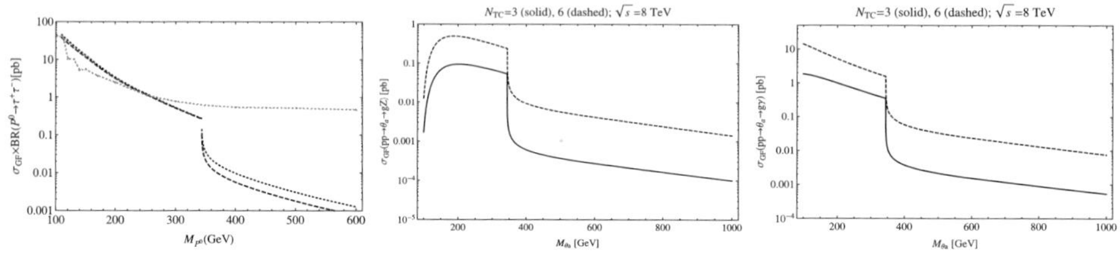

Fig. 1. The predicted cross sections for P^0 (left) and θ^a (center, right).

5. Summary

We have explicitly computed the techni-pion masses in the walking one-family model taking into account essential features of walking TC and found sizable enhancement on the QCD correction to the masses by the walking feature, while no sizable EW corrections due to the dramatic cancellation between photon and Z boson contributions: the main mass corrections are QCD and ETC originated. We further discussed the phenomenological implications to the LHC signatures, focusing on P^0 and θ_a.

Acknowledgments

This work was supported by the JSPS Grant-in-Aid for Scientific Research (S) #22224003, (C) #23540300 (K.Y.) and #23-01781 (J.J.). J. J. was also supported by the JSPS Postdoctoral Fellowships for Foreign Researchers P11781.

References

1. K. Yamawaki, M. Bando and K. Matumoto, Phys. Rev. Lett. **56**, 1335 (1986); M. Bando, T. Morozumi, H. So and K. Yamawaki, Phys. Rev. Lett. **59**, 389 (1987).
2. M. Bando, K. Matumoto and K. Yamawaki, Phys. Lett. B **178**, 308 (1986).
3. S. Matsuzaki and K. Yamawaki Phys. Lett. B **719**, 378 (2013); Phys. Rev. D **86**, 115004 (2012).
4. W. E. Caswell, Phys. Rev. Lett. **33**, 244 (1974); T. Banks and A. Zaks Nucl. Phys. B **196**, 189 (1982).
5. E. Farhi and L. Susskind, Phys. Rept. **74**, 277 (1981).
6. J. Jia, S. Matsuzaki, and K. Yamawaki Phys. Rev. D **87**, 016006 (2013).
7. The ATLAS Collaboration, ATLAS-CONF-2011-132.

Non-Abelian Dual Superconductivity in SU(3) Yang-Mills Theory due to Non-Abelian Magnetic Monopoles

Kei-Ichi Kondo

Department of Physics, Graduate School of Science, Chiba University, Chiba 263-8522, Japan
**E-mail: kondok@faculty.chiba-u.jp*

Akihiro Shibata

Computing Research Center, High Energy Accelerator Research Organization (KEK), and Graduate Univ. for Advanced Studies (Sokendai), Tsukuba 305-0801, Japan
E-mail: akihiro.shibata@kek.jp

Seikou Kato

Fukui National College of Technology, Sabae 916-8507, Japan
E-mail: skato@fukui-nct.ac.jp

Toru Shinohara

Department of Physics, Graduate School of Science, Chiba University, Chiba 263-8522, Japan
E-mail: sinohara@graduate.chiba-u.jp

We give numerical evidences for the non-Abelian dual superconductivity due to non-Abelian magnetic monopoles in SU(3) Yang-Mills theory as a mechanism for quark confinement, based on our new formulation of lattice gauge theory.

Keywords: Quark confinement; dual superconductivity; non-Abelian magnetic monopole.

1. Introduction

The dual superconductivity is the promising mechanism for quark confinement. In the first paper, we have proposed the non-Abelian dual superconductivity picture in the SU(3) Yang-Mills (YM) theory, and already presented numerical evidences for the restricted field-field dominance and the non-Abelian magnetic monopole dominance in the string tension, by applying our new formulation of YM theory to a lattice. See Fig. 1.

In this presentation, we give further numerical evidences for confirming the non-Abelian dual superconductivity for $SU(3)$ YM theory based on the second paper:

- dual Meissner effect: the measured chromoelectric flux tube between a quark and antiquark pair strongly supports the non-Abelian dual Meissner effect due to non-Abelian magnetic monopoles.

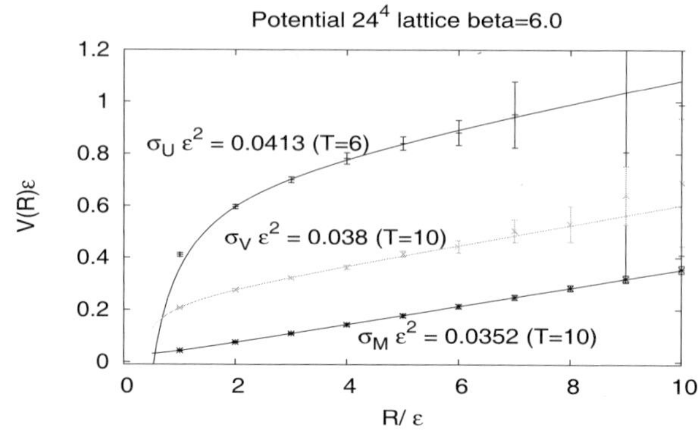

Fig. 1. $SU(3)$ quark-antiquark potentials as functions of the quark-antiquark distance R: (from above to below) (i) full potential $V_f(R)$ (red curve), (ii) restricted part $V_r(R)$ (green curve) and (iii) magnetic–monopole part $V_m(R)$ (blue curve), measured at $\beta = 6.0$ on 24^4 using 500 configurations where ϵ is the lattice spacing.

- gauge-invariant magnetic-monopole current: the nonvanishing gauge-invariant magnetic-monopole current exists around the chromoelectric field.
- type of dual superconductivity: the type of the dual superconductor is the type I in SU(3) Yang-Mills, rather than the border between the type I and II, in marked contrast to the SU(2) case.

2. Summary of the presentation

For this purpose, we have used our new formulation of $SU(3)$ YM theory on a lattice to extract the restricted field from the original $SU(3)$ YM field, which has played a dominant role in confinement of quarks in the fundamental representation, i.e., the restricted-field dominance and the non-Abelian magnetic monopole dominance in the string tension, as shown in Fig. 1.

In order to confirm the dual Meissner effect, we have measured the chromofield strength connecting a quark and an antiquark for both the original YM field and the restricted field. See Fig. 2. We have observed the dual Meissner effect in $SU(3)$ YM theory, i.e., only the chromoelectric field exists and the magnetic-monopole current is induced around the flux connecting a quark and an antiquark. See Fig. 3 and Fig. 4.

We have checked that the nonvanishing gauge-invariant magnetic-monopole current is induced around the chromoelectric field, which supports the dual picture of the ordinary superconductor exhibiting the electric current around the magnetic field. See Fig. 5.

Moreover, we have determined the type of non-Abelian dual superconductivity according to the Ginzburg-Landau (GL) theory, i.e., type I for $SU(3)$ YM theory

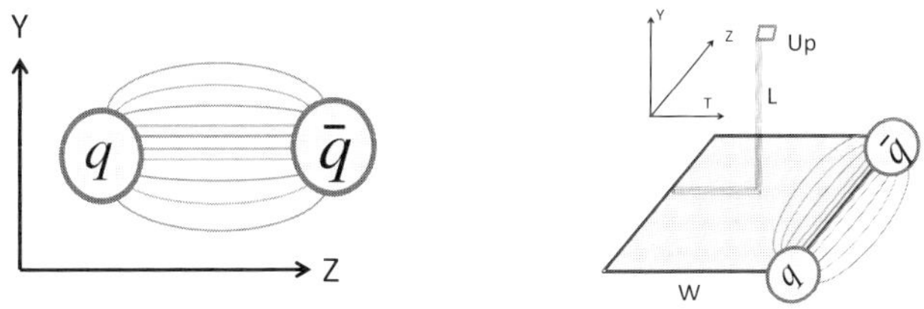

Fig. 2. (Left) The setup of measuring the chromo-flux produced by a quark–antiquark pair. (Right) The gauge-invariant connected correlator $(U_p L W L^\dagger)$ between a plaquette U and the Wilson loop W.

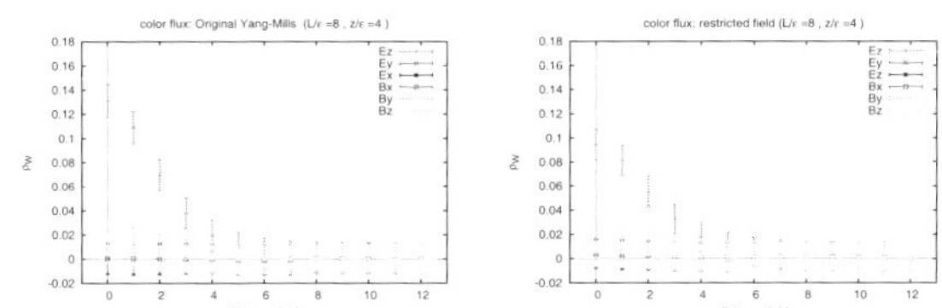

Fig. 3. Measurement of components of the chromoelectric field $\boldsymbol{E}$ and chromomagnetic field $\boldsymbol{B}$ as functions of the distance y from the z axis. (Left panel) the original $SU(3)$ YM field, (Right panel) the restricted field.

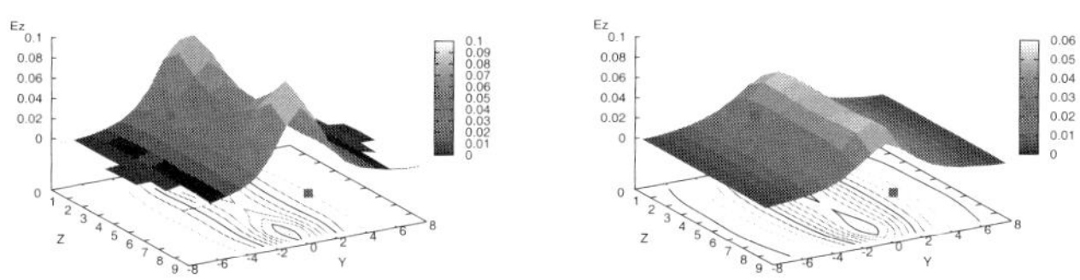

Fig. 4. The distribution in Y-Z plane of the chromoelectric field E_z connecting a pair of quark and antiquark: (Left panel) chromoelectric field produced from the original YM field, (Right panel) chromoelectric field produced from the restricted field.

with the GL parameter:

$$\kappa = 0.45 \pm 0.01. \tag{1}$$

which should be compared with the border of type I and II for the SU(2) YM theory. See Fig. 6. These features are reproduced from the restricted part, i.e., type I with

$$\kappa = 0.48 \pm 0.02. \tag{2}$$

These results confirm the non-Abelian dual superconductivity picture for quark confinement. In order to confirm the non-Abelian dual superconductivity, it will be interesting to investigate if the interactions between two chromoelectric flux tubes are attractive (expected for the type I) or repulsive (expected for the type II). The interactions should reflect the internal non-Abelian nature which depends on the

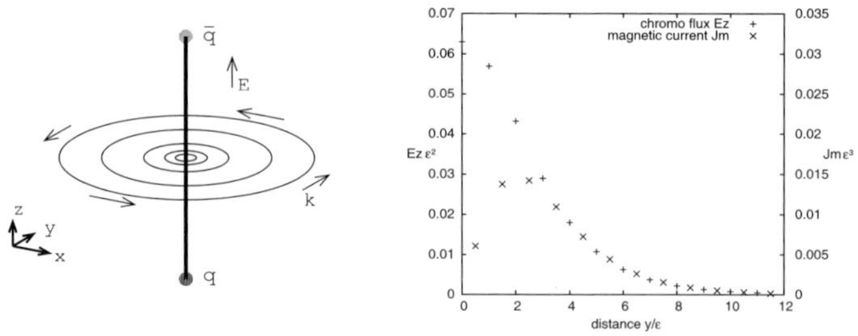

Fig. 5. The magnetic-monopole current $\mathbf{k}$ induced around the flux along the z axis connecting a quark-antiquark pair. (Left panel) The positional relationship between the chromoelectric field E_z and the magnetic current $\mathbf{k}$. (Right panel) The magnitude of the chromo-electronic current E_z and the magnetic current $J_m = |\mathbf{k}|$ as functions of the distance y from the z axis.

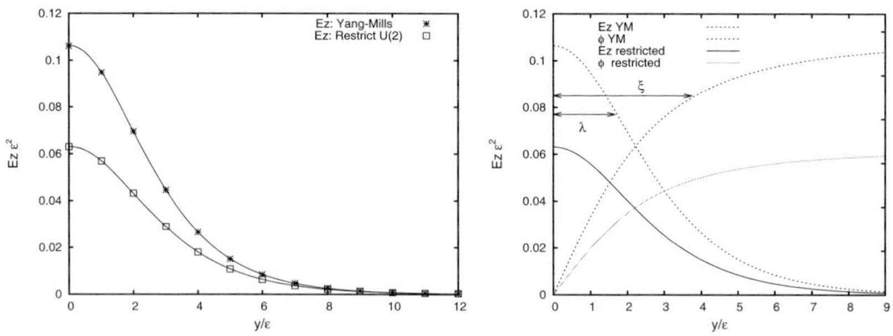

Fig. 6. (Left panel) The plot of the chromoelectric field E_z versus the distance y in units of the lattice spacing ϵ and the fitting as a function $E_z(y)$ of y. The red cross for the original $SU(3)$ field and the green square symbol for the restricted field. (Right panel) The order parameter ϕ reproduced as a function $\phi(y)$ of y, together with the chromoelectric field $E_z(y)$. Here λ is the penetration depth and ξ is the coherence length obtained according to the Ginzburg-Landau theory.

distance between two tubes. It will be also interesting to extend our result to the finite temperature case to see the fate of the dual superconductivity, and to see when and how the dual superconductivity disappears above the critical temperature.

Acknowledgement

This work is supported by Grant-in-Aid for Scientific Research (C) 24540252 from the Japan Society for the Promotion Science (JSPS), and also in part by JSPS Grant-in-Aid for Scientific Research (S) 22224003. The numerical calculations are supported by the Large Scale Simulation Program No.T11-15 (FY2011) and No.12-13 (FY2012) of High Energy Accelerator Research Organization (KEK).

References

1. K.-I. Kondo, A. Shibata, T. Shinohara, and S. Kato, arXiv:1007.2696 [hep-th], Phys. Rev. D**83**, 114016 (2011).
2. A. Shibata, K.-I. Kondo, S. Kato and T. Shinohara, arXiv:1212.6512 [hep-lat], Phys. Rev. D**87**, 054011 (2013).

Uniqueness and Significance of Weak Solution of Non-Perturbative Renormalization Group Equation to Analyze Dynamical Chiral Symmetry Breaking

Ken-Ichi Aoki, Shin-Ichiro Kumamoto and Daisuke Sato

Institute for Theoretical Physics, Kanazawa Unicersity, Kanazawa, Ishikawa 920-1192, Japan

E-mail: aoki, kumamoto, satodai@hep.s.kanazawa-u.ac.jp

We propose quite a new method of analyzing the dynamical chiral symmetry breaking in gauge theories. Starting with the non-perturbative renormalization group equation for the Wilsonian fermion potential, we define the weak solution* of it in order to mathematically authorize solutions with singularity. The weak solution is obtained uniquely and it successfully predicts the physically correct vacuum, chiral condensates, dynamical mass, through its auto-convexizing power for the effective potential. Thus it works perfectly even for the first order phase transition in the finite density QCD.

1. Introduction

Among various methods to analyze the dynamical chiral symmetry breaking, the non-perturbative renormalization group is quite effective since it may include non-ladder diagrams which cure the gauge invariance problem.[6] In the lowest order approximation, the Wilsonian effective action is expressed by a scale dependent fermion potential $V_{\mathrm{W}}(x,t)$, where $x = \int d^4x\bar{\psi}(x)\psi(x)$ is an operator variable, $t = \log(\Lambda_0/\Lambda(t))$ is the renormalization scale. This potential keeps all information of the fermionic interactions, the free energy V_0, the dynamical mass M_d, the 4-fermion interactions G *etc* as follows:

$$V_{\mathrm{W}}(x,t) = V_0(t) + M_d(t)x + G(t)\frac{x^2}{2} + \cdots . \tag{1}$$

The renormalization group equation for G (in case $M = 0$) is written as

$$\frac{dG}{dt} = -2G + \frac{1}{2\pi^2}G^2, \tag{2}$$

which makes G diverge at a finite scale t_c for upper critical initial value. This blowup nature itself is a correct behavior since it expresses the divergence of the susceptibility due to the spontaneous chiral symmetry breakdown.[1,2] However, due to this divergence we can not go beyond t_c, and there is no way to calculate infrared physical quantities such as the chiral condensate or the dynamical mass.

*The authors greatly appreciate helpful comments by Prof. Akitaka Matsumura who told us how to construct the weak solution.

2. Partial differential equation and its weak solution

Various methods have been used to bypass this singularity, e.g., the bare mass[4], auxiliary fields,[1–3] *etc.* Here we propose a new direct method of solving the renormalization group equation as a partial differential equation (PDE):[7]

$$\frac{\partial M(x,t)}{\partial t}+\frac{\partial f(M(x,t),t)}{\partial x}=0,\ M=\frac{\partial V_{\rm W}}{\partial x},\ f(M,t)=-\frac{1}{4\pi^2}e^{-4t}\ln[1+M^2e^{2t}],\quad (3)$$

where $M(x,t)$ is called the mass function. The 4-fermion interaction G corresponds to $\frac{\partial M}{\partial x}$ at the origin, and its divergence at finite $t_{\rm c}$ means that the above PDE has no global classical solution beyond $t_{\rm c}$.

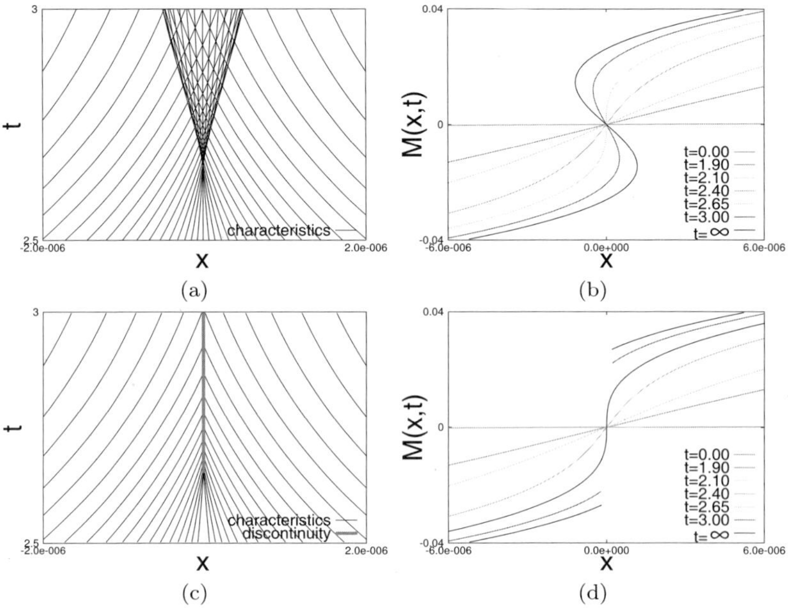

Fig. 1. NJL plot. (a) Characteristics. (b) Mass function. (c) Discontinuity. (d) Weak solution.

The expected behavior of physically meaningful solution $M(x,t)$ is plotted in Fig. 1(d), where M has a jump at the origin after $t_{\rm c}$. In order to accommodate such singular solution with discontinuity, we write down the weak version of the PDE,[5,7]

$$\int_0^{\infty}dt\int_{-\infty}^{\infty}dx\left[M\frac{\partial\varphi}{\partial t}+f(M,t)\frac{\partial\varphi}{\partial x}\right]+\int_{-\infty}^{\infty}dx\ M(x,0)\varphi(x,0)=0.\quad (4)$$

The weak solution is defined as to satisfy the above equation for any smooth and bounded test function $\varphi(x,t)$. The weak solution satisfy original PDE except for discontinuity and position of discontinuity $x=S(t)$ is controlled by the Rankine-Hugoniot (RH) condition,

$$\frac{dS(t)}{dt}[M_+-M_-]=f(M_+)-f(M_-),\quad (5)$$

where M_+, M_- are right and left limit at the discontinuity point respectively.

To obtain the weak solution, first we set up characteristic curves $x = X(t)$ representing the contour lines of M, which satisfies

$$\frac{dX(t)}{dt} = \frac{\partial f(M,t)}{\partial M}. \tag{6}$$

Where these curves are crossing each other, we must pick up one curve and introduce discontinuity according to the RH condition, and then we get the unique function $M(x,t)$. Fig. 1 shows the Nambu-Jona-Lassinio (NJL) model example ($g = 1.005g_c$) of these procedures. The contour lines are plotted in (a), where at some finite t_c they start crossing with each other. The cross section of the total contours is seen in (b), where the derivative $\frac{\partial M}{\partial x}$ diverges at the origin at some t_c.

Note that in obtaining $M(x,t)$ as in (b) there is nothing singular, and it is just a motion of 'string'. However, to get the renormalized potential we have to define $M(x,t)$ as a unique function of x. Then the RH condition determines the discontinuity as shown in (c), and the mass function in (d). This is the weak solution of the PDE and it defines global t solution uniquely.

3. Weak solution results for the physical quantities

We show results in the finite density NJL where the first order phase transition occurs. Contours and mass function are plotted in Fig. 2, where at the central region five-fold structure appears corresponding to the three-fold local minima. In the renormalization procedure, two discontinuities appear pairwisely, move towards the origin, and finally merge into one at the origin as shown in Fig. 2(a).

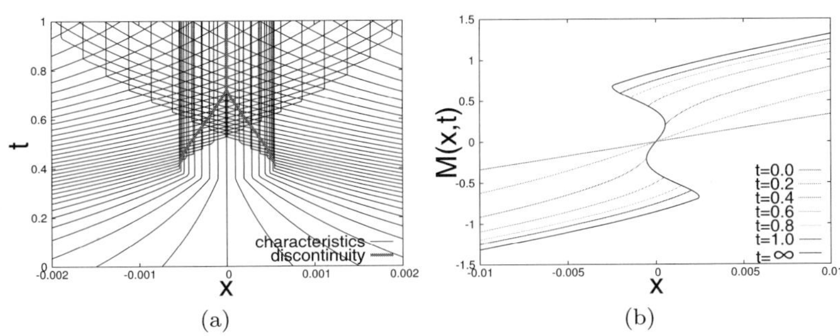

(a) (b)

Fig. 2. NJL $g = 1.7g_c$, $\mu = 0.7$. (a) Characteristics and jumps. (b) Evolution of mass function.

Snapshots in the course of renormalization are shown in Fig. 3, where the mass function $M(t,x)$, the Wilsonian fermion potential V_W and the Legendre effective potential for σ are plotted. It is astonishing that our method of weak solution uniquely determines their singularity structures and the resultant Legendre effective potential is always convexized. This means the dynamical mass and the chiral condensates are uniquely calculated, and perfectly correct in the sense that even in case there are multi local minima, the lowest free energy minimum is always chosen automatically. This feature is quite a new finding and shows powerfulness of the purely fermionic non-perturbative renormalization group and its weak solution.[7] This analysis has been applied to QCD, even with finite density or non-ladder, and proved to work perfectly to give physical quantities without any ambiguity.[6]

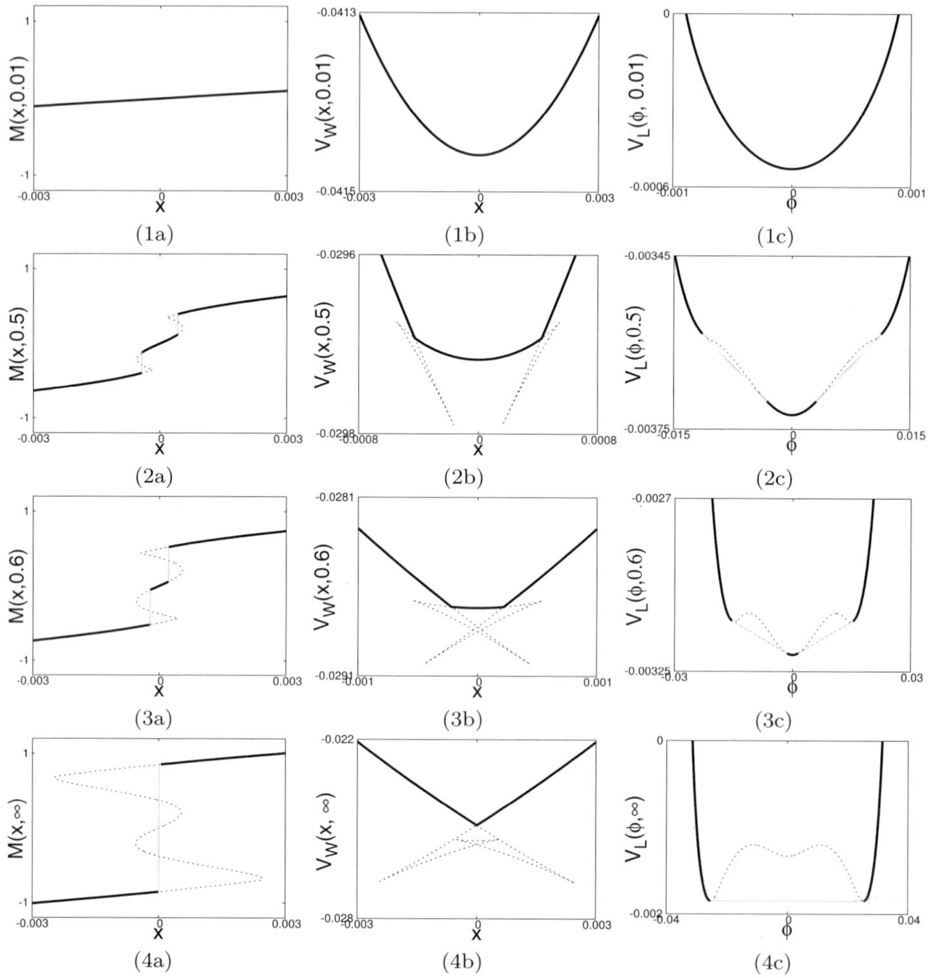

Fig. 3. Evolution of physical quantities by weak solution (NJL $g = 1.7g_c$, $\mu = 0.7$, $t = 0.01$, 0.5, 0.6, ∞). (a) Mass function. (b) Wilsonian fermion potential. (c) Legendre effective potential.

References

1. K-I. Aoki, *Proc. SCGT96*, 171 (1996):hep-ph/9706264, *Prog. Theor. Phys. Suppl.* **131**, 129 (1998), *Int. J. Mod. Phys. B* **14**, 1249 (2000).
2. K-I. Aoki, K. Morikawa, J.-I. Sumi, H. Terao and M. Tomoyose, *Prog. Theor. Phys.* **102**, 1151 (1999), *Phys. Rev. D* **61**, 045008 (2000).
3. H. Gies and C. Wetterich *Phys. Rev. D* **65**, 065001 (2001).
4. K-I. Aoki and K. Miyashita *Prog. Theor. Phys.* **121**, 875 (2009).
5. L. C. Evans, *Partial Differential Equations*, 2nd ed. (AMS, 2010).
6. K-I. Aoki and D. Sato *Prog. Theor. Exp. Phys.* **2013**, 043B04 (2013).
7. K-I. Aoki and S.-I. Kumamoto and D. Sato in preparation.

GUT Model Discrimination by Nucleon Decay via Dim. 6 Effective Int.

Nobuhiuro Maekawa[1,2,a] and Yu Muramatsu[1,b]

[1] *Department of Physics, Nagoya University, Nagoya 464-8602, Japan*
[2] *Kobayashi Maskawa Institute, Nagoya University, Nagoya 464-8602, Japan*
[a] *E-mail:maekawa@eken.phys.nagoya-u.ac.jp*
[b] *E-mail:mura@eken.phys.nagoya-u.ac.jp*

Anomalous $U(1)_A$ SUSY GUT models are attractive because they can solve most of difficulties in SUSY GUT models in a natural manner. One of the most important predictions of anomalous $U(1)_A$ SUSY GUT is that the nucleon decay via dim. 6 interactions are enhanced and the rough estimation gives $\tau(p \to \pi^0 + e^+) \sim 10^{34}$ years which is nothing but the present experimental lower bound, while nucleon decay via dim. 5 interactions are suppressed. We calculate dim. 6 effective int. and nucleon lifetimes for each decay mode in $SU(5), SO(10)$ or E_6 GUT models with various unitary matrices which diagonalize Yukawa matrices. In this calculation we use more than 50000 model points. One important observation is that the gauge group can be determined by measuring the branching ratios of nucleon. The detailed results will be given in our paper.[1]

Keywords: nucleon decay; SUSY GUT

1. Introduction

SUSY GUTs are attractive because theoretically, it can unify not only three gauge interactions into a single gauge interaction but also quarks and leptons into fewer multiplets. Moreover, experimentally, not only measured values of three gauge couplings in the SM can be explained quantitatively in the supersymmetric (SUSY) GUTs, but also the various hierarchies of masses and mixings of quarks and leptons can be understood qualitatively by the unification of one generation quarks and leptons into **10** and $\bar{\mathbf{5}}$ of $SU(5)$ if it is assumed that the **10** matters induce stronger hierarchies for Yukawa couplings than the $\bar{\mathbf{5}}$ matters. Unfortunately, several issues are known in SUSY GUT scenario. The most difficult problem is the doublet-triplet splitting problem. Another problem is the unrealistic Yukawa relations which are caused by the unification of matters.

These issues can be solved by introducing the anomalous $U(1)_A$ gauge symmetry under the natural assumption that all the interactions which are allowed by the symmetry of the theory are introduced with $O(1)$ coefficients. Note that all nonrenormalizable interactions are introduced with $O(1)$ coefficients in the anomalous $U(1)_A$ GUT scenario.[2] One important prediction in the scenario is the enhancement of nucleon decay via dimension 6 operators and the suppression of nucleon decay via dimension 5 operators. Actually the unification scale Λ_u can be given by

$$\Lambda_u \sim \lambda^{-a}\Lambda, \tag{1}$$

where $\lambda < 1$ is the ratio of the Fayet-Illiopoulos parameter to the cutoff scale Λ which is taken as the usual unification scale $\Lambda_G \sim 2 \times 10^{16}$ GeV. (In the anomalous $U(1)_A$ GUT scenario, the success of the gauge coupling unification in the minimal SUSY $SU(5)$ GUT can be naturally explained by taking $\Lambda = \Lambda_G$.[3]) Here a is the anomalous $U(1)_A$ charge of the adjoint Higgs and must be negative, and in the eq. (1) an $O(1)$ coefficient is neglected. Therefore, the unification scale becomes smaller than the usual SUSY GUT scale Λ_G, which results in the enhancement of the nucleon decay via dimension 6 interactions. The prediction is strongly dependent on the parameters λ, a, and the $O(1)$ coefficient, because the lifetime of nucleon is proportional to the unification scale to the fourth. The typical unification scale becomes 4.4×10^{15} GeV if $\lambda = 0.22$, $a = -1$, and $O(1)$ coefficient is taken to be 1. Then the prediction of the nucleon lifetime becomes around the present lower bound $\tau_P > 10^{34}$ years, which is obtained by the SuperKamiokande experiment.[4]

In this work, we calculate the partial decay widths for various decay modes of nucleon via dimension 6 interactions in the anomalous $U(1)_A$ GUT models. In the calculation of the hadron matrix elements, we use the Lattice results.[5] Since the nucleon lifetime is strongly dependent on the parameters as $\tau_P \propto \Lambda_u^4$, we focus mainly on the various ratios of partial decay widths.

2. Model ambiguities

For simplicity, we fix the unification scale $\Lambda_u = 5 \times 10^{15}$ GeV in our calculation. Since the partial decay widths are also dependent on the explicit GUT models, we will briefly mension which GUT models we consider.

(1) Firstly, we adopt $SU(5)$, $SO(10)$, and E_6 as the unified gauge gruop in our calculation. GUT Models with larger unified gauge group have more superheavy gauge multiplets which induce nucleon decay. Especially, the breaking scale of $SO(10)$ Higgs which makes the rank of the group decreasing becomes smaller than the unification scale in the anomalous $U(1)_A$ GUTs. Therefore, the effects from the $SO(10)$ superheavy gauge multiplets is expected to be comparatively large. That is one of the interesting predictions in the anomalous $U(1)_A$ SUSY GUT.
(2) Secondly, not only three **16** of $SO(10)$ but also several **10** of $SO(10)$ are introduced as matter sector in $SO(10)$ models (In E_6 models, the fundamental representation **27** includes **10** of $SO(10)$ automatically.) because the realistic quark and lepton masses and mixings can be naturally obtained.[2]
(3) Finally, we take account of the uncertainties for the Yukawa couplings by considering the various unitary mattices which diagonalize the Yukawa matrices. We just assume that the diagonalizing matrices are classified by the largeness of the

angles. We use Cabbibo-Kobayashi-Maskawa type matrix and Maki-Nakagawa-Sakata type matrix as the diagonalizing matrices. In this paper we generate 10 matrices for each diagonalizing matrix in order to consider the ambiguities of diagonalizing matrices in calculating the nucleon lifetimes. As the result, we calculate the nucleon partial decay widths for more than 50000 model points.

3. The results on nucleon decay widths

Here we just note what we understand from the detail calculation of the nucleon decay widths by taking more than 50000 model points. See our paper[1] for the detail results.

(1) First, if the unification scale is taken to be 5×10^{15} GeV, it is not surprising that the proton decay ($P\rightarrow e\pi$) will be observed in near future. The most model points predict shorter lifetime of the proton than the experimental lower bound. Of course, as noted in introduction, the absolute value of the nucleon lifetime is strongly dependent on the unification scale. For example, if $O(1)$ factor for the scale is 2, then the lifetime becomes $2^4=16$ times larger. But anyway, for experiments this must be a good news.
(2) Second, the predicted value for the decay width of $N(P)\rightarrow\nu\pi$ has less dependence on ambiguities of parameters. The point is that the decay width is obtained by the sum of the three decay widths for three neutrinos. As the result, the diagonalizing matrix for the neutrinos which is expected to have large mixings has only tiny effect because of the cancellation due to the unitarity of the matrix.
(3) Finally, the ratio $\Gamma(N\rightarrow\nu+\pi)/\Gamma(P\rightarrow e+\pi)$ is important to fix the unified gauge group especially in the anomalous $U(1)_A$ SUSY GUT. This is because this ratio becomes larger for larger rank unified gauge group. We will explain this result in more detail in the next section.

4. Model discrimination

The essential point for the importance of the ratio $\Gamma(N\rightarrow\nu+\pi)/\Gamma(P\rightarrow e+\pi)$ is as follows[a]. In $SU(5)$ GUT, the gauge bosons $X(\mathbf{\overline{3}},\mathbf{2})_{\frac{5}{6}}+h.c.$ induce the nucleon decay. If the unified gauge group is $SO(10)$, additional gauge bosons $X'(\mathbf{3},\mathbf{2})_{\frac{1}{6}}+h.c.$ which belong to $\mathbf{10}+\mathbf{\overline{10}}$ of $SU(5)$ also can induce the nucleon decay. In E_6 GUT, gauge bosons $X''(\mathbf{3},\mathbf{2})_{\frac{1}{6}}+h.c.$ also does it. An important point is that the effective interactions induced by the X bosons are included in the effective interactions $\mathbf{10}^\dagger\mathbf{10}\mathbf{10}^\dagger\mathbf{10}$ and $\mathbf{10}^\dagger\mathbf{10}\mathbf{\overline{5}}^\dagger\mathbf{\overline{5}}$, while the X' and X'' induce only $\mathbf{10}^\dagger\mathbf{10}\mathbf{\overline{5}}^\dagger\mathbf{\overline{5}}$ interactions. Here $\mathbf{10}=(q(\mathbf{3},\mathbf{2})_{\frac{1}{6}},u_R^c(\mathbf{\overline{3}},\mathbf{1})_{-\frac{2}{3}},e_R^c(\mathbf{1},\mathbf{1})_{1})$ and $\mathbf{\overline{5}}=(d_R^c(\mathbf{\overline{3}},\mathbf{1})_{\frac{1}{3}},l(\mathbf{1},\mathbf{2})_{-\frac{1}{2}})$ are matter fields. Since only $\mathbf{\overline{5}}$ includes neutrino, the ratio $\Gamma(N\rightarrow\nu+\pi)/\Gamma(P\rightarrow e+\pi)$ is

[a]The importance of the ratio has been discussed in the literature in $SO(10)$ GUT[6]

expected to be small in $SU(5)$ GUT. In $SO(10)$ or E_6 GUT models, if the contributions from X' or X'' are sizable, the ratio becomes larger. Of course, when $m_{X'}, m_{X''} >> m_X$, the contributions from the additional gauge bosons X' and X'' can be neglected, and as the result, the ratio becomes almost the same as in $SU(5)$ GUT models. However, it is interesting that in the anomalous $U(1)_A$ SUSY GUT with the Dimopoulos-Wilczek mechanism[7] to solve the doublet-triplet splitting problem, the relation $m_{X'} \sim m_X$ is required, and therefore, the ratio $\Gamma(N \to \nu + \pi)/\Gamma(P \to e + \pi)$ becomes larger than in $SU(5)$ GUT. The requirement is caused by making the solution for the doublet-triplet splitting problem natural.[2] In our calculation, the ratio becomes 0.19-0.34 in $SU(5)$, but 0.26-1.6 in $SO(10)$ and 0.36-0.95 in E_6. It is interesting that the ratio can be O(1), and therefore, if the proton decay ($P \to e + \pi$) is seen, the neutron decay ($N \to \nu + \pi$) also may be seen.

References

1. N. Maekawa and Y. Muramatsu, in preparation.
2. N. Maekawa, Prog. Theor. Phys. **106**, 401 (2001); M. Bando and N. Maekawa, Prog. Theor. Phys. **106**, 1255 (2001); N. Maekawa and T. Yamashita, Prog. Theor. Phys. **107**, 1201 (2002).
3. N. Maekawa, Prog. Theor. Phys. **107**, 597 (2002); N. Maekawa and T. Yamashita, Phys. Rev. Lett. **90**, 121801 (2003).
4. H. Nishino *et al.* [Super-Kamiokande Collaboration], Phys. Rev. D**85**, 112001 (2012); K. Abe *et al.*, arXiv:1109.3262 [hep-ex].
5. Y. Aoki, C. Dawson, J. Noaki and A. Soni, Phys. Rev. D **75**, 014507 (2007)
6. See for example, P. Langacker, Phys. Rept. **72**, 185 (1981).
7. S. Dimopoulos and F. Wilczek, NSF-ITP-82-07; M. Srednicki, Nucl. Phys. B**202**, 327(1982).

Revisiting the Calculation of the Neutron Electric Dipole Moment with the QCD Sum Rule Method*

Natsumi Nagata

Department of Physics, Nagoya University, Nagoya 464-8602, Japan
Department of Physics, University of Tokyo, Tokyo 113-0033, Japan
E-mail: natsumi@eken.phys.nagoya-u.ac.jp

We study the neutron electric dipole moment in the presence of the CP-violating operators up to the dimension five in terms of the QCD sum rules. By using input parameters obtained from the lattice simulation, we derive a conservative limit for the contributions of the CP violating operators.

Keywords: Neutron Electric Dipole Moment; QCD Sum Rules

1. Introduction

The neutron electric dipole moment (nEDM) is an interesting quantity since it provides a sensitive probe to the CP-violating sources both within and beyond the Standard Model (SM). In terms of the relativistic Lagrangian form, the interaction is written as

$$\mathcal{L}_{\rm nEDM} = -\frac{i}{2} d_n \overline{N} F_{\mu\nu} \sigma^{\mu\nu} \gamma_5 N \ , \tag{1}$$

where N and $F_{\mu\nu}$ denote the neutron field and the electromagnetic field strength tensor, respectively, and $\sigma^{\mu\nu}$ is defined by $\sigma^{\mu\nu} \equiv \frac{i}{2}[\gamma^\mu, \gamma^\nu]$. In the non-relativistic limit, it describes the interaction between the electric field and the spin of neutron. It is found that this interaction violates both parity and time-reversal invariance, which indicates the CP-violation as well. In the SM, CP-violation is induced through the Cabbibo-Kobayashi-Maskawa (CKM) matrix. Its contribution to the nEDM is, however, found to be quite suppressed since the CP-violating effect is accompanied by the flavor-changing interaction, while the nEDM is flavor conserving interaction. Recent calculations tell that the SM contribution to the nEDM lies around $d_n \sim 10^{-(31-32)}\ e\cdot$cm,[2,3] which is still below the experimental limit $|d_n| < 2.9 \times 10^{-26}\ e\cdot$cm[4] by several orders of magnitude. Thus, the nEDM provides a clean, background-free probe of the CP-violating interactions in physics beyond the SM.

*This article is based on the work with Junji Hisano, Jeong Yong Lee (Nagoya University), and Yasuhiro Shimizu (Tohoku University), which is published as Ref. 1.

From a theoretical point of view, one needs to interpret the experimental limits on the nEDM as constraints on the CP violation in the high-energy theory. To that end, we want to derive precise relations between the nEDM and the CP-violating parameters which include the CP violating effects in the more fundamental theory. In this work, we use the method of the QCD sum rules[5] in order to derive such relations.[a] Further, we extract some QCD parameters from the results obtained with the lattice simulations. As a result, we derive a conservative limit for the contributions of the CP-violating operators compared with previous calculations.[7,8]

2. Calculation

The CP-violating interactions induced in the high-energy theory affect the nEDM through the following effective Lagrangian:

$$\mathcal{L}_{\not{C}\not{P}} = -\sum_{q=u,d,s} m_q \bar{q} i\theta_q \gamma_5 q + \theta_G \frac{\alpha_s}{8\pi} G^A_{\mu\nu}\tilde{G}^{A\mu\nu} - \frac{i}{2}\sum_{q=u,d,s} d_q \bar{q} F_{\mu\nu}\sigma^{\mu\nu}\gamma_5 q - \frac{i}{2}\sum_{q=u,d,s} \tilde{d}_q \bar{q} g_s G^A_{\mu\nu} T^A \sigma^{\mu\nu}\gamma_5 q \,. \quad (2)$$

Here we write down up to dimension five operators at the QCD scale; m_q represents the quark masses; $G^A_{\mu\nu}$ is gluon field strength tensor; g_s is the strong coupling constant with $\alpha_s = g_s^2/4\pi$; $\tilde{G}^A_{\mu\nu} \equiv \frac{1}{2}\epsilon_{\mu\nu\rho\sigma}G^{A\rho\sigma}$ with $\epsilon^{0123} = +1$; T^A denotes the generators in the $SU(3)_C$ algebra. The second, third and forth terms in Eq. (2) are called the effective QCD θ term, the electric and chromoelectric dipole moments (CEDMs) for quarks, respectively. The coefficients of the CP-violating operators, θ_q, θ_G, d_q, and $\tilde{d}_q$, are all assumed to be quite small, and we keep only the terms up to the first order of these parameters. θ_q and θ_G are moved into each other via the chiral rotation, thus, the sum of them, $\bar{\theta} \equiv \sum_{q=u,d,s}\theta_q + \theta_G$ is physical parameter.

As mentioned to in the Introduction, we use the method of QCD sum rules[5] in order to relate the CP-violating parameters with the nEDM. In this method, the correlation functions of hadron fields are evaluated in terms of the operator product expansion (OPE). In the calculation, the short-distance contribution is evaluated in terms of the perturbative QCD and included into the Wilson coefficients of the OPE, while the long-distance contribution is taken into account as the vacuum expectation values (VEVs) of the quark and gluon operators. Then, by using the dispersion relations, the result is related with the sum of the contributions of the hadronic states. In the procedure, the Borel transformation is applied in order to suppress the contribution of excited states and/or continuum spectrum as well as to eliminate the subtraction terms.

At this point, we need to introduce the neutron field. We denote it as $\eta_n(x)$ hereafter. The strength of its coupling with the one particle state of neutron is

[a] An alternative approach is to exploit the chiral perturbation theory. A latest calculation based on it is given in Ref. 6. See also K. Fuyuto's article in this proceedings.

parametrized by

$$\langle\Omega|\eta_n(x)|N(\boldsymbol{p},s)\rangle = \lambda_n e^{\frac{i}{2}\alpha_n\gamma_5}\, u_n(\boldsymbol{p},s)\, e^{-ip\cdot x}\ , \tag{3}$$

where $|\Omega\rangle$ and $|N(\boldsymbol{p},s)\rangle$ indicate the vacuum and the one-particle neutron state, respectively, and $u_n(\boldsymbol{p},s)$ is an ordinary spinor wave function. The low-energy constant λ_n is to be determined later. Our goal is to extract the nEDM from the two-point function of $\eta_n(x)$ evaluated on an electromagnetic field background with the CP violating sources. The phase factor in Eq. (3), however, causes mixture between electric and magnetic dipole moment structures and makes it difficult to pick out only the EDM from the QCD sum rules. To avoid the problem we select a term independent of the phase α_n, *i.e.*, chiral invariant. Such terms are found to have the Lorentz structure of $\{\tilde{F}_{\mu\nu}\sigma^{\mu\nu}, \not{q}\}$, thus we focus on the terms proportional to this structure in the following calculation.

To carry out the OPE calculation, we express $\eta_n(x)$ by the quark fields such that the field has the same quantum numbers as neutron. The most general interpolator for neutron on the ordinary CP-even background is parametrized as $\eta_n(x) = j_1(x) + \beta j_2(x)$ with $j_1(x) = 2\epsilon_{abc}\left(d_a^T(x)C\gamma_5 u_b(x)\right)d_c(x)$ and $j_2(x) = 2\epsilon_{abc}\left(d_a^T(x)Cu_b(x)\right)\gamma_5 d_c(x)$, where the subscripts, a,b,c, denote the color indices and C is the charge conjugation matrix. We have found that for our calculation, $\beta=+1$ is an appropriate choice since this choice eliminates the sub-leading terms and simultaneously excludes the contamination of the CP-odd currents.

Now we just show the result for the OPE calculation of the correlator:

$$\Pi(q)^{(\mathrm{OPE})} = \frac{1}{16\pi^2}\langle\overline{q}q\rangle \log\left(\frac{-q^2}{\Lambda^2}\right)\{\tilde{F}\cdot\sigma, \not{q}\}\ \Theta\ , \tag{4}$$

with

$$\Theta \equiv (4e_d m_d\rho_d - e_u m_u\rho_u)\chi\overline{\theta} + (4d_d - d_u) + (\kappa - \frac{1}{2}\xi)(4e_d\tilde{d}_d - e_u\tilde{d}_u)\ , \tag{5}$$

and

$$\begin{aligned}\rho_u &= \frac{m_*}{m_u}\left[1 + \frac{m_0^2}{2\overline{\theta}}\left\{\frac{\tilde{d}_u - \tilde{d}_d}{m_d} + \frac{\tilde{d}_u - \tilde{d}_s}{m_s}\right\}\right],\\ \rho_d &= \frac{m_*}{m_d}\left[1 + \frac{m_0^2}{2\overline{\theta}}\left\{\frac{\tilde{d}_d - \tilde{d}_u}{m_u} + \frac{\tilde{d}_d - \tilde{d}_s}{m_s}\right\}\right],\end{aligned} \tag{6}$$

where $m_* \equiv m_u m_d m_s/(m_u m_d + m_d m_s + m_u m_s)$; $\langle\overline{q}q\rangle$ is the quark condensate; Λ is an arbitrary constant with mass dimension one; e_q denote the electric charges of quarks; $\langle\overline{q}\sigma_{\mu\nu}q\rangle_F = e_q\chi F_{\mu\nu}\langle\overline{q}q\rangle$, $g_s\langle\overline{q}G^A_{\mu\nu}T^Aq\rangle_F = e_q\kappa F_{\mu\nu}\langle\overline{q}q\rangle$, and $2g_s\langle\overline{q}\gamma_5\tilde{G}^A_{\mu\nu}T^Aq\rangle_F = ie_q\xi F_{\mu\nu}\langle\overline{q}q\rangle$;[9] $\langle\overline{q}g_s(G\cdot\sigma)q\rangle = -m_0^2\langle\overline{q}q\rangle$.[10] On the other hand, the correlator can be also evaluated in terms of the phenomenological model which reflects the hadron contribution. By equating these two correlators after the Borel transformation, we have

$$\lambda_n^2 d_n m_n = -\Theta\langle\overline{q}q\rangle\frac{m_n^4}{32\pi^2}\exp(2)\ . \tag{7}$$

Now all we have to do is to determine λ_n. In our work, we extract its value from the lattice simulations:[11]

$$\lambda_n = -0.0436 \pm 0.0047_{(\mathrm{stat})} \pm 0.0084_{(\mathrm{syst})} \ \mathrm{GeV}^3 \ , \tag{8}$$

In the previous work, on the other hand, this value is also evaluated by using the QCD sum rules. By comparing these values, we can find that the lattice value is several times larger than that evaluated based on the QCD sum rules. Therefore, if you use the lattice value for λ_n, you can expect that the resultant nEDM value is smaller than those in the previous literature.

By substituting the value in Eq. (8) to the sum rule in Eq. (7), we finally obtain the nEDM d_n as

$$d_n = 2.0 \times 10^{-1} \ \Theta \ , \tag{9}$$

with $\langle \bar{q}q \rangle \simeq -(262 \ \mathrm{MeV})^3$ and numerically, it follows that

$$d_n = 8.2 \times 10^{-17} \bar{\theta} \ [e \ \mathrm{cm}] + 0.79 d_d - 0.20 d_u + e(-0.30 \tilde{d}_u + 0.30 \tilde{d}_d - 0.01 \tilde{d}_s) \ . \tag{10}$$

The estimate of the error in this calculation is discussed in Ref. 1. It is found that this result is about 50% smaller than the previous results,[7,8] and thus with using the result one would obtain a conservative limit for the contributions of the CP-violating operators in physics beyond the SM.

3. Conclusion

We have evaluated the nEDM based on the method of the QCD sum rules. By using the input parameters obtained from the lattice simulation, we have derived a conservative limit for the contributions of the CP-violating operators.

Acknowledgments

The author would like to Junji Hisano, Jeong Yong Lee, and Yasuhiro Shimizu for collaboration. This work is supported by Research Fellowships of the Japan Society for the Promotion of Science for Young Scientists.

References

1. J. Hisano, J. Y. Lee, N. Nagata and Y. Shimizu, *Phys.Rev.* **D85**, p. 114044 (2012).
2. T. Mannel and N. Uraltsev, *Phys.Rev.* **D85**, p. 096002 (2012).
3. I. Khriplovich and A. Zhitnitsky, *Phys.Lett.* **B109**, p. 490 (1982).
4. C. Baker, D. Doyle, P. Geltenbort, K. Green, M. van der Grinten *et al.*, *Phys.Rev.Lett.* **97**, p. 131801 (2006).
5. M. A. Shifman, A. Vainshtein and V. I. Zakharov, *Nucl.Phys.* **B147**, 385 (1979).
6. K. Fuyuto, J. Hisano and N. Nagata, *Phys.Rev.* **D87**, p. 054018 (2013).
7. M. Pospelov and A. Ritz, *Phys.Rev.Lett.* **83**, 2526 (1999).
8. M. Pospelov and A. Ritz, *Phys.Rev.* **D63**, p. 073015 (2001).
9. B. Ioffe and A. V. Smilga, *Nucl.Phys.* **B232**, p. 109 (1984).
10. V. Belyaev and B. Ioffe, *Sov.Phys.JETP* **56**, 493 (1982).
11. Y. Aoki *et al.*, *Phys.Rev.* **D78**, p. 054505 (2008).

Localization and ABJ Wilson Loop Calculation

Keita Nii* and Shinji Hirano

Graduate School of Science, Nagoya University,
Nagoya, 464-8602, Japan
**E-mail: nii@th.phys.nagoya-u.ac.jp*

Masaki Shigemori

Kobayashi Maskawa Institute,
Nagoya, 464-8602, Japan

We study the supersymmetric Wilson loop in a 3-dimensional $\mathcal{N} = 6$ supersymmetric $U(N_1)_k \times U(N_2)_{-k}$ Chern-Simons Matter gauge theory known as ABJ theory. Using the Lens space matrix model, we obtain the exact results for the supersymmetric Wilson loops in ABJ theory. The final result is given in the form of a $\mathsf{min}(N_1, N_2)$-dimensional integral. This result contains not only perturbative effects but also non-perturbative ones. We show that these results correctly reproduces the perturabative expression for the supersymmetric Wilson loop.

Keywords: ABJ theory, Wilson loop, Localization

1. Introduction

In recent years a localization technique is often used to extract the exact results from the supersymmetric gauge theories.[1] This technique enables us to calculate the supersymmetric quantities (for example the partition function and the supersymmetric Wilson loop) even if the gauge coupling is strong or the rank of gauge group is finite. Then it is worthwhile for the studies of AdS/CFT correspondence or non-perturbative effect. Basically the result by a localization takes the form of a finite dimensional integral, more precisely a matrix eigenvalue integral. In the case of ABJ theory[2] which is 3d $\mathcal{N} = 6$ supersymmetric $U(N_1)_k \times U(N_2)_{-k}$ Chern-Simons Matter gauge theory and known as the dual description of M-theory on $AdS_4 \times S^7/Z_k$ with a discrete torsion, the 1/6 BPS Wilson loop can be calculated by using a localization technique as follows.[3]

$$\hat{W}_{\mathsf{ABJ}}(N_1, N_2) \equiv \frac{(2\pi)^{\frac{1}{2}N_1(N_1+1)+\frac{1}{2}N_2(N_2+1)}}{G_2(N_1+1)G_2(N_2+1)} W_{\mathsf{ABJ}}(N_1, N_2) \tag{1}$$

$$W^i_{\mathsf{ABJ}} \equiv \mathcal{N}_{\mathsf{ABJ}} \int \prod_{j=1}^{N_1} \frac{d\mu_j}{2\pi} \prod_{a=1}^{N_2} \frac{d\nu_a}{2\pi} \frac{\prod_{j<k}\left(2\sinh\frac{\mu_j-\mu_k}{2}\right)^2 \prod_{a<b}\left(2\sinh\frac{\nu_a-\nu_b}{2}\right)^2}{\prod_{j,a}\left(2\cosh\frac{\mu_j-\nu_a}{2}\right)^2}$$
$$\times \exp\left[-\frac{1}{2g_s}\left(\sum_j \mu_j^2 - \sum_a \nu_a^2\right) + p\mu_i\right] \tag{2}$$

where $g_s = \frac{2\pi i}{k}$ is a Chern-Simons gauge coupling and

$$\frac{(2\pi)^{\frac{1}{2}N_1(N_1+1)+\frac{1}{2}N_2(N_2+1)}}{G_2(N_1+1)G_2(N_2+1)}, \quad \mathcal{N}_{\mathsf{ABJ}} = \frac{i^{-\frac{\kappa}{2}(N_1^2-N_2^2)}}{N_1!N_2!} \tag{3}$$

are the normalization factors. The function $G_2(N_1+1)$ is a Barnes G-function.

$$G_2(n+2) = \prod_{j=1}^{n} j!, \quad (n: \text{ integer}) \tag{4}$$

The inserted factor $p\mu_i$ is due to the Wilson loop. p is the winding number. In the $p = 0$ limit this reproduces the partition function. Hirano et al.[4] further calculated the integral of the partition function using the $U(N_1)\times U(N_2)$ Lens space matrix model which becomes ABJ matrix model in a certain limit.[5] Using the same technique we calculate the above integral further.

The organization of this paper is following. In section 2, we give the $U(N_1) \times U(N_2)$ Lens space matrix model and calculate that integral exactly. In section 3, we explain how to obtain the original ABJ results from the Lens space matrix model and show the results. In section 4, we summarize our results and comment about the future direction.

2. Lens space matrix model

First we show the $U(N_1) \times U(N_2)$ Lens space matrix model for the Wilson loop.

$$\hat{W}^i_{\mathsf{Lens}}(N_1, N_2) = \frac{(2\pi)^{\frac{1}{2}N_1(N_1+1)+\frac{1}{2}N_2(N_2+1)}}{G_2(N_1+1)G_2(N_2+1)} W_i(N_1, N_2) \tag{5}$$

$$W_i(N_1, N_2) = \mathcal{N}_{\mathsf{lens}} \int \prod_{j=1}^{N_1} \frac{d\mu_j}{2\pi} \prod_{a=1}^{N_2} \frac{d\nu_a}{2\pi} \prod_{j<k}\left(2\sinh\frac{\mu_j-\mu_k}{2}\right)^2 \prod_{a<b}\left(2\sinh\frac{\nu_a-\nu_b}{2}\right)^2$$
$$\times \prod_{j,a}\left(2\cosh\frac{\mu_j-\nu_a}{2}\right)^2 \exp\left[-\frac{1}{2g_s}\left(\sum_j \mu_j^2 + \sum_a \nu_a^2\right) + p\mu_i\right] \tag{6}$$

This differs from the original ABJ Wilson loop integral in that the factor $\prod_{j,a}\left(2\cosh\frac{\mu_j-\nu_a}{2}\right)^2$ appears in the numerator. This simplifies the calculation because this integrals are purely finite number of Gaussian integrals. Then we can exactly integrate this integral and we obtain

$$\hat{W}^i_{\mathsf{Lens}}(N_1,N_2)=\mathcal{N}_{\mathsf{lens}}\frac{(2\pi)^{\frac{1}{2}N_1(N_1+1)+\frac{1}{2}N_2(N_2+1)}}{G_2(N_1+1)G_2(N_2+1)}(N_1-1)!\,N_2!\left(\frac{g_s}{2\pi}\right)^{\frac{N}{2}}$$
$$\times q^{-\frac{1}{6}N(N^2-1)-\frac{1}{2}p^2+p}(1-q)^{\frac{1}{2}N(N-1)}G_2(N+1,q)S(N_1,N_2) \quad (7)$$

$$S(N_1,N_2)=\frac{1}{N_1!}\sum_{s_1,s_2,\cdots,s_{N_1}\geq 0}\sum_{k=1}^{N_1}q^{-ps_k+pN-2p}(-1)^{\frac{1}{2}N_1(N_1-1)+\sum_{j=1}^{N_1}s_j}$$
$$\left(\prod_{j\neq k}\frac{1-q^{s_j-s_k-p}}{1-q^{s_j-s_k}}\right)\left[\prod_{j=1}^{N_1}\frac{\left(-q^{j+p\delta_{kj}}\right)_{-N}}{(q^j)_{-N}}\frac{\left(-q^{j-p\delta_{k(N_1-j+1)}}\right)_{N_2}}{(q^j)_{N_2}}\right]\prod_{j=1}^{N_1-1}\frac{1}{(q)_j^{\,2}}$$
$$\prod_{1\leq j<l\leq N_1}\left(-q^{l-j-p\delta_{kj}}\right)_1\left(-q^{l-j+p\delta_{kl}}\right)_1\prod_{j=1}^{N_1}\frac{\left(q^{s_j+1}\right)_{-N}}{\left(-q^{s_j+1+p\delta_{kj}}\right)_{-N}}$$
$$\prod_{1\leq j<l\leq N_1}\frac{\left(q^{s_l-s_j}\right)_1^{\,2}}{\left(-q^{s_l-s_j+p\delta_{kl}}\right)_1\left(-q^{s_l-s_j-p\delta_{kj}}\right)_1} \quad (8)$$

where $q=\mathrm{e}^{-\frac{2\pi i}{k}}$ and we use the q-Pochhammer symbol:

$$(a)_n\equiv\prod_{j=0}^{n-1}(1-aq^k),\qquad (a)_{-n}=\frac{1}{(aq^{-n})_n} \quad (9)$$

The original ABJ Wilson loop can be obtained by analytically continuing the N_2 into $-N_2$ in the Lens space result.[5]

$$\hat{W}^i_{\mathsf{ABJ}}(N_1,N_2')=\lim_{\epsilon\to 0}\hat{W}^i_{\mathsf{Lens}}(N_1,-N_2'+\epsilon) \quad (10)$$

3. Results

We show the analytically continued result for $N_1<N_2'$.

$$\hat{W}^i_{\mathsf{ABJ}}(N_1,N_2')=i^{-\frac{\kappa}{2}(N_1^2+N_2'^2)}\frac{(2\pi)^{\frac{N_1^2+N_2'^2}{2}}}{N_1N_1!}g_s^{\frac{N_1+N_2'}{2}}q^{-\frac{1}{2}p^2-p}(-1)^{N_1(N_2'-1)}$$
$$\times\frac{\prod_{j=1}^{N_2'-N_1-1}(q)_j\prod_{j=1}^{N_1-1}(q)_j^{\,-2}}{G_2(N_1+1)G_2(N_2'+1)}\sum_{s_1,\cdots,s_{N_1}\geq 0}\sum_{k=1}^{N_1}q^{-ps_k+pN_1-pN_2'}(-1)^{\sum_{j=1}^{N_1}s_j}$$
$$\times\prod_{j\neq k}\frac{\left(q^{s_j-s_k-p}\right)_1}{\left(q^{s_j-s_k}\right)_1}\prod_{j=1}^{N_1}\frac{\left(q^{s_j+1}\right)_{N_2'-N_1}}{\left(-q^{s_j+1+p\delta_{kj}}\right)_{N_2'-N_1}}$$
$$\times\prod_{1\leq j<l\leq N_1}\frac{\left(q^{s_l-s_j}\right)_1^{\,2}\left(-q^{l-j-p\delta_{kj}}\right)_1\left(-q^{l-j+p\delta_{kl}}\right)_1}{\left(-q^{s_l-s_j+p\delta_{kl}}\right)_1\left(-q^{s_l-s_j-p\delta_{kj}}\right)_1}$$
$$\times\prod_{j=1}^{N_1}\frac{\left(q^{j-N_2'}\right)_{N_2'-j}(q)_{j-1}\left(-q^{j+p\delta_{kj}}\right)_{N_2'-N_1}}{\left(-q^{j-p\delta_{k(N_1-j+1)}-N_2'}\right)_{N_2'}(q^j)_{N_2'-N_1}} \quad (11)$$

This result contain the summation $\sum_{s=0}^{\infty}$, so we need some regularization. Non-perturbatively we can regularize the summation by changing the $\sum_{s\geq 0}\cdots\to$

$\frac{-1}{2\pi i}\int_{-i\infty}^{i\infty}\frac{\pi ds}{\sin\pi s}\cdots$. The non-perturbative effect appears as the residue. The perturbative result is obtained as follows.

$$\begin{aligned}\hat{W}_i(1,-2) = &-\frac{\pi^{5/2}i^{-1/2}g^{5/2}}{2\sqrt{2}} - \frac{p^2\pi^{5/2}g^{7/2}}{4\sqrt{2}}i^{-1/2} - \frac{i^{-1/2}g^{9/2}\pi^{5/2}}{96} \\ &\times\left(\sqrt{2}-12\sqrt{2}p^2+3\sqrt{2}p^4\right) \\ &-\frac{\pi^{5/2}}{192}\left(7\sqrt{2}p^2-12\sqrt{2}p^4+\sqrt{2}p^6\right)i^{-1/2}g^{11/2}+O(g^{13/2}) \quad (12)\end{aligned}$$

where we use the zeta function regularization for the above summation and only show the result of the $U(1)_k \times U(2)_{-k}$ gauge group.

$$\mathsf{Li}_n(z) = \sum_{k=1}^{\infty}\frac{z^k}{k^s}, \quad \sum_{s=1}^{\infty}(-1)^s s^n = \mathsf{Li}_{-n}(-1) \quad (13)$$

4. Conclusions and discussions

We exactly calculated the supersymmetric Wilson loop in ABJ theory i.e. the $\mathcal{N}=6$ supersymmetric $U(N_1)_k \times U(N_2)_{-k}$ Chern-Simons Matter theory by using the $U(N_1)\times U(N_2)$ lens space matrix model. The result which we have obtained is useful to do the non-perturbative analysis. The original integral is a (N_1+N_2)-dimensional integral. On the other hand the one which we have obtained is a $\mathsf{min}(N_1, N_2)$ integral. We have also checked that our result from the lens space matrix model reproduces the correct perturbative result.

This result is a starting point where we do the non-perturbative analysis. For instance, we should check the Seiberg duality for the supersymmetric Wilson loop and construct the mapping rule for the supersymmetric Wilson loop. We should also apply our method to more general Chern-Simons Matter theory with fewer supersymmetries.

5. Acknowledgments

KN would like to thank Tadakatsu Sakai, Yuki Sato, Seishi Enomoto, Igari Takafumi, Norihito Shirai, Nozawa Shinsaku, Takayama Kenichi, Beye Florian, Yumi Iwasaki, Sho Hitakatsu, Masahiro Nakamatsu, Shota Kumekawa and Wakayama Koji for comments and fruitful discussions.

References

1. A. Kapustin, B. Willett and I. Yaakov, JHEP 1003, 089 (2010) [arXiv:0909.4559 [hep-th]], N. Hama, K. Hosomichi and S. Lee, JHEP 1105, 014 (2011).
2. O. Aharony, O. Bergman and D. L. Jafferis, JHEP 0811, 043 (2008) [arXiv:0807.4924 [hep-th]]
3. N. Drukker, J. Plefka and D. Young, JHEP 0811, 019 (2008) [arXiv:0809.2787 [hep-th]]
4. Awata Hidetoshi, Hirano Shinji and Shigemori Masaki, [arXiv:1212.2966[hep-th]]
5. M. Marino, P. Putrov, JHEP 1006, 011 (2010). [arXiv:0912.3074 [hep-th]]

Analysis of Spontaneous Mass Generation by Iterative Method in the Nambu-Jona-Lasinio Model and Gauge Theories

Ken-Ichi Aoki, Shinnosuke Onai and Daisuke Sato

Institute for Theoretical Physics, Kanazawa University,
Kanazawa 920-1192, Japan
E-mail: aoki, onai, satodai@hep.s.kanazawa-u.ac.jp

We propose a new iterative method to directly calculate the spontaneous mass generation due to the dynamical chiral symmetry breaking. We can conclude the physical mass definitely without recourse to any other consideration like the free energy comparison.

1. Introduction

Owing to its outstanding feature, dynamical chiral symmetry breaking phenomena have been widely studied in various fields of elementary particle physics. The standard method to discuss the spontaneous mass generation is to formulate a coupled system of self-consistent equations and find its non-trivial solution. However, those equations are no more than the *necessary* condition and it is needed to examine solutions to select correct one by using another mean e.g. by referring to the free energy of each solution. Even if it is done, there are still unclear points whether the minimal free energy solution ensures the physically correct answer.

We adopt the Nambu-Jona-Lasinio (NJL) model[1] and give a new iteration method that directly sums up an infinite number of diagrams of the standard perturbation theory in the ladder approximation. Using this method, we demonstrate that the physically meaningful result is automatically obtained with the correct critical coupling constant. The NJL model has four-fermion interactions among the massless fermions with the chiral invariance. Here we added the bare mass m_0 to the Lagrangian to make the standard perturbation theory work well,

$$\mathcal{L}_{\mathrm{NJL}} = \bar{\psi}\partial\!\!\!/\psi + \frac{2\pi^2 g}{N}\left[\left(\bar{\psi}\psi\right)^2 + \left(\bar{\psi}i\gamma_5\psi\right)^2\right] - m_0\bar{\psi}\psi, \tag{1}$$

where N is the number of fermion flavors. We consider $1/N$ leading contribution to the mass. Diagrammatically it is a sum of infinite diagrams called 'tree', where considering the fermion-antifermion pair as a single meson, the tree diagrams are defined by those without any meson loops, or in other words we regard a series of loops as a *fat* propagator.Usual method is to set up an equation satisfied by this infinite sum of diagrams. Above the critical coupling constant, there actually exists a non-trivial solution which does not vanish after zero bare mass limit. However, it is unclear that how the finite mass should come out of the infinite sum of the diagrams while each finite diagram must vanish at the zero bare mass limit. In this article we set up a method to directly sum up the infinite diagrams and show how the finite mass come out without any ambiguity.[3]

2. Iteration Method

First of all we classify diagrams in the *tree* using the node length of each diagram. Node length of a diagram is defined by the maximum number of loops in a continuous route towards the edge loop, or maximum number of nodes of *fat* propagator legs in the diagram. Fig. 1(a) shows the counting rule of node length and classification of diagrams. Here we define $M^{(n)}$ is a sum of diagrams whose node length is no greater than n.

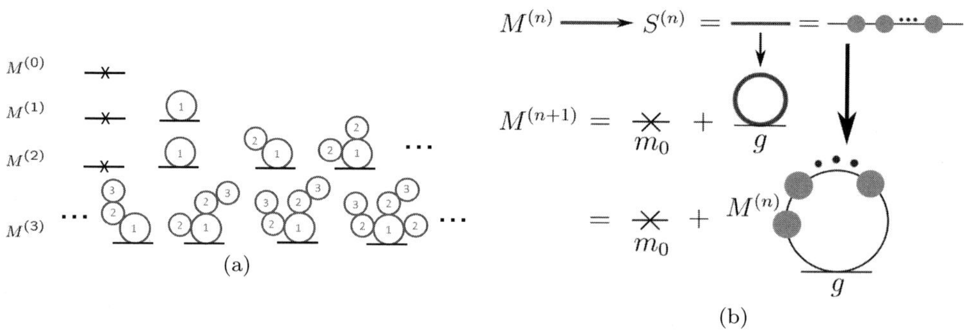

Fig. 1. Node length. (a) Definition. (b) Node length iteration.

Then we write down the iteration to calculate $M^{(n+1)}$ by using $M^{(n)}$ as shown in Fig. 1(b). The transformation function is a one loop integral and we denote it by F,

$$M^{(n+1)} = F\left(M^{(n)}\right), \quad F(M) = m_0 + gM\left(1 - M^2 \log\left(1 + M^{-2}\right)\right). \tag{2}$$

Therefore the total sum of the tree diagrams is obtained by $M^{(\infty)}$, infinitely many times of transformation of the same F.

3. Mass Generation

Iterative transformation here is best understood by a graphical method where the transformation function $y = F(x)$ and a straight line of $y = x$ are drawn as shown in Fig. 2. Each iteration process can be drawn on this figure by a successive move of points. In any case the iterative transformation finally reaches a stable fixed point. Fixed points are crossing points between $y = F(x)$ and $y = x$, and position of fixed points are shown in Fig. 3.

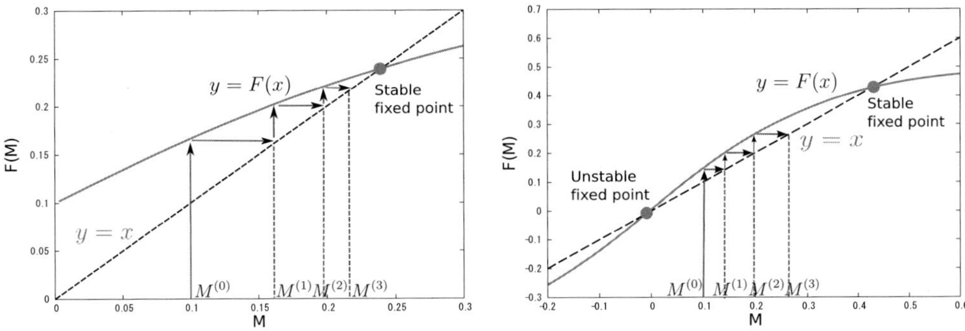

(a) (b)

Fig. 2. Iterative steps. (a) Weak coupling : g=0.7. (b) Strong coupling : g=1.5.

In the weak coupling region, there is only one fixed point near the origin which is stable, and iteration from any starting point must reach it, as shown in Fig. 2(a). When the coupling constant becomes strong, there appear pair creation of fixed points, one is stable and the other is unstable. Then there are two stable fixed points each of which has its attractive region, 'territory'. We must be careful about the initial starting point of iteration, that is, $M^{(0)}$, which must be the bare mass m_0 by definition. In Fig. 3 we set a positive value for the bare mass, then the initial point is in the territory of the right-hand side stable fixed point, as seen in Fig. 2(b). Therefore for all region of the coupling constant, the physical result is controlled by the right most stable fixed point in Fig. 3. The critical coupling constant for the change of fixed point structure depends on the bare mass and it becomes unity in the vanishing bare mass limit.

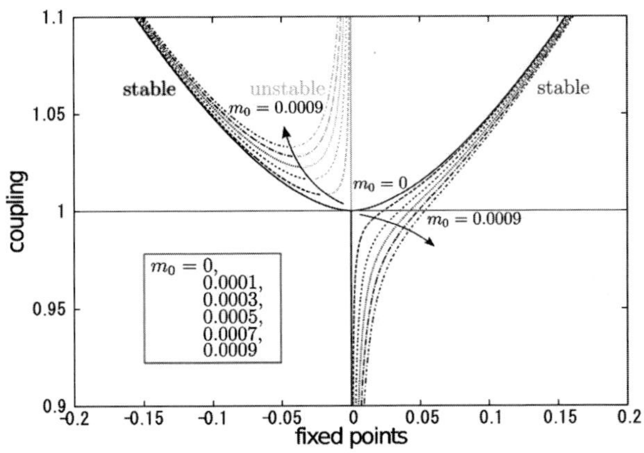

Fig. 3. Fixed point structure with the coupling constant and the bare mass.

Let's see some features of mass generation with respect to the node length n in Fig. 4. In the weak coupling case (a), the dynamical mass is generated rather quickly at low n and becomes constant, which should be called the perturbative characteristics. By decreasing the bare mass the final mass goes to zero. In the strong coupling case (b), the generation of the dynamical mass depends strongly on the bare mass, and it is mainly generated at some narrow range of node length. Decreasing the bare mass, the region of mass generating node length becomes large, but the output mass is almost constant, which means the spontaneous mass generation. It is also seen that the shape of generation curves look the same form, just displacement in the node length space. These features are readily verified by the iterative nature of our calculation well seen in Fig. 2(b).

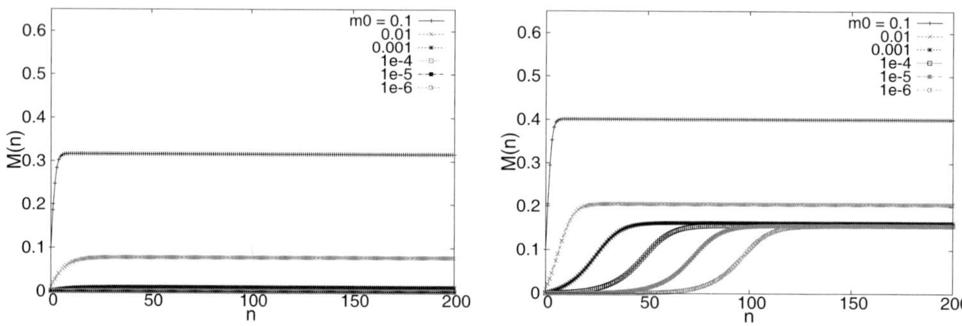

(a) (b)

Fig. 4. Mass generation by the node length iteration. (a) Weak coupling : g=0.9. (b) Strong coupling : g=1.1.

Finally we mention about gauge theories. We add all planar diagrams which are equivalent to the ladder Schwinger-Dyson equation.[2] To set up iterative method, we define ladder depth of each planar diagram which is the maximum number of gluon propagators towards the fermion propagator. We define mass function $\Sigma^{(n)}$ which contains all planar diagrams whose ladder depth is no greater than n. Then we write down the iteration transformation as follows.

$$\Sigma^{(n+1)}(x) = \mathscr{F}\left[\Sigma^{(n)}(x)\right], \tag{3}$$

$$\mathscr{F}\left[\Sigma(x)\right] = m_0 + \frac{\lambda(x)}{4x}\int_0^x \frac{y\Sigma(y)dy}{y+\Sigma^2(y)} + \int_x^{\Lambda^2} \frac{\lambda(y)\Sigma(y)dy}{4(y+\Sigma^2(y))}, \tag{4}$$

where the mass function $\Sigma(x)$ is a function of momentum squared x. The functional $\mathscr{F}$ is now an infinite dimensional map and there are infinite number of fixed point functions. Our analysis clarified that only one of fixed points is perfectly stable and reached by proper initial function $\Sigma(x) = m_0$.[3]

References

1. Y. Nambu and G. Jona-Lasinio, *Phys. Rev.* **122**, 345 (1961); **124**, 246 (1961).
2. K-I. Aoki, M. Bando, T. Kugo, M.G. Mitchard and H. Nakatani, *Prog. Theor. Phys.* **84**, 683 (1990).
3. K-I. Aoki, S. Onai and D. Sato in preparation.

Running Coupling in SU(2) with Adjoint Fermions

Jarno Rantaharju*

Riken Advanced Institute for Computational Science,
7-1-26, Minatojima-minami-machi, Chuo-ku, Kobe, Hyogo, Japan

Kari Rummukainen†

Department of Physics and Helsinki Institute of Physics,
P.O.Box 64, FI-00014 University of Helsinki, Finland

Kimmo Tuominen‡

Department of Physics, P.O.Box 35 (YFL),
FI-40014 University of Jyväskylä, Finland,
and
Helsinki Institute of Physics, P.O. Box 64,
FI-00014 University of Helsinki, Finland.

We present a measurement of the Schrödinger Functional running coupling in SU(2) lattice gauge theory with adjoint fermions. We use HEX smearing and clover improvement to reduce the discretization effects. We obtain a robust continuum limit for the step scaling, which confirms the existence of a non-trivial fixed point.

Keywords: Lattice field theory, Conformal field theory

1. The Model

We study the running of the Schrödinger Functional coupling in the SU(2) lattice field theory with 2 fermions in the adjoint representation. The model, dubbed Minimal Walking Technicolor, has been studied recently as a possible candidate for a walking technicolor theory and as a part of the ongoing mapping of the conformal window on the lattice.[1–5] We define the lattice model by the action

$$S = S_G + S_F, \tag{1}$$

where S_F is the clover improved Wilson fermion action with smeared gauge links and S_G is a partially smeared version of the Wilson plaquette action. We use hypercubic stout smearing, or HEX smearing,[6] to reduce the discretization errors and allow

*jarno.rantaharju@riken.jp
†kari.rummukainen@helsinki.fi
‡kimmo.i.tuominen@jyu.fi

simulations at larger coupling. The gauge action is defined as

$$S_G = \beta_L \sum_{x;\mu<\nu} (1-c_g)\mathcal{L}_{x,\mu}(U) + c_g\mathcal{L}_{x,\mu}(V) \tag{2}$$

$$\mathcal{L}_{x,\mu}(U) = \left(1 - \frac{1}{2}\mathrm{Tr}\left[U_\mu(x)U_\nu(x+a\hat{\mu})U_\mu^\dagger(x+a\hat{\nu})U_\nu^\dagger(x)\right]\right),$$

where V is the smeared gauge field and we choose $c_g = 0.5$.

The fermion action is given by

$$S_F = a^4 \sum_x \left[\overline{\psi}(x)(iD_W + m_0)\psi(x) + ac_{\mathrm{sw}}\overline{\psi}(x)\frac{i}{4}\sigma_{\mu\nu}F_{\mu\nu}(x)\psi(x)\right], \tag{3}$$

We expect the smearing to bring the discretization errors close to the tree-level values and choose $c_{\mathrm{sw}} = 1$. We have performed a few short measurements of the clover coefficient and find them consistent with the tree-level value, even at small values of β.

The smeared links are calculated in three sequential stout smearing steps, summing over the directions that are orthogonal to those in the previous steps.[6] We have chosen the smearing coefficients for each step to be $\alpha_1 = 0.78$, $\alpha_2 = 0.61$ and $\alpha_1 = 0.35$.

We measure the running coupling using the Shcödinger functional method.[7,8] We use boundary conditions to the temporal direction to induce a chromoelectric background field and measure the coupling by the response to a change in the background field. The boundary conditions are

$$U_\mu(\overline{x}, t=0) = e^{-i\eta\sigma_3 a/L}, \quad U_\mu(\overline{x}, t=L) = e^{-i(\pi-\eta)\sigma_3 a/L} \tag{4}$$

with σ_3 the third Pauli matrix and we choose $\eta = \pi/4$. The spatial boundary conditions are periodic for the gauge field. The fermion field is set to zero at the temporal boundaries and have twisted periodic boundary conditions to the spatial directions: $\psi(x + L\hat{i}) = exp(i\pi/5)\psi(x)$.

At the classical level the derivative of the action with respect to η is

$$\frac{\partial S^{cl.}}{\partial\eta} = \frac{k}{g_0^2},$$

where k is a function of N and η. We define the full renormalized coupling at quantum level by

$$\langle\frac{\partial S}{\partial\eta}\rangle = \frac{k}{g^2},$$

The running of the coupling is quantified by the step scaling function σ. It describes the change of the measured coupling when the linear size of the system is changed from L to sL keeping the bare coupling g_0^2 constant.

$$\Sigma(u, s, L/a) = g^2(g_0^2, sL/a)|g^2(g_0^2, L/a) = u \tag{5}$$

$$\Sigma(u, s, L/a) = \sigma(u, s) + c(u, s)a^2 \tag{6}$$

Table 1. The values of χ^2 and degrees of freedom in the interpolation in eq. 8.

	$L/a = 4$	$L/a = 6$	$L/a = 8$	$L/a = 12$	$L/a = 16$	combined
χ^2	32.77	38.02	23.71	19.27	11.48	125.2
d.o.f	4	5	5	5	5	24

We choose $s = 2$ and obtain the continuum limit from measurements at $L/a = 6$ and $L/a = 8$. The measured values of the coupling squared are shown in figure 2. We also show the lattice step scaling function $\Sigma(g^2, 2, L/a)/g^2 = g^2(g_0^2, 2L/a)/g^2(g_0^2, L/a)$ on the left side in figure 1.

The Schrödinger functional boundary conditions reduce the amount of zero modes in the system and allow simulations at zero quark mass. Since the Wilson fermion action breaks chiral symmetry and allows additive renormalization of the quark mass, we use the PCAC relation to find the value of κ where the renormalized mass vanishes. The quark mass M is defined by

$$aM(x_0) = \frac{1}{4}\frac{(\partial_0^* + \partial_0)f_A(x_0)}{f_P(x_0)}. \tag{7}$$

We define κ_c as the value of the parameter κ where the mass $aM(L/2)$ vanishes. To find κ_c we measure the mass at 3 to 7 values of κ on lattices of size $L/a = 16$ and interpolate to find where the mass is zero. In practice we achieve $aM < 0.003$.

We use reweighting to correct for the residual effect of the nonzero mass. The difference can be seen in figure 1, where we show the lattice step scaling function calculated from both the original and the reweighted measurements.

The continuum limit $\sigma(g^2, 2)$ is taken keeping the coupling g^2 equal, but the measurements have been calculated at constant g_0^2, and do not correspond to the same value of g^2 at different lattice sizes. The measurements, therefore, need to be shifted to matching values of g^2. The most economical and convenient way to achieve this is to interpolate the measurements at each lattice size L/a by fitting to a function of g_0^2. This results in a value of $g^2(g_0^2, L/a)$ over a continuous range of g_0^2.

We use the interpolating function

$$\frac{1}{g^2(g_0^2, L/a)} = \frac{1}{g_0^2}\left[\frac{1 + \sum_{i=1}^{n} a_i g_0^2}{1 + \sum_{i=1}^{m} b_i g_0^2}\right], \tag{8}$$

with $n = 4$ and $m = 2$. These values were chosen to maximize the combined P value for the fit, calculated from the sum of χ^2 and degrees of freedom for each fit.The interpolating functions are then used to calculate $\Sigma(u, 2, L/a)$ at a continuous range of u, and the continuum limit is calculated by fitting to the quadratic function in equation 6. The result is shown in figure 3.

2. Conclusions

Both the lattice results and the estimated continuum limit show a non-trivial infrared fixed point between $g^2 = 2$ and 4. This is in agreement with previous studies.

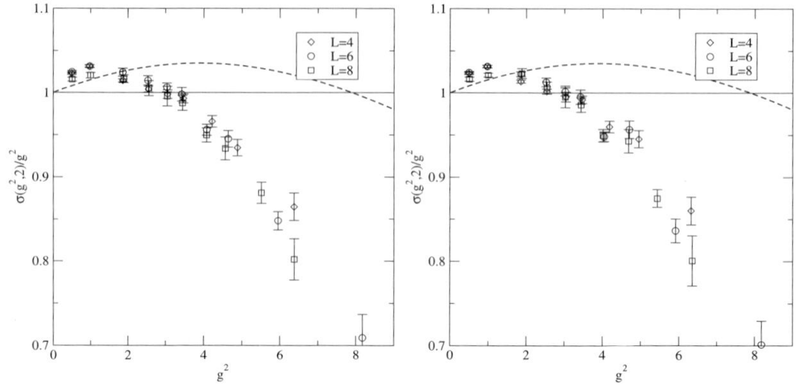

Fig. 1. The scaled lattice step scaling function $\Sigma(g^2, 2, L/a)/g^2 = g^2(g_0^2, 2L/a)/g^2(g_0^2, L/a)$. The plot on the left shows the result using the original measurements and the plot on the right using reweighted values. The black dashed line gives the continuum 2-loop perturbative result for $\sigma(g^2, 2)/g^2$.

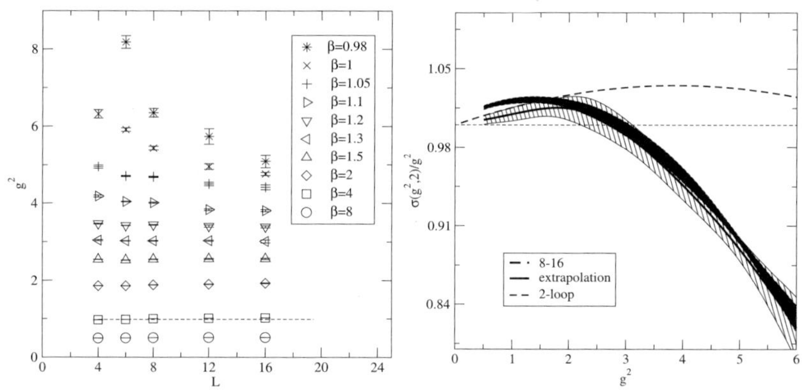

Fig. 2. On the left, the measured values of $g^2(g_0^2, L/a)$ against a/L. The black dashed line gives an example of the running in 2-loop perturbation theory at modest coupling, normalized so that it matches the measurement at $L/a = 6$. The measurements have been reweighted to zero mass. On the right, the scaled step scaling function $\sigma(g^2, 2)/g^2$ using reweighted measurements. The red line with the hashed band correspond to the continuum extrapolation using lattice sizes $L = 6$ and 8 and the black line with the green band correspond to the lattice result at the largest lattice size $L = 8$. The black dashed line gives the 2-loop perturbative result.

Furthermore, a large part of the discretization errors in the interesting region seems to be removed by the clover improvement and the HEX smearing. Smearing also reduces the computation time required to generate lattices and enables simulations at larger coupling. We plan to verify this further by repeating the measurement with an increased lattice size.

References

1. F. Bursa, L. Del Debbio, D. Henty, E. Kerrane, B. Lucini, A. Patella, C. Pica and T. Pickup *et al.*, Phys. Rev. D **84**, 034506 (2011) [arXiv:1104.4301 [hep-lat]].
2. A. J. Hietanen, J. Rantaharju, K. Rummukainen and K. Tuominen, JHEP **0905**, 025 (2009) [arXiv:0812.1467 [hep-lat]].
3. F. Bursa et. al. PoS LAT **2009**, 056 (2009) [arXiv:0910.2562 [hep-ph]].
4. A. J. Hietanen, K. Rummukainen and K. Tuominen, Phys. Rev. D **80**, 094504 (2009) [arXiv:0904.0864 [hep-lat]].
5. T. DeGrand, Y. Shamir and B. Svetitsky, Phys. Rev. D **83**, 074507 (2011) [arXiv:1102.2843 [hep-lat]].
6. S. Durr, Z. Fodor, C. Hoelbling, S. D. Katz, S. Krieg, T. Kurth, L. Lellouch and T. Lippert *et al.*, JHEP **1108** (2011) 148 [arXiv:1011.2711 [hep-lat]].
7. M. Luscher, R. Narayanan, R. Sommer, U. Wolff, P. Weisz, *Determination of the running coupling in the SU(2) Yang-Mills theory from first principles,* Nucl. Phys. Proc. Suppl. **30** (1993) 139-148.
8. M. Luscher, R. Narayanan, P. Weisz and U. Wolff, *The Schrodinger functional: A Renormalizable probe for nonAbelian gauge theories,* Nucl. Phys. B **384**, 168 (1992) [arXiv:hep-lat/9207009].
9. M. Hasenbusch and K. Jansen, Nucl. Phys. Proc. Suppl. **106**, 1076 (2002) [hep-lat/0110180].

Weak Solution Method of the Non-Perturbative Renormalization Group Equation to Describe Dynamical Chiral Symmetry Breaking and its Application to Beyond the Ladder Analysis in QCD

Ken-Ichi Aoki and Daisuke Sato

Institute for Theoretical Physics, Kanazawa University, Kanazawa 920-1192, Japan
E-mail: aoki@hep.s.kanazawa-u.ac.jp; satodai@hep.s.kanazawa-u.ac.jp

The method of non-perturbative renormalization group (NPRG) is applied to the analysis of dynamical chiral symmetry breaking (DχSB) in QCD. We show that the DχSB solution of the NPRG flow equation can be obtained without the bosonization. The solution, having the singular point, can be authorized as the weak solution of partial differential equation, and can be easily evaluated using the method of the characteristic curve. Also we show that our non-ladder extended approximation improves almost perfectly the gauge dependence of the chiral condensates.

Keywords: non-perturbative renormalization group, dynamical chiral symmetry breaking, QCD, weak solution

1. Introduction

The approach of the "non-perturbative renormalization group" (NPRG) based on Wilson's idea allows us to analyze dynamical chiral symmetry breaking (DχSB) in QCD with systematic approximations compared to the ladder Schwinger-Dyson analysis which suffers from the strong gauge dependence. The NPRG approach has been formulated as several forms of the functional differential equation, so people often call these formulations the "functional renormalization group". Among them, we adopt the flow equation derived by Wetterich.[1]

In NPRG, the 4-fermi coupling constant dynamically generated by the gauge interaction blows up at an infrared scale as a signal of DχSB. Due to its explosive behavior, the access to the broken symmetry scale below the critical scale is a nontrivial problem. In the earlier studies, the auxiliary field (bosonization),[3,4] the dynamical bosonization,[5] and the explicit breaking term such as the bare mass term[6] have been introduced to access the broken symmetry scale. However, using the weak solution method[a], we can directly access the broken symmetry scale only with the multi-fermi interactions but without any help by those methods.

[a]The authors greatly appreciate helpful comments by Prof. Akitaka Matsumura who told us how to construct the weak solution.

2. Truncated Effective Action

In order to study the dynamical chiral symmetry breaking in QCD with three-flavor massless quarks, we define the following truncated effective action:

$$\Gamma_\Lambda[\Phi] = \int_x \left\{ \frac{Z_A}{4} F^a_{\mu\nu} F^{\mu\nu}_a + \frac{1}{2\xi} (\partial_\mu A_\mu)^2 + \bar{\psi} \left(Z_\psi \partial\!\!\!/ + i g_s A\!\!\!/ \right) \psi - V(\psi, \bar{\psi}; \Lambda) \right\}, \quad (1)$$

where we use the covariant gauge with the gauge-fixing parameter ξ, and do not represent the ghost sector for simplicity. In this definition, the multi-fermi interactions are represented by $V(\psi, \bar{\psi}; \Lambda)$, which we call the fermion potential.

We substitute the effective action (1) into the Wetterich flow equation, and expand it in terms of the fields and the derivatives, so that we can obtain the coupled non-perturbative RG equations of these coupling constants. Then, lowering the cutoff scale Λ by solving the RG equations, the gauge interaction dynamically induces multi-fermi interactions. Among them, the scalar 4-fermi interaction brings about the DχSB at an intermediate cutoff scale as the Nambu–Jona-Lasinio model does. The 4-fermi coupling constant blows up at a finite infrared scale Λ_c, showing the signal of the DχSB, and consequently its RG flow cannot go beyond the critical scale Λ_c.

3. Partial Differential Equation

To go beyond the critical scale Λ_c, we solve the flow equation without expanding the fermion potential with respect to the scalar fermion-bilinear field $\sigma(\equiv \bar{\psi}\psi)$ that is the central operator for DχSB. For simplicity, the arguments of the fermion potential are limited to the scalar operator: $V(\psi, \bar{\psi}; \Lambda) \to V(\sigma; \Lambda)$. Besides, the interactions are limited to the ladder type ones, and consequently we obtain the ladder flow equation as the following partial differential equation (PDE):[3,4]

$$\partial_t V(\sigma; t) = -\eta_\psi \sigma \partial_\sigma V + \frac{\Lambda^4}{4\pi^2} \ln \left[1 + \Lambda^{-2} \left(\partial_\sigma V + (3 + \xi) \frac{C_2 \, g_s^2 \sigma}{4\Lambda^2} \right)^2 \right], \quad (2)$$

where C_2 is the second Casimir invariant of $SU(3)$. Here the dimensionless scale parameter $t(\equiv \log \Lambda_0/\Lambda)$ and the anomalous dimension of the quark field $\eta_\psi(\equiv \partial_t \log Z_\psi)$ are introduced. The PDE (2) can be numerically solved by replacing the derivatives with the finite differences. Practically, we work with the PDE for the first derivative of the fermion potential, $M(\sigma; t)(\equiv \partial_\sigma V)$, which we call the mass function. As shown in Fig. 1, the numerical solution of the mass function at $\sigma = 0$ has a finite jump below the critical scale Λ_c. This is nothing but the signal of DχSB, and the infrared value of the mass function at the limit, $\sigma \to +0$, is nothing but the dynamical mass m_{dyn}.

4. Weak Solution

While the singular behavior have numerically been evaluated, the non-analytic solution cannot be mathematically authorized as a solution of the PDE. Such a solution

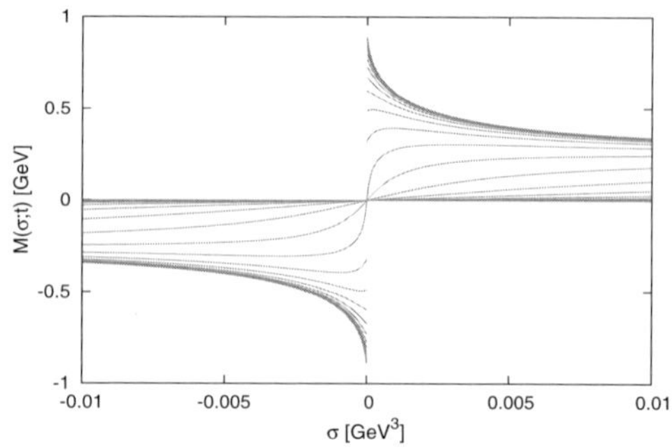

Fig. 1. RG evolution of the mass function $M(\sigma;t)$.

can be defined as the "weak solution" which satisfies the integral form of the PDE.[7] Here we skip the detail of its definition, but briefly introduce the method of the characteristic curve to evaluate the weak solution.

Representing the right-hand side of Eq. (2) as a function $F(\sigma, M; t)$, the PDE for the mass function can be written by

$$\partial_t M(\sigma;t) = -F_\sigma - \partial_\sigma M \cdot F_M, \tag{3}$$

where we use the notations, $F_\sigma \equiv \partial_\sigma F(\sigma, M;t)|_M$, $F_M \equiv \partial_M F(\sigma, M;t)|_\sigma$. Now let us consider the mass function on a characteristic curve, $\sigma = \bar{\sigma}(t;\sigma_0)$, where σ_0 is the initial position of $\bar{\sigma}$. If $\bar{\sigma}(t;\sigma_0)$ obeys the following differential equation,

$$\frac{d}{dt}\bar{\sigma}(t;\sigma_0) = F_M, \tag{4}$$

then the derivative $\partial_\sigma M$ vanishes in the PDE:

$$\frac{d}{dt}M(\bar{\sigma};t) = -F_\sigma. \tag{5}$$

Eqs. (4) and (5) are coupled ordinary differential equations equivalent to the PDE (3). The parametric curve of the mass function $M(\bar{\sigma};t)$ in terms of σ_0 can be constructed by solving them for each σ_0 as shown in Fig. 2. Below the critical scale Λ_c, the parametric curve is no longer regarded as the one-to-one mass function of σ. However, a condition of the weak solution called by "Rankine-Hugoniot condition" uniquely gives the one-to-one function shown in Fig. 1.[8]

5. Result and Discussion

Finally we show the result of the non-ladder extended flow equation, where the non-ladder corrections are added to the ladder flow equation (2).[2] In this approximation, the commutator of the generator of $SU(3)_c$ is ignored consistently for obtaining the gauge independent result.[9]

The gauge dependence of the chiral condensates $\langle\bar{\psi}\psi\rangle$, which is actually obtained by introducing the bare mass term of quarks as its source term, is shown in Fig. 3.

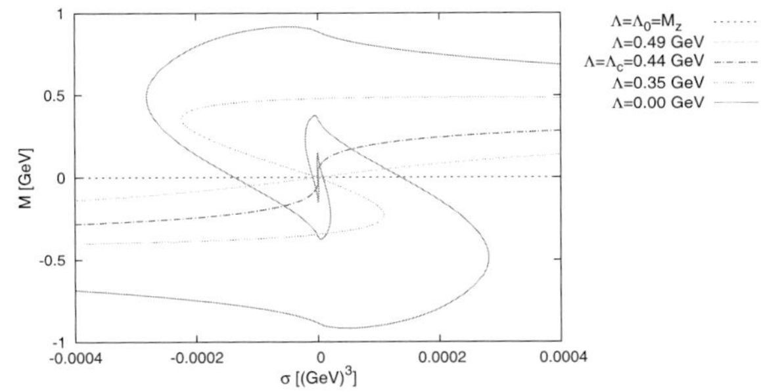

Fig. 2. Parametric curve of the mass function $M(\bar{\sigma}; t)$ in terms of σ_0.

The gauge dependence of the non-ladder extended results is greatly suppressed compared with the ladder results. Therefore we conclude that our non-ladder extended flow equation well respects the gauge independence. Here it should be noted that, in the Landau gauge ($\xi = 0$), the gauge-dependent ladder result of the chiral condensates coincides with the almost gauge-independent non-ladder extended one. This feature of the Landau gauge proves a folklore that the ladder approximation looks good particularly in the Landau gauge.

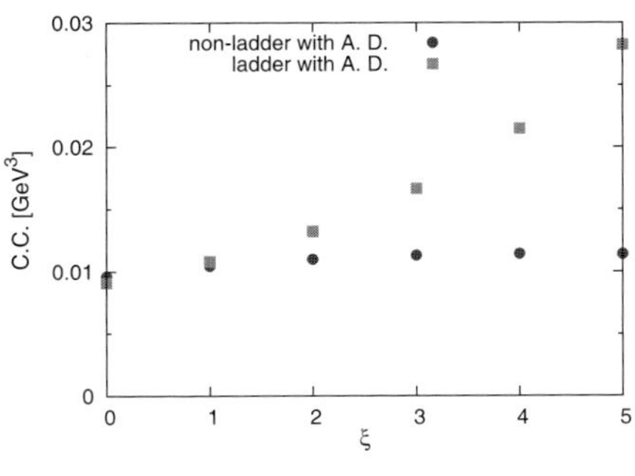

Fig. 3. The dependence of the chiral condensates $\langle\bar{\psi}\psi\rangle$ on the gauge-fixing parameter ξ.

References

1. C. Wetterich, *Phys. Lett. B* **301**, 90 (1993).
2. K-I. Aoki and D. Sato, *Prog. Theor. Exp. Phys.* **2013**, 043B04 (2013).
3. K-I. Aoki, *Proc. SCGT96*, 171 (1996):hep-ph/9706264, *Prog. Theor. Phys. Suppl.* **131**, 129 (1998), *Int. J. Mod. Phys. B* **14**, 1249 (2000).
4. K-I. Aoki, K. Morikawa, J. -I. Sumi, H. Terao and M. Tomoyose, *Prog. Theor. Phys.* **102**, 1151 (1999), *Phys. Rev. D* **61**, 045008 (2000) .
5. H. Gies and C. Wetterich, Phys. Rev. D **65**, 065001 (2002) .
6. K-I. Aoki and K. Miyashita, *Prog. Theor. Phys.* **121**, 875 (2009).
7. L. C. Evans, *Partial Differential Equations*, 2nd ed. (AMS, 2010).
8. K-I. Aoki, D. Sato and S.-I. Kumamoto, in preparation (2013).
9. K-I. Aoki, K. Takagi, H. Terao and M. Tomoyose, *Prog. Theor. Phys.* **103**, 815 (2000).

QCD Corrections to Electric Dipole Moment from Dimension-six Four-Quark Operators

Koji Tsumura

Department of Physics, Nagoya University, Nagoya 464-8602, Japan
**E-mail: ko2@eken.phys.nagoya-u.ac.jp*

Based on Ref. 1, we derive the renormalization-group equations for the CP-violating interaction up to the mass dimension six at one-loop level. In order to evaluate the scalar boson mediated electric dipole moment, all the four-quark operators are taken into account. The effect of the renormalization-group improvement by the QCD corrections can give 50 % modifications for the chromoelectric dipole moment as compared to the value predicted in the new physics scale.

Keywords: Electric dipole moment, CP violation

1. Introduction

The electric dipole moment (EDM) is an important observable to probe new physics beyond the standard model (SM) with additional CP violating sources. In the SM, Kobayashi-Maskawa (KM) phase gives CP violating phenomena. However, the contributions to the EDM of quarks, leptons, and nucleons are extremely suppressed. The recent estimation of the KM contributions to the electron EDM and the neutron EDM is found in the references.[2]

It is expected that the SM would be extended around TeV scale by more fundamental theory in order to solve many issues such as naturalness problem of the Higgs boson mass. New physics beyond the SM generically provides the new CP violating phases, and can predict observable EDM contributions.[3]

In models with extended Higgs sector such as two-Higgs-doublet model, the EDM and chromoelectric dipole moment (CEDM) are generated at the loop level, if CP symmetry is broken in the Yukawa interactions. Although the Yukawa coupling constants for lighter fermions are expected to be small, it is known that a two-loop diagram (so called the Barr-Zee diagram) can give significantly large contribution to the EDM and the CEDM.[4] This is because that the chirality flipping in the heavy-quark loops can enhance contributions, where the heavy-quark loop and the external lighter fermion lines are connected by the neutral scalar bosons via CP violating Yukawa interactions. For this kind of effective operators, it is unclear that which renormalization scale should be taken for the strong gauge coupling constant α_s for CEDM, where the CEDM is QCD analogue of the EDM (an external photon is replaced by a gluon line).

In this proceeding paper, we present improved renormalization-group equations (RGEs) at one-loop level by considering effective CP-violating operators up to mass dimension six. Using the one-loop RGEs, we evaluate the impact of the QCD corrections to the CEDM assuming new CP violating Yukawa interaction with a scalar ϕ. We found that the effect of the QCD correction to the CEDM can be larger than 50% for $m_\phi \gtrsim 100$ GeV as compared the value calculated in the new physics scales.

2. RGE improvement

In this section, we present the one-loop RGEs for the Wilson coefficients for the CP-violating operators up to the dimension six. The operator bases are defined as

$$\begin{aligned}\mathcal{L}_{\rm CPV} = & \sum_{i=1,2,4,5}\sum_{q} C_i^q(\mu)\mathcal{O}_i^q(\mu) + C_3(\mu)\mathcal{O}_3(\mu) \\ & + \sum_{i=1,2}\sum_{q'\neq q} \widetilde{C}_i^{q'q}(\mu)\widetilde{\mathcal{O}}_i^{q'q}(\mu) + \frac{1}{2}\sum_{i=3,4}\sum_{q'\neq q} \widetilde{C}_i^{q'q}(\mu)\widetilde{\mathcal{O}}_i^{q'q}(\mu)\,, \end{aligned} \tag{1}$$

where the sum of q runs not only light quarks but also heavy ones. The effective operators are defined as

$$\begin{aligned}\mathcal{O}_1^q &= -\frac{i}{2}m_q\overline{q}\, eQ_q(F\cdot\sigma)\gamma_5 q\,, \\ \mathcal{O}_2^q &= -\frac{i}{2}m_q\overline{q}\, g_s(G\cdot\sigma)\gamma_5 q\,, \\ \mathcal{O}_3 &= -\frac{1}{6}g_s f^{ABC}\epsilon^{\mu\nu\rho\sigma}G^A_{\mu\lambda}G^B{}_\nu{}^\lambda G^C_{\rho\sigma}\,, \end{aligned} \tag{2}$$

and

$$\begin{aligned}\mathcal{O}_4^q &= \overline{q_\alpha}q_\alpha\overline{q_\beta}\, i\gamma_5 q_\beta\,, \\ \mathcal{O}_5^q &= \overline{q_\alpha}\sigma^{\mu\nu}q_\alpha\overline{q_\beta}\, i\sigma_{\mu\nu}\gamma_5 q_\beta\,, \\ \widetilde{\mathcal{O}}_1^{q'q} &= \overline{q'_\alpha}q'_\alpha\overline{q_\beta}\, i\gamma_5 q_\beta\,, \\ \widetilde{\mathcal{O}}_2^{q'q} &= \overline{q'_\alpha}q'_\beta\overline{q_\beta}\, i\gamma_5 q_\alpha\,, \\ \widetilde{\mathcal{O}}_3^{q'q} &= \overline{q'_\alpha}\sigma^{\mu\nu}q'_\alpha\overline{q_\beta}\, i\sigma_{\mu\nu}\gamma_5 q_\beta\,, \\ \widetilde{\mathcal{O}}_4^{q'q} &= \overline{q'_\alpha}\sigma^{\mu\nu}q'_\beta\overline{q_\beta}\, i\sigma_{\mu\nu}\gamma_5 q_\alpha\,, \end{aligned} \tag{3}$$

where m_q are masses for quark q, $F_{\mu\nu}$ and $G^A_{\mu\nu}(A = 1$–$8)$ are the electromagnetic and gluon field strength tensors, g_s is the strong coupling constant ($\alpha_s = g_s^2/4\pi$), $F\cdot\sigma \equiv F_{\mu\nu}\sigma^{\mu\nu}$, $G\cdot\sigma \equiv G^A_{\mu\nu}\sigma^{\mu\nu}T^A$, and $\tilde{G}^A_{\mu\nu} \equiv \frac{1}{2}\epsilon_{\mu\nu\rho\sigma}G^{A\rho\sigma}$ with $\sigma^{\mu\nu} = \frac{i}{2}[\gamma^\mu, \gamma^\nu]$ and $\epsilon^{0123} = +1$. The matrix T^A denotes the generators in the $\mathrm{SU}(3)_C$ algebra, and f^{ABC} is the structure constant.

We refer dimension five operators $\mathcal{O}_1^q$ and $\mathcal{O}_2^q$ as EDM and CEDM operators. The RGEs for the Wilson coefficients of these operators are given as follows,

$$\mu\frac{\partial}{\partial\mu}\mathbf{C} = \mathbf{C}\boldsymbol{\Gamma}, \tag{4}$$

where the Wilson coefficients are written in a column vector as

$$\mathbf{C} = (C_1^q, C_2^q, C_3, C_4^q, C_5^q, \widetilde{C}_1^{q'q}, \widetilde{C}_2^{q'q}, \widetilde{C}_1^{qq'}, \widetilde{C}_2^{qq'}, \widetilde{C}_3^{q'q}, \widetilde{C}_4^{q'q}). \tag{5}$$

The anomalous dimension matrix is calculated at one-loop level as

$$\mathbf{\Gamma} = \begin{bmatrix} \frac{\alpha_s}{4\pi}\gamma_s & \mathbf{0} & \mathbf{0} \\ \frac{1}{(4\pi)^2}\gamma_{sf} & \frac{\alpha_s}{4\pi}\gamma_f & \mathbf{0} \\ \frac{1}{(4\pi)^2}\gamma'_{sf} & \mathbf{0} & \frac{\alpha_s}{4\pi}\gamma'_f \end{bmatrix}, \tag{6}$$

where each sub-matrix is given in Ref. 1. The anomalous dimensions for the dimension-five operators are calculated in the reference,[5] and that for the three-gluon operator $\mathcal{O}_3$ is calculated.[6] The operator mixings among the dimension-five operators and the three-gluon operator are found in the reference.[7] We newly derive the anomalous dimension sub-matrices γ_{sf}, γ_{sf}, γ_f and $\gamma_{f'}$ related to the four-quark operators. Note that the operator mixings γ_{sf} and γ'_{sf} are generated at $\mathcal{O}(\alpha_s^0)$.

3. Numerical results

We here consider a (CP-even) real scalar field ϕ which has CP violating Yukawa interactions with quarks as

$$\mathcal{L}_\phi = 2^{1/4}G_F^{1/2}m_q\overline{q_\alpha}(f_S^q + i\,f_P^q\gamma_5)q_\alpha\phi, \tag{7}$$

After integrating the real scalar out, the Wilson coefficients of the four-quark operators are generated at tree level as

$$\begin{aligned} C_4^q(m_\phi) &= \sqrt{2}G_F\frac{m_q^2}{m_\phi^2}f_S^q f_P^q\,, \\ \widetilde{C}_1^{q'q}(m_\phi) &= \sqrt{2}G_F\frac{m_q m_{q'}}{m_\phi^2}f_S^q f_P^{q'}\,, \\ \widetilde{C}_1^{qq'}(m_\phi) &= \sqrt{2}G_F\frac{m_q m_{q'}}{m_\phi^2}f_S^{q'} f_P^q\,. \end{aligned} \tag{8}$$

Since $\widetilde{C}_1^{q'q}$ and $\widetilde{C}_1^{qq'}$ are non-zero, C_2^q are generated at two-loop level via the operator mixings with $\widetilde{C}_4^{q'q}$ and $\widetilde{C}_4^{qq'}$. Using the RGEs for the Wilson coefficients, we obtain

$$C_2^q = \frac{\alpha_s}{8\pi^3}\frac{m_{q'}}{m_q}\left(\ln\frac{m_\phi}{m_{q'}}\right)^2\left[\widetilde{C}_1^{q'q} + \widetilde{C}_1^{qq'}\right]. \tag{9}$$

The result is consistent with the calculation of the Barr-Zee diagram.

In Fig. 1, we evaluate the impact of the QCD corrections to the Wilson coefficient $C_2^q(\mu) = \tilde{d}_q(\mu)/m_q$ for the CEDM operator. The numerator is evaluated by solving full RGEs numerically and including the effect of the running strong gauge coupling constant $\alpha_s(\mu)$, while the denominator is evaluated with constant strong gauge coupling at m_b scale. Thus, the ratio shows the improvement of the Wilson coefficient by our new RGEs. The effect is significantly large and more than 50% for $m_\phi \gtrsim 100$ GeV.

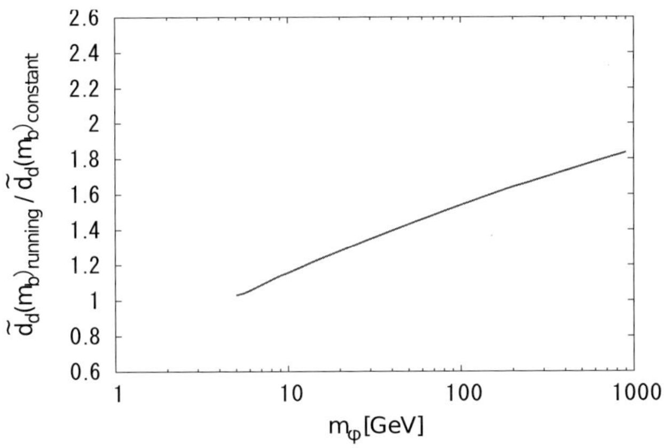

Fig. 1. The ratio of the CEDM for a down quark at m_b scale, where the running strong gauge coupling constant and the full RGE are numerically evaluated in the numerator.

4. Summary

We have studied the renormalization-group improvement of the Wilson coefficient for the EDM and the CEDM by QCD corrections. We have derived the one-loop RGEs including all the (flavor-conserving) CP violating operators up to dimension in the operator bases. Assuming a new scalar field ϕ with the CP violating Yukawa interaction, we have presented a new evaluation of the quark EDM. We found that the effect of the RGE improvement by the QCD corrections on the CEDM for a down quark at m_b scale can be larger than 50% for $m_\phi \gtrsim 100$ GeV as compared to the value predicted in the new physics scale.

References

1. J. Hisano, K. Tsumura, M. J. S. Yang and, Phys. Lett. B **713**, 473 (2012) [arXiv:1205.2212 [hep-ph]].
2. T. Mannel, N. Uraltsev and , Phys. Rev. D **85**, 096002 (2012) T. Mannel, N. Uraltsev and , JHEP **1303**, 064 (2013).
3. See for reviews, M. Pospelov and A. Ritz, Annals Phys. **318**, 119 (2005).
4. S. M. Barr and A. Zee, Phys. Rev. Lett. **65**, 21 (1990) [Erratum-ibid. **65** (1990) 2920].
5. M. A. Shifman, A. I. Vainshtein and V. I. Zakharov, Phys. Rev. D **18**, 2583 (1978) [Erratum-ibid. D **19**, 2815 (1979)].
6. J. Dai and H. Dykstra, Phys. Lett. B **237**, 256 (1990), E. Braaten, C. S. Li and T. C. Yuan, Phys. Rev. D **42**, 276 (1990).
7. G. Boyd, A. K. Gupta, S. P. Trivedi and M. B. Wise, Phys. Lett. B **241**, 584 (1990).

Analysis of the Scalar Tetraquarks on the Lattice

Masayuki Wakayama[a] and Chiho Nonaka[b]

[ab] *Department of physics, Nagoya University, Furo-cho, Chikusa-ku, Nagoya 464-8602, Japan*
[b] *Kobayashi-Maskawa Institute for the Origin of Particles and the Universe (KMI), Nagoya University, Furo-cho, Chikusa-ku, Nagoya 464-8602, Japan*
[a] *E-mail: wakayama@hken.phys.nagoya-u.ac.jp* [b] *E-mail: nonaka@hken.phys.nagoya-u.ac.jp*

We study the possibility that the scalar mesons exist as tetraquark states. The energy shift of two pions as a function of spatial lattice size makes a distinction between bound states and scattering states. We perform a calculation of two pion scattering amplitudes, using quenched $N_f = 2$ Wilson fermion and plaquette/Iwasaki gauge actions. We obtain the indication that the tetraquark states in the case of the $I = 0, 2$ channels are no bound states. And we find that the bound energy depends strongly on m_π rather than m_π/m_ρ.

Keywords: Tetraquarks; Scalar meson; Lattice; Proceedings; World Scientific Publishing.

1. Introduction

Since many light scalar mesons were found in experiments,[1] a lot of theoretical studies have been devoted to investigate their states.[2–5] For the structure of light scalar mesons, in addition to conventional two-quark states from the quark model, several possibilities are proposed ; four-quark states and so on.

Following the procedure which was proposed by Alford and Jaffe,[3] we explore the existence of four-quark states in the scalar channel on larger lattice with a finer lattice spacing. We also investigate the dependence of bound energies for four-quark states on m_π/m_ρ and on m_π.

2. Four-quark states from Lattice QCD

We calculate four-quark correlators in $N_f = 2$ lattice QCD under the assumption that four-quark states of scalar mesons are consisted of bound states of the two

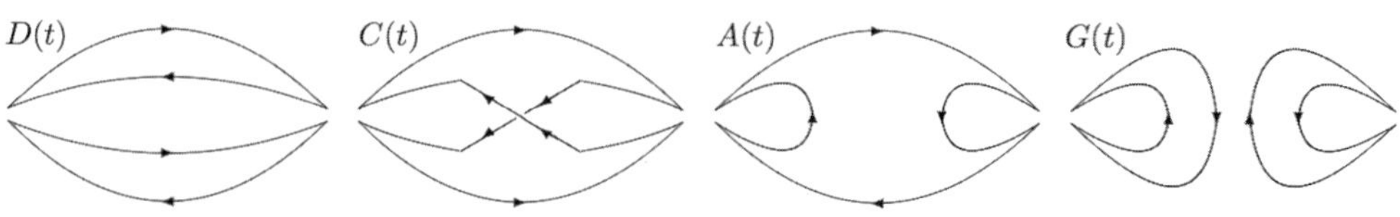

Fig. 1. The diagrams for four-quark correlators.

pions. We employ the procedure which was proposed by Alford et al.[3] The four-quark correlators of $I = 0, 2$ channels are given by,

$$J_{I=0}(t) = D(t) + \frac{1}{2}C(t) - 3A(t) + \frac{3}{2}G(t)\ , \quad J_{I=2}(t) = D(t) - C(t)\ , \tag{1}$$

where $D(t)$, $C(t)$, $A(t)$ and $G(t)$ are corresponding to the diagrams in Fig. 1. To evaluate the four-quark states clearly, we carry out the calculation in the quenched approximation where two-quark, multi-quark and glueball states do not mix with four-quark states as intermediate states. We drop the contribution of the diagrams for $A(t)$ and $G(t)$, ignoring the two-quark annihilation and the vacuum channels. To obtain the bound energy δE_I for the four-quark states, we construct the ratio of $J_I(t)$ and the pion correlator $P(t)$ and fit it to an exponential at large t,

$$R_I(t) \ = \ \frac{J_I(t)}{(P(t))^2} \underset{t\to\infty}{\longrightarrow} \frac{Z_I}{Z_\pi^2} \exp\left(-\delta E_I\, t\right) + \cdots\ . \tag{2}$$

We discuss the possibility that the four-quark states exist as bound states from the N_L dependence of the bound energies.

If a four-quark state is a bound state, δE_I is negative and would approach to a negative constant in the large N_L region. On the other hand if a four-quark state is a scattering state, it is expected that δE_I obeys the scattering formula of a two-particle state in a cube box of size L with periodic boundary condition in Ref. 6,

$$\delta E_I = E_I - 2m_\pi = \frac{T_I}{L^3}\left[1 + 2.8373\left(\frac{m_\pi T_I}{4\pi L}\right) + 6.3752\left(\frac{m_\pi T_I}{4\pi L}\right)^2\right] + \mathcal{O}\left(L^{-6}\right), \tag{3}$$

where E_I is the total energy, and T_I is the scattering amplitude which can be written by the scattering length a_I as $T_I = -4\pi a_I/m_\pi$. From Eq.(3), we find that δE_I is proportional to L^{-3} in the region where the physical spatial lattice size $L = N_L a$ (a is the lattice spacing) is enough large.

3. Results

In Ref. 3, they observed that the $I = 2$ channel is clearly the scattering state but found the possibility that the $I = 0$ channel is the four-quark bound state. However to get the conclusive result for the four-quark bound state in the $I = 0$ channel, further calculations on a larger lattice with finer lattice spacing are required.

We calculate the four-quark correlators on a larger lattice with $N_f = 2$ Wilson fermion using configurations which are produced with plaquette/Iwasaki gauge actions. The lattice spacings are shown in Table 1. We impose the Dirichlet boundary condition in the temporal direction on the quark fields.

The results for $I = 0$ channel are shown in Fig. 2. In Fig. 2 (a), the quark masses are set as the values of m_π/m_ρ become close to those in Ref. 3 (See Table 1). In spite of the same values of m_π/m_ρ, the values of m_π are different among them, which comes from the lattice artifact due to the large lattice spacings. In Fig. 2 (b), the bound energies are obtained under the same values of $m_\pi \sim 370$ MeV. We

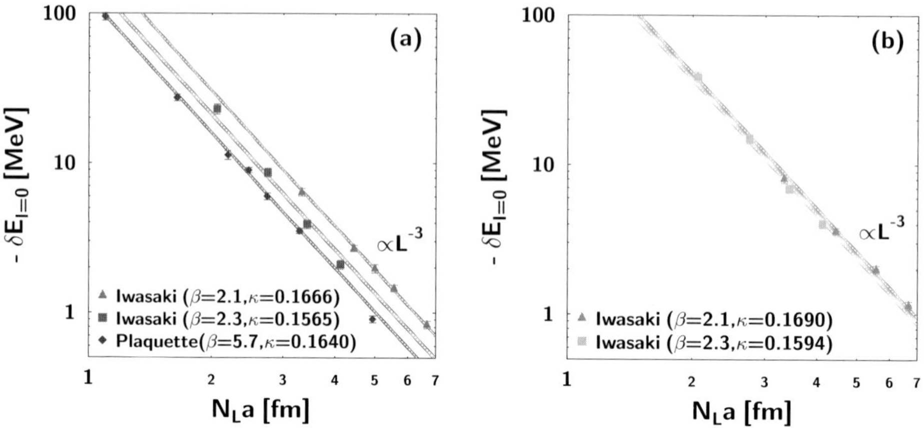

Fig. 2. The physical lattice size $N_L a$ dependence of $\delta E_{I=0}$. In left figure (a) (right figure (b)), the values of m_π/m_ρ (m_π) are fixed. In both figures the lines are proportional to L^{-3}.

find that in all cases of Figs.2 (a) and (b) the symbols are on the L^{-3} lines, which is contradicted with the result of Ref. 3. We can not observe the suggestion of the existence of the bound state in the $I = 0$ channel.

In Fig. 2 (a) (Fig. 2 (b)) the three (two) symbols which have the same values of m_π/m_ρ (m_π) are on the different (same) L^{-3} lines. It implies that δE_I depends strongly on m_π rather than m_π/m_ρ, which is deduced from $T_I = -4\pi a_I/m_\pi$. Since the scattering amplitude is proportional to the inversion of m_π, the absolute value of δE_I decreases with the increasing amount of m_π.

In Ref. 3 they insisted that they found the deviation from the L^{-3} line (the black dashed line in Fig. 3) in the L dependence of $\delta E_{I=0}$. However we can give another explanation to it. Because of the m_π differences among data plots, we can draw the three different L^{-3} lines instead of one dashed L^{-3} line in Fig. 3. It means that there are no bound states in the $I = 0$ channel.

4. Summary

We have investigated a four-quark state as a bound state from the spatial lattice size dependence of the bound energies for a four-quark state. We found that the four-quark states in the cases of $I = 0, 2$ channels exist as no bound states, i.e., all bound energies for the four-quark states as a function of L are on the L^{-3} lines, which is contradicted with the previous work.[3] The symbols which have the same values of m_π/m_ρ (m_π) are on the different (same) L^{-3} lines. This results suggest that the absolute value of bound energy depends directly on m_π rather than m_π/m_ρ.

This work is a staring point of study of four-quark states in the light scalar mesons. To reach the conclusive results for the light scalar meson states, further investigation is needed; full QCD calculation with light quark masses on larger lattice size, application of four-quark interpolators and so on.

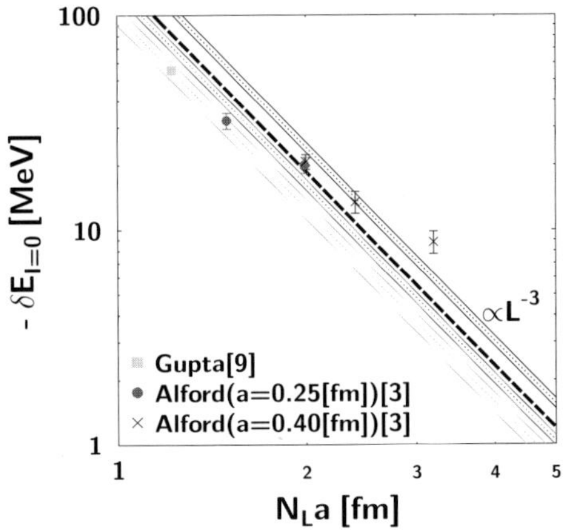

Fig. 3. The $N_L a$ dependence of $\delta E_{I=0}$. The lines are proportional to L^{-3}. The color difference corresponds to m_π difference. The black dashed line was drawn in Fig.5 of Ref.3.

Table 1. The values of m_π/m_ρ and m_π. In the upper table results of Ref. 3 are shown. In the lower table results of our calculation are shown.

data	a [fm]	m_π/m_ρ	m_π [MeV]
Gupta *et al.*[9]	0.0762(8)[10]	——	940(30)
Alford *et al.* ($a = 0.25$ [fm])	0.249(5)	0.756(5)	840(11)
Alford *et al.* ($a = 0.40$ [fm])	0.400(4)	0.756(4)	790(6)
Iwasaki ($\beta = 2.1, \kappa = 0.1666$)	0.278(3)[7]	0.736(3)	451(6)
Iwasaki ($\beta = 2.3, \kappa = 0.1565$)	0.172(4)[7]	0.746(6)	593(16)
Plaquette ($\beta = 5.7, \kappa = 0.1640$)	0.140(4)[8]	0.748(4)	754(12)
Iwasaki ($\beta = 2.1, \kappa = 0.1690$)	0.278(3)[7]	0.646(4)	368(5)
Iwasaki ($\beta = 2.3, \kappa = 0.1594$)	0.172(4)[7]	0.545(7)	374(11)

Acknowledgments

This work is supported in part by Nagoya University Global COE Program (G07). Numerical calculations were performed on the cluster system "φ" at KMI. Grant-in-Aid for Young Scientists (B) (22740156), Grant-in-Aid for Scientific Research (S) (22224003) and the Kurata Memorial Hitachi Science and Technology Foundation.

References

1. J. Beringer *et al.* [Particle Data Group Collaboration], Phys. Rev. D **86**, 010001 (2012).
2. R. L. Jaffe, Phys. Rev. D **15**, 267 (1977); R. L. Jaffe, Phys. Rev. D **15**, 281 (1977).
3. M. G. Alford and R. L. Jaffe, Nucl. Phys. B **578**, 367-382 (2000) [hep-lat/0001023].
4. T. Kunihiro *et al.* [SCALAR Collaboration], Phys. Rev. D **70**, 034504 (2004)
5. S. Prelovsek *et al.*, Phys. Rev. D **82**, 094507 (2010) [arXiv:1005.0948 [hep-lat]].
6. M. Lüscher, Commun. Math. Phys. **105**, 153 (1986)
7. P. de Forcrand *et al.* [QCD-TARO Collaboration], hep-lat/9806008.
8. M. Fukugita *et al.*, Phys. Rev. D **52**, 3003 (1995).
9. R. Gupta, A. Patel and S. R. Sharpe, Phys. Rev. D **48**, 388 (1993) [hep-lat/9301016].
10. C. T. H. Davies *et al.*, Phys. Rev. D **56**, 2755 (1997) [hep-lat/9703010].

Non-Perturbative Study of 16-Flavor QCD

Yasumichi Aoki[a], Tatsumi Aoyama[a], Masafumi Kurachi[a], Toshihide Maskawa[a], Kei-ichi Nagai[a], Hiroshi Ohki[a], Akihiro Shibata[b], Koichi Yamawaki[a], Takeshi Yamazaki[a]*

(LatKMI Collaboration)

[a] *Kobayashi-Maskawa Institute for the Origin of Particles and the Universe (KMI), Nagoya University, Nagoya, 464-8602, Japan*

[b] *Computing Research Center, High Energy Accelerator Research Organization (KEK), Tsukuba 305-0801, Japan*

It is well known that the SU(3) gauge theory with the fundamental 16-flavor fermion is governed by a non-trivial infrared fixed point in the 2-loop perturbation theory, while the theory has not been well investigated by non-perturbative lattice gauge simulations. We investigate properties of 16-flavor QCD by lattice simulation with highly improved actions(HISQ/tree) at wide range of the lattice spacing. We present the Polyakov loop in the spatial direction at several values of β at fixed lattice size and fermion mass. From the result we find that the finite size effect might be large at $\beta = 12$, where the lattice bare coupling is almost same as the coupling at the perturbative fixed point. We discuss the finite-size hyperscaling of the mass of pseudoscalar meson to estimate the mass anomalous dimension γ, while the results might have large finite size effect. It is encouraging that the estimated γ is consistent with the perturbative value, although our estimate has large uncertainty.

Keywords: Lattice Gauge Theories, Conformal Window

1. Introduction

One of the attractive models to explain the electroweak symmetry breaking beyond the standard model is the walking technicolor model, where the symmetry breaking results through the dynamics of strongly coupled systems. In this model, the (approximate) conformal symmetry of the strongly-coupled gauge theory plays an important role, so that it is an urgent task to understand features of such theories in non-perturbative calculations.

For this purpose, through Monte Carlo simulation of lattice gauge theory we investigate conformal signals in a gauge theory, which is expected to have the infrared fixed point. We choose the 16-flavor SU(3) gauge theory as such a theory, since in the 2-loop perturbation calculation this theory has the infrared fixed point. This work should be a test study of more critical case, such as the $N_f = 8$ and 12 SU(3) gauge theories, which we investigated.[1,2]

*Presenter

As a signal of the infrared conformality from the numerical lattice simulation, we focus on scaling behavior of physical quantities, called finite-size hyperscaling[3,4] (FSHS). While several lattice works studied the $N_f = 16$ SU(3) gauge theory,[5–9] the scaling behavior has not been studied.

In the previous studies[10] the anomalous dimension γ was estimated from simple physical quantities, such as the mass m_π of pseudoscalar meson (π) and its decay constant, through the FSHS at $\beta \equiv 6/g^2 \leq 5$. We found that the γ's obtained from the different physical quantities are reasonably consistent at each β, but the values are more than 10 times larger than the expected value of the two-loop perturbation theory. However, since the estimated γ decreases as the β increases, we expected that the γ would become compatible with the perturbative value at larger β. Thus, we started simulation at larger β than the previous works, and chose $\beta = 12$ where the lattice bare coupling g^2 is close to the one of the perturbative infrared fixed point. In this report we present the results of the Polyakov loop at $\beta = 12$ to investigate the phase and finite size effect at this β, and also preliminary analysis to estimate γ from FSHS of m_π.

2. Method

We employ a variant of the highly improved staggered quark action[11] called HISQ/tree. This action further suppresses the taste breaking in the QCD simulations compared to the Asqtad or stout-smeared staggered fermions.[12] We use this action for the many-flavor simulation to minimize the discretization effects which potentially compromise the behavior of the physical quantities at non-zero lattice spacings. Using this setup, we carry out simulations with various parameter sets for the fermion mass m_f and volumes at the bare coupling β.

The FSHS,[3,4] which is an extension of the hyperscaling,[13] is a scaling of the physical quantities, such as hadron mass m_H or decay constant, on a finite volume $L^3 \times T$ with the fixed L/T at small m_f. In this scaling, m_H is described by a function $f(x)$, $Lm_H = f(x)$, where x is the scaling variable $x = Lm_f^{\frac{1}{1+\gamma_*}}$ with γ_* being the mass anomalous dimension at the infrared fixed point. This scaling reproduces the original hyperscaling, $m_H \propto m_f^{\frac{1}{1+\gamma_*}}$, in the infinite volume limit. From this FSHS of the physical quantity we will estimate γ.

3. Results

Before starting the FSHS analysis at $\beta = 12$, we investigate the Polyakov loop for study of finite size effect, because it is naively expected that finite size effect would be very large at larger β, in other words smaller lattice spacing. The left panel of Fig. 1 shows the scatter plots of the Polyakov loop in the x direction, P_x, at $\beta = 5, 8,$ and 12 with the lattice size $18^3 \times 32$ and $m_f = 0.3$. The P_x goes away from the origin as the β increases. This tendency is considered that finite size effect becomes larger at larger β, since P_x in the infinite volume limit behaves like the one at $\beta = 5$

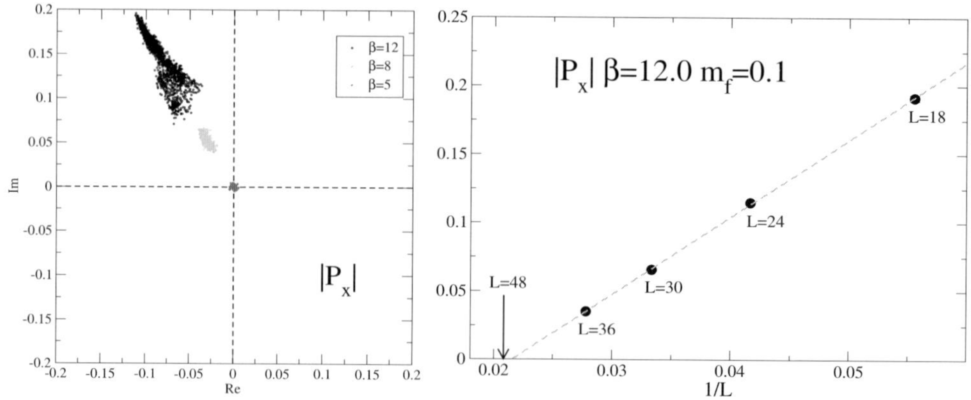

Fig. 1. Left panel is scatter plot of Polyakov loop in x direction, P_x, at $\beta = 5$, 8, and 12 with $L^3 \times T = 18^3 \times 32$ at $m_f = 0.3$. Right panel is inverse spatial size dependence of $|P_x|$ at $\beta = 12$, $m_f = 0.1$ with $T/L = 3$. Dashed line is linear extrapolation to $|P_x| = 0$.

in the panel. We also study the L dependence of $|P_x|$. The right panel of Fig. 1 shows that the $|P_x|$ roughly behaves as $1/L$ at $\beta = 12$ with fixed $T/L = 3$ and $m_f = 0.1$, and would vanish at $L \sim 48$. It is an important future work to perform simulations on $L \geq 48$, and investigate the finite size dependence of the P_x as well as the FSHS.

While finite size effect may be large in the current data, we perform a rough estimate of the γ from the FSHS of m_π at $\beta = 12$. The data are obtained on $L = 18, 24$, and 30 with the fixed ratio $T/L = 3$ at $m_f = 0.03$–0.3. Figure 2 shows the FSHS analysis with two values of γ using the data for the various L and m_f. A nice alignment of the data is observed in both γ values. This means that the data obey the FSHS with those values of γ. We estimate γ from the analysis, and obtain γ =0.0–0.15. This estimate might include the large finite size effect discussed in the above, so that study of the finite size dependence should be necessary. It is, however, encouraging that the range of the estimated value covers the perturbative result in contrast to the results at the smaller values of β.

4. Summary and outlook

We have investigated the property of the 16-flavor SU(3) gauge theory by the numerical simulation of the lattice gauge theory. We have found that the data at $\beta = 12$ might have large finite size effect, because the Polyakov loop in the spatial direction has non-zero value at larger β in $L = 18$. We have estimated that the Polyakov loop would vanish in $L \geq 48$ from its L dependence. It is important to calculate physical quantities on the large volumes where the finite size effect would be subleading. Although the finite size effect may be large, we have performed a trial analysis to estimate γ from the FSHS, and obtained γ =0.0-0.15. It is encouraging that the perturbative result is included in the estimated range.

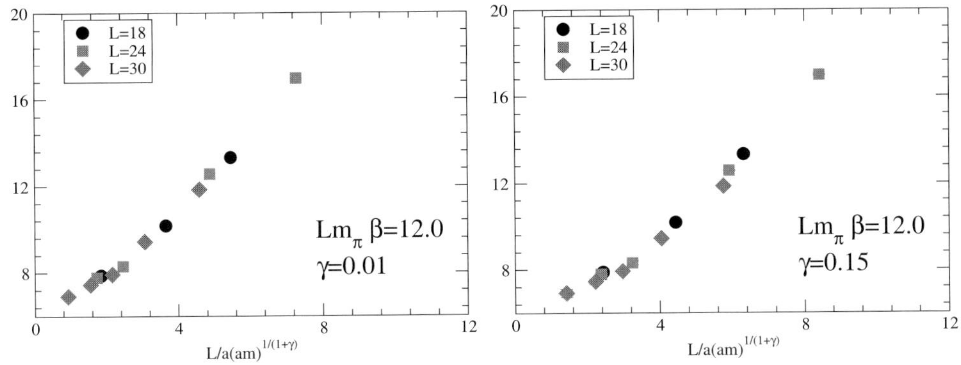

Fig. 2. Test of FSHS with different values of γ. Different symbols denote data on different lattice sizes.

Acknowledgments

Numerical simulation has been carried out on the supercomputer system φ at KMI in Nagoya university, and on the computer facilities of the Research Institute for Information Technology at Kyushu University. This work is supported by the JSPS Grant-in-Aid for Scientific Research (S) No.22224003, (C) No.23540300 (K.Y.) and (C) No.21540289 (Y.A.), and also by Grants-in-Aid of the Japanese Ministry for Scientific Research on Innovative Areas No. 23105708 (T.Y.).

References

1. Y. Aoki, T. Aoyama, M. Kurachi, T. Maskawa, K.-i. Nagai *et al.* (LatKMI), *Phys.Rev.* **D86**, 054506 (2012).
2. Y. Aoki, T. Aoyama, M. Kurachi, T. Maskawa, K.-i. Nagai *et al.* (LatKMI), arXiv:1302.6859[hep-lat] (2013).
3. T. DeGrand, *Phys.Rev.* **D80**, 114507 (2009).
4. L. Del Debbio and R. Zwicky, *Phys. Rev.* **D82**, 014502 (2010).
5. P. H. Damgaard, U. M. Heller, A. Krasnitz and P. Olesen, *Phys. Lett.* **B400**, 169 (1997).
6. Y. Iwasaki, K. Kanaya, S. Kaya, S. Sakai and T. Yoshie, *Phys. Rev.* **D69**, 014507 (2004).
7. Z. Fodor, K. Holland, J. Kuti, D. Nogradi and C. Schroeder, *Phys. Lett.* **B681**, 353 (2009).
8. A. Hasenfratz, *Phys. Rev.* **D82**, 014506 (2010).
9. K.-I. Ishikawa, Y. Iwasaki, Y. Nakayama and T. Yoshie, arXiv:1301.4785[hep-lat] (2013).
10. Y. Aoki, T. Aoyama, M. Kurachi, T. Maskawa, K.-i. Nagai *et al.* (LatKMI), *PoS* **LATTICE2012**, 029 (2012).
11. E. Follana *et al.*, *Phys. Rev.* **D75**, 054502 (2007).
12. A. Bazavov, T. Bhattacharya, M. Cheng, C. DeTar, H. Ding *et al.* (HotQCD), *Phys.Rev.* **D85**, 054503 (2012).
13. V. A. Miransky, *Phys. Rev.* **D59**, 105003 (1999).

WORKSHOP TIME TABLE

4 Dec. 2012

	Chair: K. Yamawaki	
9:00 - 9:45	Y. Iwasaki (KEK)	Conformal Window and Correlation Functions in Lattice Conformal QCD
9:45 - 10:30	J. Kuti (U. C. San Diego)	Can the Higgs impostor hide close to the conformal window ?
	Chair: Y. Aoki	
10:50 - 11:15	L. Del Debbio (Univ. of Edinburgh)	IR fixed points from lattice simulations
11:15 - 11:40	A. Hasenfratz (Univ. of Colorado Boulder)	Reaching the chiral limit in many flavor systems
11:40 - 12:05	N. Yamada (KEK)	Exploring many flavor QCD with Wilson quark action
12:05 - 13:30	Group Photo and Lunch	
	Chair: V. A. Miransky	
13:30 - 13:55	E. Neil (Fermilab)	From Lattice Strong Dynamics to Phenomenology
13:55 - 14:20	H. Ohki (KMI, Nagoya Univ.)	Lattice study of conformality in twelve-flavor QCD
14:20 - 14:45	K.-i. Nagai (KMI, Nagoya Univ.)	Exploring walking behavior in SU(3) gauge theory with 8 HISQ quarks
14:45 - 15:10	E. Rinaldi (KMI, Nagoya Univ.)	The scalar spectrum of many-flavour QCD
	Chair: M. Hayakawa	
15:40 - 16:05	K. Miura (Laboratori Nazionali di Frascati)	Thermodynamic Lattice Study for (Pre-)conformal Dynamics in Strongly Flavored QCD-Like Theory
16:05 - 16:30	X.-Y. Jin (RIKEN AICS)	Lattice QCD with 12 Quark Flavors: A Careful Scrutiny
16:30 - 16:55	K. Tuominen (Univ. of Jyvaskyla)	On gauge theory phase diagrams at zero and finite temperature
18:00 - 20:00	Poster Session and Welcome Party	

5 Dec. 2012

	Chair: M. Tanabashi	
9:00 - 9:45	O. Jinnouchi (Tokyo Institute of Technology)	Latest ATLAS results on Higgs and BSM physics
9:45 - 10:30	E. H. Simmons (Michigan State Univ.)	Coloron Models and LHC Phenomenology
	Chair: M. Tanabashi	
10:50 - 11:35	R. E. Shrock (Stony Brook Univ.)	Studies on the UV to IR Evolution of Gauge Theories and Quasiconformal Behavior
11:35 - 12:00	R. S. Chivukula (Michigan State Univ.)	Vector Boson Signals of Electroweak Symmetry Breaking
	Chair: N. Maekawa	
13:30 - 14:15	F. Sannino (CP3-Origins & DIAS)	Technicolor after the Higgs Discovery
14:15 - 15:00	S. Matsuzaki (Kyoto Sangyo Univ.)	125 GeV Techni-dilaton at the LHC
	Chair: K. Tobe	
15:30 - 15:55	M. Piai (Swansea Univ.)	Holographic Technidilaton and LHC searches
15:55 - 16:20	D-K. Hong (Pusan National Univ.)	Composite Higgs and Techni-dilaton at LHC
16:20 - 16:45	S. Aoki (Univ. of Tsukuba)	Hadronic interactions from lattice QCD
18:00 - 20:00	Banquet	

6 Dec. 2012

	Chair: M. Harada	
9:00 - 9:45	B. Mueller (Duke Univ.)	Quark-Gluon Plasma: From RHIC to LHC
9:45 - 10:10	C. Nonaka (Nagoya Univ.)	Dynamical model based on hydrodynamics for relativistic heavy ion collisions
10:10 - 10:35	Y. Akamatsu (KMI, Nagoya Univ.)	In-medium QCD forces at high temperature
	Chair: K.-I. Kondo	
10:55 - 11:20	M. Harada (Nagoya Univ.)	Holographic Mean-Field Theory for Baryon Many-Body Systems
11:20 - 11:45	H. K. Lee (Hanyang Univ.)	Dilatons for dense hadronic matter
11:45 - 12:10	Y.-L. Ma (Nagoya Univ.)	Baryonic matter from hidden local symmetry induced from holographic models
	Chair: T. Sakai	
13:30 - 13:55	M. Okawa (Hiroshima Univ.)	Twisted reduction in large N QCD with adjoint Wilson fermions
13:55 - 14:20	M. Hanada (KEK)	Large-Nc gauge theory and chiral random matrix theory
14:20 - 14:45	A. Shibata (KEK)	Non-Abelian dual superconductivity in SU(3) Yang-Mills theory due to non-Abelian magnetic monopoles
	Chair: K. Yamawaki	
15:15 - 16:00	K. Konishi (Univ. of Pisa and INFN, Sezione di Pisa)	New Confinement Phases from Singular SQCD Vacua
16:00 - 16:45	M. Shifman (Univ. of Minnesota)	Non-Abelian Strings in Supersymmetric Yang-Mills
17:30 - 19:30	Free Discussion and Group Dinner	

The most interesting models
Sextet model close to CW?
Close to conformal window large
the identification of conformal and
new algorithmic developments have
fermion matrix inversion, improved
Scaler spectrum from disconnected
Lattice BSM community is "theory

SCGT 12
KMI-GCOE Workshop on
Strong Coupling Gauge Theories
in the LHC Perspective

4th of July Fireworks, 2012

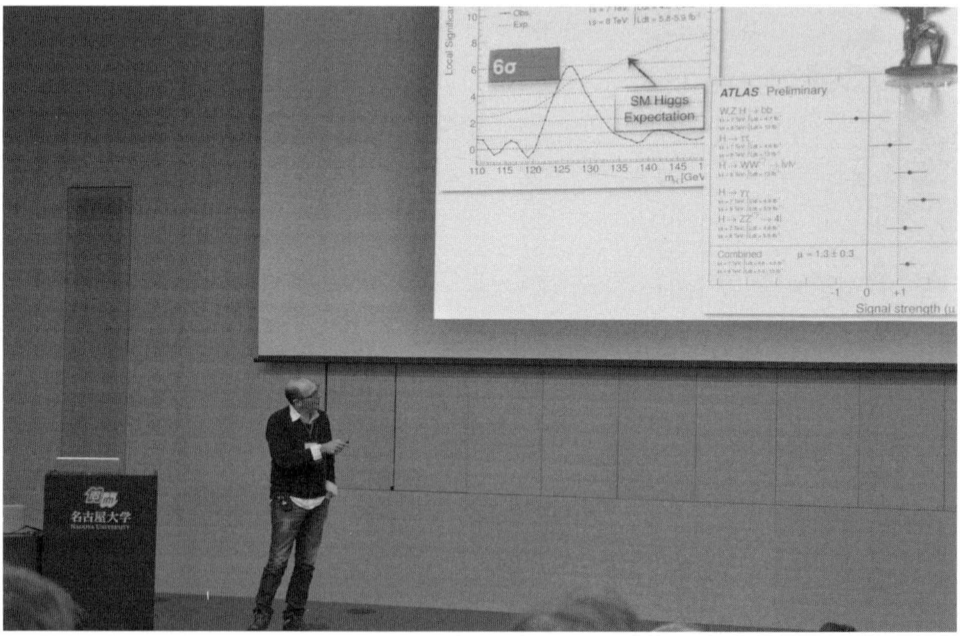
6σ
SM Higgs
Expectation
ATLAS Preliminary
Combined
μ = 1.3 ± 0.3
Signal strength
名古屋大学

the TD is favored, or

名古屋大学
NAGOYA UNIVERSITY

名古屋大学
NAGOYA UNIVERSITY

名古屋大学
NAGOYA UNIVERSITY

Oscillating couplings
SCGT 12
KMI-GCOE Workshop

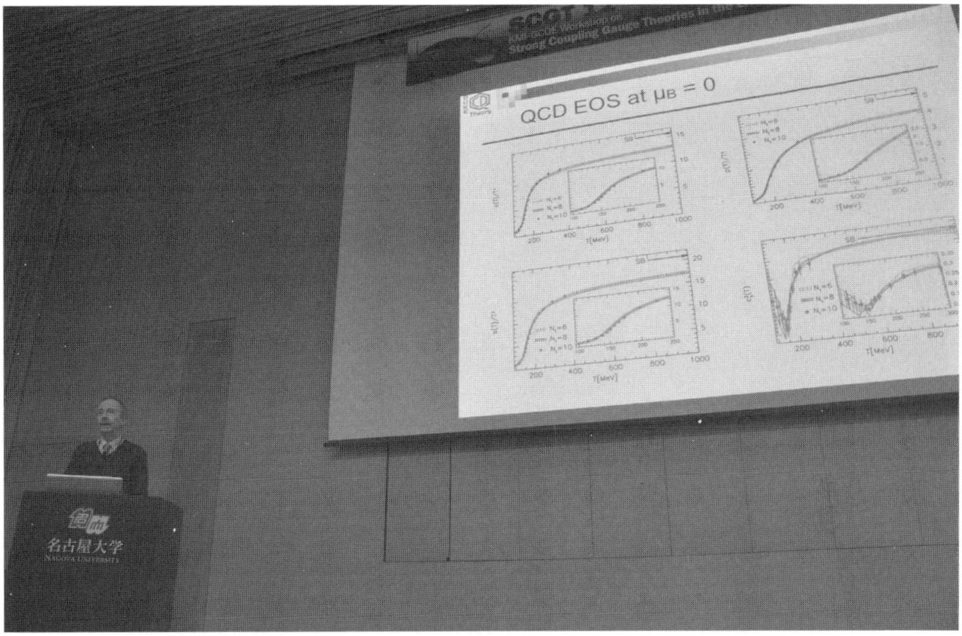
QCD EOS at μB = 0
名古屋大学
NAGOYA UNIVERSITY

rks
hotino
Global symmetry G;
string;
n the world sheet.
M. Shifman 10
SCGT 12
KMI-GCOE Workshop on
Strong Coupling Gauge Theories
in the LHC Perspective

LIST OF PARTICIPANTS

Abe, Tomohiro	Tsinghua University
Akamatsu, Yukinao	KMI, Nagoya University
Aoki, Sinya	University of Tsukuba
Aoki, Yasumichi	KMI, Nagoya University
Aoyama, Tatsumi	KMI, Nagoya University
Araki, Takeshi	Maskawa Institute, Kyoto Sangyo University
Bamba, Kazuharu	KMI, Nagoya University
Chivukula, R Sekhar	Michigan State University
Del Debbio, Luigi	University of Edinburgh
Fukano, Hidenori	University of Jyvaskyla
Fuyuto, Kaori	Nagoya University
Grinstein, Benjamin	University of California, San Diego
Hanada, Masanori	KEK
Harada, Masayasu	Nagoya University
Hasenfratz, Anna	University of Colorado Boulder
Hashimoto, Michio	Chubu University
Hayakawa, Masashi	Nagoya University
He, BingRan	Nagoya University
Hill, Christopher	Fermilab
Hirochi, Kyosuke	Nagoya University
Hisano, Junji	Nagoya University
Hong, Deog-Ki	Pusan National University
Hoshino, Hironori	Nagoya University
Iijima, Toru	KMI, Nagoya University
Ikemura, Takumi	Nagoya University
Ishida, Hiroyuki	Maskawa Institute, Kyoto Sangyo University
Iso, Satoshi	KEK

Iwasaki, Yoichi	KEK
Jia, Junji	KMI, Nagoya University
Jin, Xiao-Yong	RIKEN AICS
Jinnouchi, Osamu	Tokyo Institute of Technology
Karliner, Marek	Tel Aviv University
Katsuragawa, Taishi	Nagoya University
Kikukawa, Yoshio	University of Tokyo
Kitano, Ryuichiro	Tohoku University
Kondo, Kei-Ichi	Chiba University
Konishi, Kenichi	University of Pisa and INFN, Sezione di Pisa
Kugo, Taichiro	YITP, Kyoto University
Kumamoto, Shin-Ichiro	Kanazawa University
Kurachi, Masafumi	KMI, Nagoya University
Kuti, Julius	University of California, San Diego
Kuwahara, Takumi	Nagoya University
Lee, Hyun Kyu	Hanyang University
Ma, Yong-Liang	Nagoya University
Maekawa, Nobuhiro	KMI, Nagoya University
Maru, Nobuhito	Keio University
Maskawa, Toshihide	KMI, Nagoya University
Matsuzaki, Shinya	Maskawa Institute, Kyoto Sangyo University
Miransky, Vladimir	Western University
Miura, Kohtaroh	Laboratori Nazionali di Frascati
Motohiro, Yuichi	Nagoya University
Mueller, Berndt	Duke University
Muramatsu, Yu	Nagoya University
Nagai, Kei-ichi	KMI, Nagoya University
Nagai, Ryo	Nagoya University

Nagata, Natsumi	Nagoya University
Nakajima, Hideo	Utsunomiya University
Neil, Ethan	Fermilab
Nii, Keita	Nagoya University
Nishihara, Hiroki	Nagoya University
Nojiri, Shin'ichi	KMI, Nagoya University
Nonaka, Chiho	Nagoya University
Nozawa, Shinsaku	Nagoya University
Ohki, Hiroshi	KMI, Nagoya University
Okawa, Masanori	Hiroshima University
Onai, Shinnosuke	Kanazawa University
Piai, Maurizio	Swansea University
Rantaharju, Jarno	RIKEN AICS
Rinaldi, Enrico	KMI, Nagoya University
Sakai, Tadakatsu	Nagoya University
Sannino, Francesco	CP3-Origins & DIAS
Sato, Daisuke	Kanazawa University
Shaposhnikov, Mikhail	Ecole Polytechnique Federale de Lausanne
Shibata, Akihiro	KEK
Shifman, Mikhail	University of Minnesota
Shigemori, Masaki	KMI, Nagoya Univ
Shrock, Robert	Stony Brook University
Simmons, Elizabeth	Michigan State University
Sogami, Ikuo	Maskawa Institute, Kyoto Sangyo University
Takeuchi, Tatsu	Virginia Tech and Kavli IPMU
Tanabashi, Masaharu	KMI, Nagoya University
Terao, Haruhiko	Nara Women's University
Tobe, Kazuhiro	Nagoya University

Toda, Kouichi	Toyama Prefectural University
Tsumura, Koji	Nagoya University
Tuominen, Kimmo	University of Jyvaskyla
Umetsu, Koichiro	Kyoto Sangyo University
Wakayama, Masayuki	Nagoya University
Yamada, Norikazu	KEK
Yamawaki, Koichi	KMI, Nagoya University
Yamazaki, Takeshi	KMI, Nagoya University
Yoda, Hiroshi	Nagoya University

AUTHOR INDEX